CYCLOBUTADIENE
and Related Compounds

ORGANIC CHEMISTRY

A SERIES OF MONOGRAPHS

Edited by

ALFRED T. BLOMQUIST

Department of Chemistry, Cornell University, Ithaca, New York

1. Wolfgang Kirmse. CARBENE CHEMISTRY, 1964
2. Brandes H. Smith. BRIDGED AROMATIC COMPOUNDS, 1964
3. Michael Hanack. CONFORMATION THEORY, 1965
4. Donald J. Cram. FUNDAMENTALS OF CARBANION CHEMISTRY, 1965
5. Kenneth B. Wiberg (Editor). OXIDATION IN ORGANIC CHEMISTRY, PART A, 1965; PART B. *In preparation.*
6. R. F. Hudson. STRUCTURE AND MECHANISM IN ORGANO-PHOSPHORUS CHEMISTRY, 1965
7. A. William Johnson. YLID CHEMISTRY, 1966
8. Jan Hamer (Editor). 1,4-CYCLOADDITION REACTIONS, 1967
9. Henri Ulrich. CYCLOADDITION REACTIONS OF HETEROCUMULENES, 1967
10. M. P. Cava and M. J. Mitchell. CYCLOBUTADIENE AND RELATED COMPOUNDS, 1967

IN PREPARATION

Reinhard W. Hoffmann. Dehydrobenzene and Cycloalkynes

CYCLOBUTADIENE
and Related Compounds

M. P. CAVA and M. J. MITCHELL

*Chemistry Department,
Wayne State University,
Detroit, Michigan*

WITH A CHAPTER ON THEORY BY

H. E. SIMMONS AND A. G. ANASTASSIOU

*Central Research Laboratories,
E. I. du Pont de Nemours & Co., Inc.
Wilmington, Delaware*

1967

ACADEMIC PRESS   New York and London

COPYRIGHT © 1967, BY ACADEMIC PRESS INC.
ALL RIGHTS RESERVED.
NO PART OF THIS BOOK MAY BE REPRODUCED IN ANY FORM,
BY PHOTOSTAT, MICROFILM, OR ANY OTHER MEANS, WITHOUT
WRITTEN PERMISSION FROM THE PUBLISHERS.

ACADEMIC PRESS INC.
111 Fifth Avenue, New York, New York 10003

United Kingdom Edition published by
ACADEMIC PRESS INC. (LONDON) LTD.
Berkeley Square House, London W.1

LIBRARY OF CONGRESS CATALOGUE CARD NUMBER: 66-16434

PRINTED IN THE UNITED STATES OF AMERICA

"We see a little, presume a great deal, and so jump to a conclusion."

JOHN LOCKE

Preface

The aim of this monograph is to present in readily accessible form all of the information available on four-membered carbocyclic compounds having only trigonally hybridized carbon atoms in the ring. Cyclobutadiene is the most elusive and the most celebrated compound of this type, but biphenylene, benzocyclobutadiene, dimethylenecyclobutene, and the rest are in many ways of equal interest and importance. All of these compounds have been investigated experimentally in recent years and all are treated fully in the present volume.

The guidelines which we followed were: (1) the text should be as comprehensive as possible; (2) the organization should be highly formal in order to serve the specialized needs of workers in the field; and (3) critical commentary should be introduced wherever necessary to resolve conflicting reports in the literature.

In attempting to make the monograph as comprehensive as possible, we have made an exhaustive survey of all of the usual reference sources, including *Chemical Abstracts, Beilsteins Handbuch der Organischen Chemie*, and the *Zentralblatt*, and, in addition, we have made a reference-by-reference investigation of all previous bibliographies on the subject and have solicited pertinent preprints and reprints from workers in the field.

Each of the cyclobutadienoid ring systems cited above has been treated in a separate chapter which has been subdivided into standard sections dealing with specific topics, e.g., methods of preparation, chemistry, physical properties. While it is true that a rigid organizational scheme of this type presents a fragmented view of the subject to the uninitiated reader, it has the overriding advantage in our opinion of better serving the needs of current workers in the field, most of whom require an easily consulted comprehensive reference work rather than a qualitative review. The text on cyclobutadiene has been divided into three chapters, of which the first deals with cyclobutadiene and substituted cyclobutadienes, the second with cyclobutadiene–metal complexes, and the third with the divalent ions of cyclobutadiene. In addition, cyclobutene-3,4-dione and its analogs and 3, 4-dimethylenecyclobutene and its analogs have been termed "cyclobutadienequinones" and "methylene analogs of cyclobutadienequinone," respectively, and have been assigned to separate chapters. The text on benzocyclobutadiene has been apportioned among four chapters (Benzocyclobutadiene, 1, 2-Benzocyclobutadienequinone, Methylene Analogs of 1,2-Benzocyclobutadienequinone, and Higher Aromatic Analogs of Benzocyclobutadiene) and the text on biphenylene has been divided into two chapters (Biphenylene and The Benzobiphenylenes).

An alphabetical bibliography has been provided at the end of each chapter in which related works, e.g., preliminary communications and their matching papers, have been brought together into groups bearing collective reference numbers. Such works are designated individually by the letters a, b, c, etc., and are cited in the text either by a number alone (which indicates that the information in question is to be found in all references of the group bearing that number) or by a number and one or more letters (which indicates that the information occurs only in the particular reference or references cited). In this way, the reader can determine at a glance whether he should consult the preliminary communication, the full paper, or both in order to obtain further information on a particular subject.

Recent years have seen a remarkable increase in the number of papers published in the field of cyclobutadiene chemistry. While this increase has been gratifying to us as researchers, it has also been a source of considerable difficulty to us as authors; on more than one occasion we have had to make substantial changes in the manuscript in order to incorporate a newly published paper. Fortunately, many of our colleagues were able to send us preprints and personal communications prior to publication, and this privileged information helped us to keep the number of revisions to a minimum. But even so, we found it necessary in the end to set an arbitrary cut-off date (January 1, 1964) beyond which no new material was to be incorporated into the manuscript. Of course, all papers that reached us in preprint form before that date were incorporated, even though the published papers did not appear until later. Finally, in an effort to update the contents, we added an appendix containing abstracts of papers published in 1964 and 1965. Because of the haste with which the appendix was compiled, a few papers have probably escaped detection, but we believe that the preponderant majority of papers have been included and have been adequately abstracted. The information contained in the appendix, though informally organized, is retrievable through the index.

We are indebted to more than two hundred of our colleagues for sending us reprints, preprints, and personal communications during the time when the manuscript was in progress. Their help was of inestimable value and saved us many hours of library work. We owe a debt of gratitude to Professor G. W. Griffin for critically reading the entire manuscript and to Dr. E. R. Atkinson, Dr. Margareta Avram, Professor J. E. H. Hancock, Professor P. M. Maitlis, Dr. K. W. Ratts, and Dr. R. P. Stein for reading selected portions of it and offering advice. We also wish to express our thanks to Dr. Leonard T. Capell,

Mr. John F. Stone, and Mr. Donald F. Walker of the Chemical Abstracts Service for their advice on nomenclature; to our former colleagues of the Chemistry Department of The Ohio State University for many helpful discussions; and to the librarians of the Chemical Abstracts Service, the Batelle Memorial Institute, and the McPherson Chemistry Library of The Ohio State University for their unstinting aid.

M. P. CAVA

February, 1967 M. J. MITCHELL

Contents

Preface	vii

Chapter 1. Cyclobutadiene — 1

A. History	3
B. Generation of Cyclobutadiene and Substituted Cyclobutadienes	11
C. Unsuccessful Approaches to the Cyclobutadienes	26
D. Evidence of the Triplet Diradical Character of the Cyclobutadienes	43
E. Chemistry of Cyclobutadiene and the Substituted Cyclobutadienes	49
References	84

Chapter 2. The Cyclobutadiene–Metal Complexes — 88

A. History	89
B. Preparation of Cyclobutadiene–Metal Complexes	90
C. Cyclobutadiene–Metal Complexes as Transient Intermediates	100
D. Unsuccessful Approaches to the Cyclobutadiene–Metal Complexes	104
E. Chemistry of the Cyclobutadiene–Metal Complexes	106
F. Physical Properties of the Cyclobutadiene–Metal Complexes	112
References	119

Chapter 3. Cyclobutadiene Divalent Ions — 122

A. History	123
B. Cyclobutadiene Dication	123
C. Cyclobutadiene Dianion	126
References	127

Chapter 4. Cyclobutadienequinone — 128

A. History	128
B. Synthesis of Cyclobutadienequinones	129
C. Unsuccessful Attempts to Prepare Cyclobutadienequinones	137
D. Chemistry of the Cyclobutadienequinones	138
E. Physical Properties of the Cyclobutadienequinones	150
References	156

Chapter 5. Methylene Analogs of Cyclobutadienequinone — 157

A. History	158
B. Dimethylenecyclobutenes	158
C. Tetramethylenecyclobutanes	168

D. Trimethylenecyclobutanone and the Dimethylenecyclobutanediones	173
E. Condensed Methylenecyclobutene Aromatic Systems	178
References	179

Chapter 6. Benzocyclobutadiene — 180

A. History	180
B. Generation of Benzocyclobutadiene and Substituted Benzocyclobutadienes	182
C. Unsuccessful Approaches to the Benzocyclobutadienes	187
D. Chemistry of Benzocyclobutadiene	192
References	217

Chapter 7. Benzocyclobutadienequinone — 219

A. History	220
B. Synthesis of Benzocyclobutadienequinone	220
C. Unsuccessful Attempts to Prepare Benzocyclobutadienequinone	223
D. Physical Properties of Benzocyclobutadienequinone	223
E. Chemistry of Benzocyclobutadienequinone	225
References	230

Chapter 8. Methylene Analogs of 1,2-Benzocyclobutadienequinone — 232

A. History	232
B. 2-Methylenebenzocyclobutenones	232
C. 1,2-Dimethylenebenzocyclobutenes	235
References	241

Chapter 9. Higher Aromatic Analogs of Benzocyclobutadiene — 242

A. History	243
B. Naphtho[b]cyclobutadiene	243
C. Phenanthro[l]cyclobutadiene	252
References	254

Chapter 10. Biphenylene — 255

A. History	255
B. Synthesis of Biphenylenes	256
C. Unsuccessful Approaches to the Biphenylenes	272
D. Proof of Structure of Biphenylene	276
E. Chemistry of Biphenylene	278

F. Physical Properties of Biphenylene	285
G. Chemical and Physical Properties of Substituted Biphenylenes	291
References	313

Chapter 11. **The Benzobiphenylenes** — 317

A. History	318
B. Synthesis of Benzo- and Dibenzobiphenylenes	321
C. Unsuccessful Approaches to the Benzobiphenylenes	334
D. Chemical and Physical Properties of the Benzo- and Dibenzobiphenylenes	335
E. Higher Aromatic Analogs of the Benzobiphenylenes	364
F. Heterocyclic Benzobiphenylenes	364
References	366

Chapter 12. **Theoretical Aspects of the Cyclobutadiene Problem** — 368

H. E. Simmons and A. G. Anastassiou

A. The π-Electronic States of Cyclobutadiene	369
B. Some Conclusions and Comparisons	386
C. Jahn–Teller Considerations	388
D. Substituted Cyclobutadienes, Condensed Aromatic Cyclobutadienes, and Cyclobutadiene–Metal Complexes	393
References	414
Addendum	418

Appendix. **Abstracts of Papers Appearing in the Literature in 1964–1965** — 423

Author Index	465
Subject Index	480

CHAPTER 1

Cyclobutadiene

Cyclobutadiene is the simplest member of the series of fully conjugated cyclic polyenes having the general composition $(C_2H_2)_n$, of which benzene has long been the best-known example. In the era before the development of molecular orbital theory, organic chemists quite naturally assumed that cyclobutadiene would be a stable aromatic system with a chemistry paralleling that of benzene. As a result, several attempts were made around the turn of the century to prepare cyclobutadiene and substituted cyclobutadienes, and a number of compounds were assigned cyclobutadiene structures on the basis of rather limited experimental evidence. These attempts were entirely unsuccessful and none of the proposed cyclobutadiene structures has withstood the test of reinvestigation; to this day, neither cyclobutadiene nor a substituted cyclobutadiene has been isolated and characterized directly. Furthermore, modern molecular orbital theory contradicts the earlier ingenuous assumption of stability and predicts that, unlike the $4n+2$ π-electron systems, cyclobutadiene, which is a $4n$ π-electron system, should be a highly unstable species having zero delocalization energy and exhibiting the properties of a triplet diradical.* This prediction is in line with what is known about the chemistry of cyclobutadiene and it leaves little hope that a stable cyclobutadiene can be synthesized.

Despite this unencouraging prospect (or perhaps because of it) chemists have sought in recent years to synthesize cyclobutadienoid systems in which the

*See, however, the Addendum to Chapter 12 for a recent prediction by Dewar of a singlet ground state for cyclobutadiene.

instability of the four-membered system would be diminished through some special structural or electronic feature. This approach has met with a certain degree of success, inasmuch as it has resulted in the preparation of biphenylene (*1*), tetramethylcyclobutadienenickel chloride (*2*), and 1,2-diphenylnaphtho-[*b*]cyclobutadiene (*3*), all of which are isolable and may be viewed as cyclobutadienoid systems. It must be noted, however, that the major contributors to the resonance hybrid of both biphenylene and diphenylnaphthocyclobutadiene are dimethylenecyclobutene structures rather than cyclobutadiene structures and that the carbocycle of the metal complex has little or no cyclobutadienoid character.

However, it is not impossible that a simple cyclobutadiene, substituted with carefully chosen groups, might prove to have a singlet ground state and hence be isolable. A particularly interesting idea along these lines has been proposed by Roberts,[128] who has suggested that the cyclobutadiene system might be appreciably stabilized by a pair of neighboring conjugative substituents, one of which would be electron-releasing and the other of which would be electron-attracting.

In the last decade the problem of synthesizing a stable cyclobutadiene derivative has come to be recognized as a challenge of classical proportions. It has also become increasingly apparent that we are only now beginning to understand the chemistry of cyclobutadiene and that this understanding might well portend the development of a new and more sophisticated phase of cyclobutadiene studies. Perhaps one of the most significant pieces of research in the field was published only recently by Watts, Fitzpatrick, and Pettit, in which the formation of free cyclobutadiene is described unambiguously for the first time.*,[148]

---

* Pettit's work demonstrates that, although cyclobutadiene is an extremely unstable substance, it has, nevertheless, a detectable lifetime. This work, which was published after the manuscript of Chapter 1 had been prepared, is described in ref. 148 and in the Appendix.

Various aspects of cyclobutadiene chemistry have been reviewed by Avram et al.,[13] Baker and McOmie,[15] Criegee,[45] Gelin,[76] Murata,[115] Nenitzescu et al.,[117] Vogel,[144] and Vol'pin.[146]

A. History

The first attempt to synthesize compounds of the cyclobutadiene series appears to have been made by W. H. Perkin, Jr., who tried to prepare "tetrenedicarboxylic acid" [cyclobutadiene-1,2-dicarboxylic acid (*1*)], which he likened to phthalic acid, and "tetrenecarboxylic acid" [cyclobutadienecarboxylic acid, (*6*)], the formal cyclobutadienoid analog of benzoic acid.*,[120] Treatment of 1,2-dibromocyclobutane-1,2-dicarboxylic acid (*2*) with various bases gave only 2-bromocyclobutenecarboxylic acid (*3*), while treatment of the dimethyl ester (*5*) with the same reagents gave only cyclobutene-1,2-dicarboxylic acid (*4*), and not the expected cyclobutadiene diacid (*1*). Similarly, 2-bromocyclobutenecarboxylic acid (*3*) failed to give cyclobutadienecarboxylic acid (*6*) on treatment with aqueous or alcoholic solutions of potassium hydroxide.

Some ten years later (1905) an attempt was made by Willstätter and Schmädel to synthesize the parent hydrocarbon itself.[156] Thus, 1,2-dibromocyclobutane (*8*), prepared from cyclobutene (*7*), was dehydrobrominated by treatment (a) with quinoline at high temperatures, which gave only a little butadiene and a red nitrogen-containing high molecular weight material, and (b) with potassium hydroxide at 100°–105°, which gave 1-bromocyclobutene

---

* For several incorrect cyclobutadiene structure assignments antedating Perkin's work, see p. 6 *ff*. An earlier attempt to prepare cyclobutadiene from crotonaldehyde [A. Kekulé, *Ann.*, **162**, 77 (1872)] failed almost at the outset and is not included in the present discussion.

[(8) → (9)] rather than the desired product, cyclobutadiene (10). Under forcing conditions (powdered potassium hydroxide, 210°), the dibromide gave acetylene; cyclobutadiene seemed to be a possible intermediate in this reaction.

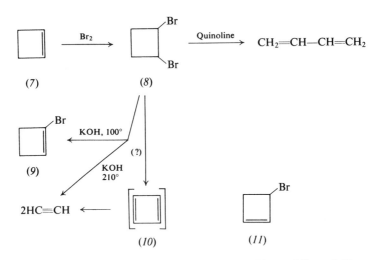

In discussing these results some years later, Favorskii and Favorskaya concurred in the belief that the critical intermediate in the formation of acetylene was cyclobutadiene (10), but they suggested that Willstätter had represented the structure of the unsaturated bromide incorrectly and that the latter was in fact an allyl bromide (11) rather than a vinyl bromide (9).[65] However, this proposal overlooks the fact that Willstätter and Schmädel successfully proved the structure of their monobromide to be (9) by oxidizing it to succinic acid.[156] [Oxidation of (11) would be expected to give α-bromosuccinic acid.]

Another early approach to the cyclobutadiene system, this one via 1,3-disubstituted cyclobutanes, was recorded by one of Willstätter's students, Roman Malachowski, in a dissertation entitled "Zur Kenntnis der Cyclobutanderivate."[106] The starting material in this unsuccessful and otherwise unpublished work was presumed to be cyclobutane-1,3-dicarboxylic acid (12), but it is now known that the diacid was in fact a methylcyclopropanedicarboxylic acid[58]; consequently, Malachowski's efforts to prepare 1,3-cyclobutadienedicarboxylic acid were doomed to failure from the very outset.

Willstätter was probably motivated in his attempts to synthesize cyclobutadiene, as he was later in his successful synthesis of cyclooctatetraene,[157] by a desire to test the general validity of Thiele's theory of partial valence.[142] (Extrapolation of this simple theory suggests that cyclobutadiene and cyclooctatetraene should be stable aromatic analogs of benzene.) It is of interest to note, however, that Willstätter apparently attached little importance to his attempted synthesis of cyclobutadienes, since he mentioned this work only in passing when he wrote his autobiography many years later.[154]

During the period 1915–1950, little work appeared in the literature concerning attempts to synthesize cyclobutadiene and its derivatives. It is probable that a number of investigations of at least a preliminary nature were initiated, but it appears that none of these was sufficiently fruitful to warrant publication. Investigations of this nature were carried out in the laboratories of Erwin Ott (Münster) and E. P. Kohler (Harvard) and their scope and purpose can be reconstructed at least in part from the dissertations of the students of these men. Thus, Stiebler[138] and Knoche,[92] at Münster, attempted to synthesize a number of cyclobutane derivatives potentially suitable as precursors of cyclobutadiene and, at Harvard, Jones[88] explored potential routes to some 1,3-disubstituted cyclobutanes, but was unable to prepare a suitable precursor of cyclobutadiene.

The most significant efforts of this era were carried out by E. R. Buchman and his co-workers (Schlatter, Reims, Deutsch, Silberman, Chemerda, and Howton) at the California Institute of Technology.[34, 35, 36, 37] Buchman attempted to prepare cyclobutadiene by applying the Hofmann degradation to various quaternary ammonium hydroxides of the cyclobutane series but his efforts were without success.[34] Thus, the thermal decomposition of *trans*-1,2-cyclobutanebis(trimethylammonium) hydroxide (*13*) afforded no hydrocarbon product and only cyclobutanone and basic products could be detected. A similar study using *cis*-1,3-cyclobutanebis(trimethylammonium) hydroxide (*14*) gave evidence of the formation of traces of vinylacetylene and cyclobutene only, and the thermal decomposition of 2-cyclobutenyltrimethylammonium hydroxide (*15*) gave negative results similar to those obtained with the 1,3-bis-(quaternary base). A diolefin believed to be methylenecyclobutene was detected in the thermolysis of (2-methylenecyclobutyl)trimethylammonium hydroxide (*16*).

The recent era of intensive investigation into the preparation and properties of the cyclobutadienes dates from the proposal of Longuet-Higgins and Orgel in 1956 that the cyclobutadienes should form stable complexes with some of the transition metals.[102] The proposal was subsequently substantiated through the preparation of the first cyclobutadiene–metal complex by Criegee and Schröder in 1959.[53]

1. ERRONEOUS CYCLOBUTADIENE STRUCTURES

From time to time, incorrect cyclobutadiene structures have been assigned to various reaction products. Considerable effort has been devoted to elucidating the structures of these compounds, but since they are not cyclobutadienes and have only a peripheral bearing on the subject, they are discussed here only briefly.*

(1) In 1890, Liebermann and Bergami obtained a compound having the empirical formula $(C_9H_6O)_2$, through cyclization of truxillic acid (17) with concentrated sulfuric acid, which they named "truxone" [(17) → (18)].[99] Reduction of truxone with hydriodic acid gave a hydrocarbon of the empirical formula $(C_9H_6)_n$ which they named "truxene" and which on oxidation with chromic acid gave a carbonyl-containing compound having the empirical formula $(C_9H_4O)_n$, to which they gave the name "truxenequinone." Liebermann and Bergami did not fix a value for n and merely reported that the low solubility and high melting point of truxene led them to suspect that n was greater than 2. This was subsequently confirmed by Hausmann, who prepared truxene by the acid-catalyzed trimerization of 1-indanone and correctly assigned to it the structure of tribenzylenebenzene (19).[85] In addition, truxenequinone was found to be identical with the tribenzoylenebenzene which had been obtained by Gabriel and Michael from phthalylacetic acid [(21) → (20)][73] and which had been shown by them to give sym-triphenylbenzene on destructive distillation with lime [(20) → (22)].[73] However, the conversion of truxone, a dimeric compound,† into tribenzylenebenzene, a trimeric compound, was too unusual a transformation for easy acceptance; this led Liebermann to propose an incorrect trimeric structure for truxone (23)[98] and, what is more regrettable, it caused Kipping to propose an equally incorrect dimeric cyclobutadiene structure for truxene, viz., 5,10-dihydrocyclobuta[1,2-a:3,4-a']diindene (24).[90] Kipping carried out some 28 ebulliscopic molecular weight determinations on truxene, and even though the results were impressively in

---

* Several cyclobutadiene structures proposed by Limpricht, Schiff, and others at a time when organic structure theory was in its infancy are omitted from the discussion; cf. "hydroxycyclobutadiene" for furan (ref. 100), "2-hydroxycyclobutadiene-1-carboxaldehyde" for furfuraldehyde (ref. 131), etc.

† Relative to cinnamic acid.

favor of the trimeric structure (*19*),[89] he rejected it in favor of its dimeric cognate (*24*) and proposed a similar structure, 5,10-cyclobuta[1,2-*a*:3,4-*a'*]diindenequinone (*25*) for truxenequinone. Finally, the correctness of the trimeric structures for truxene and truxenequinone, i.e., (*19*) and (*20*), was demonstrated conclusively by Stobbe[140] and confirmed by several other researchers.[38]

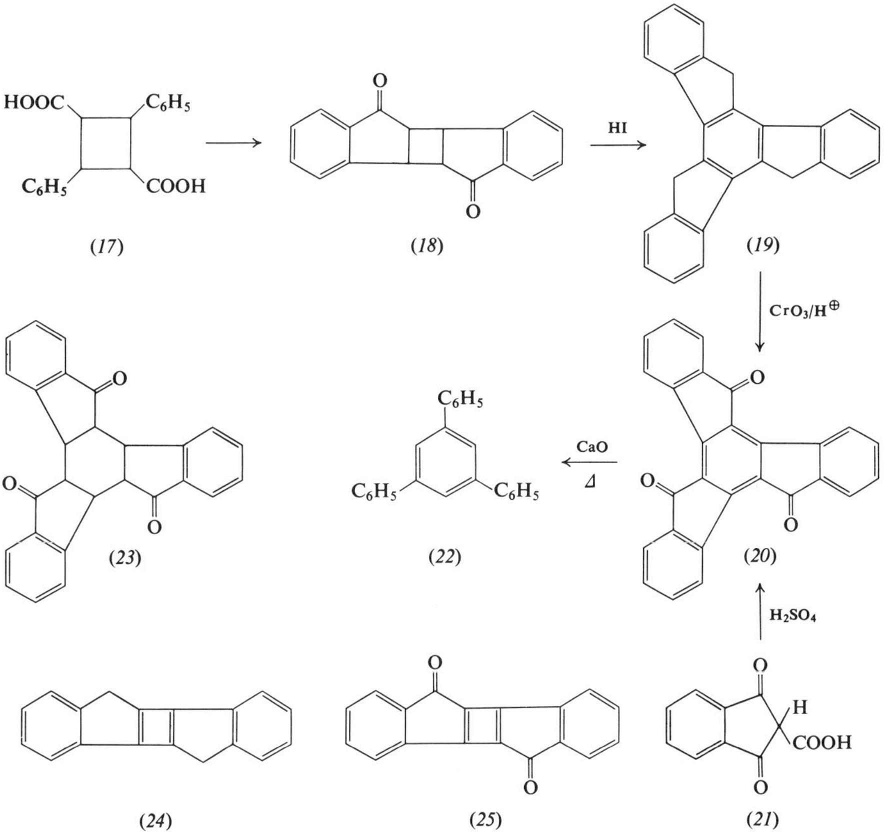

(2) Manthey[107] proposed an erroneous cyclobutadiene structure, 2,4-diphenylcyclobutadiene-1,3-dicarboxylic acid (*26*), for "triphenyltrimesic acid," a condensation product of phenylpropiolic acid.[95] His proposal was supported by Lanser who, however, appears to have preferred the isomeric structure, 3,4-diphenylcyclobutadiene-1,2-dicarboxylic acid (*27*).[96] Lanser prepared the silver salt, monomethyl ester, monethyl ester, anhydride, and imide of "triphenyltrimesic acid" and the ammonium salt and the silver salt of its monoamide, all of which he described erroneously as derivatives of (*27*).[96] Ruhemann and Merriman obtained "triphenyltrimesic acid" [to which

they also assigned structure (27)] as a product of the reaction of phenylpropiolyl chloride with ethyl sodiomalonate.[130] In addition, Ruhemann and Merriman claimed that they had prepared the anhydride of "triphenyltrimesic acid" [to which they assigned structure (28)] by the reaction of phenylpropiolyl chloride with acetylacetone in the presence of pyridine. "Triphenyltrimesic acid" has since been shown to be 1-phenylnaphthalene-2,3-dicarboxylic acid (29).[111, 139]

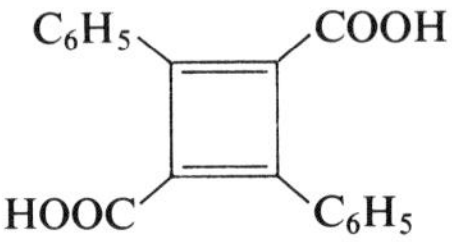

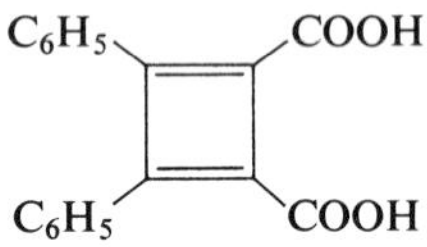

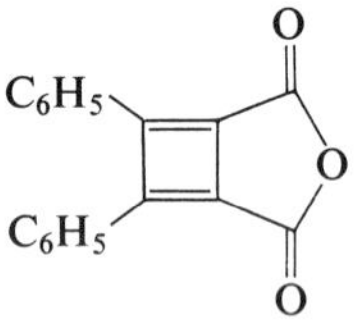

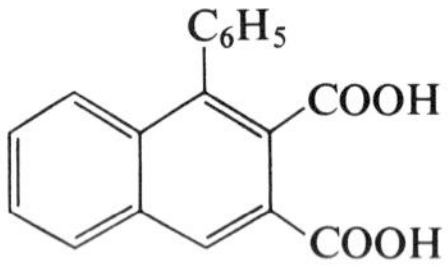

(3) Ruhemann and Merriman believed that a product which they had obtained by treating a cyclopentenedione derivative [(30), obtained in several steps from phenylpropiolic acid] with sodium carbonate was 2-acetyl-3-methyl-4-phenylcyclobutadienecarboxylic acid (31) and that an acid cyclization product of this substance was 1-acetyl-2-methyl-7H-cyclobuta[a]inden-7-one (32).[130] Structure (32) was said to be in accord with the ultraviolet spectrum of the cyclization product,[122] but, in fact, both (31) and (32) were subsequently

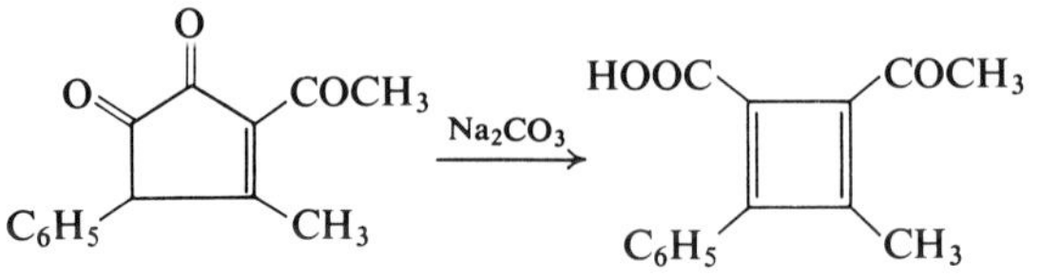

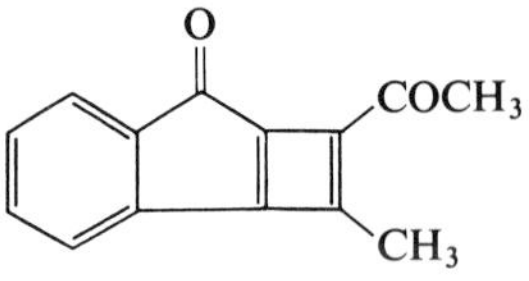

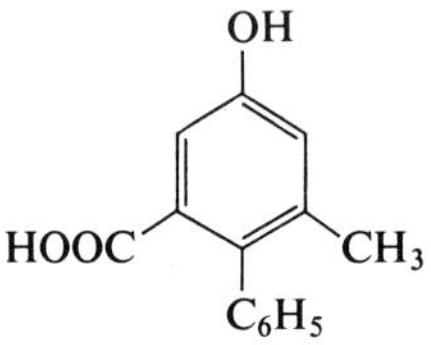

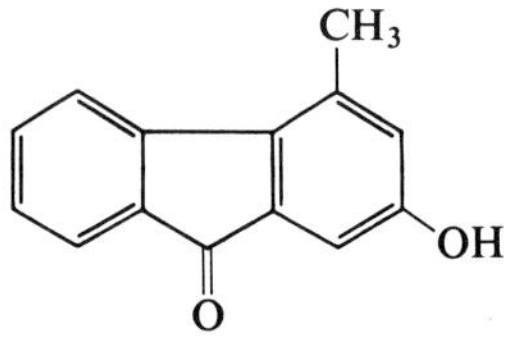

shown to be incorrect. The two products are 5-hydroxy-3-methyl-2-phenylbenzoic acid (*33*) and 2-hydroxy-4-methylfluorenone (*34*), respectively.[81,147]

(4) A compound obtained by Gastaldi and Cherchi as a by-product in the condensation of equimolar amounts of ethyl benzoate and acetophenone in the presence of sodium ethoxide, and also by heating acetophenone with sodium ethoxide, was described at first as 1,3-diphenylcyclobutadiene (*35*)[75] but it was later identified by the same authors as 3,5-diphenyltoluene (*36*).[74]

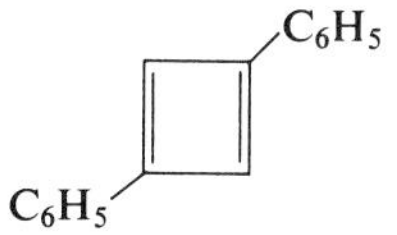

(*35*) 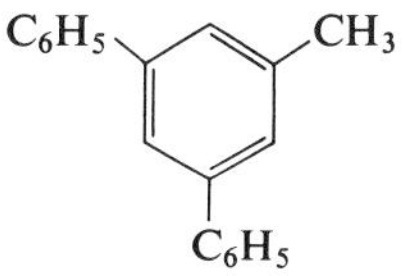   (*36*)

(5) Brass and Mosl proposed cyclobutadienoid structure (*37*) for a ketone which they obtained either by treating dithiadiene derivative (*38*) with mineral acid or by heating 2,3-dichloroindenone (*39*) with copper.[24] Reduction of the ketone afforded a hydrocarbon to which they assigned the corresponding cyclobutadiene structure, cyclobuta[1,2-*a*:4,3-*a'*]diindene (*40*), which is isomeric with the incorrect dimeric structure (*24*) suggested by Kipping for truxene. Brass and Mosl's compounds were subsequently shown to be tribenzoylenebenzene [(*20*), for (*37*)] and tribenzylenebenzene [(*19*), for (*40*)], respectively.[38,132,140]

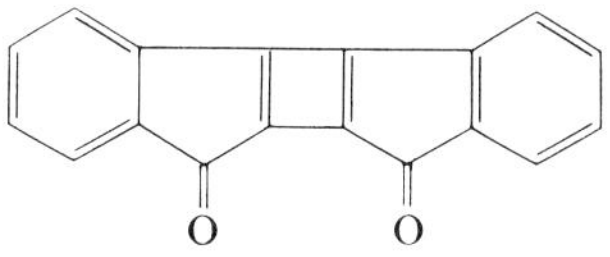

(*37*) 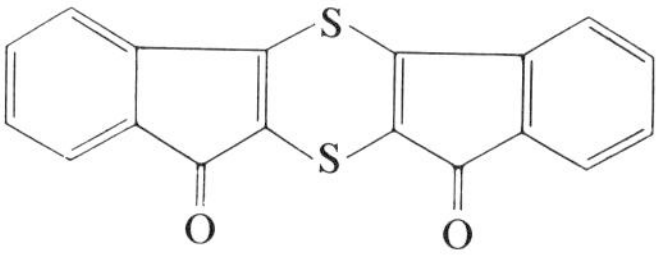   (*38*)

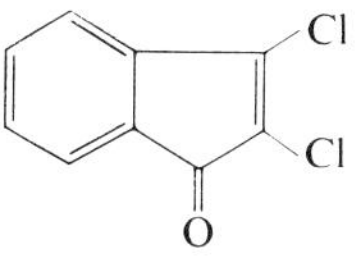

(*39*) 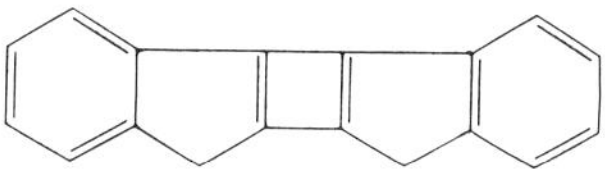   (*40*)

(6) A hydrocarbon, $C_{16}H_{12}$, obtained by treating 2-amino-1-phenylethanol with mineral acid, was reported at first to be 1,3-diphenylcyclobutadiene (*35*)[40]; subsequently, it was shown to be 2-phenylnaphthalene,[39,124] which was almost certainly produced via phenylacetaldehyde.[124]

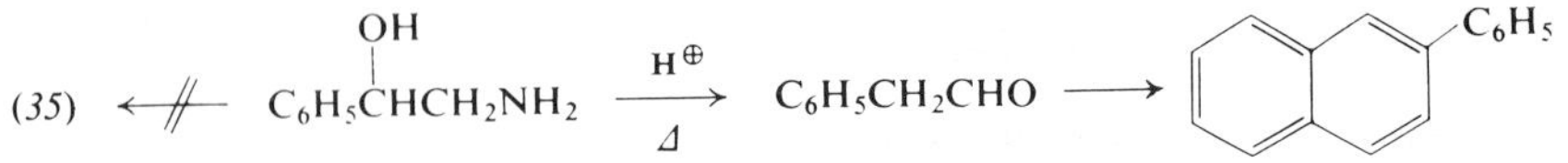

(35) ↚ C₆H₅CH(OH)CH₂NH₂ →[H⊕][Δ] C₆H₅CH₂CHO → [naphthalene-C₆H₅]

(7) Several cyclobutadiene structures, which were assigned on the basis of minimal evidence, remain to be challenged. Thus, Störmer and Biesenbach converted phenylacetaldehyde into a liquid hydrocarbon to which they assigned the structure of 1,3-diphenylcyclobutadiene.[141] Dixit assigned cyclobutadiene structures to several compounds [(44a)–(46b)] which he obtained by the transformation of three compounds believed to be the corresponding cyclobutenones [(41)–(43)].[59] The latter were prepared by the reaction of phenol, anisole, or 2-methylanisole with acetonedicarboxylic acid, but their identity is no more certain than that of the alleged cyclobutadienes.

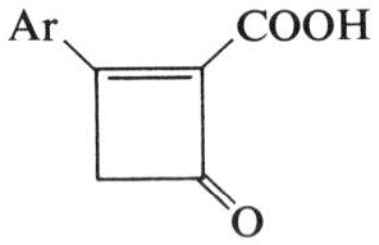

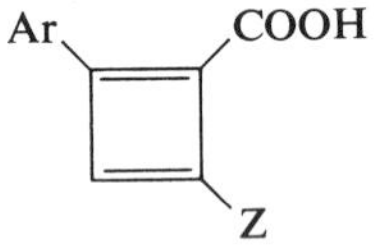

(41): Ar = p-HOC₆H₄
(42): Ar = p-MeOC₆H₄
(43): Ar = 3-Me-4-MeOC₆H₃

(44a): Ar = p-HOC₆H₄; Z = Cl
(44b): Ar = p-AcOC₆H₄; Z = AcO
(45a): Ar = p-MeOC₆H₄; Z = Cl
(45b): Ar = p-MeOC₆H₄; Z = AcO
(46a): Ar = 3-Me-4-MeOC₆H₃; Z = Cl
(46b): Ar = 3-Me-4-MeOC₆H₃; Z = AcO

A side-product obtained in unspecified yield in the reaction of phenylpropiolyl chloride with dry ethyl sodiomalonate was assumed to have either structure (47) or (48).[130] The compound is probably 2,2-dicarbethoxy-4-phenylbenz[f]indane-1,3-dione (49). [See also compound (29).]

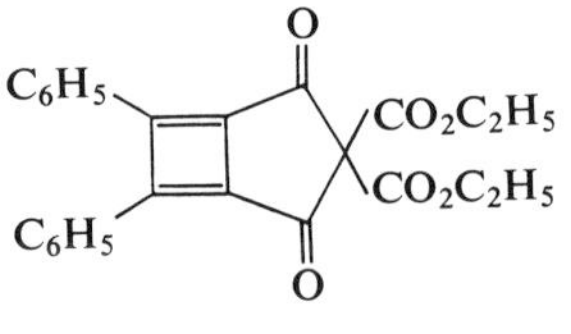

(47)

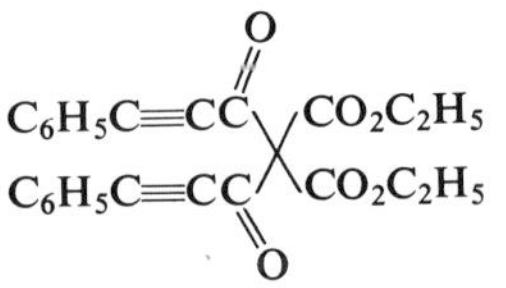

(48)

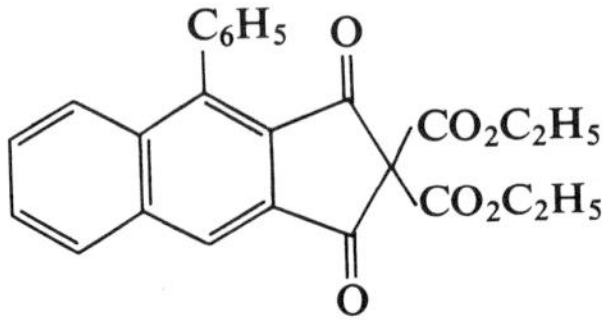

(49)

B. Generation of Cyclobutadiene and Substituted Cyclobutadienes

Neither cyclobutadiene nor a substituted cyclobutadiene has been isolated and characterized directly. Although many reactions are known in which a cyclobutadiene is probably generated as a transient intermediate, the question of whether or not such a cyclobutadiene intermediate is actually produced in any given case can usually be answered only on the basis of indirect evidence, i.e., through the isolation of a stable product which can reasonably be viewed as a primary, or even as a secondary, transformation product of a cyclobutadiene.* The product isolated can be a cyclobutadiene dimer, a Diels–Alder adduct, a cleavage product of a cyclobutadiene, or a cyclobutadiene–metal complex; or it can be a product arising from any one of these primary transformation products through further chemical change or rearrangement.

At the present time, it is not possible to make a clear and meaningful distinction between those reactions in which cyclobutadiene intermediates are actually generated and those in which such intermediates merely appear to have been generated. As a result, *all* reactions that give at least one product formally attributable to a cyclobutadiene intermediate have had to be included in the present section.† A more critical evaluation of these reactions, together with arguments to show why some of them probably do not proceed via true cyclobutadiene intermediates, is reserved for Section D. In the interim, the reader would do well to view each case as it is presented, as lying somewhere in a continuum of probability ranging from the certain to the unlikely: Some cases will be found to be difficult or even impossible to rationalize without invoking a cyclobutadiene intermediate; others will be less clear-cut and, in these, a cyclobutadiene intermediate will sometimes be merely the most attractive, rather than the most compelling, of several alternatives.

Reactions that produce cyclobutadiene–metal complexes are discussed in Chapter 2.

1. Dehalogenation

When *all-trans*-1,2,3,4-tetrabromocyclobutane (*1*) is treated with lithium amalgam in ether,[9] or when *cis*-3,4-dichlorocyclobutene (*2*) is treated with sodium amalgam in ether,[5] the syn dimer (*4*) of cyclobutadiene (*3*) is formed. In contrast, the dehalogenation of (*2*) with *lithium* amalgam gives the anti dimer, which is probably not formed via cyclobutadiene (see Section E, 1, a).

---

* For the first example of the unambiguous detection of cyclobutadiene, see Abstract 1.23 in the Appendix.

† For a discussion of a number of reactions believed *not* to proceed via cyclobutadiene intermediates, see Section C, "Unsuccessful Approaches to the Cyclobutadienes."

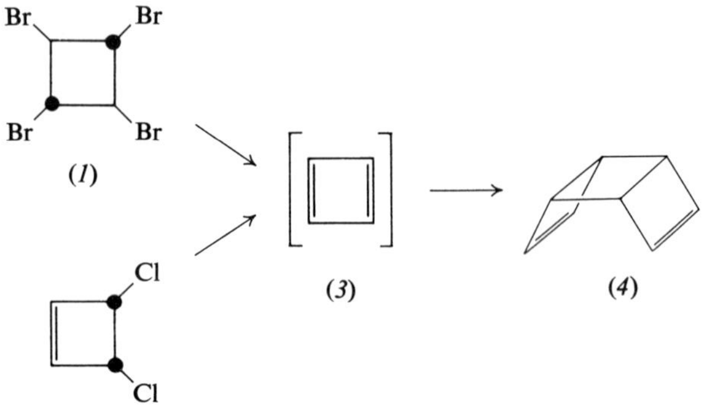

(1) (2) (3) (4)

The dehalogenation of *cis*-3,4-dichloro-1,2,3,4-tetramethylcyclobutene (*5a*) with lithium amalgam in ether[49] or furan,[49, 52] sodium amalgam in ether,[52] lithium in liquid ammonia,[1] zinc–copper couple in various solvents,[49] *n*-butyllithium in ether,[1] activated zinc dust,[20] or potassium in the vapor phase[135] results in the formation of the syn dimer (*7*) of tetramethylcyclobutadiene (*6*).

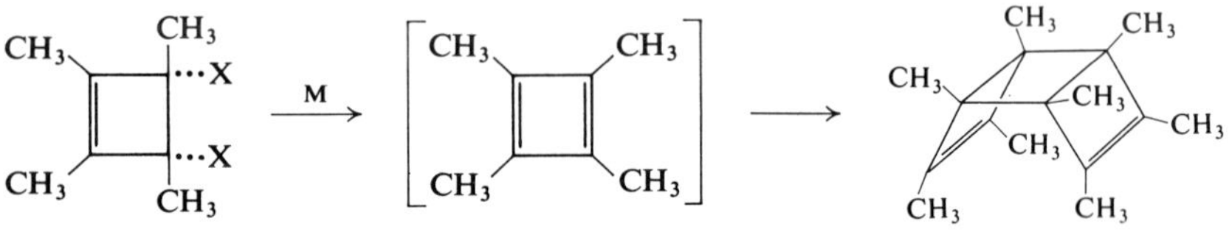

(*5a*): X=Cl
(*5b*): X=Br
(*5c*): X=I

(*6*) (*7*)

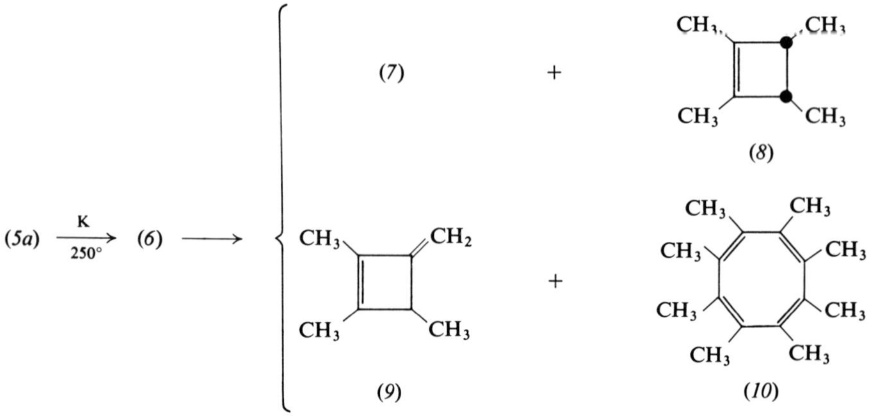

(*8*) (*9*) (*10*)

B. GENERATION OF CYCLOBUTADIENE

[The dehalogenation of the dichloride (*5a*) with potassium vapor gives in addition two disproportionation products, (*8*) and (*9*), of tetramethylcyclobutadiene (*6*) and a rearrangement product (*10*) of its syn dimer (*7*)[135] (see also Section D).] The reaction of *cis*-3,4-dibromo-1,2,3,4-tetramethylcyclobutene (*5b*) with lithium amalgam in ether,[52] and the reaction of *cis*-3,4-diiodo-1,2,3,4-tetramethylcyclobutene (*5c*) with zinc[93] also afford the syn dimer (*7*) of tetramethylcyclobutadiene.

The formation of tetramethylcyclobutadiene as a transient intermediate in the dehalogenation of *cis*-3,4-dichloro-1,2,3,4-tetramethylcyclobutene with lithium amalgam, zinc, etc., is supported by the results of trapping experiments in which adducts of tetramethylcyclobutadiene were formed. Thus, dehalogenation of the dichloride (*5a*) with activated zinc dust in the presence of 2-butyne and dimethyl acetylenedicarboxylate gave hexamethylbenzene (*12a*) and dimethyl tetramethylphthalate (*12b*), respectively.*,[20]

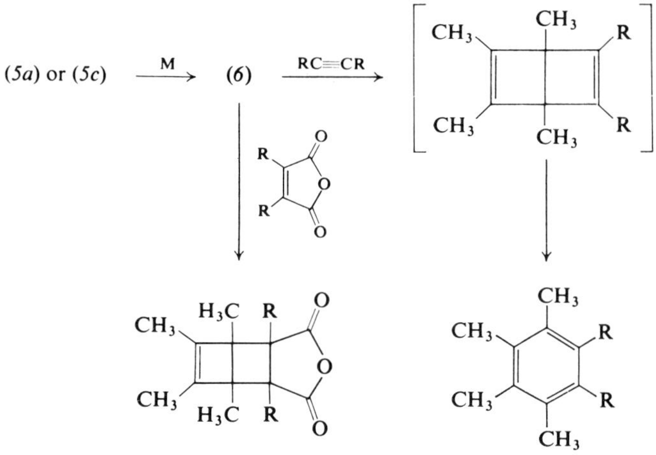

$$(5a) \text{ or } (5c) \xrightarrow{M} (6) \xrightarrow{RC\equiv CR} \left[\begin{array}{c}\text{tetramethyl cyclobutadiene-alkyne adduct}\end{array}\right]$$

(*11a*): R = H
(*11b*): R = CH$_3$
(*11c*): R = Br

(*12a*): R = CH$_3$
(*12b*): R = CO$_2$CH$_3$

The dehalogenation of *cis*-3,4-diiodo-1,2,3,4-tetramethylcyclobutene (*5c*) with mercury also probably generates tetramethylcyclobutadiene, as shown by trapping experiments in which adducts of maleic anhydride,[45] dimethylmaleic anhydride,[45] and dibromomaleic anhydride[56] (*11a*, *b*, and *c*, respectively) were formed.

A key synthetic intermediate in all work with tetramethylcyclobutadiene is

---

* For a cautionary note to the effect that tetramethylcyclobutadiene may not be produced under all circumstances in the dehalogenation of (*5a*) see Section E, 2, c and ref. 45.

cis-3,4-dichloro-1,2,3,4-tetramethylcyclobutene (*5a*). The dichloride was first prepared in 1952 by the reaction of 2-butyne with chlorine or thionyl chloride.[136] Since then, the method of preparation has been improved to such an extent that a yield of 50% of the dichloride can be obtained in a one-step operation in which 2-butyne is treated with chlorine at $-20°$ in the presence of boron trifluoride.[51] The method appears to be limited to the synthesis of the tetramethyl compound (*5a*) inasmuch as all attempts to prepare dichlorocyclobutenes substituted with other groups have been totally unsuccessful.[45] The dibromide (*5b*) and the diiodide (*5c*) are available from the dichloride by standard methods.[127]

The dehalogenation of 3,4-dibromo-1,2,3,4-tetraphenylcyclobutene (*13*) with phenyllithium yields an unstable dimer (*15*) of tetraphenylcyclobutadiene (*14*), probably the syn dimer.[69] (See also Section E, 1, f.)

(*13*) (*14*) (*15*)

2. Dehydrohalogenation

Heating 1,2-dibromocyclobutane (*16*) with powdered potassium hydroxide at 210° gives acetylene,[13, 156] which appears to be formed through generation and subsequent fissioning of cyclobutadiene (*3*). However, since no other example of acetylene formation from a cyclobutadiene has been encountered, the fissioning reaction, and hence the generation of cyclobutadiene, must be viewed as unlikely. (See also Section F, 5, h.)

In contrast, dehydrobromination of the dibromide (*16*) with boiling quinoline gives a small amount of butadiene and a red nitrogen-containing polymer.[156] It is possible that cyclobutadiene (*3*) is actually generated in this case and that it abstracts two hydrogen atoms from quinoline (or from the starting material) to give cyclobutene which then undergoes thermal isomerization to give butadiene.

When a mixture of 1,2- and 1,3-dibromocyclobutane (obtained by the aluminum bromide-catalyzed isomerization of 1,2-dibromocyclobutane) was dehalogenated with solid potassium hydroxide at 110°, vinylacetylene was obtained in 36.6% yield.[8] Although the product may be viewed as a rearranged cleavage product of cyclobutadiene, in all likelihood it was formed via 1-bromo-1,3-butadiene (*18*).[8]

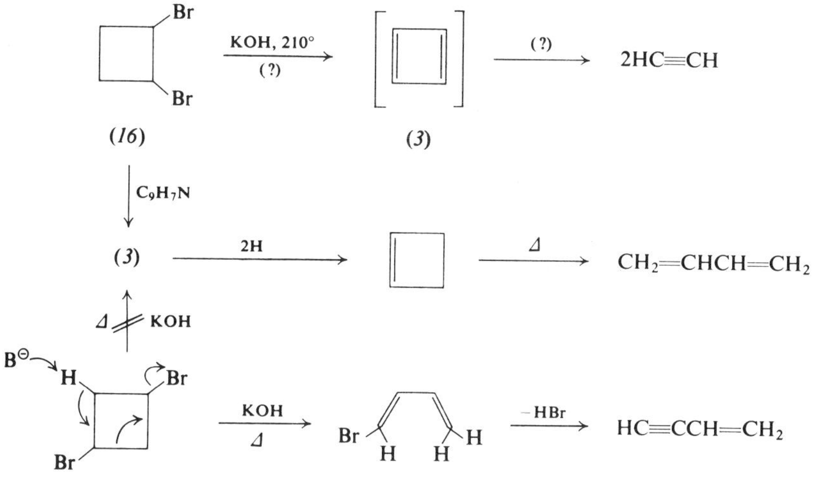

B. GENERATION OF CYCLOBUTADIENE

The "dehydrohalogenation" of 1,2-dichloro-*all-trans*-1,2,3,4-tetraphenylcyclobutane (*19*) with a variety of strong bases (phenyllithium, phenylsodium, or lithium diethylamide) gives 1,2,3,4-tetraphenylbutadiene (*20*) in high yield, possibly via tetraphenylcyclobutadiene (*14*).[134] However, it is probable that the reaction proceeds by *dehalogenation* of the dichloride (*19*) to give tetraphenylcyclobutene, followed by rearrangement of the latter to give tetraphenylbutadiene. (See also Section E, 5, a.)

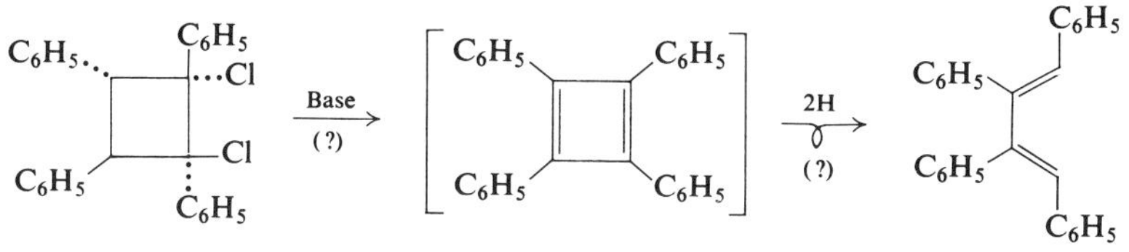

A dimer (*29*) of 1-fluoro-2,3,4-triphenylcyclobutadiene (*28*) was obtained by treating 2,4-dichloro-3,3-difluoro-1-phenylcyclobutene (*21*) with phenyllithium, presumably through formation and subsequent dehydrohalogenation of intermediate (*25*).[91] The latter was in fact formed in good yield when the proper stoichiometric amounts of dihalide (*21*) and phenyllithium were combined, and it was found to undergo dehydrohalogenation in the expected manner when treated with phenyllithium [(*25*) → (*28*) → (*29*)].

Two other tetrahalo-1-phenylcyclobutenes and the compounds obtained from them by treatment with appropriate stoichiometric amounts of phenyllithium were used to synthesize the same dimer (29) of 1-fluoro-2,3,4-triphenylcyclobutadiene (28) via the common intermediate (25). Thus, dimer (29) was obtained from 4,4-dichloro-3,3-difluoro-1-phenylcyclobutene (24) by treatment with excess phenyllithium [(24) → (25) → (28) → (29)].[116] Similarly, 2-chloro-1,4-diphenyl-3,3-difluorocyclobutene (27), which is undoubtedly the first intermediate in the transformation of (24), also gave the dimer (29) of fluorotriphenylcyclobutadiene when treated with excess phenyllithium.[91,116] Finally, 4-chloro-1-phenyl-3,3,4-trifluorocyclobutene (23) reacted analogously with phenyllithium, as did 1,4-diphenyl-2,3,3-trifluorocyclobutene (26).[116]

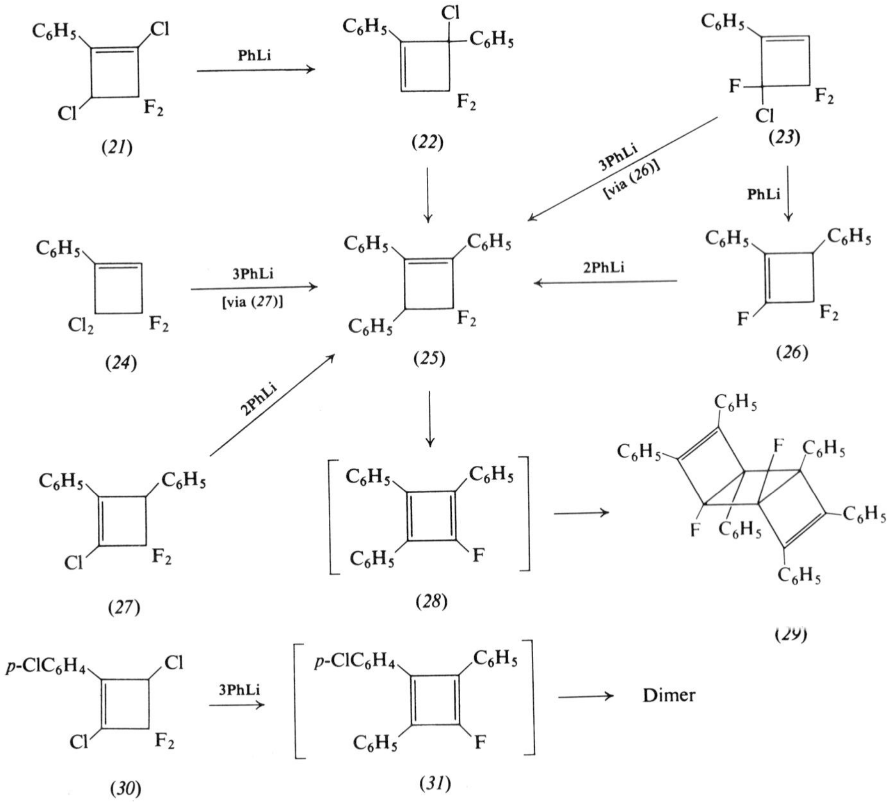

In a reaction analogous to the conversion of (21) into dimer (29), 2,4-dichloro-3,3-difluoro-1-(p-chlorophenyl)cyclobutene (30) reacted with excess phenyllithium to give a dimer of 1-(p-chlorophenyl)-2,4-diphenyl-3-fluorocyclobutadiene (31).[116]

The dehydrohalogenation of 1-(1'-cyclohexenyl)-2,4-dichloro-3,3-difluorocyclobutene (32) with triethylamine at 180° yields 1,3-dichloro-2-fluoro-

5,6,7,8-tetrahydronaphthalene (*34*), which is presumably a rearrangement product of 1-(1'-cyclohexenyl)-2,4-dichloro-3-fluorocyclobutadiene (*33*).[133] The same product is formed by dehydrohalogenation of 1-(1'-cyclohexenyl)-4,4-dichloro-3,3-difluorocyclobutene.

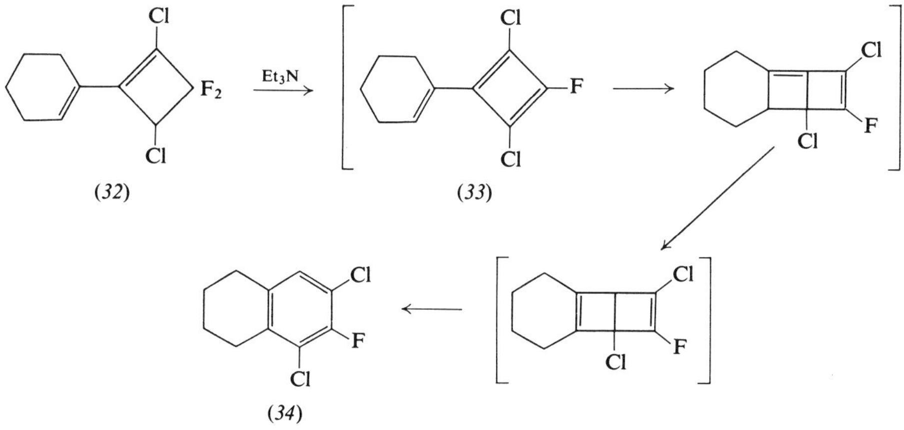

3. Acetylene Dimerization

Cyclobutadienes appear to be generated from acetylenes under certain circumstances, i.e., if the acetylenic compound is subjected to ultraviolet irradiation or if the olefinic linkage is naturally strained. Thus, cyclohexyne (*36*), a highly strained acetylene which is known only as a transient intermediate, undergoes tetramerization to yield the dimer (*40*) of 1,2,3,4,5,6,7,8-octahydrobiphenylene (*37*).[158] [The cyclohexyne was generated in tetrahydrofuran solution by the action of magnesium or lithium amalgam on 1,2-dibromocyclohexene (*35a*) or 1-bromo-2-fluorocyclohexene (*35b*).] A side product of the reaction, formed in small yield (3–9%), is dodecahydrotriphenylene (*38*). The side product (*38*) may be viewed as the rearrangement product of a Diels–Alder adduct (*39*) of octahydrobiphenylene and cyclohexyne, but substantiating evidence is lacking (see Section E, 2, c).

Similarly, 3,3,6,6-tetramethylcyclohexyne (*42*) dimerizes to give the cyclobutadiene derivative, 1,1,4,4,5,5,8,8-octamethyl-1,2,3,4,5,6,7,8-octahydrobiphenylene (*45*). [The cyclohexyne was generated by the reaction of 3,3,6,6-tetramethylcyclohexane-1,2-dione dihydrazone (*41*) with silver oxide in benzene or, better, by the reaction of 3,3,6,6-tetramethyl-1,2-dibromocyclohexene (*43*) with sodium in ether at room temperature.] However, in contrast to octahydrobiphenylene, the octamethyl analog undergoes rearrangement [(*45*) → (*44a*) (or *b*)] and hydrogen abstraction [(*45*) → (*46a*) (or *b*)] rather than dimerization.[2]

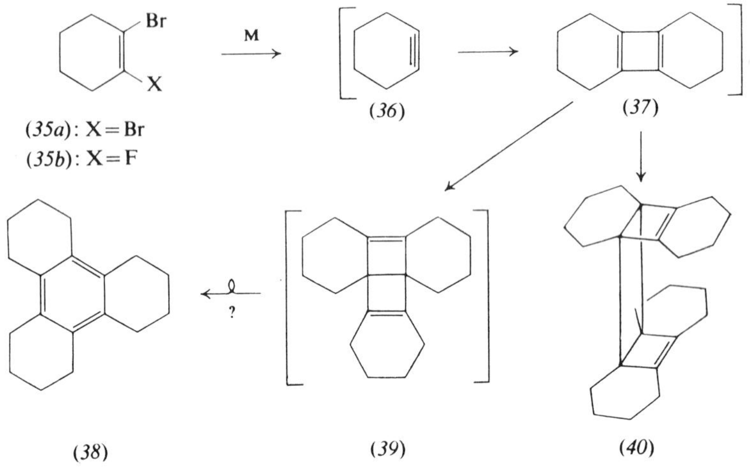

Br
X
M
(35a): X=Br
(35b): X=F
(36)
(37)
?
(38)
(39)
(40)

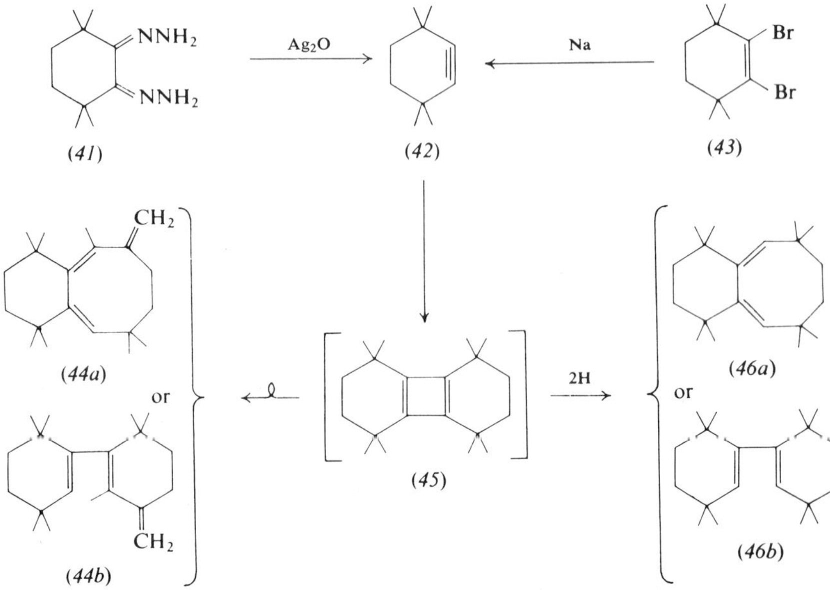

NNH2
NNH2
Ag2O
Na
Br
Br
(41)
(42)
(43)
CH2
(44a)
or
CH2
(44b)
2H
(45)
(46a)
or
(46b)

Prolonged ultraviolet irradiation of diphenylacetylene in hexane affords octaphenylcyclooctatetraene (*47*), which can be viewed as a rearrangement product of tetraphenylcyclobutadiene dimer (*15*).[33] The other products of the reaction are 1,2,3-triphenylazulene (*49*), 1,2,3-triphenylnaphthalene (*49a*), and hexaphenylbenzene (*49b*), each of which can be attributed to tetraphenylcyclobutadiene (see also Section E, 2, c and Section E, 3, b).

B. GENERATION OF CYCLOBUTADIENE

$$2C_6H_5C{\equiv}CC_6H_5 \xrightarrow{h\nu} \begin{bmatrix} \text{tetraphenylcyclobutadiene} \end{bmatrix}$$
(14)

(49) 1,2,3-triphenylazulene + (49a) + (49b) + (47) octaphenylcyclooctatetraene

Octaphenylcyclooctatetraene is also formed when diphenylacetylene is treated with phenylmagnesium bromide.[143] It is possible that an unstable cyclobutadiene–magnesium complex is involved in this transformation. (But see also Chapter 2.)

Irradiation of acetylene gives cyclooctatetraene in trace amounts, presumably via cyclobutadiene.[94]

4. RING CLOSURE OF BUTADIENES

Butadienes suitably substituted with labile or reactive groups in the 1,4-positions can be made to undergo ring closure under the proper reaction conditions to give cyclobutadienes. Thus far, however, the method has been applied to the generation of tetraphenylcyclobutadiene only.

Octaphenylcyclooctatetraene (47), which may be viewed as the rearrangement product of the syn dimer (15) of tetraphenylcyclobutadiene (14), is obtained (a) by thermolysis of (4-bromo-1,2,3,4-tetraphenyl-*cis,cis*-1,3-butadienyl)dimethyltin bromide (48) in triglyme at about 130°,[67] (b) by reaction of 1,4-dilithiotetraphenylbutadiene (48a) with cupric bromide in refluxing 1,2-dimethoxyethane,[25] and (c) by thermolysis of 1,4-diiodotetraphenylbutadiene (48b) in 2-ethoxyethanol at 200°.[25] The reaction of the dilithio compound (48a) with the diiodo compound (48b) gives a compound believed to be the syn dimer (15) of tetraphenylcyclobutadiene, which isomerizes readily to octaphenylcyclooctatetraene (47).[69] It is possible that the reaction between (48a) and (48b) proceeds by metal–halogen interchange to give an iodo-lithio intermediate (48c) which undergoes ring closure to give tetraphenylcyclobutadiene, etc.

When 1-bromo-4-iodo-1,2,3,4-tetraphenylbutadiene (48d) is treated with activated zinc in refluxing tetrahydrofuran for 18 hours, 1,2,3,4-tetraphenylbutadiene, 1,2,3-triphenylazulene (49) and a hydrocarbon $C_{28}H_{22}$ of unknown structure are produced.[66] The suggestion has been made that triphenylazulene arises by way of tetraphenylcyclobutadiene (*14*),[66] but proof is lacking.

Ultraviolet irradiation of (4-iodo-1,2,3,4-tetraphenyl-*cis,cis*-1,3-butadienyl)-dimethyltin iodide (*48e*) for 8 hours in benzene gave only triphenylazulene.*[, 66]

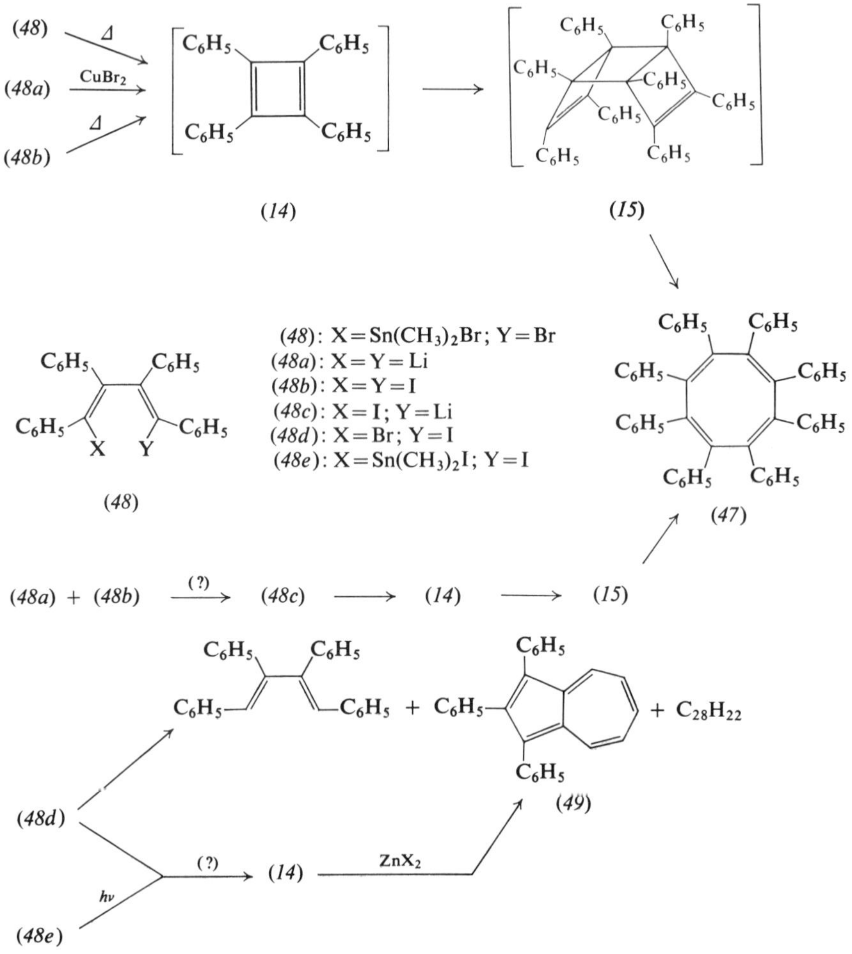

(48)
(48a)
(48b)
Δ
CuBr2
(14)
(15)
(47)
(48): X=Sn(CH3)2Br; Y=Br
(48a): X=Y=Li
(48b): X=Y=I
(48c): X=I; Y=Li
(48d): X=Br; Y=I
(48e): X=Sn(CH3)2I; Y=I
(48a) + (48b) (?) (48c) (14) (15)
+ C28H22
(49)
(48d)
(48e)
hν
(?)
ZnX2

The formation of tetraphenylcyclobutadiene in the thermolysis of (4-bromo-1,2,3,4-tetraphenyl-*cis,cis*-1,3-butadienyl)dimethyltin bromide is supported by the formation of adducts with dimethyl acetylenedicarboxylate, diethyl fumarate, diethyl maleate, and dimethyl maleate.[67] (See Section E, 2, c.)

* But the formation of triphenylazulene is not necessarily indicative of the intermediacy of tetraphenylcyclobutadiene. See Section C, 8, a and Section E, 3, b.

5. HOFMANN ELIMINATION AND RELATED METHODS

Butadiene is formed in small yield when 1,3-bis(dimethylamino)cyclobutane dimethiodide (*50a*) is heated with aqueous potassium hydroxide or with silver oxide suspended in methanol.[13] Butadiene is also obtained when solid 1,3-bis(dimethylamino)cyclobutane dimethohydroxide (*50b*) is subjected to dry pyrolysis, ostensibly via the sequence cyclobutadiene → cyclobutene → butadiene.[13]

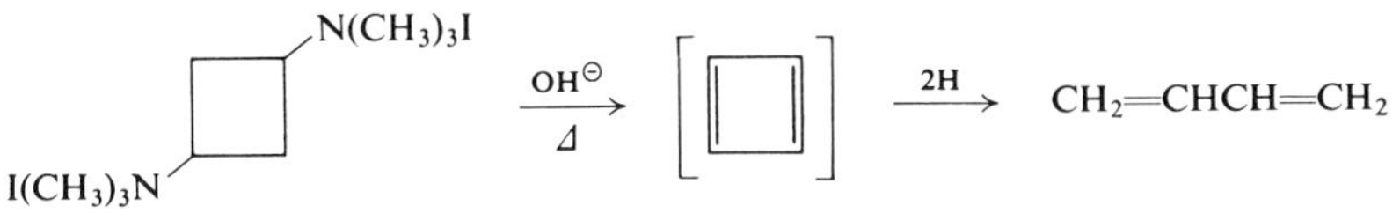

(*50a*): X=I
(*50b*): X=OH

Similarly, the reaction of 1,3-diphenyl-2,4-bis(dimethylamino)cyclobutane dimethiodide (*51*) with lithium diethylamide in furan at −5° yields a compound believed to be *trans*-1,3-diphenylbutadiene (*53*), together with two dimers

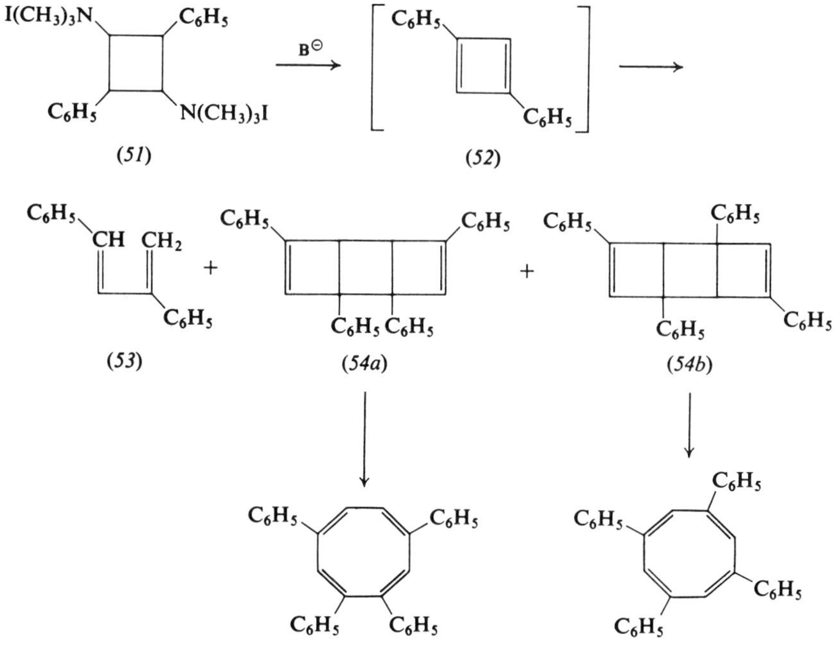

[(54a) and (54b)] of 1,3-diphenylcyclobutadiene (52).[151] If the elimination reaction is carried out with potassium *tert*-butoxide in *tert*-butyl alcohol, 1,3,4,6- and 1,3,5,7-tetraphenylcyclooctatetraene are obtained [(55a) and (55b), respectively]. The cyclooctatetraenes are believed to be formed by thermal rearrangement of the corresponding 1,3-diphenylcyclobutadiene dimers, (54a)[153] and (54b).

6. REVERSE DIELS–ALDER REACTION

The pyrolysis of several compounds which are formal adducts of cyclobutadiene yields rearrangement products and, in smaller yield, butadiene. Both the rearrangement products and the butadiene have been ascribed to cyclobutadienoid intermediates but it now appears likely that the rearrangement products were not formed in this way (see also Section D and Section E, 2, b).

Pyrolysis of dimethyl tricyclo[4.2.2.0^{2,5}]deca-3,7,9-triene-7,8-dicarboxylate (56) yields butadiene, dimethyl phthalate (57),[42] dimethyl 2,6-naphthalate (58),[13,42] dimethyl 3,4-dihydro-2,6-naphthalate (59),[42] and dimethyl 1,2-dihydro-2,6-naphthalate (60).[42] Similarly, pyrolysis of tricyclo[4.2.2.0^{2,5}]deca-3,7,9-triene (61) gives butadiene, benzene, naphthalene, 1,2-dihydronaphthalene (62), and tetralin[118]; 2a,3,10,10a-tetrahydro-3,10-ethenocyclobuta[b]-anthra-4,9-quinone (63) gives butadiene, anthraquinone, and benz[a]anthra-7,12-quinone (64)[13]; 4,7-dimethoxy-2a,3,8,8a-tetrahydro-3,8-ethenocyclobuta[b]naphthalene (65a) yields butadiene, 1,4-dimethoxynaphthalene, and 1,4-dimethoxyphenanthrene (66a)[13]; the 4,7-diacetoxy analog (65b) of (65a) yields butadiene and the corresponding naphthalene and phenanthrene (66b) derivatives.[13]

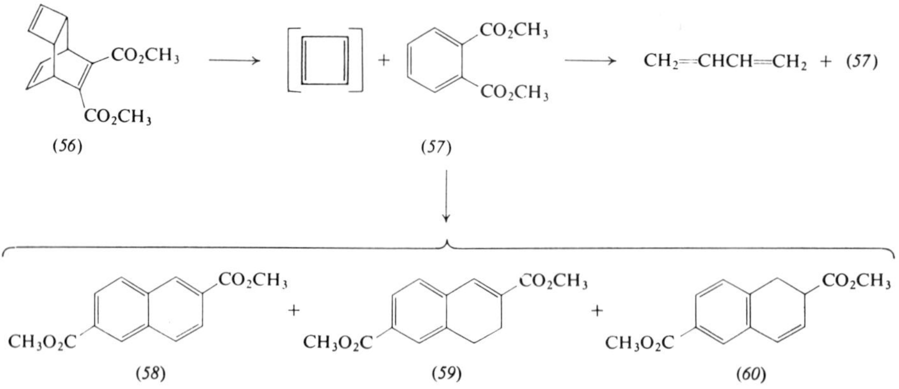

B. GENERATION OF CYCLOBUTADIENE 23

(61) → CH$_2$=CHCH=CH$_2$ + [naphthalene-like structures] (62)

(63) → CH$_2$=CHCH=CH$_2$ + [anthraquinone] + [benz[a]anthraquinone] (64)

(65a): R = CH$_3$
(65b): R = CH$_3$CO

→ CH$_2$=CHCH=CH$_2$ + [dimethoxynaphthalene] + [dimethoxyphenanthrene]

(66a): R = CH$_3$
(66b): R = CH$_3$CO

7. Decomposition of Cyclobutadiene–Metal Complexes*

Tetramethylcyclobutadienenickel chloride (67) decomposes at 190° under 0.01 mm pressure to give octamethylcyclooctatetraene (10), tetramethylbutadiene, and an almost inseparable liquid mixture of two isomeric C$_{16}$H$_{24}$ hydrocarbons, (68a) and (68b).[45] All four products may be viewed as having been derived from tetramethylcyclobutadiene. [The cyclooctatetraene probably arises by valence isomerization of the syn dimer of tetramethylcyclobutadiene, (7) ⤳ (10)[135]; tetramethylbutadiene very likely results from tetramethylcyclobutadiene by hydrogen abstraction, followed by thermal isomerization of the intermediate tetramethylcyclobutene[135]; the C$_{16}$H$_{24}$ isomeric hydrocarbons are known to be formed by acid-catalyzed rearrangement of 1,2,3,4,5,6,7,8-octamethylbicyclo[4.2.0]oct-2,4,7-triene (70), which is itself formed in good yield by the valence isomerization of *syn*- and *anti*-octamethyltricyclooctadiene, (7) and (69), respectively, and octamethylcyclooctatetraene (10).[55]] A methylenebicyclo[5.1.0]octadiene structure (71) was tentatively proposed for the C$_{16}$H$_{24}$ liquid mixture[53a] before it was realized that it is composed of the two isomeric hydrocarbons (68a) and (68b).[55] Thermolysis of the triphenylphosphine complex of tetramethylcyclobutadienenickel chloride [C$_8$H$_{12}$NiCl$_2$·P(C$_6$H$_5$)$_3$] gives the same four products; in contrast, thermolysis of the phenanthroline complex (C$_8$H$_{12}$NiCl$_2$·C$_{12}$H$_8$N$_2$)[121] gives *anti*-1,2,3,4,5,6,7,8-octamethyltricyclo[4.2.0.0^{2,5}]octa-3,7-diene (69), 3-methyl-

---
* Reactions that produce cyclobutadiene–metal complexes are discussed in Chapter 2.

ene-1,2,4-trimethylcyclobutene (9), as well as the $C_{16}H_{24}$ liquid hydrocarbon mixture (68a and b).[45, 46] The anti tricyclo compound (69) is also formed when tetramethylcyclobutadienenickel chloride is decomposed in boiling water.[54] It is not at all certain that tetramethylcyclobutadiene is an intermediate in the formation of (69).

1,2,3,4-Tetramethylcyclobutadienenickel chloride reacts with cyclopentadienylsodium[50] to give a mixture of two isomeric π-cyclobutenylnickel–cyclopentadiene complexes,[47] possibly via tetramethylcyclobutadiene (see Chapter

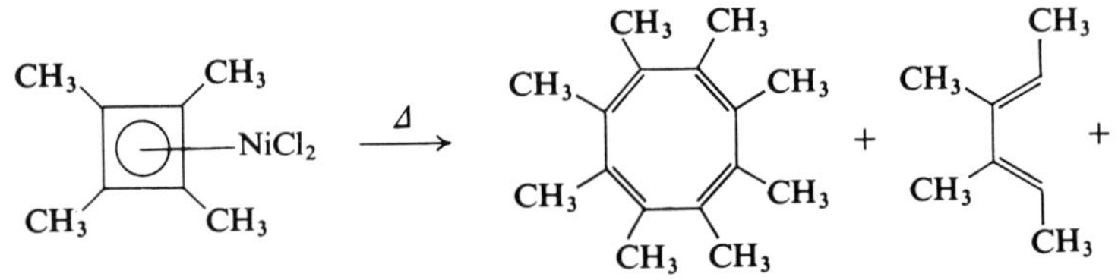

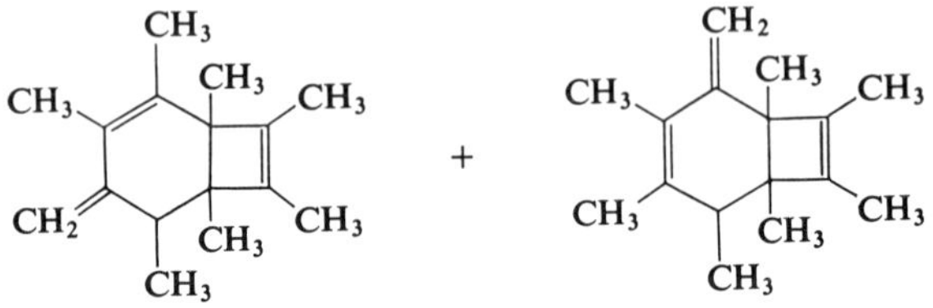

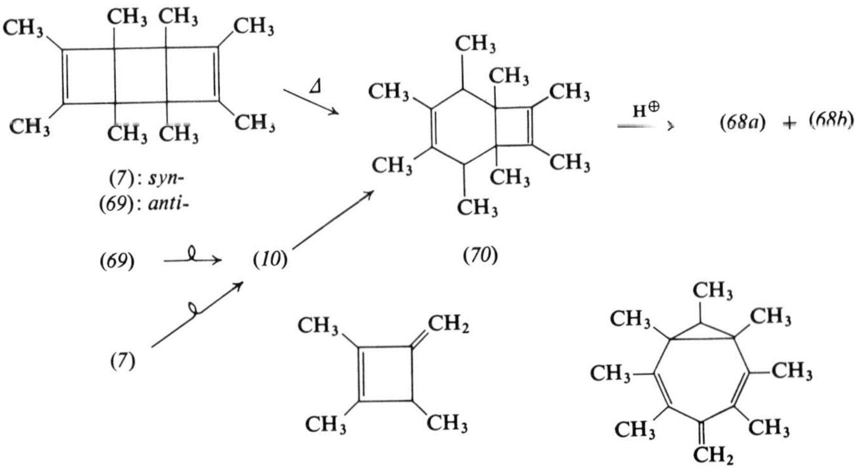

2, Section C). Treating an aqueous solution of tetramethylcyclobutadiene-nickel chloride with sodium nitrite gives cis-1,2,3,4-tetramethylcyclobutene-3,4-diol in 50% yield, again, possibly via tetramethylcyclobutadiene.[53]

The dry thermolysis of tetraphenylcyclobutadienepalladium chloride (72) affords mainly 1,4-dichloro-1,2,3,4-tetraphenylbutadiene (48f),[58] but it is not known whether or not tetraphenylcyclobutadiene (14) is an intermediate in this transformation. Treating tetraphenylcyclobutadienepalladium chloride with triphenylphosphine or trimethyl phosphite generates tetraphenylcyclobutadiene which is fairly stable in the absence of air, as shown by the persistence of its green color for several hours in solution and the formation ultimately of the dimer, octaphenylcyclooctatetraene (47), in 70% yield.[105]

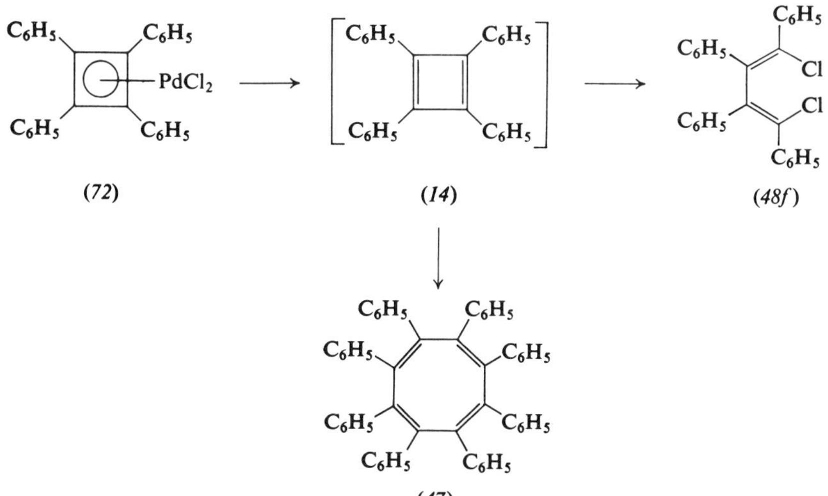

8. Thermolysis of a Heterocyclopentadiene

Octaphenylcyclooctatetraene (47) is produced in the thermal decomposition of a compound believed to be tetraphenylmercurole (73), possibly by way of tetraphenylcyclobutadiene.[25]

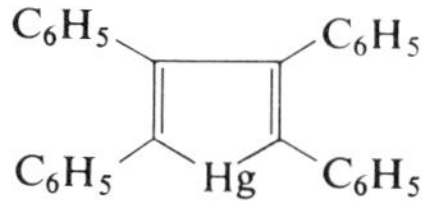

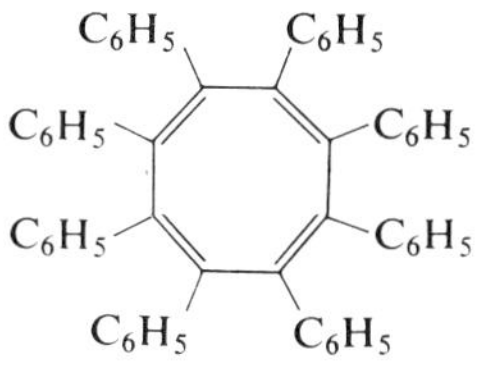

C. Unsuccessful Approaches to the Cyclobutadienes*

In the previous section, those reactions were discussed in which cyclobutadiene intermediates are generated (or at least appear to be generated) as evidenced by the formation of secondary products (dimers, adducts, etc.) attributable to such intermediates. In the present section, those reactions are described that lead to patently negative, rather than to positive or merely inconclusive, results.

1. By Repositioning an Unsaturated Linkage

A number of unsuccessful attempts have been made to generate substituted cyclobutadienes by shifting or repositioning an unsaturated linkage. In one of the simplest of these, it was found that phenylcyclobutadienequinone (*1*) could not be reduced to the corresponding cyclobutadienoid hydroquinone (*2*) under a variety of conditions.[137] (See also Chapter 4, Section D, 1, d.)

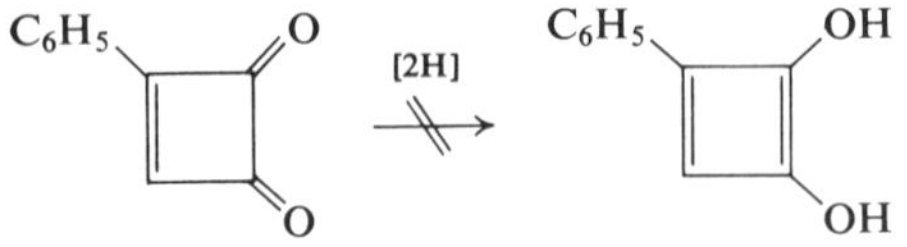

In a related approach, ultraviolet irradiation of a mixture of diphenylcyclobutadienequinone (*4*) and stilbene afforded only a dimer of (*4*) rather than the desired cyclobutadienoid 1,4-dioxene (*3*).[21]

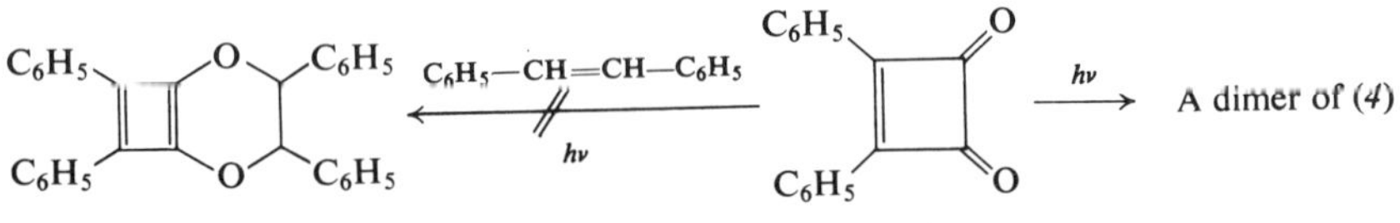

Several unsuccessful attempts have been made to generate a cyclobutadiene by the Diels–Alder addition of a dienophile to a dimethylenecyclobutene. Thus, the reaction of either 1,2-diphenyl-3,4-dimethylenecyclobutene (*7*)[23] or 1,2-dimethyl-3,4-dimethylenecyclobutene (*8*)[46, 79] with tetracyanoethylene affords cycloaddition products (*9*) and (*10*) rather than the desired normal Diels–Alder adducts (*5*) and (*6*). (See also Chapter 5, Section B, 4, b.)

* See also Section A.

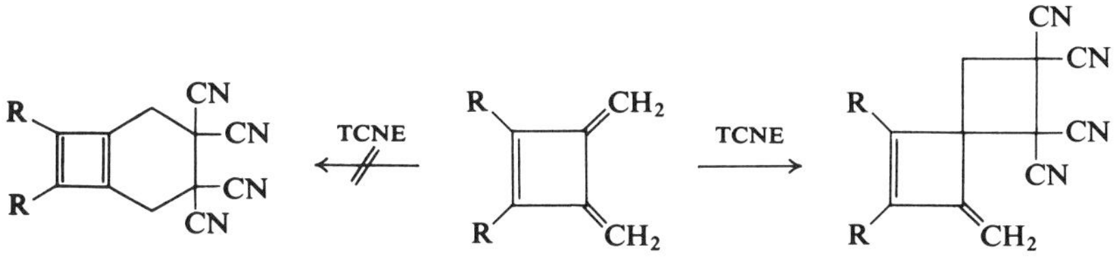

C. UNSUCCESSFUL APPROACHES TO THE CYCLOBUTADIENES

(5): R=C$_6$H$_5$
(6): R=CH$_3$

(7): R=C$_6$H$_5$
(8): R=CH$_3$

(9): R=C$_6$H$_5$
(10): R=CH$_3$

Evidence was sought unsuccessfully for the enolization of 2,4-dichloro-3-phenylcyclobutenone (12) to 1,3-dichloro-2-hydroxy-4-phenylcyclobutadiene (11).[129] An attempt to form the cyclobutadienoid enol acetate (13) by treating the cyclobutenone (12) with isopropenyl acetate was also unsuccessful.[129] Similarly, no evidence could be found for the formation of the bis(enol) (14) of methylketene dimer.[159]

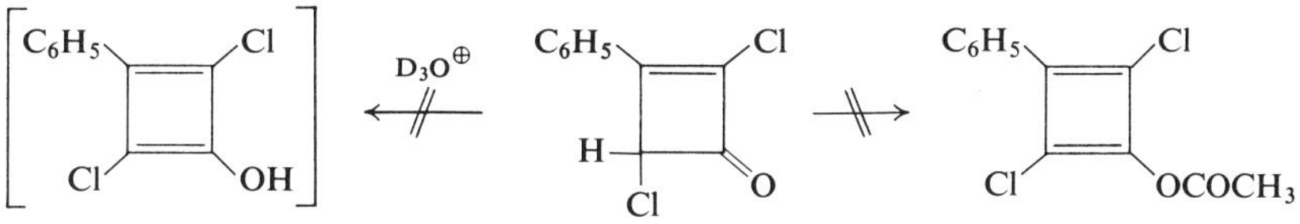

(11) (12) (13)

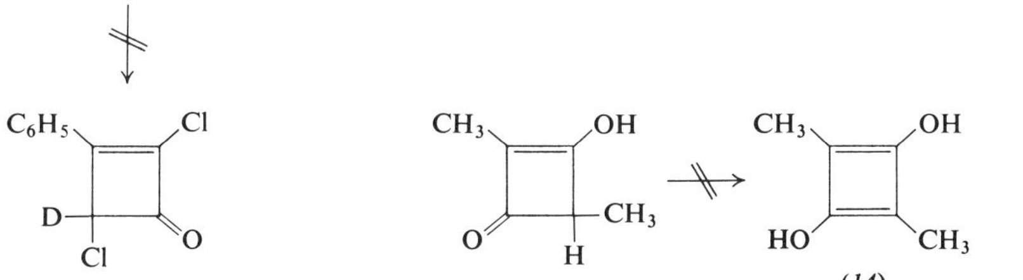

(14)

Simple molecular orbital calculations indicate that the cyclobutadienyl-carbinyl radical, cation, and anion (15a, b, and c, respectively) should have a substantial delocalization energy (1.59 β). However, all attempts to generate such cyclobutadienylcarbinyl species by the removal of a hydrogen atom, a hydride ion, or a proton from the 4-position of methylenecyclobutene were unsuccessful.[3]

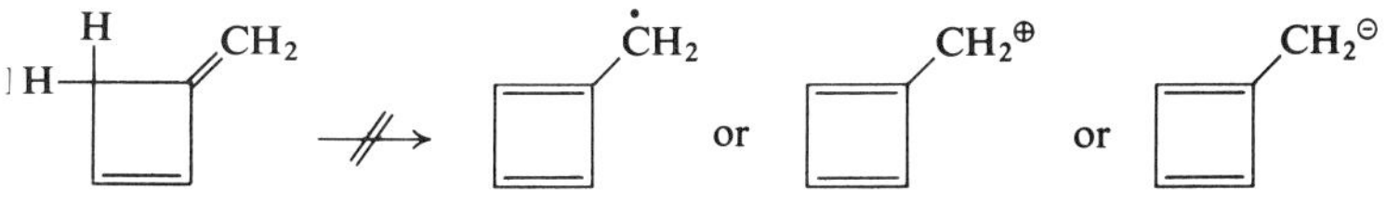

(15a) (15b) (15c)

An unsuccessful attempt was made to synthesize 2-chloro-4-nitro-3-phenyl-cyclobutenone (16), a compound in which enolization to the corresponding cyclobutadiene (17) might be particularly favored because of hydrogen bonding in the enol. However, treatment of the starting material, 4,4-dichloro-3-phenylcyclobutenone (18), with sodium nitrite gave a compound believed to be N-hydroxy-1-chloro-2-phenylmaleimide (19) rather than the desired α-nitroketone (16). The nitroketone (16) may be a transient intermediate in the formation of compound (19).[29]

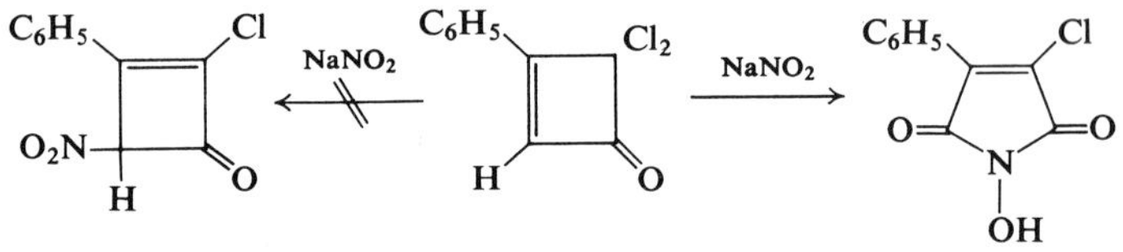

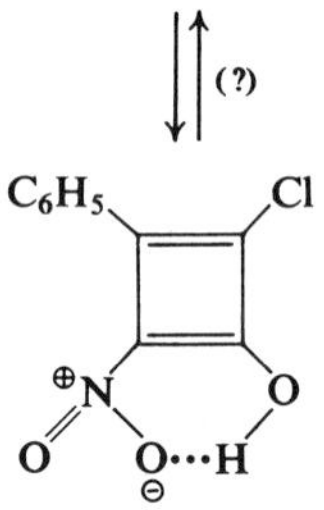

2. By Dehydrohalogenation of Halocyclobutenes

A number of halocyclobutenes failed to afford characterizable dehydro-halogenation products attributable to a cyclobutadiene intermediate. For example, attempts were made to generate 1-fluoro-2,3,4-triphenylcyclo-butadiene (21) by dehydrofluorination of 3,3-difluoro-1,2,4-triphenylcyclo-butene (20) with (a) triethylamine, (b) sodium methoxide, and (c) phenyl-magnesium bromide; unlike phenyllithium, these reagents failed to give a dimer of (21) and, hence, failed to generate (21) from (20).[116]

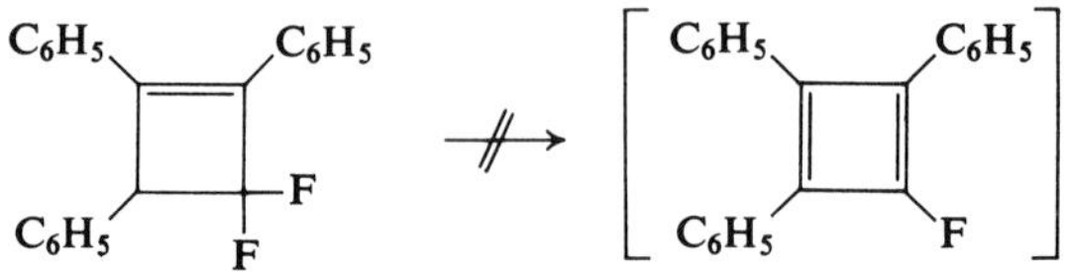

C. UNSUCCESSFUL APPROACHES TO THE CYCLOBUTADIENES

Dehydrohalogenation of 2-chloro-3,3-difluoro-1-vinylcyclobutene (23) gave polymeric material. No definite product attributable to 1-chloro-2-fluoro-4-vinylcyclobutadiene (22) was detected.[78] Attempts to dehydrofluorinate the related 2-chloro-3,3-difluoro-1-phenylcyclobutene (25) to 1-chloro-2-fluoro-4-phenylcyclobutadiene (24) failed because the compound (25) proved to be impervious to the common strong base reagents.[133]

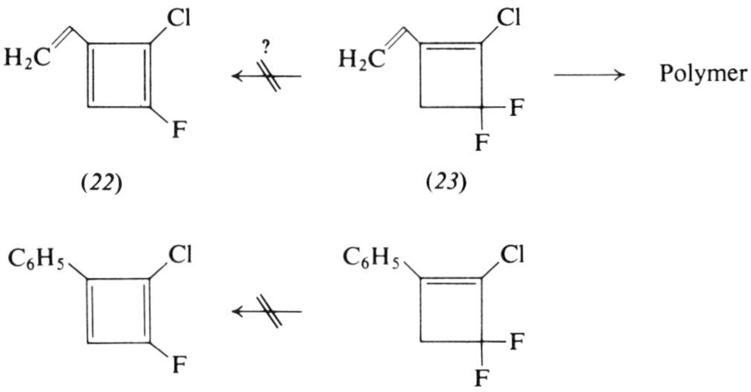

An attempt to prepare a substituted cyclobutadiene stabilized electronically by a pair of "push–pull" substituents was unsuccessful, as shown below. Thus, the reaction of 2-chloro-3,3-difluoro-4-nitro-1-phenylcyclobutene (26) with a number of secondary amines failed to follow the desired course, (26) → (27) → (28) → (31). Instead, the cyclobutene (26) reacted with the amines to give the corresponding 1-phenyl-1-dialkylamino-2-chloro-3-fluoro-4-nitro-1,4-butadienes (32). The reaction course apparently involves an S_N2 displacement of fluoride from (26), followed by spontaneous ring cleavage of the resulting cyclobutene intermediate (29).[29]

The resistance of cyclobutene (26) to dehydrofluorination (and hence to cyclobutadiene formation) is confirmed by its reaction with butyllithium to give a *stable* lithium salt (30) from which the deuterated starting material (33) could be obtained by treatment with deuterium oxide.[30]

In a somewhat different approach to the problem, an attempt was made to synthesize a cyclobutadiene electronically similar to (31) by a different route. 1-Phenyl-2-nitro-3,3-difluorocyclobutane (34) was treated with dimethylamine to give 1-dimethylamino-2-nitro-3-phenylcyclobutene (39) via intermediates (35) and (36). Bromination of enamine (39) afforded a dibromide of probable structure (38). However, all attempts to dehydrohalogenate the dibromide to 1-bromo-2-dimethylamino-3-nitro-4-phenylcyclobutadiene (37) led only to uncharacterizable amorphous products.[30]

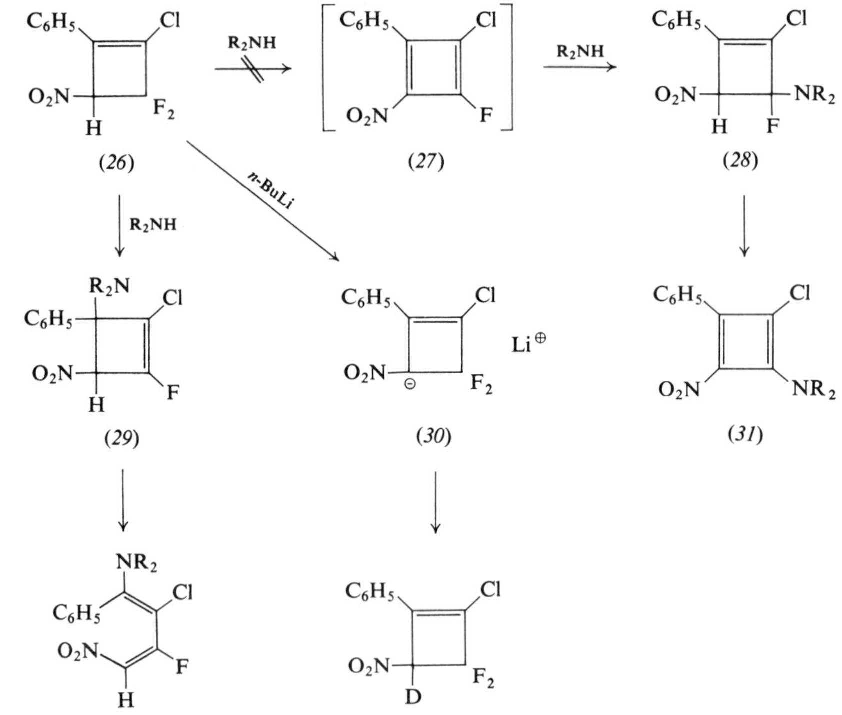

(32): R=(CH₃), C₂H₅, C₄H₉, or
R₂=(CH₂)₄

One of the stereoisomers (*41a*) of 1,3-dinitro-2,4-diphenyl-3-bromocyclobutene (prepared from a β-nitrostyrene photodimer) gave only the ring-cleavage product, 1,3-dinitro-2,4-diphenyl-1-bromo-1,4-butadiene (*40*), on

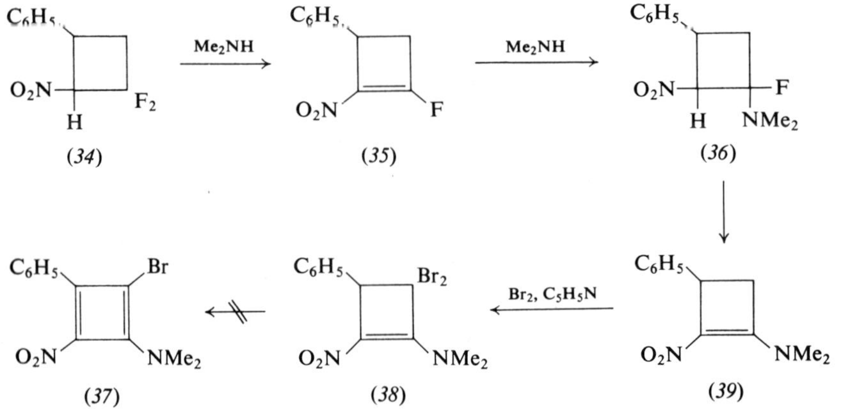

C. UNSUCCESSFUL APPROACHES TO THE CYCLOBUTADIENES

treatment with triethylamine.[123] Similarly, dehydrobromination of 1,3-dibromo-1,3-dinitro-2,4-diphenylcyclobutane (*43*) (also prepared from a β-nitrostyrene photodimer) afforded a mixture of the cyclobutene (*41a*) and its stereoisomer (*41b*) and no product attributable to 1,3-dinitro-2,4-diphenylcyclobutadiene (*42*) was detected.[114] The use of bases other than triethylamine in the dehydrobromination of (*43*) gave no better results.[114, 123]

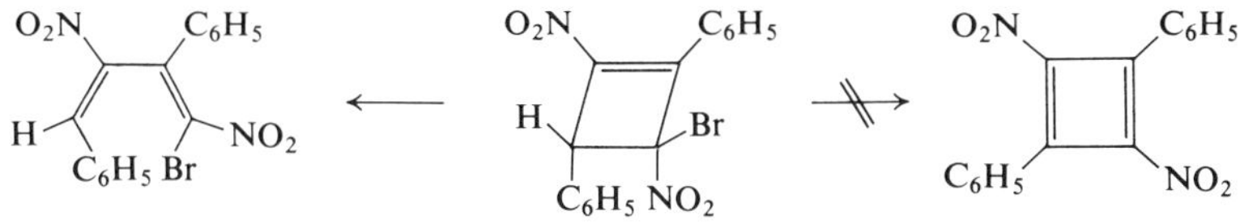

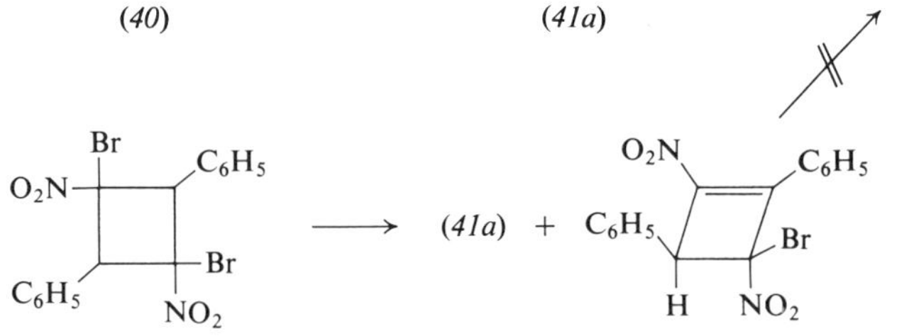

An early attempt to prepare cyclobutadiene-1,3-dicarboxylic acid (*46*) from a compound believed to be cyclobutane-1,3-dicarboxylic acid (*45*) by bromination followed by dehydrobromination has been recorded.[106] The failure of this attempt is due to the fact that the alleged cyclobutanedicarboxylic acid (prepared by base treatment of ethyl α-chloropropionate) is actually 1-methylcyclopropane-1,2-dicarboxylic acid (*44*).[58]

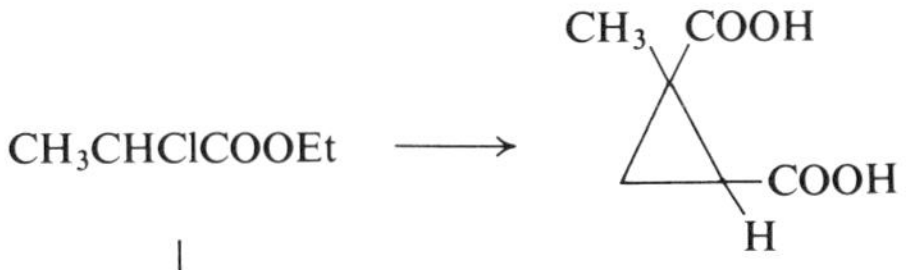

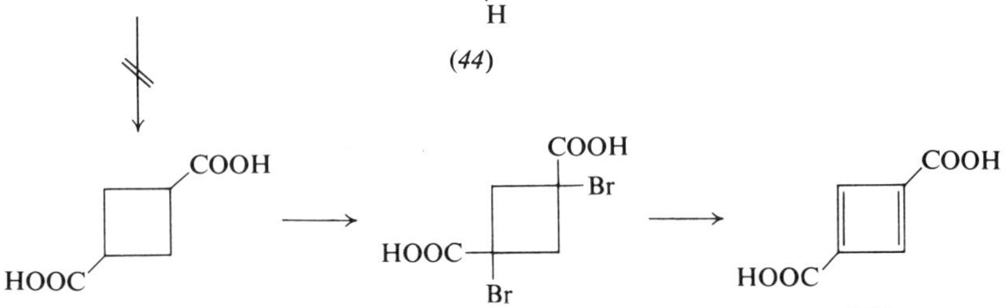

An attempt to aromatize 1,2,3,4-tetraphenylcyclobutane with N-bromosuccinimide afforded 1,2,3,4-tetraphenylbutadiene instead. The dehydrohalogenation of 1,2(and 3)-dibromocyclobutane with solid potassium hydroxide at 110° gives vinylacetylene, formed in all likelihood via a noncyclobutadienoid intermediate [64] (see Section B, 2).

3. By Dehalogenation

An attempt to prepare cyclobutadiene-1,2,3,4-tetracarboxylic acid (*49*), starting from a compound presumed to be a tetrabromodibenzotricyclooctadiene (*47*) foundered at the outset when it was found that ozonolysis of (*47*) failed to give a key intermediate, 1,2,3,4-tetrabromocyclobutane-1,2,3,4-tetracarboxylic acid (*48*).[70] The reason for this failure is now apparent: X-ray analysis of the starting material has demonstrated that it is 3,4,7,8-tetrabromo-1,2,5,6-dibenzocyclooctatetraene (*50*) and not (*47*).[110]

cis,trans,cis-Cyclobutane-1,2,3,4-tetracarboxylic acid (*51*) could not be converted into a bromo derivative suitable for the preparation of cyclobutadiene-1,2,3,4-tetracarboxylic acid (*49*).[48]

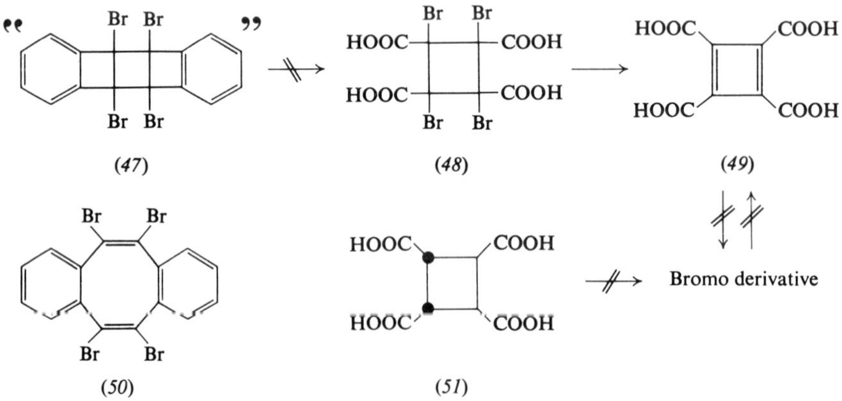

The following reagents were found to be ineffective in the dehalogenation of *cis*-3,4-dichloro-1,2,3,4-tetramethylcyclobutene (*52*) to 1,2,3,4-tetramethylcyclobutadiene (*54*): (a) sodium iodide in acetone, (b) sodium in dioxane, (c) magnesium in ether, and (d) magnesium–magnesium iodide in ether.[49]

The dehalogenation of *cis*-3,4-diiodo-1,2,3,4-tetramethylcyclobutane (*53*) with mercury or silver powder gave air-sensitive products of unknown structure rather than the expected dimer of tetramethylcyclobutadiene (*54*). Doubts have been expressed concerning the generation of tetramethylcyclobutadiene in

C. UNSUCCESSFUL APPROACHES TO THE CYCLOBUTADIENES   33

these reactions; these doubts receive support from the isolation of an iodine-containing cyclobutene (55) formed in the reaction of diiodide (53) with mercury in the presence of 3-methyl-2-penten-4-one.[93] [But see also Section B, 1 for the formation of Diels–Alder adducts of (54) with other dienophiles under these conditions.]

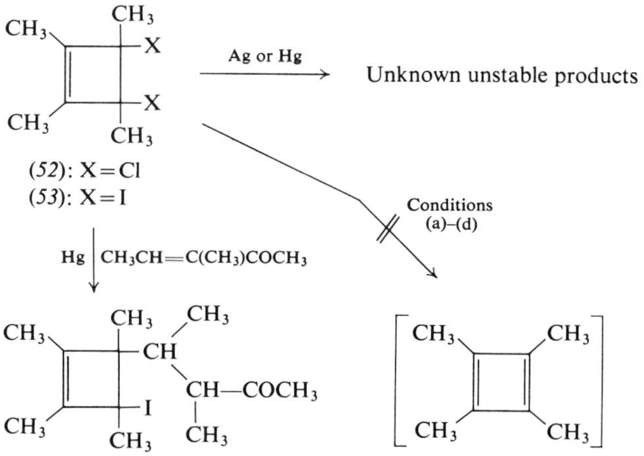

(52): X=Cl
(53): X=I

Ag or Hg → Unknown unstable products

Conditions (a)–(d)

Hg | $CH_3CH=C(CH_3)COCH_3$

(55)   (54)

3,4-Dichloro-1,2,3,4-tetrafluorocyclobutene (57), prepared from 1,4-dichloro-1,2,3,4-tetrafluoro-1,4-butadiene, and 3,4-diiodo-1,2,3,4-tetrafluorocyclobutene (56) have been subjected to various dehalogenation conditions, but no evidence for the generation of 1,2,3,4-tetrafluorocyclobutadiene (58) has been found in any of these reactions.[63, 84] Thus, attempted dehalogenation of dichloride (57) with (a) zinc dust in dioxane at 120°–180°[84] and (b) with lithium amalgam in ether[63] afforded only a small amount of hydrogenated product in the first case and an amorphous gum in the second. Dichloride (57) was converted smoothly into diiodide (56) with sodium iodide in acetone, probably by way of an S_N2' process rather than by elimination–addition.[63] Treatment of the diiodide with zinc–copper couple in ether gave an iodine-free polymer which appears to contain cyclobutene residues. Similarly, diiodide (56) eliminated iodine under both photolytic and thermolytic conditions, but no product definitely attributable to 1,2,3,4-tetrafluorocyclobutadiene (58) was isolated.[63] Reaction of diiodide (56) with phenyllithium under a variety of conditions afforded only amorphous brown products and no evidence pointing to the formation of the desired product, 1,2-difluoro-3,4-diphenylcyclobutadiene (59) was obtained.[63]

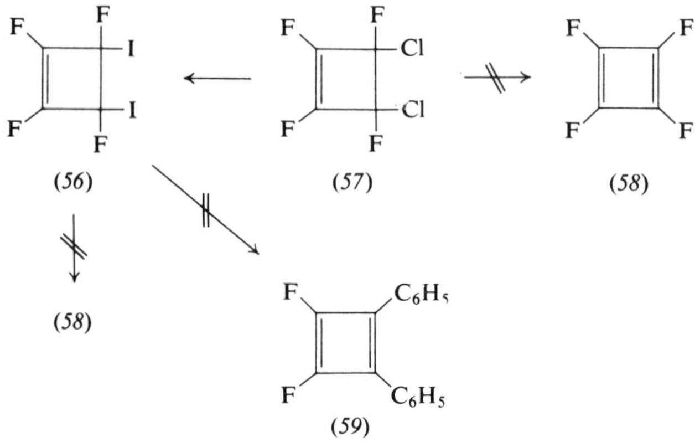

4. BY HOFMANN ELIMINATION AND RELATED METHODS

Subjecting 1,2-bis(dimethylamino)cyclobutane dimethohydroxide (60) to the conditions of the Hofmann elimination gave 1,2-bis(dimethylamino)cyclobutane (61) and three oxygenated compounds attributed to the intermediary formation of 1-dimethylaminocyclobutene.[37] Similarly, subjecting 1,3-bis(dimethylamino)cyclobutane dimethobromide (62) to the conditions of Wittig's modification of the Hofmann elimination [i.e., treating (62) with phenyllithium, cyclopentadienyllithium, or cyclopentadienylpotassium] gave, not cyclobutadiene, but a hydrocarbon, $C_{10}H_{10}$, believed to be 1(or 3)-phenylcyclobutene [(64a) or (64b)].[12] The origin of (64) is obscure [(62) → (63) → (64a) or (64b)?], but it is unlikely that cyclobutadiene was involved in its formation. When the dimethobromide was subjected to the conditions of the Robiant–Wittig modification of the Hofmann reaction (methylene bromide, followed by phenyllithium), the only product was a polymeric substance.[12] The Cope–Bumgardner method was equally unsuccessful: pyrolysis of the bis(amine oxide) (66) gave no identifiable product other than dimethylhydroxylamine.[12]

Similarly, subjecting 4-(dimethylamino)-1,3-diphenylcyclobutene methohydroxide to Hofmann elimination gave only carbonyl-containing products rather than products attributable to 1,3-diphenylcyclobutadiene (67).[151] 3-Acetoxycyclobutene (prepared by treating cyclobutene with selenium oxide in acetic anhydride) was investigated as a potential synthetic precursor of cyclobutadiene, but the results were inconclusive.[108]

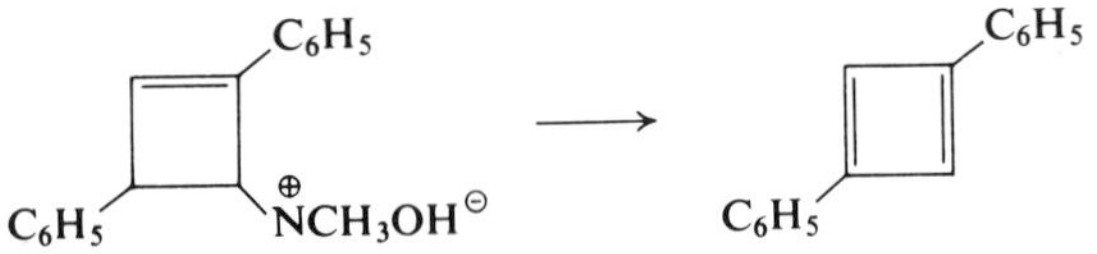

C. UNSUCCESSFUL APPROACHES TO THE CYCLOBUTADIENES

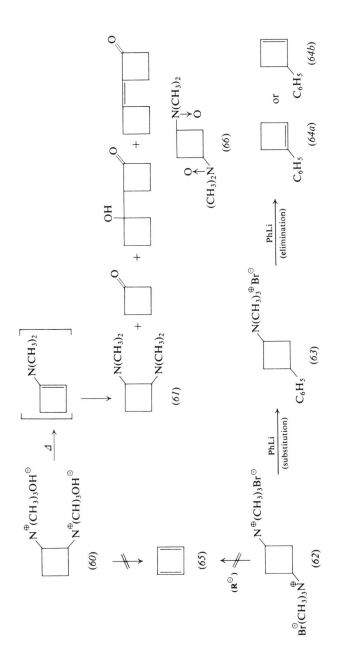

5. By the Dimerization of Acetylenes

Several cyclobutadiene structures have been proposed by Dunicz as intermediates in the photochemical polymerization of acetylene.[61] (See also Section B, 3.)

When it is heated under autogenous pressure, perfluoro-2-butyne forms a white crystalline product to which a tetrameric, cyclo[4.2.0.0^{2,5}]octadiene structure (68) was originally assigned.[31] Because the assigned structure may be viewed as that of a dimer of tetrakis(trifluoromethyl)cyclobutadiene, it was of more than routine interest. Further investigation of the product led Ekström to propose a cyclooctatetraene structure (69),[62, 63] but both tetrameric structures were subsequently rejected in favor of a trimeric benzenoid structure (70).[32, 83]

Because the acetylenic linkages in compounds of type (71, m=1 or 2) might be expected to undergo valence tautomerism to the corresponding cyclobutadiene structures (72, m=1 or 2) with a minimum decrease in entropy, the cyclododecadiyne homolog of (71, m=2) was prepared and examined as a possible precursor to (72, m=2), but no evidence for the formation of the latter could be obtained.[44] (But see also the formation of biphenylene from 1,3,7,9-cyclododecatetrayne, Chapter 10, Section B, 6.) Similar cyclobutadienoid structures, i.e., (72, m=3) and (72, m=4) were considered and rejected for the two dimeric hydrocarbons formed in the dehalogenation of 1-bromo-2-chlorocycloheptene (73, m=3) and 1-bromo-2-chlorocyclooctene (73, m=4) with sodium amide in liquid ammonia.[17] The hydrocarbons were identified as the dimethylenecyclobutane structures (76, m=3) and (76, m=4), which were undoubtedly formed via cyclic allenes (75, m=3, and 75, m=4), and not via the corresponding cycloalkynes (74, m=3, and 74, m=4).

When the two halides (73, m=3) and (73, m=4) were dehalogenated with sodium in ether, the only products isolated were cycloheptene and cyclooctene, respectively. On the other hand, when 1,2-dibromocyclohexene was dehalogenated under the same conditions, the product was a complex mixture from

C. UNSUCCESSFUL APPROACHES TO THE CYCLOBUTADIENES 37

which only dodecahydrotriphenylene could be isolated (Favorsky, 1912).[*,64] Similarly, treatment of 1,2-dibromocyclohexene with magnesium or lithium amalgam failed to give cyclohexyne or a cyclobutadiene dimeride.[†,90] The dehalogenation of 3-methyl-1,2-dibromocyclohexene[60] and the pyrolysis of silver 2-chlorocyclohexenecarboxylate[82] also failed to afford such products.

Attempts were made to prepare difluoroacetylene by several methods.[86] In the absence of any evidence to the contrary, it was reasonable to believe that difluoroacetylene might be made to undergo dimerization to give 1,2,3,4-tetrafluorocyclobutadiene, which might be somewhat stabilized by the four electron-withdrawing fluorine atoms.[86] Interestingly enough, the preparation of difluoroacetylene has been claimed in two patents[77,113] and in one of these the isolation of a fluorocarbon, C_4F_4, possibly tetrafluorocyclobutadiene (58) was reported.[112,113] The fluorocarbon was detected by its mass spectrum in the mixture resulting from the decomposition of difluoroacetylene (prepared by thermolysis of difluoromaleic anhydride); the compound (58?) boils at approximately 0° and decomposes in turn to products of unknown structure (polymers?) on standing in the gaseous state at room temperature for several hours.[113]

---

* When the dehalogenation of 1,2-dibromocyclohexene with sodium in ether was carried out in the presence of 1,3-diphenylisobenzofuran, an adduct of cyclohexyne was formed (Gwynn, 1962, ref. 82). Since it has been shown that cyclohexyne can dimerize to a cyclobutadienoid intermediate (Wittig and Mayer, 1963, ref. 158), it seems likely that such an intermediate might well have been generated unknowingly by Favorsky as early as 1912. Furthermore, since Willstätter's work (1905, ref. 156) probably did not result in the formation of cyclobutadiene, it is not impossible that Favorsky, and not Willstätter, was the first to generate a cyclobutadiene.

† However, for a successful application of this reaction, see Section B, 3.

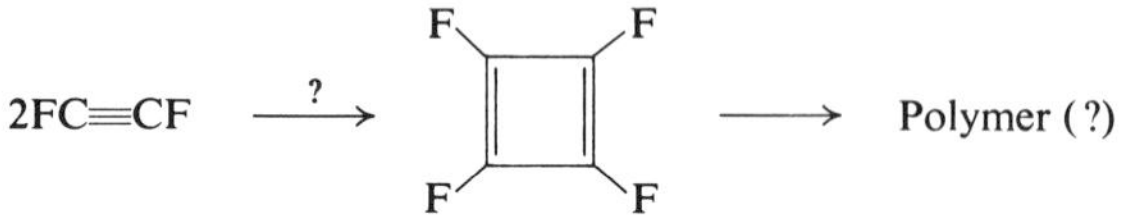

$$\text{(F-substituted maleic anhydride)} \xrightarrow{\Delta} FC\equiv CF + FC\equiv CCOF + 2CO_2 + CO$$

$$2FC\equiv CF \xrightarrow{?} \underset{(58)}{\text{(tetrafluorocyclobutadiene)}} \longrightarrow \text{Polymer (?)}$$

6. By Ring Closure of Butadienes and Related Compounds

When 1,4-dibromo-1,2,3,4-tetraphenylbutadiene (77) was treated with lithium amalgam, a hydrocarbon of unknown structure ($C_{28}H_{22}$) was formed instead of a dimer of tetraphenylcyclobutadiene.[66] Similarly, when either 1-bromo-4-iodo-1,2,3,4-tetraphenylbutadiene (78) or the corresponding dibromide (77) was treated with butyllithium, 1,2,3,4-tetraphenylbutadiene (79) was obtained by way of its 1,4-dilithio derivative.[66] The reaction of bromoiodide (78) with zinc in tetrahydrofuran afforded a mixture of (79), the unknown hydrocarbon, m.p. 148°–149°, and a trace of 1,2,3-triphenylazulene (80),[66] which may have been formed via a cyclobutadiene. (Compare Section B, 3 and Section C, 8.)

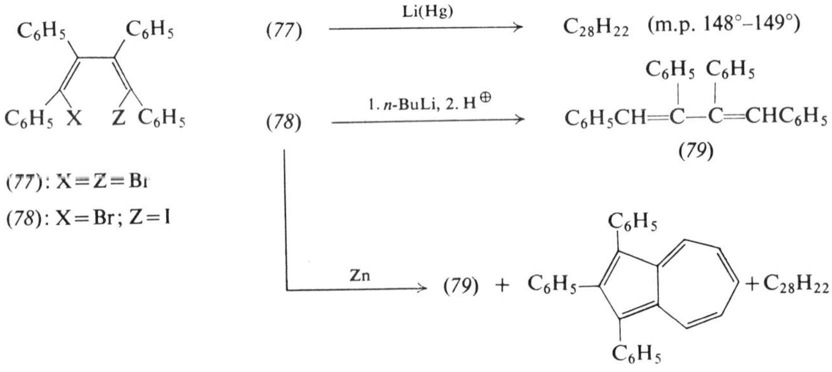

$$(77) \xrightarrow{Li(Hg)} C_{28}H_{22} \text{ (m.p. 148°–149°)}$$

$$(78) \xrightarrow{1.\ n\text{-BuLi},\ 2.\ H^{\oplus}} C_6H_5CH=\underset{\underset{C_6H_5}{|}}{C}-\underset{\underset{C_6H_5}{|}}{C}=CHC_6H_5 \quad (79)$$

(77): X = Z = Br
(78): X = Br; Z = I

$$\xrightarrow{Zn} (79) + (80) + C_{28}H_{22}$$

(80) = 1,2,3-triphenylazulene

An unsuccessful attempt was made to generate 1,2,3,4-tetraphenylcyclobutadiene (81) by the pyrolysis of a hydrocarbon originally believed to be "octaphenylcubane"[143] but subsequently shown to be octaphenylcyclooctatetraene (82).[119]

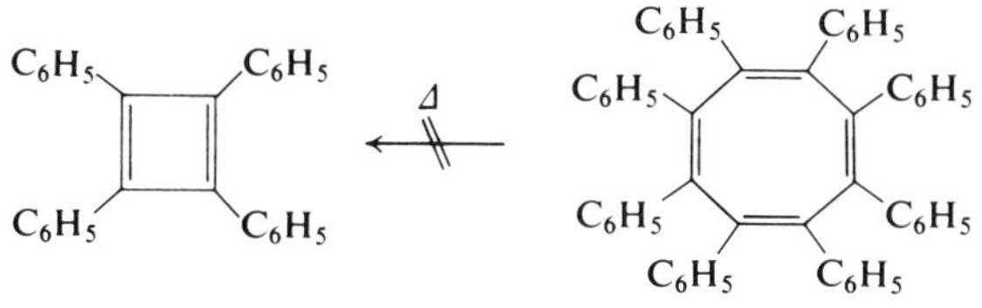

C. UNSUCCESSFUL APPROACHES TO THE CYCLOBUTADIENES

(81) ⇄̸ (82)

7. BY A REVERSE DIELS–ALDER REACTION

The generation of cyclobutadiene by the thermolysis of three formal adducts of cyclobutadiene, (83),[125] (84),[125] and (85),[126] was attempted without success by Reppe. The work was subsequently repeated by Nenitzescu and co-workers, who were able to isolate butadiene, which is attributable to cyclobutadiene.[13] (See also Section B, 6.)

Similarly, thermolysis of (86) failed to give a product attributable to tetramethylcyclobutadiene,[20] but since the other expected thermolysis fragment of the molecule, dimethyl 3,4,5,6-tetramethylphthalate, was isolated, it is possible that tetramethylcyclobutadiene was generated in the reaction.

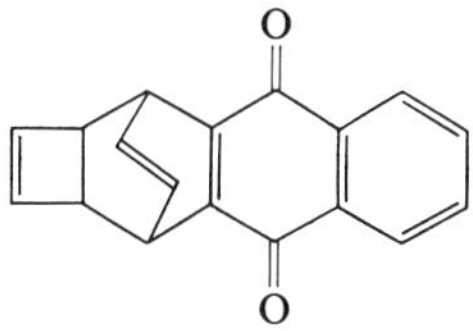

(83) 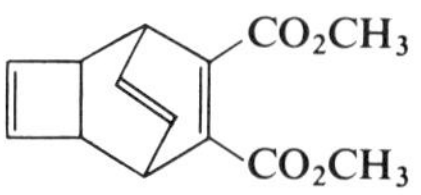   (84) 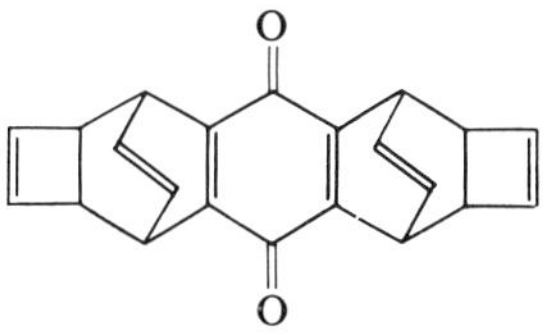   (85)

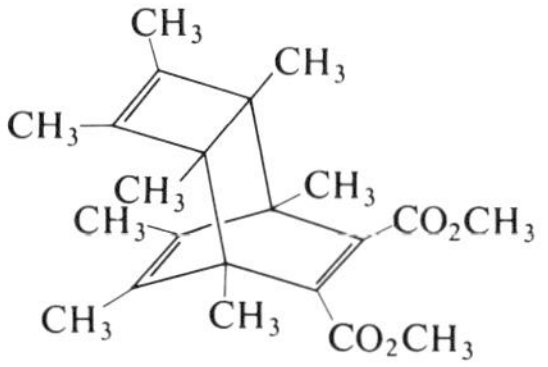

(86)

8. BY REARRANGEMENT REACTIONS

a. *Of Cyclopropenylcarbinyl Cations*

Up to the present time, attempts to generate cyclobutadienes by the ring-enlargement of cyclopropenylcarbinyl cations have not been successful. Thus, when triphenylcyclopropenyl bromide (87) was treated with phenyldiazomethane, 1,2,3-triphenylazulene (80) was formed by the path indicated

1. CYCLOBUTADIENE

C. UNSUCCESSFUL APPROACHES TO THE CYCLOBUTADIENES 41

$(87) \rightarrow (88) \rightarrow (91) \rightarrow (93) \rightarrow (80)$, and the desired product, tetraphenylcyclobutadiene (81), was not formed.*,[26,30] A key intermediate in this transformation, cyclopropylcarbinyl cation (91), is also formed in the dehydration of (2,3-diphenyl-2-cyclopropenyl)diphenylcarbinol (89). The course of this dehydration is shown in the sequence $(89) \rightarrow (90) \rightarrow (91)$, etc.[27] The other product of the reaction, 1,2,4-triphenylnaphthalene (92), is also formed, undoubtedly, via cyclobutenylcarbonium ion (90).

Treating diphenylacetylene with 2,4-dinitrobenzenesulfenyl chloride gave triphenylazulene (80) in 25% yield, together with a small amount of a white crystalline compound, $C_{28}H_{20}$, m.p. 154°–155°, the identity of which is not known.[4] [The latter is neither 1,2,3-triphenylnaphthalene[4] (prepared by treating diphenylacetylene with lithium metal) nor 1,2,4-triphenylnaphthalene.[27]] Balaban has suggested that the hydrocarbon is tetraphenylcyclobutadiene,[16] but this has been disputed by Breslow.[27]

1,2,3-Triphenylazulene (80) was also formed in the photodimerization of diphenylacetylene, but in the latter case it appears much more likely that tetraphenylcyclobutadiene was actually an intermediate in the reaction,

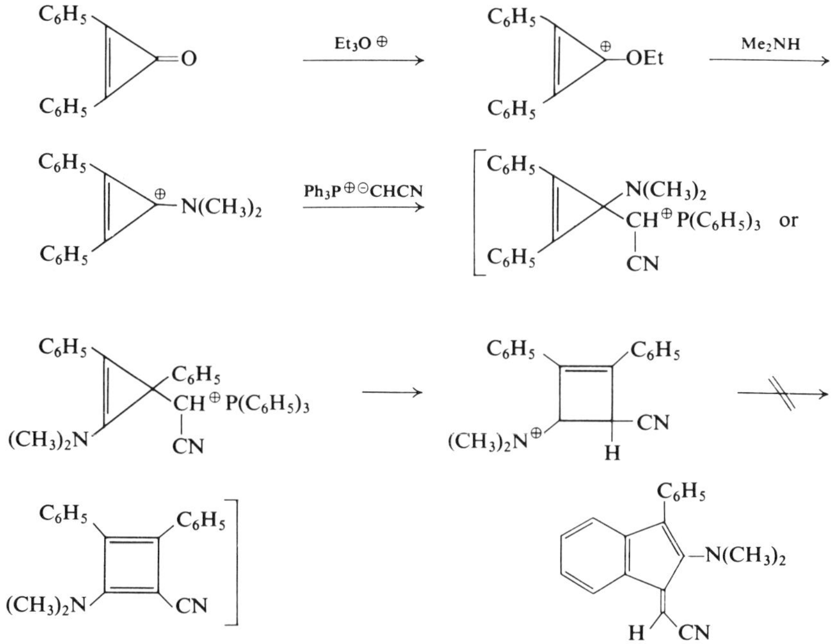

* A small amount of pentaphenylcyclopentadiene was also formed, presumably by attack of excess phenyldiazomethane on cyclobutenylcarbonium ion (88).

because the dimer of tetraphenylcyclobutadiene was also isolated (see Section B, 3).

An ingeniously contrived attempt to prepare a "push–pull" stabilized cyclobutadiene through rearrangement of a cyclopropenylcarbinyl cation is outlined on p. 40. The reaction sequence failed to give the desired product, 1-cyano-2-dimethylamino-3,4-diphenylcyclobutadiene (94) and gave an indene derivative (95) instead.[28] It seems unlikely that cyclobutadiene (94) is an intermediate in the formation of the indene (95).

An attempt to prepare either 1,2-diphenylcyclobutadiene (67a) or 1,3-diphenylcyclobutadiene (67) via diphenyltetrahedrane (97) failed. Thus, photolysis of (2,3-diphenyl-2-cyclopropenyl)diazomethane (96) afforded diphenylacetylene as the only isolable product.[152]

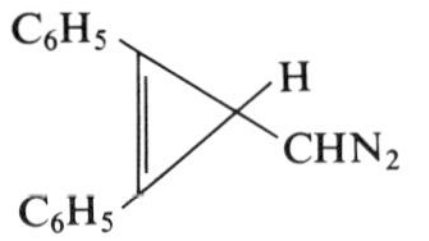

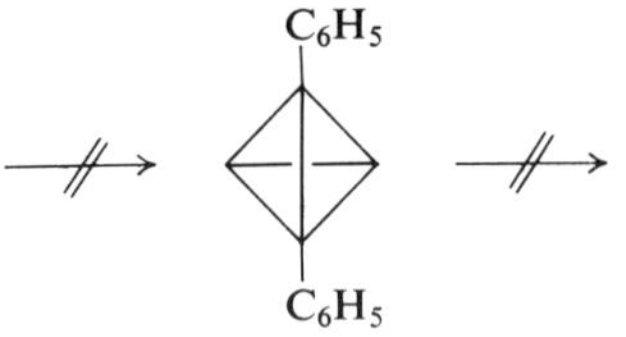

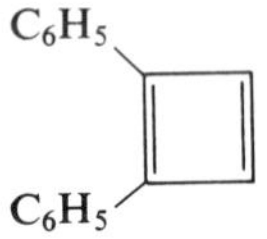

b. Of 1,1,3,3-Tetrasubstituted Cyclobutane-2,4-bis(carbenes) and Cyclobutane-2,4-bis(carbonium) Ions

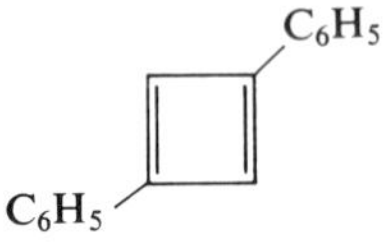

Attempts to convert 1,1,3,3-tetramethyl-2,4-cyclobutanediol (98) into tetramethylcyclobutadiene (54) by treating the former with various dehydrating agents failed and no characterizable products were isolated.[18] In related work, 1,1,3,3-tetramethyl-2,4-cyclobutanedione ditosylhydrazone (99) was treated with base under carbene-producing conditions but the product isolated was not a compound attributable to tetramethylcyclobutadiene. Rather, it was a tetramethylbutatriene (100) possibly formed by the pathway shown below.[103]

9. FROM PRESUMED CYCLOBUTADIENE–METAL COMPLEXES

The reaction of aqueous sodium chloride with a compound assumed to be the cyclobutadiene–silver nitrate complex (101)[7, 10, 72, 109] gave a hydrocarbon C_8H_8, which was shown to be syn-tricyclo[4.2.0.0^{2,5}]octa-3,7-diene (103),[9a, 10]

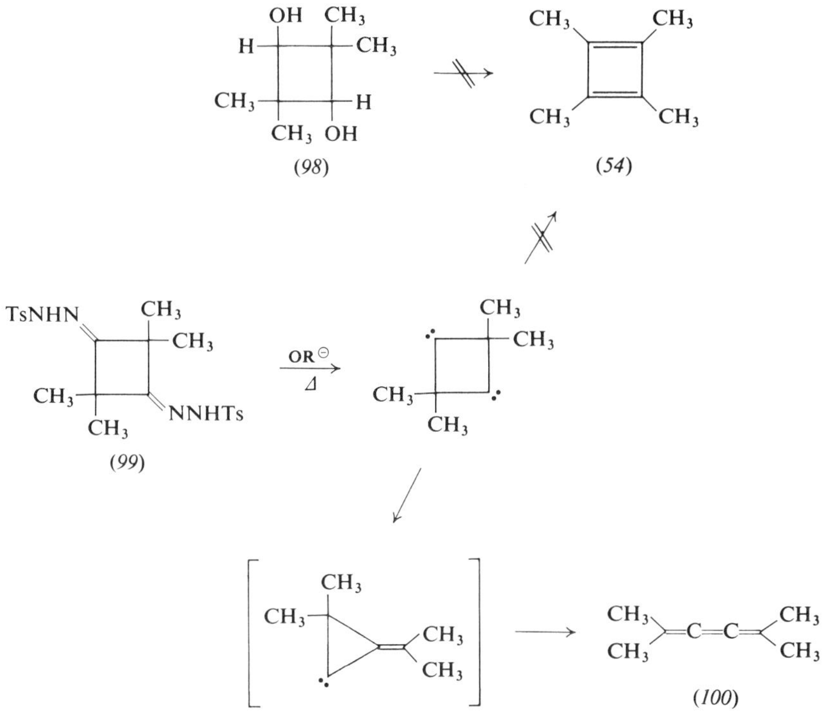

D. TRIPLET DIRADICAL CHARACTER

apparently a dimer of cyclobutadiene. However, a recent NMR study of the silver nitrate complex has demonstrated that it is actually a tricyclooctadiene–silver nitrate complex (*102*)[6] and hence, the presumed generation of cyclobutadiene in this transformation must now be discounted. (See also Chapter 2, Section D.)

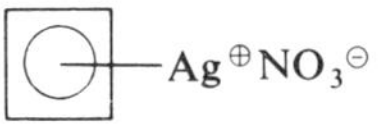

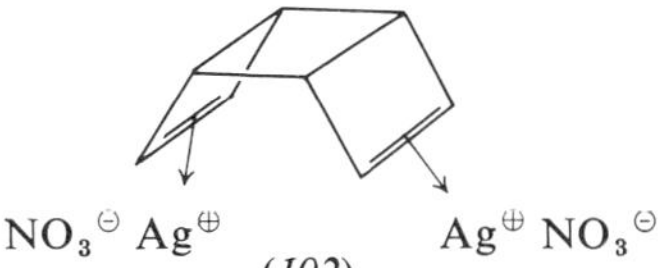

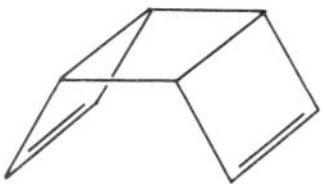

10. BY DOERING'S DECARBOXYLATION METHOD

Attempts to prepare cyclobutadiene by the lead dioxide decarboxylation of *cis,trans,cis*-1,2,3,4-cyclobutanetetracarboxylic bis(anhydride) were unsuccessful.[80]

D. Evidence of the Triplet Diradical Character of the Cyclobutadienes

Simple MO treatment of cyclobutadiene predicts a triplet ground state and zero delocalization energy for the molecule, and this prediction is in good agreement with much of its observed chemical behavior. The calculation of π-electron orbitals is of course based on two assumptions: (a) planarity of the molecule and (b) complete delocalization of the π electrons. If these assumptions are correct, cyclobutadiene must be a square molecule (all bonds equal in length) and all four carbon atoms must have identical reactivities. Like cyclooctatetraene (also a $4n$ π-electron system), cyclobutadiene should be capable of achieving enhanced stability by gaining two electrons to form a dianion or by losing two electrons to form a dication. (See Chapter 3.) Another way in which the molecule could achieve enhanced stability is by becoming a rectangle of alternating single (long) bonds and double (short) bonds in which the π electrons would be localized. (See also Chapter 12.)

At the present time, the little that is known about the chemistry of the cyclobutadienes indicates that they can react as triplet diradicals. Thus, when tetramethylcyclobutadiene (2) was generated by the action of potassium vapor on 3,4-dichloro-1,2,3,4-tetramethylcyclobutene (1) in the gaseous phase at approximately 250°, a hydrocarbon mixture was obtained which consisted mostly of 4-methylene-1,2,3-trimethylcyclobutene (3) and the syn dimer of tetramethylcyclobutadiene (4), together with smaller amounts of octamethyl-cyclooctatetraene (5) and cis-1,2,3,4-tetramethylcyclobutene (6)[135] (total yield of hydrocarbons, 35%). Variation of the concentration of dichloride (1) over a thirtyfold range did not cause an appreciable change in the ratio of the concentrations of the dimer (5) and the disproportionation product (3). Since the formation of the dimer is obviously a bimolecular process, the formation of the disproportionation product must also be bimolecular, otherwise the ratio of the two product concentrations would not have remained constant. This constitutes compelling evidence of the triplet diradical character of tetramethylcyclobutadiene since (a) it is unlikely that a cyclobutadiene–metal complex could have been formed under the experimental conditions described

D. TRIPLET DIRADICAL CHARACTER

and (b) a rapid bimolecular disproportionation of the type encountered [(2) → (3)] is compatible only with the assumption that the critical intermediate, tetramethylcyclobutadiene, is a triplet species.

Further support for the triplet diradical character of tetramethylcyclobutadiene (2) was obtained by generating it in the gaseous phase in the presence of triplet methylene.[135] Thus, reaction of dichloride (1) and a threefold excess of dibromomethane with potassium in the vapor phase led to a reduction in the amount of dimeric and disproportionation products formed such that the total yield of (3), (4), (5), and (6) amounted to only 10%. The major product (34% of the mixture) was 3,4-dimethylene-1,2-dimethylcyclobutene (7), a compound which should be formed readily with conservation of spin from the two triplet species [CH$_2$(↓↓) and (2)] in a concerted bimolecular process. The concerted nature of this reaction was demonstrated by generating (2) from dichloride (1) and potassium vapor in the presence of excess methyl radicals (from methyl bromide). Under these reaction conditions, the product mixture which was formed contained none of the triene (7), but consisted of the usual mixture of (3), (4), (5), and (6), along with two coupling products, 4-methylene-1,2,3,3-tetramethylcyclobutene (8) and 1,2,3,3,4,4-hexamethylcyclobutene (9).[135]

Tetraphenylcyclobutadiene (11) is generated by the thermal decomposition of (4-bromo-1,2,3,4-tetraphenyl-*cis,cis*-1,3-butadienyl)dimethyltin bromide (10) in triglyme at 130°[67] or by treatment of tetraphenylcyclobutadienepalladium chloride (16) with triphenylphosphine,[41, 105] trimethyl phosphite,[105] or ethylenebis(diphenylphosphine) [(C$_6$H$_5$)$_2$PCH$_2$CH$_2$P(C$_6$H$_5$)$_2$][41] in benzene or chloroform. Solutions of tetraphenylcyclobutadiene in triglyme and benzene are colored green,[41, 67, 105] those in chloroform are colored red.*,[41] Because of the elevated temperatures necessary to decompose the butadienyltin compound (10), the green color is fleeting in triglyme. On the other hand, the color persists for a considerable time when tetraphenylcyclobutadiene is generated at room temperature in benzene by the palladium complex method.[105]

* After the manuscript of this chapter had been prepared, a report appeared in which the color of the chloroform solution was said to be green also (see the Appendix).

When tetraphenylcyclobutadiene is generated by either of the above methods in the presence of oxygen (itself a triplet), cis-dibenzoylstilbene (15) is formed, presumably by way of a four-membered peroxide intermediate (12). Similarly, when tetraphenylcyclobutadiene (11) is generated by thermolysis of the butadienyltin compound (10) in the presence of thiophenol (a good radical scavenger), a compound believed to be 1-thiophenoxy-1,2,3,4-tetraphenylbutadiene (14) is formed, probably by way of the corresponding cyclobutene intermediate (13). Other radical reagents (i.e., nitric oxide and benzoyl peroxide) also react with the thermolysis product of the butadienyltin compound (10), but the reaction products have not been characterized.*,[67]

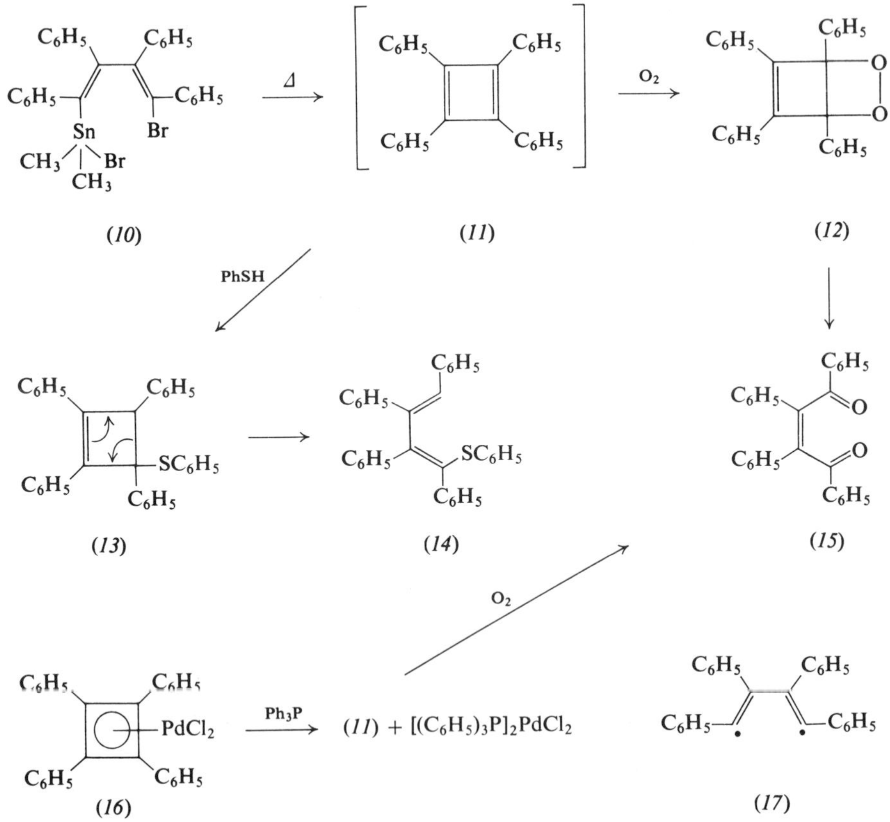

The possibility that the unstable intermediate in these reactions is not tetraphenylcyclobutadiene but its butadienyl valence tautomer (17) cannot be

---

* For the reaction of tetraphenylcyclobutadiene with nickelous bromide to form the metal complex, see Chapter 2.

D. TRIPLET DIRADICAL CHARACTER

completely excluded at this time. However, such a structure would be expected to have less resonance stabilization than the triplet tetraphenylcyclobutadiene molecule.

Butadiene (25) is formed in a number of pyrolytic reactions in which cyclobutadiene (22) would be expected to be generated. Thus, butadiene was obtained by (a) the dehydrobromination of 1,2-dibromocyclobutane (18) with boiling quinoline [13, 156]; (b) the pyrolysis of several formal Diels–Alder adducts of cyclobutadiene, i.e., 4,7-dimethoxy-2a,3,8,8a-tetrahydro-3,8-ethenocyclobuta[b]naphthalene (19),[13] dimethyl tricyclo[4.2.2.0^{2,5}]deca-3,7,9-triene-7,8-dicarboxylate (20),[13, 18] and 2a,3,10,10a-tetrahydro-3,10-ethenocyclobuta[b]anthra-4,9-quinone (21)[13]; and (c) the pyrolysis of 1,3-bis(dimethylamino)cyclobutane dimethohydroxide (22).[13] It is possible that in all of these reactions, cyclobutadiene is formed, undergoes hydrogen abstraction to form cyclobutene, and that the latter undergoes thermal cleavage to give butadiene (25).*

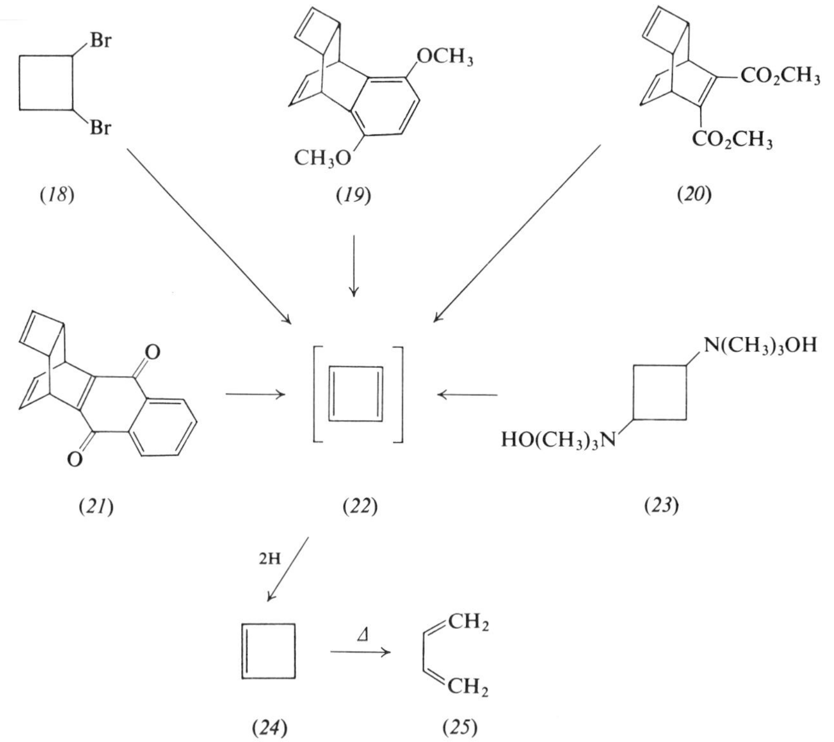

* This interpretation perhaps assumes a greater instability for cyclobutene than is actually the case. Thus, cyclobutene can be prepared in almost quantitative yield by the thermolysis of dimethyl tricyclo[4.2.0.0^{2,5}]deca-7,9-diene-7,8-dicarboxylate (ref. 43) and in good yield (40–60%) by the Hofmann degradation of N,N-dimethylammonium compounds (ref. 155).

In a similar reaction, Hofmann degradation of 1,3-diphenyl-2,4-bis(dimethylamino)cyclobutane dimethiodide (26) with lithium diethylamide afforded, as a minor product, a compound believed to be *trans*-1,3-diphenylbutadiene (31).[151] Diene (31) probably arises from 1,3-diphenylcyclobutadiene (27) by way of 1,3-diphenylcyclobutene (28). Two dimers, (29) and (30), of diphenylcyclobutadiene are the major products of this reaction.

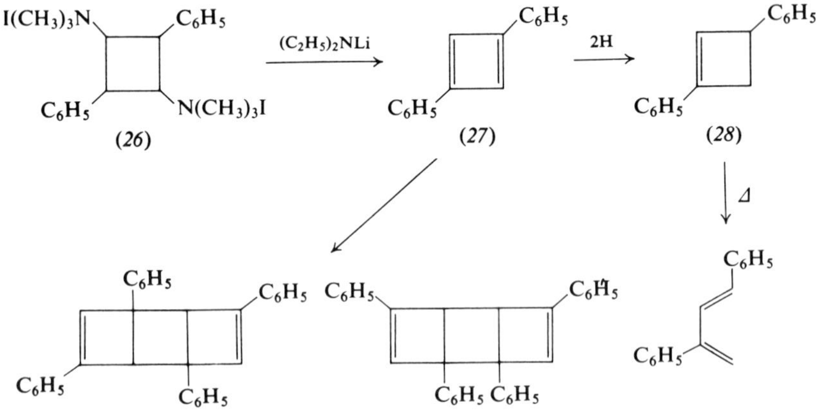

A hydrocarbon, $C_{20}H_{34}$, tentatively assigned structure (37a or b), is one of the products formed when 3,3,6,6-tetramethylcyclohexyne (33) is generated either by oxidation of hydrazone (32) or by dehalogenation of dibromide (34).[2] The mechanism of formation of (37a or b) would appear to involve dimerization of the cycloalkyne (33) to a cyclobutadiene derivative (36), followed by hydrogen abstraction from the solvent to give a cyclobutene (35a or b) and thermal cleavage of the latter to give the butadiene (37a or b). The cyclobutene (35a or b) probably rearranges with unusual ease because of internal steric strain.

It should be borne in mind that the difference in energy between the singlet states and the lowest triplet states in some cyclobutadienes might be small and that this proximity might result in the formation of thermally populated triplet states which would be the source of much of the apparent dualism in the chemical behavior of these compounds.*

---

* For recent experimental evidence of the existence of singlet cyclobutadiene at 0°, see Abstract 1.24 in the Appendix.

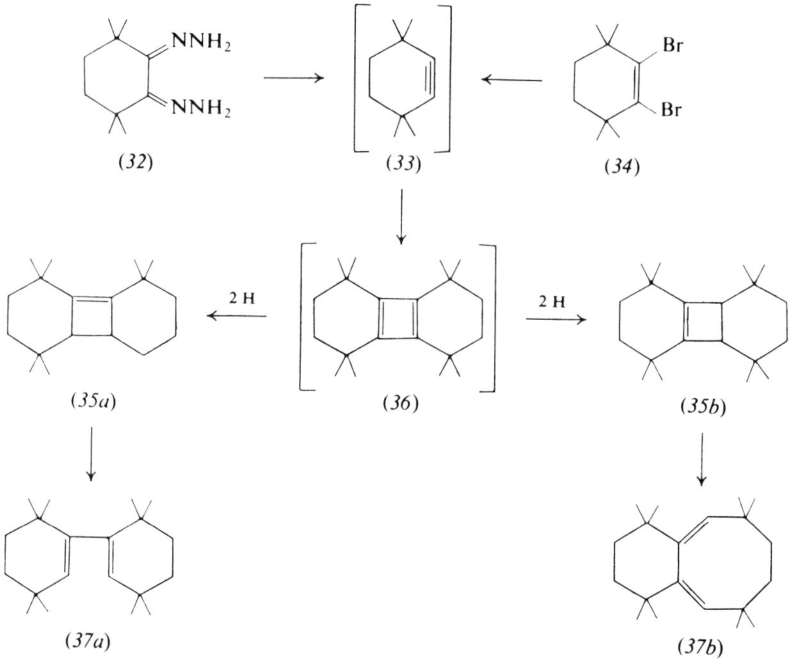

E. CHEMISTRY OF CYCLOBUTADIENE

E. Chemistry of Cyclobutadiene and the Substituted Cyclobutadienes*

At the present time, the cyclobutadienes are known only as transient intermediates and the study of their chemistry is made difficult by this fact. However, in many instances the cyclobutadienes can be isolated in the form of stable transformation products such as dimers or adducts (or even as the rearrangement products of dimers or adducts) from which the mode of reaction of the cyclobutadienes can usually be reconstructed. It must be borne in mind, however, that partly because of the versatility of the cyclobutadienes in being able to undergo a variety of transformations (dimerization, internal rearrangement, free-radical addition, adduct formation, etc.) and partly because of a dualism inherent in some of these transformations (the reaction of cyclobutadiene with an olefin may be viewed with equal justification as either a Diels–Alder reaction or a cycloaddition reaction), any interpretation of the chemistry of the cyclobutadienes contains an element of uncertainty and is necessarily speculative to a greater or lesser degree. Consequently, it is not at all unlikely that many of the interpretations in the present section will have to be modified in the future as additional experimental evidence is accumulated.

* For the diradical chemistry of the cyclobutadienes, see the preceding section and Section E, 5, a.

1. Dimerization Reactions

A number of reactions are known that produce cyclobutadiene dimers (i.e., tricyclooctadienes or products of their thermal rearrangement) under conditions which should be conducive to the generation of monomeric cyclobutadienes. The dimerization of a triplet cyclobutadiene species by cycloaddition is a process which undoubtedly is energetically favorable and leads to the formation of two new σ bonds without requiring spin inversion. Maximum overlap of π orbitals should be energetically favorable in the transition state of the dimerization (as it is in the Diels–Alder reaction), resulting in the formation of syn tricyclooctadienes (A). The limited experimental evidence so far available gives some support to this prediction, even though several dimers are known which have the opposite, or anti configuration (B). The factors which could play a role in determining the stereochemical course of dimerization are discussed in Section E, 1, a and in Section E, 1, c.

(A) Syn dimer   (B) Anti dimer

a. Of Cyclobutadiene

syn-Tricyclo[4.2.0.0²,⁵]octa-3,7-diene (6a) was formed when all-trans-1,2,3,4-tetrabromocyclobutane (1) was treated with lithium amalgam in ether,[9] or when cis-3,4-dichlorocyclobutene (7) was treated with sodium amalgam in ether.[5] Dimer (6a) is a liquid which is best isolated as its silver nitrate complex [$C_8H_8 \cdot 2AgNO_3$ (11)], from which it can easily be regenerated by treatment with aqueous sodium chloride solution.*,[9] At temperatures above 100°, (6a) rearranges rapidly to cyclooctatetraene (12).[5]

It is remarkable that the configuration of the dehalogation product of cis-3,4-dichlorocyclobutene (7) varies with the alkali metal employed. Thus,

---

* When it was first prepared from tetrabromide (1), the silver nitrate complex (11) was believed to be $C_4H_4 \cdot AgNO_3$, and the rather unstable hydrocarbon regenerated from it appeared to be cyclobutadiene itself (ref. 9). This presumed isolation of cyclobutadiene was subsequently widely quoted; more recent work has shown that the complex has the molecular formula $C_8H_8 \cdot 2AgNO_3$ (ref. 5), consequently the earlier misconception, naturally magnified by several reviewers, must be rectified. A side product of the dehalogenation of (1) with lithium amalgam is an organomercury complex of unknown composition (see Chapter 2).

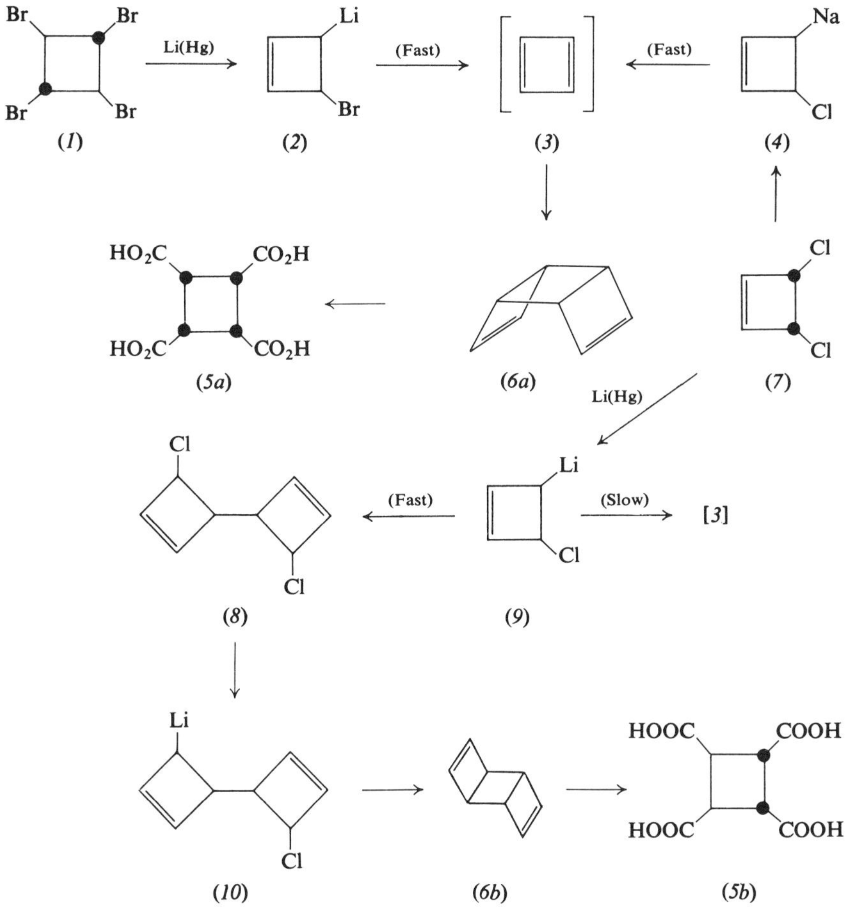

E. CHEMISTRY OF CYCLOBUTADIENE

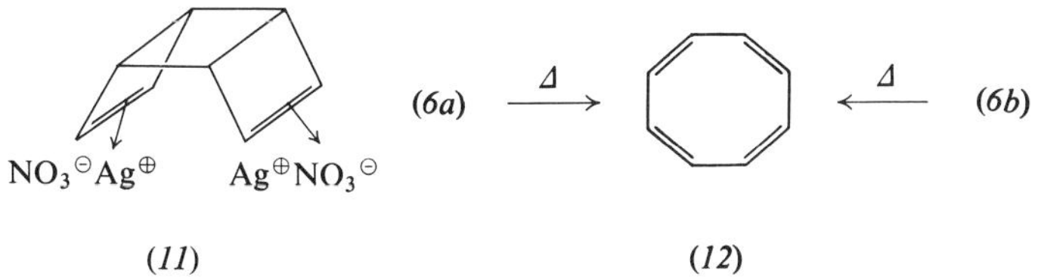

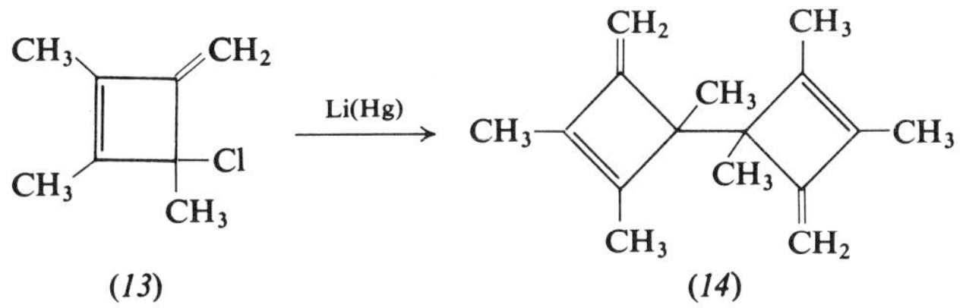

although (7) reacts with sodium amalgam to afford the syn dimer (6a), it reacts with lithium amalgam to give the corresponding anti dimer (6b).[5] Like its isomer, (6b) rearranges to cyclooctatetraene (12) on heating.[5] The interesting suggestion has been made that, while the reaction of the dichlorocyclobutene (7) with sodium probably proceeds by way of cyclobutadiene (3), its reaction with lithium might involve a sequence of two Wurtz reactions (7) → (9) → (8) → (10) → (6b). Whether a specific metallated halocyclobutene will collapse to the cyclobutadiene (affording the syn dimer) or undergo Wurtz coupling (leading to the anti dimer) may well be a function of both the metal and the halogen present. Thus, the postulated bromolithio intermediate (2) in the reaction of tetrabromide (1) with lithium should collapse more rapidly than the corresponding chlorolithio compound (9), thereby giving rise to the observed syn dimer (6a).[5] Experimental support for a Wurtz-type mechanism is found in the reaction of 3-chloro-4-methylene-1,2,3-trimethylcyclobutene (13) with lithium amalgam, which gives the dimeric hydrocarbon (14).[45]

For other reactions of cyclobutadiene, see Section D, Sections E, 2, a and b, and Table 1.1.

b. Of Tetramethylcyclobutadiene

The syn dimer (19a) of tetramethylcyclobutadiene was formed when cis-3,4-dichloro-1,2,3,4-tetramethylcyclobutene (15) was treated with lithium amalgam,[20, 49, 52] sodium amalgam,[52] lithium in liquid ammonia,[1] zinc–copper couple,[49] n-butyllithium,[1] activated zinc,[20, 45] or gaseous potassium.[135] The reaction of the corresponding dibromide (16) with lithium amalgam[52] and the reaction of the related diiodide (17) with zinc[93] also afford the syn dimer. It is likely that tetramethylcyclobutadiene (18) is an intermediate in these reactions.*

None of the above reactions produced any of the anti dimer. However, the one reaction which was conducted at an elevated temperature, viz., the reaction of the dichloride (15) with potassium vapor, also afforded a small amount of octamethylcyclooctatetraene (20),[135] probably formed by thermal isomerization of the syn dimer. Although the independent conversion of the syn dimer (19a) into tetramethylcyclooctatetraene has not been reported, the thermolysis of both (19a) and (20) to give the same valence tautomer, 1,2,3,4,5,6,7,8-octamethylbicyclo[4.2.0]octa-2,5,7-triene (22),[55] is known and makes the conversion of (19a) into (20) seem highly probable.

The anti dimer (19b) of tetramethylcyclobutadiene has been obtained only by the thermal decomposition of tetramethylcyclobutadienenickel chloride (23)[54] and tetramethylcyclobutadienenickel phenanthroline.[45] In the absence

---
* For a discussion of the difficulties attendant on any attempt to demonstrate *with certainty* the intermediacy of tetramethylcyclobutadiene in such reactions, see ref. 45.

E. CHEMISTRY OF CYCLOBUTADIENE

of any evidence to the contrary, it is reasonable to assume that tetramethylcyclobutadiene per se is not an intermediate in the formation of the anti dimer.

The configuration of both dimers has been confirmed by ozonolysis experiments.[45] Thus, the anti dimer gave *cis-trans-cis*-1,2,3,4-tetraacetylcyclobutane

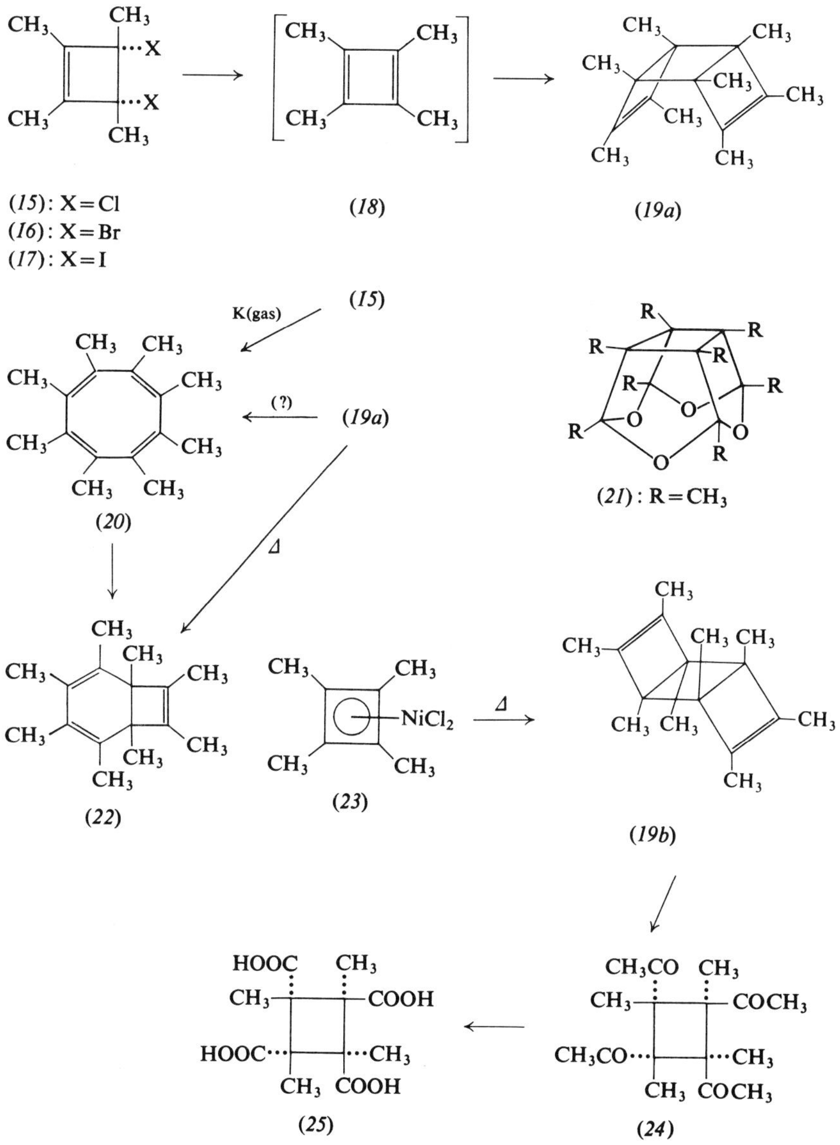

(24) on ozonolysis, which was subsequently converted into the known cis-trans-cis-tetracarboxylic acid (25). The syn dimer, on the other hand, afforded a remarkable internal ketal (21).

c. Of 1,2,3,4,5,6,7,8-Octahydrobiphenylene

A tetramer of cyclohexyne (28) is one of the products formed when 1,2-dibromocyclohexene (26) or 1-fluoro-2-bromocyclohexene (27) is treated with magnesium or lithium amalgam.[158] The first intermediate formed in the reaction is cyclohexyne (28), which apparently undergoes dimerization to give octahydrobiphenylene (29), a cyclobutadiene structure. A second dimerization step, this one involving the octahydrobiphenylene, gives the observed product,

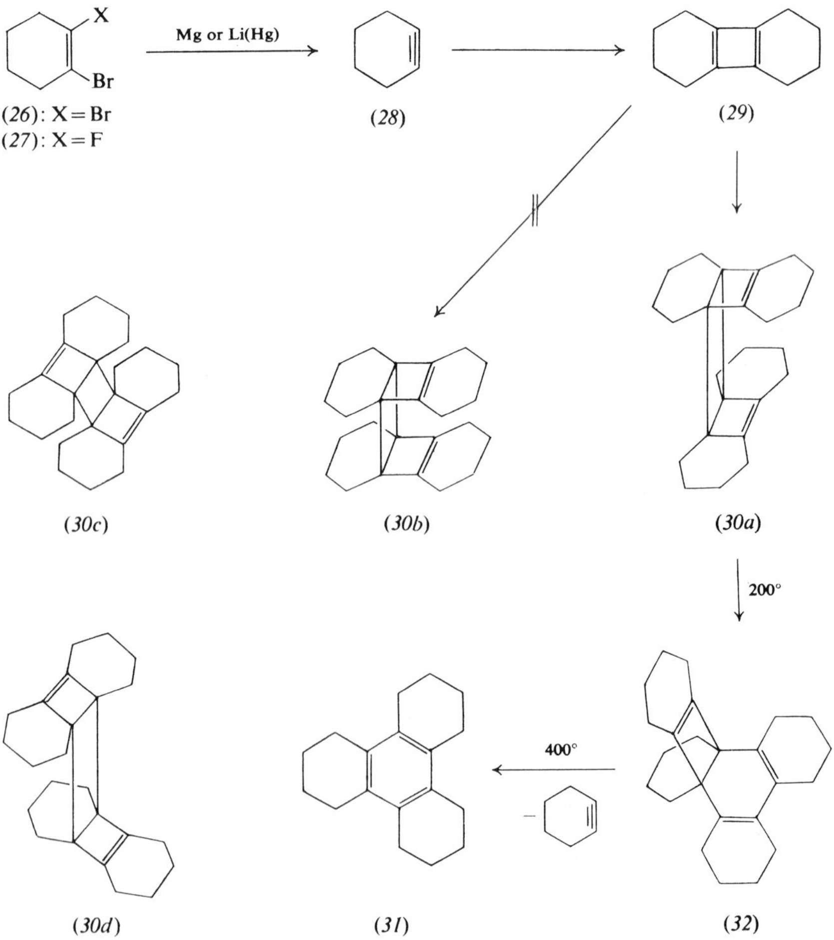

1,2,3,4,5,6,7,8,9,10,11,12-dodecahydro-8b,12b-butanobenzo[3′,4′]cyclobuta-[1′,2′:3,4]cyclobuta[1,2-e]biphenylene (30a). The assigned structure (30a) is supported by the fact that the product can be transformed in two stages into dodecahydrotriphenylene. Thus, at 200° the product undergoes cyclobutene ring-cleavage to give the corresponding butadiene (32); at 400°, the latter loses the elements of cyclohexyne to give dodecahydrotriphenylene [(32) → (31)]. The formation of dodecahydrotriphenylene as the end product of the pyrolysis eliminates the isomeric structures (30b) and (30c) [but not (30d)] from consideration.

The formation of the "perpendicular" dimer (30a) of octahydrobiphenylene rather than the "parallel" dimer (30b) can be explained on the basis of steric effects: Parallel dimerization involves eight pairs of nonbonded hydrogen interactions whereas perpendicular dimerization involves only four.* Of course, this explanation assumes that a syn (30a) and not an anti dimer (30d) is formed. Experimental evidence is not available to support this assumption, but it seems a reasonable one to make in view of the known syn configuration of the dimer of tetramethylcyclobutadiene.

d. Of 1,3-Diphenylcyclobutadiene

Generation of 1,3-diphenylcyclobutadiene (34) by the reaction of 1,3-diphenyl-2,4-bis(dimethylamino)cyclobutane dimethiodide (33) with lithium diethylamide at room temperature leads to the formation of both a "head-to-head" dimer (35a) and a "head-to-tail" dimer (35b) of the cyclobutadiene.[151] The dimers are thermally unstable and rearrange easily to the corresponding cyclooctatetraenes, viz., 1,2,4,7-tetraphenylcyclooctatetraene (36a) and 1,3,5,7-tetraphenylcyclooctatetraene (36b), respectively. Irradiation of the former (36a) afforded its valence tautomer (38), (the reaction is thermally reversible), which on further irradiation gave p-terphenyl (39) and diphenylacetylene. In addition, tautomer (38) was transformed in several steps into a diketone (37) which was synthesized by an independent route.[150]

No direct evidence regarding the configuration of dimers (35a) and (35b) is available as yet. However, by analogy with the normal dimers of cyclobutadiene and tetramethylcyclobutadiene and also on the basis of mechanistic grounds, the syn configuration seems likely.†

Subjecting dimethiodide (33) to Hofmann elimination in refluxing tert-butyl

---

* In this connection it is interesting to note that the sterically hindered analog of (29), 1,1,4,4,5,5,8,8-octamethyl-1,2,3,4,5,6,7,8-octahydrobiphenylene, appears to undergo hydrogen abstraction and rearrangement reactions but not dimerization. For a discussion of the possibility that double-bond fixation (π-electron localization) might play a role in determining the stereochemical course of dimerization of (29), see ref. 158.

† White and Dunathan prefer the alternative anti configuration for dimers (35a and b) (ref. 151).

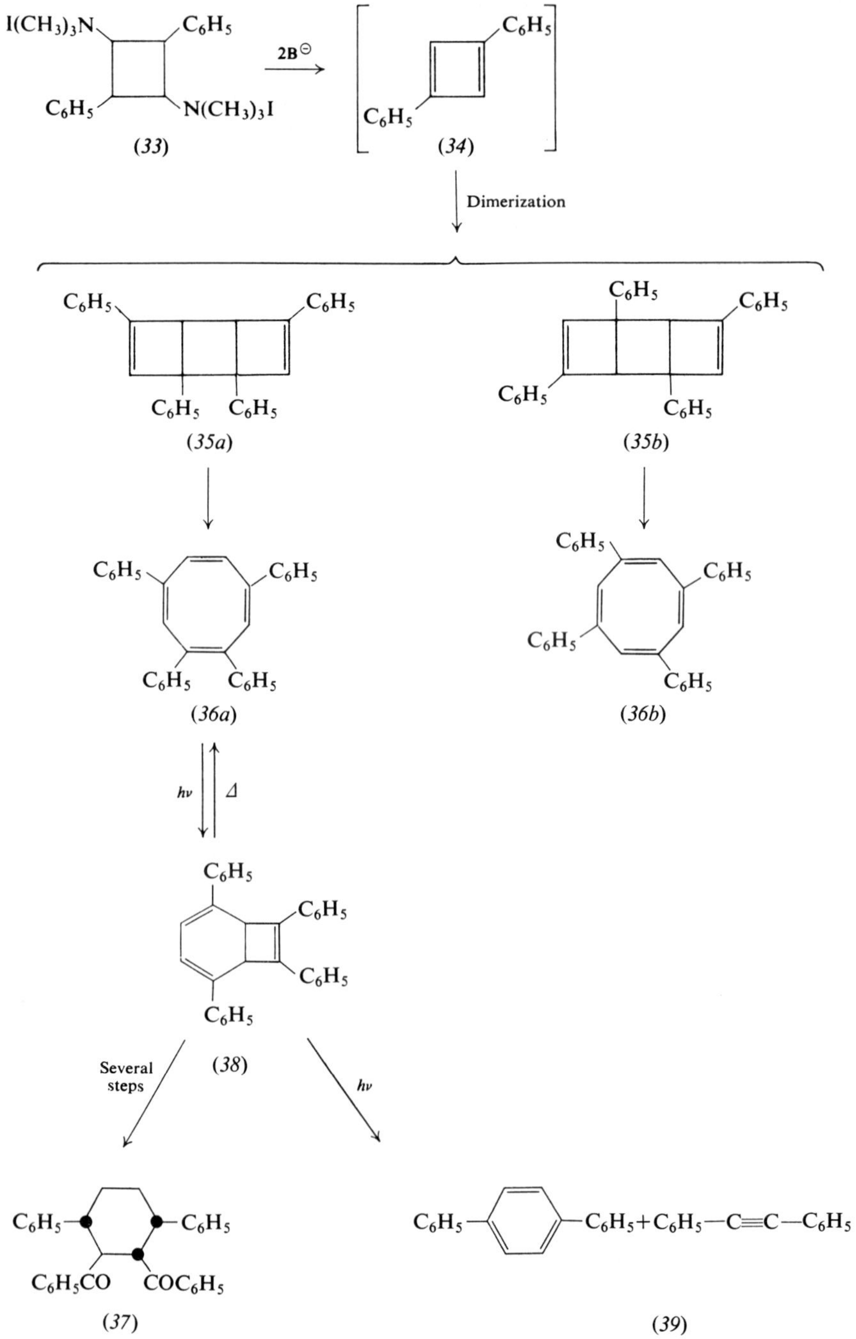
I(CH3)3N
C6H5
C6H5
N(CH3)3I
(33)
2B⊖
C6H5
C6H5
(34)
Dimerization
(35a)
(35b)
(36a)
(36b)
hν
Δ
(38)
Several steps
hν
C6H5CO
COC6H5
(37)
C6H5—C≡C—C6H5
(39)

E. CHEMISTRY OF CYCLOBUTADIENE 57

alcohol affords directly a mixture of the two cyclooctatetraenes (*36a* and *b*).[151] The latter are undoubtedly formed via the two diphenylcyclobutadiene dimers (*35a* and *b*), which are demonstrably thermolabile. [The half-life of dimer (*35b*) in solution at 25° is 12 hours, that of (*35a*) is only somewhat longer.]

e. Of 1-Fluoro-2,3,4-triphenylcyclobutadiene

A single dimer of 1-fluoro-2,3,4-triphenylcyclobutadiene was isolated when 4,4-difluoro-1,2,3-triphenylcyclobutene (*40*) was treated with phenyllithium.*,[116] In this reaction, it is assumed that dehydrofluorination of (*40*) takes place to give the cyclobutadiene (*41*); "head-to-head" dimerization of (*41*) affords the observed dimer, (*42*). Attempts to determine the structure and configuration of the dimer by chemical means were unsuccessful,[116] but an X-ray crystallographic study showed that it is the "head-to-head" dimer and that it has the anti configuration as shown in structure (*42*).[71] The factors responsible for the formation of the anti dimer rather than the syn dimer are

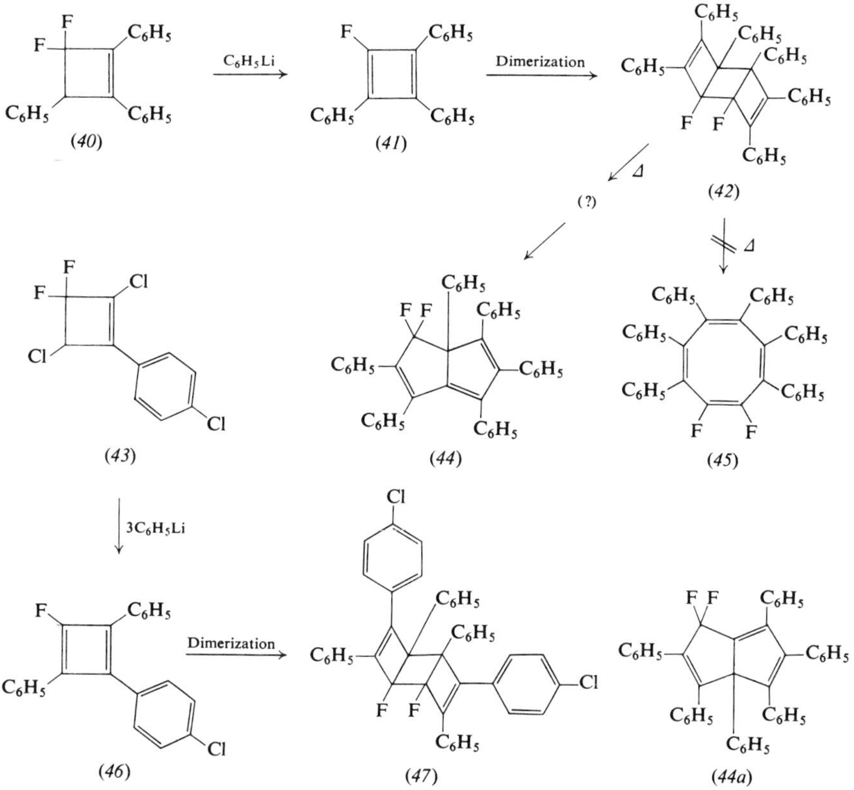

* The same dimer was obtained from the reaction of excess phenyllithium with several other halogenated phenylcyclobutenes. Since compound (*40*) was shown to be an intermediate in all cases, details are not repeated here. (See Section B, 2 and refs. 91 and 111.)

by no means clear, but the fluoro substituent may play an important role.

The dimer (42) is relatively stable thermally, becoming yellow at about 180° and melting with decomposition at about 200° to give an isomer which is not the expected cyclooctatetraene derivative (45) but rather a rearrangement product (44).[116] This behavior contrasts sharply with the rearrangement of the dimers of 1,3-diphenylcyclobutadiene and tetraphenylcyclobutadiene, which readily afford the corresponding cyclooctatetraenes. An isomeric structure (44a) was originally considered for the rearrangement product of dimer (42),[116] but an X-ray diffraction study subsequently demonstrated that the correct structure is (44).[87]

In similar work, a dimer of 1-fluoro-3-(p-chlorophenyl)-2,4-diphenylcyclobutadiene (46) was prepared by the action of excess phenyllithium on 1,3-dichloro-2-(p-chlorophenyl)-4,4-difluorocyclobutene (43).[116a] The chemical and spectral properties of the dimer (47) match those of dimer (42), consequently it may also be assigned the "head-to-head" structure and the anti configuration (47).

f. Of Tetraphenylcyclobutadiene

The reaction of 3,4-dibromo-1,2,3,4-tetraphenylcyclobutene (48) with phenyllithium affords an unstable colorless hydrocarbon (characterized only spectrally and by its X-ray powder diffraction pattern) which isomerizes to octaphenylcyclooctatetraene (56),* even in the solid state at room temperature.[69] The hydrocarbon must be an octaphenyltricyclooctadiene, probably the syn isomer (52), formed by dimerization of tetraphenylcyclobutadiene (49). In its thermal instability, dimer (52) resembles the corresponding tricyclooctadiene dimers (35a) and (35b) of 1,3-diphenylcyclobutadiene (34). The same unstable dimer (52), is formed by the reaction of 1,4-dilithio-cis,cis-tetraphenylbutadiene (53) with 1,4-diiodo-cis,cis-tetraphenylbutadiene (54).[69] In the latter reaction, it is probable that (53) and (54) first undergo metal–halogen interchange to give the mixed lithio-iodo compound (50); intramolecular cyclization of (50) would afford the observed dimer (52) via tetraphenylcyclobutadiene (49).

In view of the thermal lability of octaphenyltricyclooctadiene (52), it is not surprising that octaphenylcyclooctatetraene (56), and not octaphenyltricyclooctadiene is the product formed in more vigorous reactions in which tetraphenylcyclobutadiene (49) is believed to be generated. Thus, hydrocarbon

---

* The compound now known to be octaphenylcyclooctatetraene (56) was believed for a time to be octaphenylcubane (51). The strongest evidence in favor of the cubane structure was the absence of an expected C=C stretching band in the Raman spectrum of the compound (ref. 69). More recently, an octaphenyltricyclooctadiene structure (52) has also been proposed (ref. 143), but the cyclooctatetraene structure (56) has been established unequivocally through X-ray crystallographic analysis (ref. 101).

(56) is formed in the following diverse reactions: (a) ultraviolet irradiation of diphenylacetylene (57)[33]; (b) treatment of diphenylacetylene with phenylmagnesium bromide[143]; (c) thermolysis of (4-bromo-1,2,3,4-tetraphenyl-cis,cis-1,3-butadienyl)dimethyltin bromide (59)[68]; (d) thermolysis of 1,4-diiodotetraphenylbutadiene (54)[25]; (e) treatment of 1,4-dilithiotetraphenylbutadiene (53) with cupric bromide[25]; (f) demetallation of tetraphenylcyclobutadienepalladium chloride (58) with triphenylphosphine or trimethyl phosphite[105]; and (g) thermal decomposition of a compound believed to be tetraphenylmercurole (55).[25]

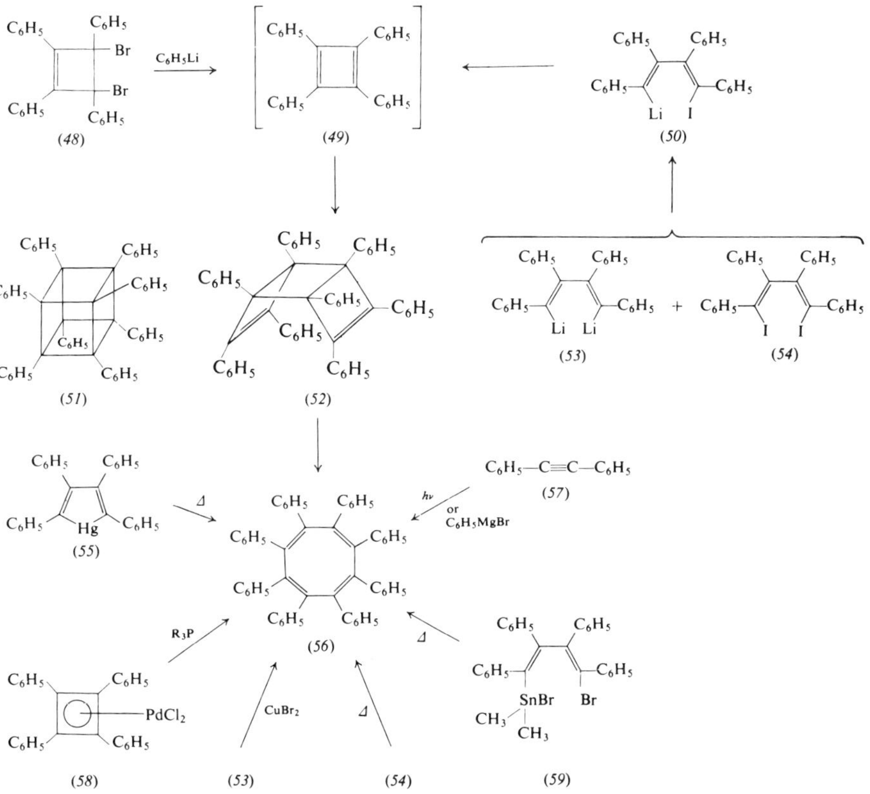

2. Olefin Addition Reactions

The cyclobutadienes appear to have no reactivity as dienophiles in the Diels–Alder reaction $(B) \not\to (A)$.* A number of reactions have been observed,

* But see Abstract 1.24 in the Appendix.

however, in which a cyclobutadiene intermediate, acting as a diene, has been trapped by a dienophile, $(B) \rightarrow (D)$. It should be noted that such reactions may also be viewed as cycloadditions, since the usual distinction between cycloaddition and Diels–Alder addition is meaningless in the case of a cyclobutadiene reacting with an olefin.

In view of its triplet character, a cyclobutadiene cannot be expected to react with an olefin in a concerted manner, because the initially formed diradical intermediate (C) must undergo spin inversion prior to ring closure. The apparent wide variation in the ease of addition of cyclobutadienes to olefins may be largely a function of the degree of stability of the diradical (C) formed initially. Both the rate of olefin addition and cyclobutadiene dimerization are, of course, influenced by the degree and type of substitution on the cyclobutadiene ring.*

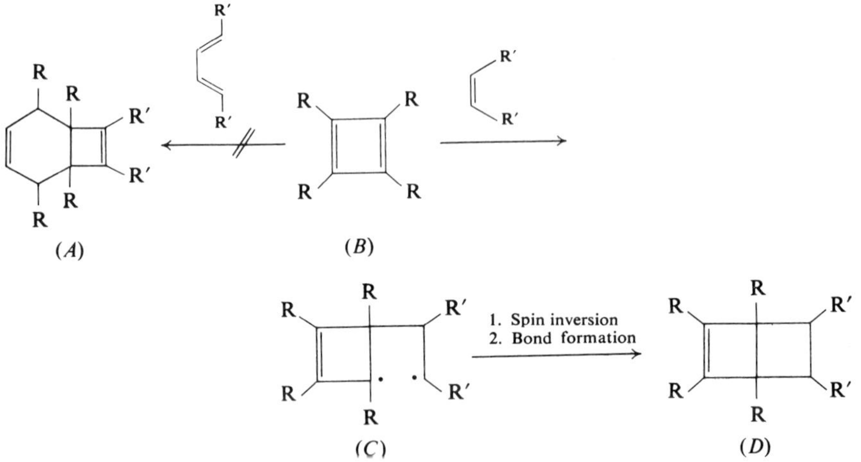

a. Attempted Addition of the Cyclobutadienes to Dienes

Unsuccessful attempts have been made to effect reaction between the members of each of the following cyclobutadiene–diene pairs: cyclobutadiene and cyclopentadiene; tetramethylcyclobutadiene and cyclopentadiene [52] or furan [49, 52]; 1,3-diphenylcyclobutadiene and furan [151]; and others.†

---

* Cyclopentadienone, an unstable molecule which reacts readily with dienophiles and which is isoelectronic with cyclobutadiene, also undergoes dimerization in preference to addition to other dienes (ref. 57).

† For a complete list, see Table 1.1.

E. CHEMISTRY OF CYCLOBUTADIENE   61

b. Apparent Addition of Cyclobutadiene to Aromatic Rings

Several compounds which are formal adducts of cyclobutadiene [e.g., *(60)*][11] yield pyrolysis products containing all of the carbon atoms in the original skeletons [e.g., *(63)*, *(64)*, *(65)*]. A mechanism involving the generation and readdition of cyclobutadiene has been proposed (see p. 62), but since the mechanism is based on several misconceptions, it is not entirely plausible. The following modification *which does not involve a cyclobutadienoid intermediate* is probably more accurate*: Thermolytic cleavage of the cyclobutene ring of *(60)* gives the corresponding butadiene *(61)*. Readdition of the strained butadiene bridge of *(61)*, this time to the substituted double bond, gives an intramolecular Diels–Alder product *(62)* which, in turn, undergoes thermal isomerization to give a cyclodecapentaene *(66)*. The cyclodecapentaene *(66)*, which is common to all three mechanisms, gives the observed products through its more stable conformer, *(66a)*. The products are dimethyl 3,4-dihydro-2,6-naphthalate *(63)*,[42] dimethyl 1,2-dihydro-2,6-naphthalate *(64)*,[42] and dimethyl 2,6-naphthalate *(65)*.[13, 42] [Compound *(63)* was originally assigned the structure of dimethyl 4a,8a-dihydro-2,6-naphthalate.[13]]

The earlier cyclobutadiene-type mechanism was founded in part on the misconception that the starting material *(60)* had the syn configuration *(60a)*. According to this mechanism, the four-membered ring would dissociate to form a cyclobutadienoid moiety or transition species *(67)* which would immediately collapse to form the isomeric structure *(68)*. Thermal ring-opening of *(68)* would lead to the common intermediate, *(66)*, and thence to the observed products.

For other examples of the reverse Diels–Alder reaction (Alder–Rickert reaction) which produce "adducts" of cyclobutadiene, see Section B, 6. The reaction also produces butadiene (Section D).

c. Addition of the Cyclobutadienes to Dienophiles

Either primary or secondary reaction products were isolated when cyclobutadienes were generated in the presence of dienophiles. Thus, dehalogenation of *cis*-3,4-dichloro-1,2,3,4-tetramethylcyclobutene *(15)* with zinc in the presence of 2-butyne or dimethyl acetylenedicarboxylate afforded hexamethylbenzene *(75)* and dimethyl tetramethylphthalate *(76)*, respectively; the bicyclic "Dewar-benzene" structures, *(69)* and *(70)*, are undoubtedly formed as unstable intermediates in these reactions.[20]

---

* Another modification, also not involving cyclobutadiene as an intermediate but having several objectionable features nevertheless, has been proposed (ref. 118). The transformation of *(60)* into *(61)* proposed by the present authors violates the recently formulated Woodward–Hoffman rules, hence, it is probably more accurate to write *(61)* as the butadienyl diradical. For a similar case, see the transformation of *(30a)* into *(32)* in Section E,1,c.

1. CYCLOBUTADIENE

Similarly, the reaction of cis-1,2-diiodo-1,2,3,4-tetramethylcyclobutene (17) with mercury in the presence of maleic anhydride or dimethylmaleic anhydride afforded the isolable but thermally labile adducts, 1,2,3,4-tetramethylbicyclo[2.2.0]hexa-2-ene-5,6-dicarboxylic anhydride (72) and 1,2,3,4,5,6-hexamethylbicyclo[2.2.0]hexa-2-ene-5,6-dicarboxylic anhydride (73),[45] respectively, which attests to the intermediacy of tetramethylcyclobutadiene.*

The use of dibromomaleic anhydride as the dienophile in the mercury dehalogenation of (17) leads to the formation of adduct (74). Conversion of this adduct into the corresponding dimethyl ester, followed by debromination of the latter with a zinc–copper couple affords the crystalline "Dewar-benzene" derivative (71) which rearranges to dimethyl tetramethylphthalate (76) on heating.[56] The isolation of (71) and its facile conversion into (76) add support to the postulated intermediacy of tetramethylcyclobutadiene (18).

When 1,2,3,4,5,6,7,8-octahydrobiphenylene (29) is generated by the

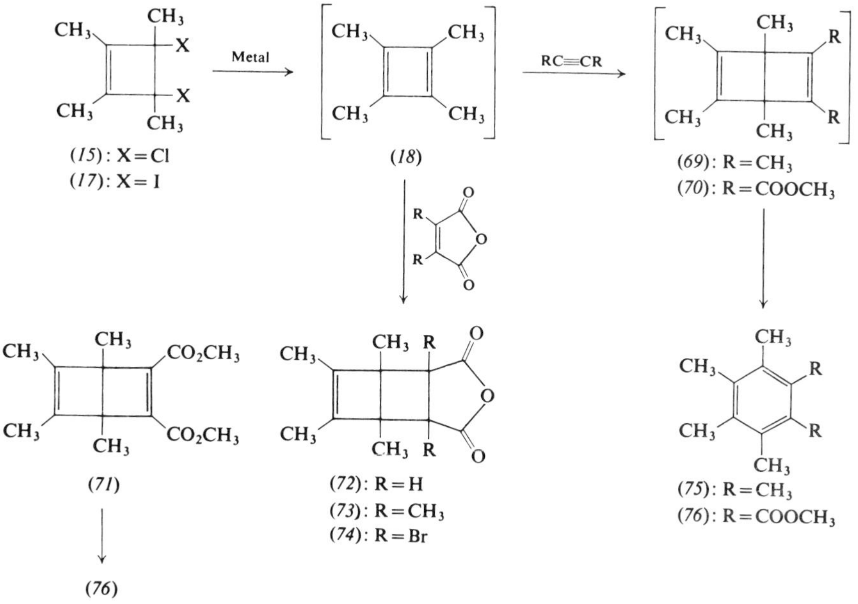

---

* For evidence tending to discredit the tetramethylcyclobutadiene mechanism, see ref. 45. In this connection, it is interesting to note that the dehalogenation of cis-3,4-dichloro-1,2,3,4-tetramethylcyclobutene with lithium amalgam (rather than with zinc) in the presence of 2-butyne gives only the syn dimer of tetramethylcyclobutadiene (rather than the adduct).

dimerization of cyclohexyne [itself formed by the dehalogenation of 1,2-dibromo- or 1-bromo-2-fluorocyclohexene, (26) or (27) → (28)] a side-product, dodecahydrotriphenylene (78) is formed in small yield (3–9%).[158] It has been suggested that the side product is formed by a Diels–Alder reaction between the octahydrobiphenylene, which is a substituted cyclobutadiene, and excess cyclohexyne [(29) → (77) → (78)], but substantiating evidence is lacking.

Similarly, in the photodimerization of diphenylacetylene, hexaphenylbenzene (80) is produced[33]; it is not inconceivable that this product is formed via a Diels–Alder reaction between tetraphenylcyclobutadiene and diphenylacetylene [(8) → (79) → (80)] rather than by simple trimerization of diphenylacetylene.*

Secondary adducts of tetraphenylcyclobutadiene (49) have been obtained by the thermolysis of (4-bromo-1,2,3,4-tetraphenyl-*cis*,*cis*-1,3-butadienyl)-dimethyltin bromide (59) in the presence of various dienophiles.[67] Thus,

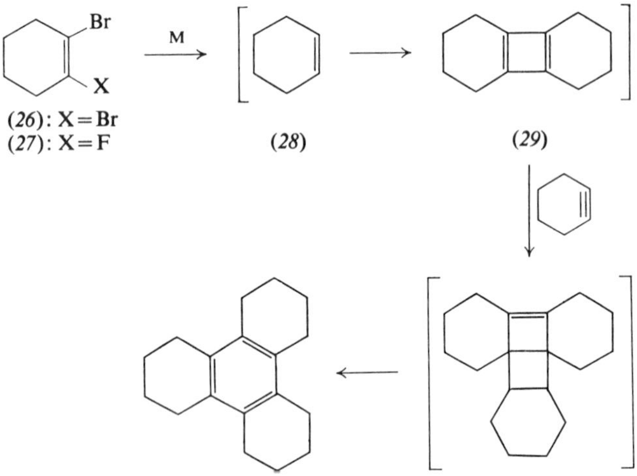

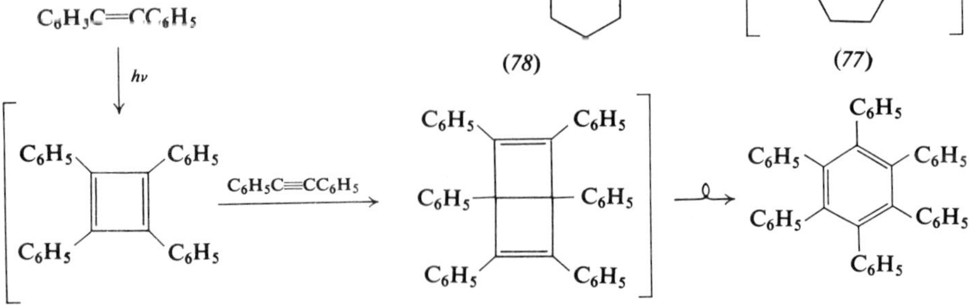

---

* For the formation of substituted benzenes from the corresponding acetylenes in the presence of transition metal catalysts, see Chapter 2.

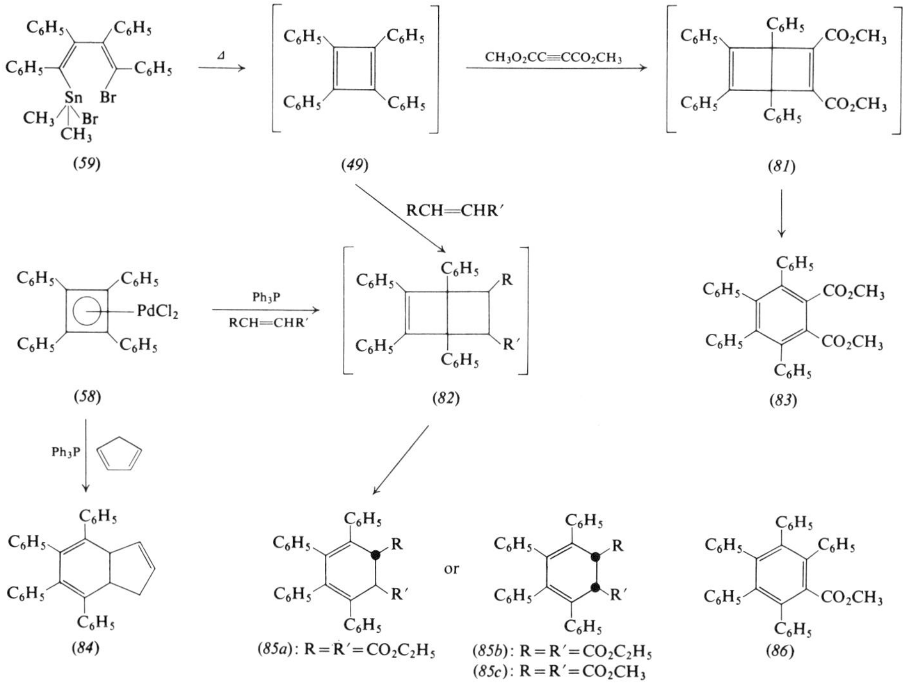

E. CHEMISTRY OF CYCLOBUTADIENE 65

dimethyl acetylenedicarboxylate, diethyl fumarate, diethyl maleate, and dimethyl maleate afford, respectively, the adducts, dimethyl tetraphenylphthalate (*83*), diethyl 3,4,5,6-tetraphenyl-*trans*-1,2-dihydrophthalate (*85a*), diethyl 3,4,5,6-tetraphenyl-*cis*-1,2-dihydrophthalate (*85b*), and dimethyl 3,4,5,6-tetraphenyl-*cis*-1,2-dihydrophthalate (*85c*). Bicyclohexene derivatives (*82*) are presumed to be intermediates in these reactions.

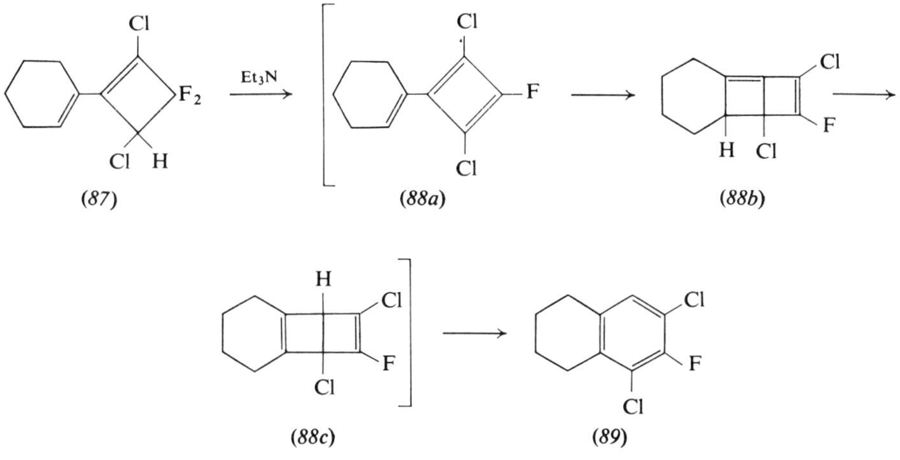

Similarly, tetraphenylcyclobutadiene [generated by the reaction of tetraphenylcyclobutadienepalladium chloride (58) with triphenylphosphine] reacts as a diene with methyl phenylpropiolate or cyclopentadiene to give the expected adducts, (86) and (84), respectively.[41]

An unusual reaction has been recorded in which the intramolecular addition of a cyclobutadiene nucleus to an unsaturated side chain appears to occur. Thus, dehalogenation of 1-(1'-cyclohexenyl)-2,4-dichloro-3,3-difluorocyclobutene (87) with triethylamine gives 1,3-dichloro-2-fluoro-5,6,7,8-tetrahydronaphthalene (89), presumably via 1-(1'-cyclohexenyl)-2,4-dichloro-3-fluorocyclobutadiene [(88a) → (88b) → (88c) → (89)].[133]

3. Skeletal Rearrangement Reactions of the Cyclobutadienes*

A few reactions are known in which skeletal rearrangement of a cyclobutadiene appears to occur. However, experimental support for the rearrangement mechanism is still meager.

a. Of 1,1,4,4,5,5,8,8-Octamethyl-1,2,3,4,5,6,7,8-octahydrobiphenylene

Generation of 3,3,6,6-tetramethylcyclohexyne (90) by either of two methods (see Table 1.1) affords, as one of three products, a hydrocarbon $C_{20}H_{32}$ tentatively assigned structure (92a or b) which appears to be a rearrangement product of the corresponding isomeric cyclobutadiene (91).[2] The mechanism of formation of (92a or b) from (91) is as yet obscure.

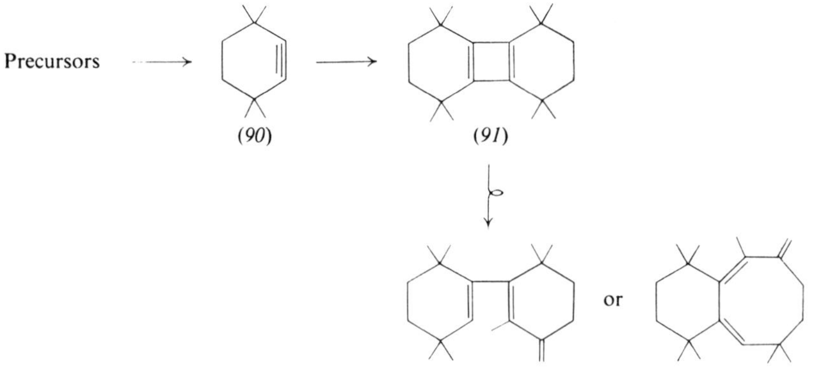

b. Of Tetraphenylcyclobutadiene

The photolysis of diphenylacetylene affords several products in low yield, one of which [octaphenylcyclooctatetraene (56)], may be taken as evidence of the generation of tetraphenylcyclobutadiene (49).[33] Two other products

* For a rearrangement reaction involving the transfer of hydrogen atoms only, see tetramethylcyclobutadiene, Section D.

isolated from the photolysis, viz., 1,2,3-triphenylazulene (*97*) and 1,2,3-triphenylnaphthalene (*98*), are not normal products of tetraphenylcyclobutadiene (*49*) but may well be its photorearrangement products. This view receives indirect support from the fact that the organotin compounds (*59*) and (*94*), which afford cyclobutadiene dimer (*56*) on heating, give triphenylazulene (*97*) on ultraviolet irradiation.[66] The photolysis of tetraphenylcyclobutadiene may be assumed to proceed via an excited state, e.g. (*49a*), to give a highly strained intermediate (*96*) which could give rise to both the azulene (*97*) and the naphthalene (*98*) by electron redistribution.

The formation of triphenylazulene in the dehalogenation of bromoiodide (*95*) with zinc[66] may proceed by way of a Lewis-acid-promoted rearrangement of tetraphenylcyclobutadiene (*49*). The mechanism of this transformation has not been firmly established and it is important to realize that the formation of triphenylazulene is not in itself proof of the intermediacy of tetraphenylcyclobutadiene. (See Sections C, 6 and C, 8.)

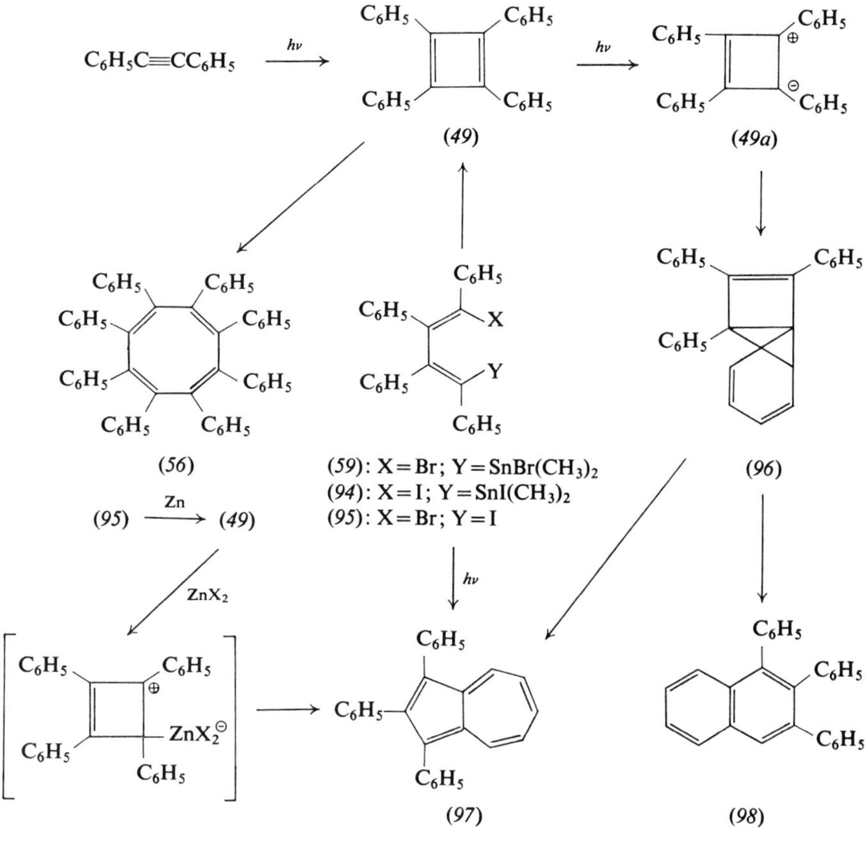

4. Formation of Stable Metal Complexes

The cyclobutadienes can react with certain transition metals or their salts to form stable cyclobutadiene–metal complexes. The preparation and properties of the complexes are discussed in detail in Chapter 2.

5. Hypothetical Reactions

a. Hydrogen Abstraction

Under certain reaction conditions cyclobutadiene and 1,3-diphenylcyclobutadiene are believed to abstract two hydrogen atoms from the reaction solvent (or from the starting material) to give butadiene and 1,3-diphenylbutadiene, formed via the corresponding cyclobutenes.* The reaction of *trans*-1,2-dichloro-*all-trans*-1,2,3,4-tetraphenylcyclobutane (*99*) with strong bases (phenyllithium, phenylsodium, or lithium amide) resembles these reactions to the extent that the corresponding butadiene, *cis,cis*-1,2,3,4-tetraphenylbutadiene (*101*) m.p. 181°–183°, is produced.[134] However, tetraphenylcyclobutadiene (*49*) is almost certain not to be an intermediate in this transformation, since neither tetraphenylcyclobutadiene dimer nor its readily isolated rearrangement product (*52*) and (*56*), respectively, was formed. A more likely mechanism for the formation of the butadiene (*101*) involves *dehalogenation* of the dichloride (*99*) by base to give *trans*-1,2,3,4-tetraphenylcyclobutene (*100*) and isomerization of the latter to give the butadiene, (*100*) → (*101*).[134]

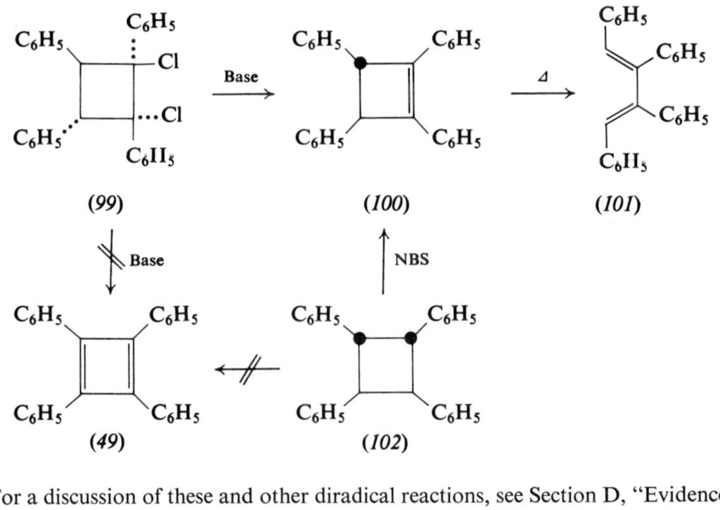

---

* For a discussion of these and other diradical reactions, see Section D, "Evidence of the Triplet Diradical Character of the Cyclobutadienes".

E. CHEMISTRY OF CYCLOBUTADIENE

An attempted aromatization of *cis,trans,cis*-1,2,3,4-tetraphenylcyclobutane (*102*) to tetraphenylcyclobutadiene (*49*) with *N*-bromosuccinimide gave *cis,cis*-1,2,3,4-tetraphenylbutadiene (*101*), m.p. 183°–184°.[14]

b. Fissioning of Cyclobutadiene

Willstätter suggested that the formation of acetylene in the dehydrohalogenation of 1,2-dibromocyclobutane with potassium hydroxide is due to the fissioning of cyclobutadiene.[156] Several authors have noted that no other example of the formation of acetylene of a substituted acetylene from a cyclobutadiene is known and that for this reason the fissioning of cyclobutadiene must be viewed as a hypothetical process for which there is no obvious driving force.[13, 104, 145, 158]

c. Valence Tautomerization of Cyclobutadiene

Lipscomb has pointed out that one way in which cyclobutadiene could achieve enhanced stability is through rearrangement to the three-dimensional tetrahedral structure commonly called "tetrahedrane" (*103*).*,[101] At the present time, there is no evidence to indicate that cyclobutadiene tends to achieve stability in this way. Furthermore, the strain energy of tetrahedrane has been estimated to be approximately 90 kcal/mole, which compares unfavorably with the value of 50 kcal/mole calculated for the strain energy of cyclobutadiene.[149]

Beesley and Thorpe were the first to lay claim to the synthesis of a tetrahedrane derivative, viz., 1-methyltricyclo[1.1.0.0^{2,4}]butane-2,3,4-tricarboxylic

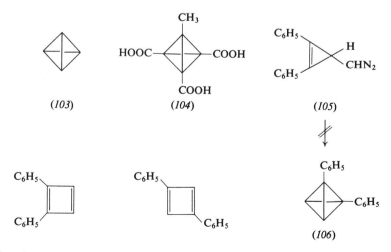

* "Tricyclo[1.1.0.0^{2,4}]butane," according to *Chemical Abstracts* nomenclature.

acid (*104*),[19] but since their work has been found to be unreproducible,[97] their claim must be discounted.

More recently, an attempt was made to detect the formation of 1,2-diphenyltricyclo[1.1.0.0^{2,4}]butane (*106*) or 1,2- or 1,3-diphenylcyclobutadiene or a transformation product derived from these compounds in the photolysis of (2,3-diphenyl-2-cyclopropenyl)diazomethane (*105*).[150] However, the only isolable product proved to be diphenylacetylene.

d. Polymerization

The free radical bromination of cyclobutene with *N*-bromosuccinimide is accompanied by polymer formation,[37] which has been attributed to the generation and subsequent polymerization of cyclobutadiene.[8]

TABLE 1.1

Cyclobutadienes[a]

Part A. Generation of cyclobutadienes		Part B. Formation of dimers, adducts, etc., of cyclobutadienes			
Method of generation[b]	Reactants	Reaction conditions and trapping agents	Product[c]	Yield (%)	Ref.
Cyclobutadiene					
1	*cis*-3,4-Dichlorocyclobutene and sodium amalgam	Ether, room temp.; followed by AgNO$_3$	I (Isolated as II)	46–51	5
	all-trans-1,2,3,4-Tetrabromocyclobutane and lithium amalgam	Ether, room temp.; followed by AgNO$_3$	I (Isolated as II)	47	5
2	1,2-Dibromocyclobutane and quinoline	Reflux	Butadiene	Small	156
3	Acetylene	UV irradiation, gas phase	Cyclooctatetraene	Trace	94
5	1,3-Bis(dimethylamino)cyclobutane dimethiodide and				
	(1) potassium hydroxide	Aqueous medium, heat	Butadiene	Small	13
	(2) silver oxide	Methanol	Butadiene	Small	13
	1,3-Bis(dimethylamino)cyclobutane dimethohydroxide	Thermolysis (dry), atm. pressure	Butadiene	Small	13
6	4,7-Diacetoxy-2a,3,8,8a-tetrahydro-3,8-ethenocyclobuta[*b*]naphthalene	Thermolysis in the melt	Butadiene[d]	—	13
	4,7-Dimethoxy-2a,3,8,8a-tetrahydro-3,8-ethenocyclobuta[*b*]naphthalene	Thermolysis in the melt	Butadiene[d]	—	13
	Dimethyl tricyclo[4.2.0.0^{2,5}]deca-3,7,9-triene-7,8-dicarboxylate	Thermolysis in the melt	Butadiene[d]	—	13

TABLE 1.1—continued

Part A. Generation of cyclobutadienes   Part B. Formation of dimers, adducts, etc., of cyclobutadienes

Method of generation[b]	Reactants	Reaction conditions and trapping agents	Product[c]	Yield (%)	Ref.
	2a,3,10,10a-Tetrahydro-3,10-ethenocyclobuta[b]anthra-4,9-quinone	Thermolysis in the melt	Butadiene[d]	—	13
	Tricyclo[4.2.2.0²,⁵]deca-3,7,9-triene	Thermolysis at the boiling point	Butadiene[d]	—	118
	1-(1′-Cyclohexenyl)-2,4-dichloro-3-fluorocyclobutadiene				
2	1-(1′-Cyclohexenyl)-4,4-dichloro-3,3-difluorocyclobutene via 1-(1′-cyclohexenyl)-2,4-dichloro-3,3-difluorocyclobutene	Triethylamine, 180°, 36 hours	III	41	133
	1-(1′-Cyclohexenyl)2,4-dichloro-3,3-difluorocyclobutene	Triethylamine, 180°, 36 hours	III	—	133
	1,3-Diphenylcyclobutadiene				
5	1,3-Bis(dimethylamino)-2,4-diphenylcyclobutane dimethiodide and				
	(1) lithium diethylamide	Ether, 0°, N₂	IV and V	10.3 and 4.2	151
		Furan, −5°	IV and V	22 (IV)	151
		tert-Butyl alcohol, reflux	VI and VII	17 and 12	151
	(2) potassium tert-butoxide	tert-Butyl alcohol, room temp.	IV and V	Low	151
		tert-Butyl alcohol and			
		(1) diphenylacetylene	—	—	151
		(2) anthracene	—	—	151
		(3) 2,3-dimethylbutadiene	—	—	151

E. CHEMISTRY OF CYCLOBUTADIENE

1-Fluoro-2,3,4-triphenylcyclobutadiene

2	2-Chloro-3,3-difluoro-1,4-diphenyl-cyclobutene and phenyllithium (via 3,3-difluoro-1,2,4-triphenyl-cyclobutenone)	—	VIIIa	—	91, 116
	4-Chloro-1-phenyl-3,3,4-trifluoro-cyclobutene and phenyllithium (via 3,3-difluoro-1,2,4-triphenyl-cyclobutene)	Ether, 0°, N_2	VIIIa	$15^{d,e}$	116
	2,4-Dichloro-3,3-difluoro-1-phenyl-cyclobutene and phenyllithium (via 3,3-difluoro-1,2,4-triphenyl-cyclobutene)	Ether, 0°, N_2	VIIIa	17	116
	4,4-Dichloro-3,3-difluoro-1-phenyl-cyclobutene and phenyllithium (via 3,3-difluoro-1,2,4-triphenyl-cyclobutene)	—	VIIIa	—	116
	3,3-Difluoro-1,2,4-triphenylcyclo-butene and phenyllithium	Ether, 0°, N_2	VIIIa	—	116
	1,4-Diphenyl-2,3,3-trifluorocyclo-butene and phenyllithium (via 3,3-difluoro-1,2,4-triphenylcyclobutene)	—	VIIIa	—	116

1-Fluoro-3-(p-chlorophenyl)-2,4-diphenylcyclobutadiene

2	2,4-Dichloro-3,3-difluoro-1-(p-chlorophenyl)cyclobutene and phenyllithium (via 3,3-difluoro-1,2,4-tris(p-chlorophenyl)cyclo-butene)	—	VIIIb	—	116a

TABLE 1.1—continued

Part A. Generation of cyclobutadienes		Part B. Formation of dimers, adducts, etc., of cyclobutadienes			
Method of generation[b]	Reactants	Reaction conditions and trapping agents	Product[c]	Yield (%)	Ref.
1,2,3,4,5,6,7,8-Octahydrobiphenylene					
3	1,2-Dibromocyclohexene and (1) lithium amalgam (via cyclohexyne)	THF, room temp., N_2	IX[d]	23	158
	(2) magnesium (via cyclohexyne)	THF, room temp., N_2	IX[d]	14	158
	1-Bromo-2-fluorocyclohexene and lithium amalgam or magnesium (via cyclohexyne)	THF, room temp., N_2	IX	—	158
1,1,4,4,5,5,8,8-Octamethyl-1,2,3,4,5,6,7,8-octahydrobiphenylene					
3	1,2-Dibromo-3,3,6,6-tetramethyl-cyclohexene and sodium (via 3,3,7,7-tetramethylcyclohexyne)	Ether	Xa (or b) and XIa (or b)	—	2
	3,3,6,6-Tetramethylcyclohexane-1,2-dione dihydrazone and silver oxide	Benzene	Xa (or b) and XIa (or b)	—	2
Tetramethylcyclobutadiene					
1	*cis*-3,4-Dibromo-1,2,3,4-tetramethyl-cyclobutene and lithium amalgam	Ether, 5 hours, with cooling	XII	81	52
	cis-3,4-Dichloro-1,2,3,4-tetra-methylcyclobutene and (1) *n*-butyllithium	Ether, cyclopentadiene	XII	87	52
		Ether, 30°	XII[d]	—	1

Compound/Reagent	Conditions	Products	Yield (%)	Ref.
(2) lithium	Liquid ammonia	XII[a]	—	1
(3) lithium amalgam	Ether, room temp., 70 hours	XII	90	47
	Furan	XII	66	49, 52
	2-Butyne	XII	—	20
(4) potassium	Vapor phase	XII, XIII, XIV, XV	—	135
(5) sodium amalgam	Ether	XII	35	52
(6) zinc–copper couple	Various solvents	XII	Low	49
(7) activated zinc dust	2-Butyne	XVIa	20	20
	Dimethyl acetylenedicarboxylate	XVIb	18	20
cis-3,4-Diiodo-1,2,3,4-tetramethylcyclobutene and				
(1) mercury	—	XVII	—	45
	Cyclopentadiene	XVII	—	45
	Dibromomaleic anhydride	XVIIIa	67	56
	Furan	XVII	—	45
	Maleic anhydride	XVIIIb	—	45
	Dimethylmaleic anhydride	XVIIIc	—	45
(2) zinc	—	XII	—	93
7 Tetramethylcyclobutadienenickel chloride and				
(1) heat	Dry thermolysis at 190°, 0.01 mm	XV, XIXa and b, XX	—	45
(2) sodium nitrite	Aqueous medium	XXI	50	53
(3) cyclopentadienylsodium	THF, 22°	XXIIa (or b)	81.6	47, 50
Tetramethylcyclobutadienenickel chloride–triphenylphosphine	Thermolysis	XV, XIXa and b, XX	—	—
Tetramethylcyclobutadienenickel chloride–phenanthroline	Thermolysis	XIV, XIX	—	45, 46

Tetraphenylcyclobutadiene

1 3,4-Dibromo-1,2,3,4-tetraphenylcyclobutene and phenyllithium	Ether, low temp.	XXIII	80	69

Table 1.1 (continued)

Part A. Generation of cyclobutadienes		Part B. Formation of dimers, adducts, etc., of cyclobutadienes			
Method of generation[b]	Reactants	Reaction conditions and trapping agents	Product[c]	Yield (%)	Ref.
3	Diphenylacetylene	Hexane, 1 week's irradiation	XXIV[c]	15.1	33
			XXV	6	
			XXVI	17.2	
	Diphenylacetylene and phenyl Grignard	Ether or THF and xylene, reflux, N_2	XXV	10	143
4	1-Bromo-4-iodo-1,2,3,4-tetraphenyl-butadiene and activated zinc dust	THF, reflux, 18 hours	XXIV	Trace	66
			XXVII	40	
	(4-Bromo-1,2,3,4-tetraphenyl-cis,cis-1,3-butadienyl) dimethyltin bromide	UV irradiation, C_6H_6, 8 hours	XXIV	—	66
		Thermolysis, 130° ± 10°, in triglyme or in the melt	XXV	85	67
		Thermolysis, 130° ± 10°, triglyme, ferric bromide	Tetraphenylfuran, benzal-1,2-diphenyl-indene, etc.	—	67
		Thermolysis, benzoyl peroxide	A mixture	—	67
		Thermolysis, diethyl fumarate	XXVIIIa	—	67
		Thermolysis, diethyl maleate	XXVIIIb	Quantitative	67
		Thermolysis, dimethyl acetylene-dicarboxylate	XXIX	—	67
		Thermolysis, dimethyl maleate	XXVIIIc	—	67
		Thermolysis, nitric oxide	A mixture	—	67
		Thermolysis, oxygen	cis-Dibenzoylstilbene	80	67
		Thermolysis, thiophene	A mixture	—	67
		Thermolysis, thiophenol	XXXa	40	67

E. CHEMISTRY OF CYCLOBUTADIENE

1,4-Diiodotetraphenylbutadiene	Thermolysis, 2-ethoxyethanol, 200°	XXV	—	25
1,4-Dilithiotetraphenylbutadiene and cupric bromide	Dimethoxyethane, reflux	XXV	15	25
1,4-Dilithiotetraphenylbutadiene and 1,4-diiodotetraphenylbutadiene		XXIII	40	69
(4-Iodo-1,2,3,4-tetraphenyl-cis,cis-1,3-butadienyl) dimethyltin iodide	UV irradiation, C_6H_6, 8 hours	XXIV	30	66
7 Tetraphenylcyclobutadiene-palladium chloride and				
(1) heat	Thermolysis, 0.1 mm (free flame)	XXXb	26	22
(2) trimethylphosphite	Benzene or chloroform	XXV	—	105
(3) triphenylphosphine	Benzene, reflux, N_2	XXV	70	41, 105
	Benzene, reflux, air	Tetraphenylfuran	12	105
	Cyclopentadiene	XXXI	—	41
	Methyl propiolate	XXXII	—	41
(4) ethylenebis(diphenylphosphine)	Benzene or chloroform	XXV	—	41
8 Tetraphenylmercurole	Thermolysis	XXV	—	25

Part C. Structural formulas and physical properties of the dimers, adducts, etc., of cyclobutadiene (I, II), 1-(1′-cyclohexenyl)-2,4-dichloro-3-fluorocyclobutadiene (III), 1,3-diphenylcyclobutadiene (IV-VII), 1-fluoro-2,3,4-triphenylcyclobutadiene (VIIIa), 1-fluoro-3-(p-chloro-diphenyl)cyclobutadiene (VIIIb), 1,2,3,4,5,6,7,8-octahydrobiphenylene (IX), 1,1,4,4,5,5,8,8-octamethyl-1,2,3,4,5,6,7,8-octahydrobiphenylene (X, XI), 1,2,3,4-tetramethylcyclobutadiene (XII–XXIII), and tetraphenylcyclobutadiene (XXII–XXXIII).

I

syn-Tricyclo[4.2.0.0^{2,5}]octa-3,7-diene, colorless oil, b.p. 45° (40 mm); IR spectrum (ref. 5a). Vol. steam.

II

$NO_3^{\ominus} Ag^{\oplus}$ $Ag^{\oplus} NO_3^{\ominus}$

I-silver nitrate complex, m.p. 138°–140° (ref. 5).

III

1,3-Dichloro-2-fluoro-5,6,7,8-tetrahydronaphthalene; m.p. 68.1°–68.4° (from EtOH).

Table 1.1 (*continued*)

IV

1,3,5,7 - Tetraphenyltricyclo[4.2.0.0^{2,5}]octa-3,7-diene, yellow needles, m.p. 133°–134°; λ_{max}^{Et2O} 262 mμ (log ϵ = 4.46); IR: 12.10, 13.10, 13.85 μ; δ = 7.42, 7.32, 6.43, 3.87.

V

1,2,4,7-Tetraphenyltricyclo[4.2.0.0^{2,5}]octa-3,7-diene; λ_{max}^{Et2O} 261 mμ (log ϵ = 4.20); IR: 7.60, 8.35 μ; δ = 7.42, 7.26, 6.70, 3.80.

VI

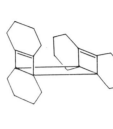

1, 3, 5, 7 - Tetraphenylcyclooctatetraene; white needles, m.p. 192°–193° (from EtOH); λ_{max}^{EtOH} 261 mμ (log ϵ = 4.73); IR: 6.15, 11.60 μ; δ = 7.37, 6.70.

VII

1, 2, 4, 7 - Tetraphenylcyclooctatetraene; yellow needles, m.p. 133°–134°; λ_{max}^{hept} 263 (log ϵ = 4.66), 325 mμ (3.81). IR: 6.15, 7.46, 9.85, 11.35 μ; δ = 7.33, 6.88, 6.61.

VIII

VIIIa (R = C$_6$H$_5$): *anti*-1,2-Difluoro-3,4,5,6,-7,8-hexaphenyltricyclo[4.2.0.0^{2,5}]octa - 3, 7-diene; m.p. 200° (decomp.); λ_{max} 227 (ϵ = 53,100), 288 mμ (28,200).

VIIIb (R = *p*-ClC$_6$H$_3$): *anti*-1,2-Difluoro-3, 4, 5, 6, 7, 8 - hexakis(*p*-chlorophenyl)tricyclo[4.2.0.0^{2,5}]octa-3,7-diene.

IX

syn-1, 2, 3, 4, 5, 6, 7, 8, 9, 10, 11, 12 - Dodecahydro-8b, 12b-butanobenzo[3', 4']cyclobuta[1', 2' : 3, 4]cyclobuta[1, 2 - *e*]biphenylene; needles, m.p. 132°–133° (from EtOH).

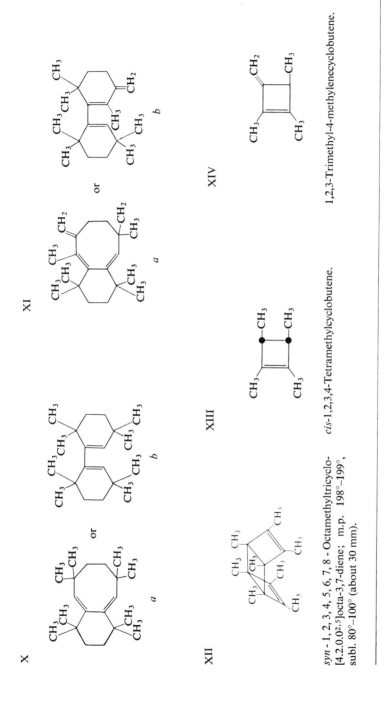

Table 1.1 (*continued*)

XV

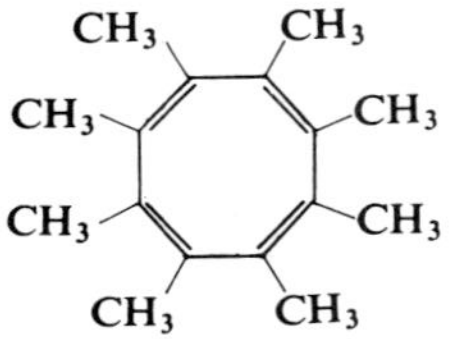

Octamethylcyclooctatetraene.

XVI

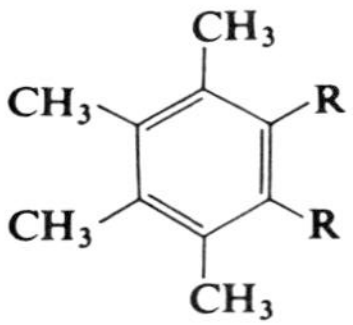

XVIa ($R=CH_3$): Hexamethylbenzene.
XVIb ($R=CO_2CH_3$): Dimethyl tetramethylphthalate.

XVII

A colorless solution from which a dark tar separates on exposure to air.

XVIII

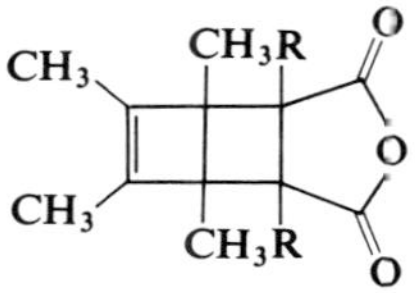

XVIIIa ($R=Br$): 5,6-Dibromo-1,2,3,4-tetramethylbicyclo[2.2.0]hexa-2-ene-5,6-dicarboxylic anhydride.

XVIIIb ($R=H$): 1,2,3,4-Tetramethylbicyclo[2.2.0]hexa-2-ene-5,6-dicarboxylic anhydride.

XVIIIc ($R=CH_3$): 1,2,3,4,5,6-Hexamethylbicyclo[2.2.0]hexa-2-ene-5,6-dicarboxylic anhydride.

XIX

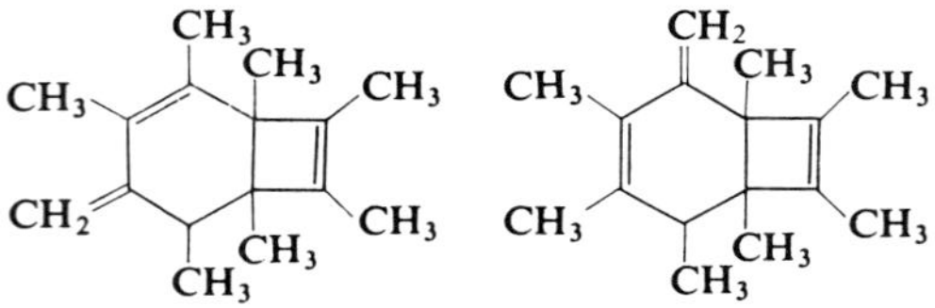

a *b*

E. CHEMISTRY OF CYCLOBUTADIENE

XX

CH₃CH=C(CH₃)—C(CH₃)=CHCH₃

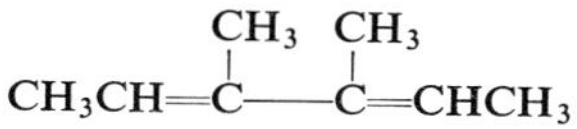

3,4-Dimethyl-2,4-hexadiene.

XXI

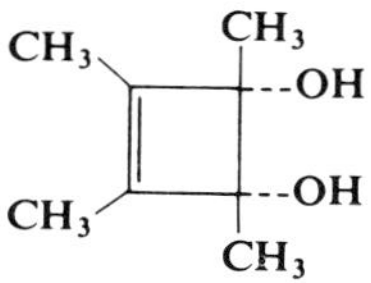

cis-1,2,3,4-Tetramethylcyclobutene-3,4-diol.

XXII

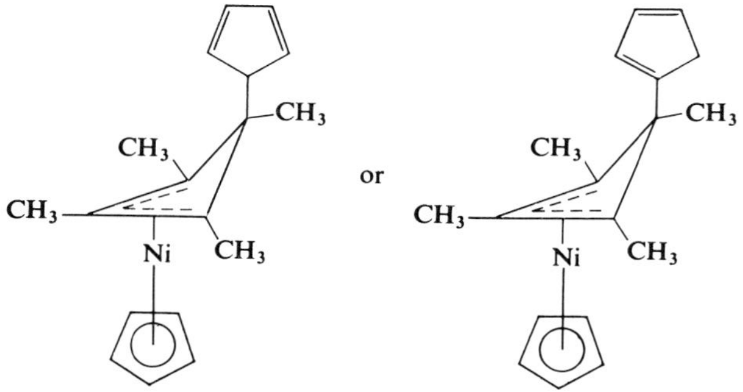

XXIII

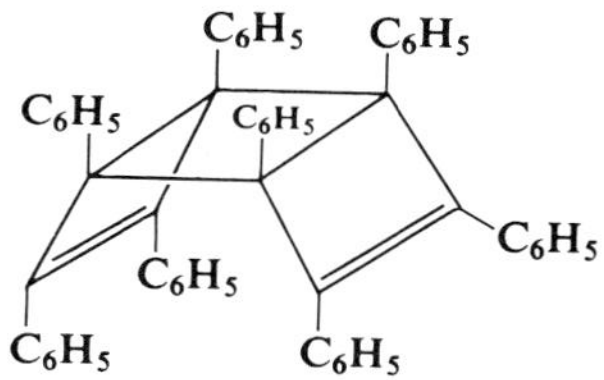

syn-1, 2, 3, 4, 5, 6, 7, 8-Octaphenyltricyclo[4.2.0.0²,⁵]octa-3,7-diene (?); thermolabile.

Table 1.1 (continued)

XXIV

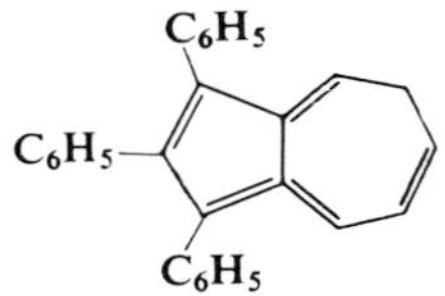

1,2,3-Triphenylazulene; m.p. 216°–217°.

XXV

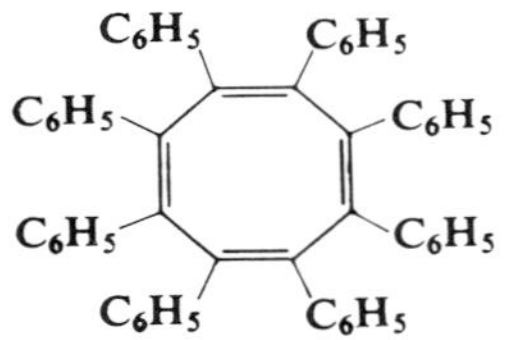

1,2,3,4,5,6,7,8-Octaphenylcyclooctatetraene; m.p. 426°–428° (from diphenyl ether); sublimes.

XXVI

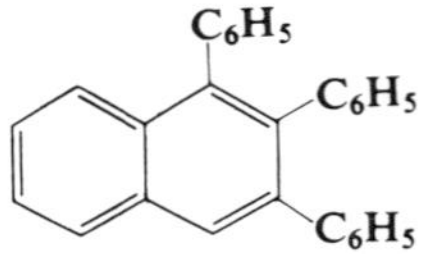

1,2,3-Triphenylnaphthalene; m.p. 151°–153°.

XXVII

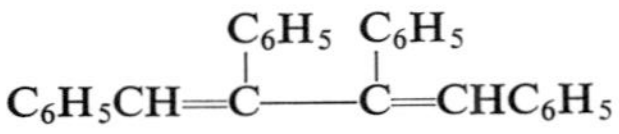

1,2,3,4-Tetraphenylbutadiene.

XXVIII

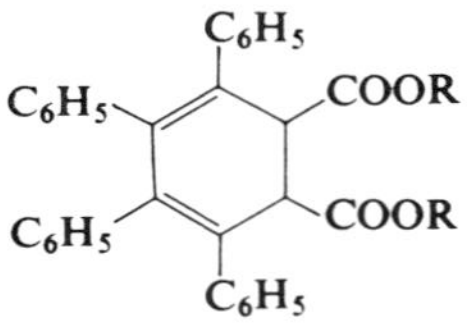

XXVIIIa (R=C_2H_5): Diethyl 3,4,5,6-tetraphenyl-*trans*-1,2,-dihydrophthalate.

XXVIIIb (R=C_2H_5): Diethyl 3,4,5,6-tetraphenyl-*cis*-1,2-dihydrophthalate, m.p. 149°–150°.

XXVIIIc (R=CH_3): Dimethyl 3,4,5,6-tetraphenyl-*cis*-1,2-dihydrophthalate.

XXIX

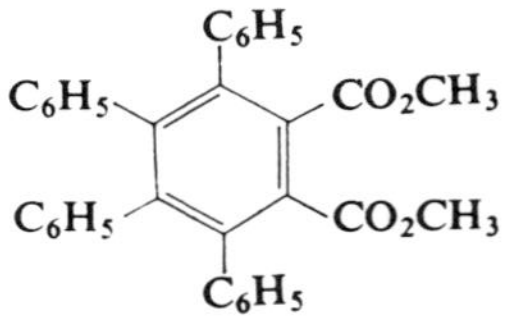

Dimethyl 3,4,5,6-tetraphenylphthalate.

E. CHEMISTRY OF CYCLOBUTADIENE

XXX

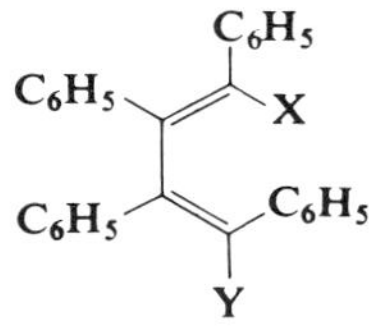

XXXa (X=C$_6$H$_5$S; Y=H): 1-Thiophenoxy-1,2,3,4-tetraphenylbutadiene (?); m.p. 132°–133°.

XXXb (X=Y=Cl): 1,4-Dichloro-1,2,3,4-tetraphenylbutadiene; m.p. 183°.

XXXI

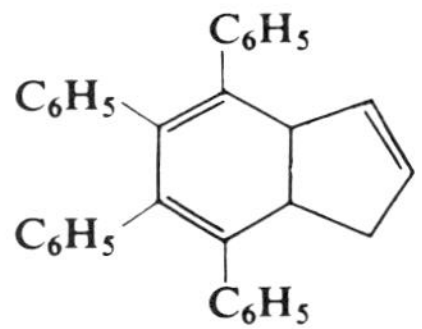

2,3,4,5-Tetraphenylbicyclo[4.3.0]nona-2,4,7-triene; m.p. 184°–187°, λ_{max}^{EtOH} 312 mμ (ϵ=9450).

XXXII

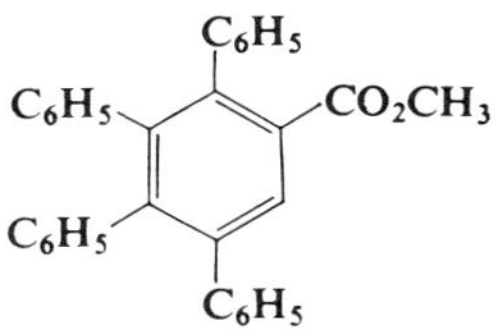

Methyl 2,3,4,5-tetraphenylbenzoate; m.p. 355°–360°.

---

[a] For cyclobutadienes isolated as transition metal complexes, see Table 2.2.

[b] Method 1 is "Dehalogenation," Section 1, B, 1; Method 2 is "Dehydrohalogenation," Section 1, B, 2; Method 3 is "Acetylene Dimerization," Section 1, B, 3; Method 4 is "Ring Closure of Butadienes," Section 1, B, 4; Method 5 is "Hofmann Elimination and Related Methods," Section 1, B, 5; Method 6 is "Reverse Diels–Alder Reactions," Section 1, B, 6; Method 7 is "Decomposition of Cyclobutadiene–Metal Complexes," Section 1, B, 7; and Method 8 is "By Thermolysis of a Heterocyclopentadiene," Section 1, B, 8.

[c] Products represented by Roman numerals are listed in Part C of this table, p. 77ff.

[d] Formed via a cyclobutadiene intermediate along with at least one other product formed via a noncyclobutadiene intermediate.

[e] Calculated from weight data recorded in the literature.

REFERENCES

1. Adam, W., *Tetrahedron Letters* **No. 21**, 1387 (1963).
2. Applequist, D. E., and Gwynn, D. E., private communication.
3. Applequist, D. E., and Roberts, J. D., *J. Am. Chem. Soc.* **78**, 4012 (1956).
4. Assony, S. J., and Kharasch, N., *J. Am. Chem. Soc.* **80**, 5978 (1958).
5. (a) Avram, M., Dinulescu, I. G., Marica, E., Mateescu, G., Sliam, E., and Nenitzescu, C. D., *Ber.* **97**, 382 (1964); (b) Nenitzescu, C. D., Avram, M., Marica, E., Dinulescu, I. G., and Mateescu, G., *Angew. Chem.* **75**, 88 (1963).
6. Avram, M., Fritz, H. P., Keller, H. J., Kreiter, C. G., Mateescu, G., McOmie, J. F. W., Sheppard, N., and Nenitzescu, C. D., *Tetrahedron Letters* 1611 (1963).
7. Avram, M., Fritz, H. P., Keller, H., Mateescu, G., McOmie, J. F. W., Sheppard, N., and Nenitzescu, C. D., *Tetrahedron* **19**, 187 (1963).
8. (a) Avram, M., Marica, E., and Nenitzescu, C. D., *Acad. Rep. Populare Romine Studii Cercetari Chim.* **7**, 155 (1959); [*Chem. Abstr.* **54**, 8664f (1960)]; (b) *ibid.*, *Rev. Chim. Acad. Rep. Populaire Roumaine* **4**, 253 (1959).
9. (a) Avram, M., Marica, E., and Nenitzescu, C. D., *Ber.* **92**, 1088 (1959); (b) Avram, M., Marica, E., Pogany, J., and Nenitzescu, C. D., *Angew. Chem.* **71**, 626 (1959).
10. Avram, M., Mateescu, G., Dinulescu, I. G., Marica, E., and Nenitzescu, C. D., *Tetrahedron Letters* **No. 1**, 21 (1961).
11. Avram, M., Mateescu, G., and Nenitzescu, C. D., *Ann.* **636**, 174 (1960).
12. (a) Avram, M., and Nenitzescu, C. D., *Rev. Chim. Acad. Rep. Populaire Roumaine* **4**, 265 (1959); (b) *ibid.*, *Acad. Rep. Populare Romine Studii Cercetari Chim.* **7**, 169 (1959).
13. Avram, M., Nenitzescu, C. D., and Marica, E., *Ber.* **90**, 1857 (1957).
14. Baker, W., Hilpern, J. W., and McOmie, J. F. W., *J. Chem. Soc.* 479 (1961).
15. Baker, W., and McOmie, J. F. W., in Ginsburg, D., "Non-Benzenoid Aromatic Compounds," Interscience, New York, (1959), p. 43.
16. Balaban, A. T., *Tetrahedron Letters* **No. 5**, 14 (1959).
17. Ball, W. J., and Landor, S. R., *J. Chem. Soc.* 2298 (1962).
18. Bartlett, P. D., and Holmes, E., private communication.
19. (a) Beesley, R. M., and Thorpe, J. F., *J. Chem. Soc.* 591 (1920); (b) *ibid.*, *Proc. Chem. Soc.* **29**, 346 (1913).
20. Berkoff, C. E., Cookson, R. C., Hudec, J., and Williams, R. O., *Proc. Chem. Soc.* 312 (1961).
21. Blomquist, A. T., and LaLancette, E. A., *J. Am. Chem. Soc.* **84**, 220 (1962).
22. Blomquist, A. T., and Maitlis, P. M., *J. Am. Chem. Soc.* **84**, 2329 (1962).
23. (a) Blomquist, A. T., and Meinwald, Y. C., *J. Am. Chem. Soc.* **81**, 667 (1959); (b) *ibid.*, *J. Am. Chem. Soc.* **79**, 5316 (1957).
24. Brass, K., and Mosl, G., *Ber.* **59**, 1266 (1926).
25. Braye, E. H., Hübel, W., and Caplier, I., *J. Am. Chem. Soc.* **83**, 4406 (1961).
26. Breslow, R., in deMayo, P. (ed.), "Molecular Rearrangements," Interscience, New York (1963), p. 276.
27. Breslow, R., and Battiste, M., *J. Am. Chem. Soc.* **82**, 3626 (1960).
28. Breslow, R., and Eicher, T., private communication.
29. Breslow, R., and Kivelevich, D., unpublished work; see Kivelevich, D., *Dissertation Abstr.* **22**, 1015 (1962).
30. Breslow, R., and Mitchell, M. J., unpublished work cited in ref. 26.
31. Brown, Henry C., *J. Org. Chem.* **22**, 1256. (1957).
32. Brown, Henry C., Gewanter, H. L., White, D. M., and Woods, W. G., *J. Org. Chem.* **25**, 634 (1960).
33. Büchi, G., Perry, C. W., and Robb, E. W., *J. Org. Chem.* **27**, 4106 (1962).

34. Buchman, E. R., *Abstr. ACS Meeting*, Sept. 13, 1954, p. 9–O. (This abstract is a summary of attempts to prepare cyclobutadiene by Buchman and his co-workers.)
35. Buchman, E. R., and Howton, D. R., *J. Am. Chem. Soc.* **70**, 3510 (1948).
36. Buchman, E. R., and Howton, D. R., *J. Am. Chem. Soc.* **70**, 2517 (1948).
37. Buchman, E. R., Schlatter, M. J., and Reims, A. O., *J. Am. Chem. Soc.* **64**, 2701 (1942).
38. Carlsohn, H., *Ber.* **60**, 473 (1927). See also Fischer, W. M., and Cirulis, A., *Ber.* **65**, 1852 (1932) and Jonescu, M., *Ber.* **60**, 1228 (1927).
39. Chatterjee, A., and Chaudhury, B., *Naturwiss.* **49**, 420 (1962).
40. Chatterjee, A., Srimany, S. K., and Chaudhury, B., *J. Chem. Soc.* 4576 (1961).
41. Cookson, R. C., and Jones, D. W., *Proc. Chem. Soc.* 115 (1963).
42. Cookson, R. C., and Marsden, J., *Chem. Ind. (London)* 21 (1961).
43. Cope, A. C., Haven, A. C., Jr., Ramp, E. E., and Trumbull, E. R., *J. Am. Chem. Soc.* **74**, 4867 (1952).
44. Cram, D. J., and Allinger, N. L., *J. Am. Chem. Soc.* **78**, 2518 (1956).
45. (a) Criegee, R., *Angew. Chem.* **74**, 703 (1962); *ibid.*, *Rev. Chim. Acad. Rep. Populaire Roumaine* **7**, 771 (1962); (b) *ibid.*, *Bull. Soc. Chim. (France)* 1 (1965).
46. Criegee, R., Dekker, J., Engel, W., Ludwig, P., and Noll, K., *Ber.* **96**, 2362 (1963).
47. Criegee, R., Förg, F., Brune, H.-A., and Schönleber, D., *Ber.* **97**, 3461 (1964).
48. Criegee, R., and Höver, H., *Ber.* **93**, 2521 (1960).
49. Criegee, R., and Louis, G., *Ber.* **90**, 417 (1957).
50. Criegee, R., and Ludwig, P., *Ber.* **94**, 2038 (1961).
51. Criegee, R., and Moschel, A., *Ber.* **92**, 2181 (1959).
52. Criegee, R., and Noll, K., *Ann.* **627**, 1 (1959).
53. (a) Criegee, R., and Schröder, G., *Ann.* **623**, 1 (1959); (b) *ibid.*, *Angew. Chem.* **71**, 70 (1959).
54. Criegee, R., Schröder, G., Maier, G., and Fischer, H.-G., *Ber.* **93**, 1553 (1960).
55. Criegee, R., Wirth, W.-D., Engel, W., and Brune, H. A., *Ber.* **96**, 2230 (1963).
56. Criegee, R., and Zanker, F., *Angew. Chem.* **76**, 716 (1964).
57. De Puy, C. H., and Lyons, C. E., *J. Am. Chem. Soc.* **82**, 631 (1960).
58. Deutsch, D. H., and Buchman, E. R., *Experientia* **6**, 462 (1950).
59. Dixit, V. M., *J. Univ. Bombay* **4**, 153 (1935). [*Chem. Abstr.* **30**, 5569[8] (1936); *Chem. Zentr.* 4557 (1936).]
60. Domnin, N. A., *Zhur. Obshch. Khim.* **9**, 1983 (1939); see also *ibid.*, *Zhur. Obshch. Khim.* **8**, 851 (1938) and the references cited therein.
61. (a) Dunicz, B. L., *J. Chem. Phys.* **12**, 37 (1944); (b) *ibid.*, *J. Am. Chem. Soc.* **63**, 2461 (1941).
62. Ekström, B., *Ber.* **92**, 749 (1959).
63. Ekström, B., Ph.D. diss., Karlsruhe (1958).
64. Favorskii, A., and Boshovsky, W., *Ann.* **390**, 122 (1912); see also Favorskii, A., *Zhur. Obshch. Khim.* **6**, 720 (1936).
65. Favorskii, A. E., and Favorskaya, T. A., *J. Russ. Phys. Chem. Soc.* **54**, 310 (1922).
66. Freedman, H. H., *J. Org. Chem.* **27**, 2298 (1962).
67. Freedman, H. H., *J. Am. Chem. Soc.* **83**, 2195 (1961).
68. Freedman, H. H., *J. Am. Chem. Soc.* **83**, 2194 (1961).
69. Freedman, H. H., and Petersen, D. R., *J. Am. Chem. Soc.* **84**, 2837 (1962).
70. Freund, G., Diplomarbeit, Karlsruhe (1960).
71. Fritchie, C., Jr., and Hughes, E. W., *J. Am. Chem. Soc.* **84**, 2257 (1962).
72. Fritz, H. P., McOmie, J. F. W., and Sheppard, N., *Tetrahedron Letters* No. **26**, 35 (1960).
73. Gabriel, S., and Michael, A., *Ber.* **10**, 1551 (1877).
74. Gastaldi, C., and Cherchi, F., *Gazz. Chim. Ital.* **45B**, 251 (1915).

75. Gastaldi, C., and Cherchi, F., *Gazz. Chim. Ital.* **44A**, 282 (1914).
76. Gelin, S., *Chim. Mod.* **8**, 241 (1963).
77. Gochenour, C. I., *U.S. Patent* 2,546,997 (to Hooker Electrochemical Co.), April 3, 1951.
78. Graham, W. H., unpublished work cited in ref. 133.
79. Griffin, G. W., and Petersen, L. I., *J. Am. Chem. Soc.* **85**, 2268 (1963).
80. Griffin, G. W., Vellturo, A. F., and Furukawa, K., *J. Am. Chem. Soc.* **83**, 2725 (1961).
81. Grundmann, C., and Litten, E., *Ber.* **85**, 261 (1952).
82. Gwynn, D. E., *Dissertation Abstr.* **23**, 437 (1962).
83. Harris, J. F., Jr., Harder, R. J., and Sausen, G. N., *J. Org. Chem.* **25**, 633 (1960).
84. Haszeldine, R. N., and Osborne, J. E., *J. Chem. Soc.* 3880 (1955).
85. Hausmann, J., *Ber.* **22**, 2019 (1889).
86. Henne, A. L., and Walton, T. R., unpublished work; see Walton, T. R., Ph.D. diss., The Ohio State Univ. (1960). [Walton, T. R., *Dissertation Abstr.* **21**, 2120 (1961).]
87. Hughes, E. D., and Beineke, T. A., private communication.
88. Jones, R. C., Ph.D. diss., Harvard (1941).
89. Kipping, F. S., *J. Chem. Soc.* 480 (1894).
90. Kipping, F. S., *J. Chem. Soc.* 269 (1894).
91. Kitahara, Y., Caserio, M. C., Scardiglia, F., and Roberts, J. D., *J. Am. Chem. Soc.* **82**, 3106 (1960).
92. Knoche, R., Ph.D. diss., Münster (1925).
93. Kristinson, H., Diplomarbeit, Karlsruhe (1962), cited in ref. 45.
94. Kuri, Z., and Shida, S., *Bull. Chem. Soc. Japan* **25**, 116 (1952).
95. Lanser, T., *Ber.* **32**, 2478 (1899).
96. Lanser, T., and Halvorsen, B. F., *Ber.* **35**, 1407 (1902).
97. Larson, H. O., and Woodward, R. B., *Chem. Ind.* (*London*) 193 (1959).
98. Liebermann, C., and Bergami, O., *Ber.* **23**, 317 (1890).
99. Liebermann, C., and Bergami, O., *Ber.* **22**, 782 (1889).
100. Limpricht, H., *Ber.* **2**, 211 (1869).
101. Lipscomb, W. N., *Tetrahedron Letters*, **No. 18**, 20 (1959).
102. Longuet-Higgins, H. C., and Orgel, L. E., *J. Chem. Soc.* 1969 (1956).
103. Maier, G., private communication.
104. Maitlis, P. M., and Games, M. L., *Chem. Ind.* (*London*) 1624 (1963).
105. Maitlis, P. M., and Stone, F. G. A., *Proc. Chem. Soc.* 330 (1962).
106. Malachowski, R., Ph.D. diss., Eidgenössichen technischen Hochschule (1911).
107. Manthey, W., *Ber.* **33**, 3081 (1900).
108. Mariella, R. P., Ph.D. diss., Carnegie Inst. Techn. (1945).
109. Martologue, N., and Mumuianu, D., *Rev. Chim. Acad. Rep. Populaire Roumaine* **6**, 303 (1961).
110. McDonald, S. G. G., private communication.
111. (a) Michael, A., *Ber.* **39**, 1908 (1906); (b) Michael, A., and Bucher, A., *Am. Chem. J.* **20**, 89 (1898).
112. Middleton, W. J., private communication.
113. Middleton, W. J., *U.S. Patent* 2,831,835 (to E. I. du Pont de Nemours and Co., Inc.), April 22, 1958.
114. (a) Miller, D. B., and Shechter, H., *Abstr. 133rd ACS Meeting*, p. 79-N (1958); (b) Miller, D. B., *Dissertation Abstr.* **19**, 1981 (1958).
115. Murata, I., *Kagaku No Ryoiki* **17**, 383 (1963). [*Chem. Abstr.* **59**, 12649h (1963).]
116. (a) Nagarajan, K., Caserio, M. C., and Roberts, J. D., *J. Am. Chem. Soc.* **86**, 449 (1964); (b) *ibid.*, *Rev. Chim. Acad. Rep. Populaire Roumaine* **7**, 1109 (1962).

REFERENCES 87

117. Nenitzescu, C. D., Avram, M., Marica, E., Maxim, M., and Dinu, D., *Acad. Rep. Populaire Romine Studii Cercetari Chim.* **7**, 481 (1959).
118. Nenitzescu, C. D., Avram, M., Pogany, I. I., Mateescu, G. D., and Farcasiu, M., *Acad. Rep. Populare Romine Studii Cercetari Chim.* **11**, 7 (1963).
119. Pawley, G. S., Lipscomb, W. N., and Freedman, H. H., *J. Am. Chem. Soc.* **86**, 4725 (1964).
120. (a) Perkin, W. H., Jr., *J. Chem. Soc.* 950 (1894); (b) Perkin, W. H., Jr., *Ber.* **26**, 2243 (1893).
121. Pfrommer, J. F., Ph.D. diss., Karlsruhe (1961).
122. Purvis, J. E., *J. Chem. Soc.* 107 (1911).
123. Ranganathan, S., *Dissertation Abstr.* **23**, 3637 (1963).
124. Ranganathan, S., and Shechter, H., *J. Org. Chem.* **27**, 2947 (1962).
125. Reppe, W., "Neue Entwicklungen auf dem Gebiete der Chemie des Acetylens und Kohlenoxyds," Springer Verlag, Berlin (1949), p. 87.
126. Reppe, W., Schlichting, O., Klager, K., and Toepel, T., *Ann.* **560**, 1 (1948).
127. Riemschneider, R., and Becker, U., *Monatsh.* **90**, 524 (1959).
128. Roberts, J. D., quoted in *Chem. Soc. Special Publication No. 12*, 111 (1958).
129. Roberts, J. D., Kline, G. B., and Simmons, H. E., Jr., *J. Am. Chem. Soc.* **75**, 4765 (1953).
130. Ruhemann, S., and Merriman, R. W., *J. Chem. Soc.* 1383 (1905).
131. Schiff, H., *Ann.* **6** (Suppl.), 1 (1868).
132. Seka, R., and Lackner, L., *Monatsh.* **74**, 212 (1943).
133. Sharts, C. M., and Roberts, J. D., *J. Am. Chem. Soc.* **83**, 871 (1961).
134. Shechter, H., Link, W. J., and Tiers, G. V. D., *J. Am. Chem. Soc.* **85**, 1601 (1963).
135. Skell, P. S., and Petersen, R. J., *J. Am. Chem. Soc.* **86**, 2530 (1964).
136. Smirnov-Zamkov, I. W., *Doklady Akad. Nauk S.S.S.R.* **83**, 869 (1952).
137. Smutny, E. J., and Roberts, J. D., *J. Am. Chem. Soc.* **77**, 3420 (1955).
138. Stiebler, P., Ph.D. diss., Münster (1922).
139. Stobbe, H., *Ber.* **40**, 3372 (1907).
140. Stobbe, H., and Zschoch, F., *Ber.* **60**, 457 (1927).
141. Störmer, R., and Biesenbach, T., *Ber.* **38**, 1965 (1905).
142. Thiele, J., *Ann.* **306**, 87 (1899).
143. Tsutsui, M., *Chem. Ind.* (*London*) 780 (1962).
144. Vogel, E., *Angew. Chem.* **74**, 829 (1962).
145. Vogel, E., *Angew. Chem.* **72**, 4 (1960).
146. Vol'pin, M. E., *Uspekhii Khimii* **29**, 147 (1960). [*Russ. Chem. Rev.* **29**, 142 (1960).]
147. Wasserman, E. R., Ph.D. diss., Radcliffe College (1948).
148. Watts, L., Fitzpatrick, J. D., and Pettit, R., *J. Am. Chem. Soc.* **87**, 3253 (1965).
149. Weltner, W., Jr., *J. Am. Chem. Soc.* **75**, 4224 (1953).
150. White, E. H., private communication.
151. White, E. H., and Dunathan, H. C., *J. Am. Chem. Soc.* **86**, 453 (1964).
152. White, E. H., and Maier, G., private communication.
153. White, E. H., and Stern, R. L., *Abstr. of Papers 141st Meeting Amer. Chem. Soc.*, Washington, D.C., March, 1962, p. 7-D.
154. Willstätter, R., "Aus Meinem Leben," Verlag Chemie, Weinheim (1949), p. 49.
155. Willstätter, R., and Bruce, J., *Ber.* **40**, 3979 (1907). See also Heisig, G. B., *J. Am. Chem. Soc.* **63**, 1698 (1941).
156. Willstätter, R., and Schmädel, W. von, *Ber.* **38**, 1992 (1905).
157. Willstätter, R., and Waser, E., *Ber.* **44**, 3428 (1911).
158. Wittig, G., and Mayer, U., *Ber.* **96**, 342 (1963).
159. Woodward, R. B., and Small, G., Jr., *J. Am. Chem. Soc.* **72**, 1297 (1950).

CHAPTER 2

The Cyclobutadiene–Metal Complexes

$$\overset{4}{\underset{3}{\boxed{\ominus}}}\overset{1}{\underset{2}{{}}}\text{—ML}_n$$

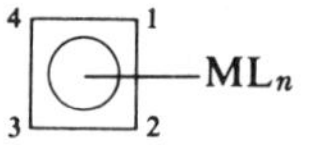

Much of current progress in cyclobutadiene chemistry owes its origins to the synthesis of the first authentic cyclobutadiene–metal complex in 1959 by Criegee and Schröder. The fact that Criegee's complex (tetramethylcyclobutadienenickel chloride) was both isolable and stable led to a resurgence of interest in the field and this in turn led to the discovery in rapid succession of several other cyclobutadiene–metal complexes. Of course, the cyclobutadiene–metal complexes are not true cyclobutadienes from the point of view of molecular orbital theory, but are simply systems resulting from the overlap of favorably oriented metal d orbitals with the antibonding π orbitals of the four-membered ring. They are related to cyclobutadiene in much the same way that the more familiar five-membered metallocenes (e.g., ferrocene and nickelocene) are related to the cyclopentadienyl radical. Indeed, very recent work by Pettit and co-workers has shown that cyclobutadieneiron tricarbonyl undergoes a series of aromatic substitution reactions paralleling those of ferrocene.*

It should be noted that the formation of a cyclobutadiene–metal complex, e.g., $(1) \rightarrow (2)$, is not in itself proof of the intermediacy of the corresponding cyclobutadiene and that the chemical transformation of one cyclobutadiene–metal complex into another, e.g., $(2) \rightarrow (3)$, also need not involve a free cyclobutadiene intermediate. On the other hand, cyclobutadiene–metal complexes have been demetallated in a variety of reactions, some of which certainly give rise to transient cyclobutadienes. The use of such complexes as

---

* See the Appendix and refs. 23 and 64 at the end of this chapter.

precursors in the generation of cyclobutadienes of varied structure is certain to become widespread in the future and to contribute greatly to the development of a systematic cyclobutadiene chemistry.

$$2C_6H_5C{\equiv}CC_6H_5 \xrightarrow{(PhCN)_2 \cdot PdCl_2}$$ [tetraphenylcyclobutadiene]–PdCl$_2$ $\xrightarrow{Fe(CO)_5}$ [tetraphenylcyclobutadiene]–Fe(CO)$_3$

(1) (2) (3)

Various aspects of the chemistry of the cyclobutadiene–metal complexes have been reviewed by Coates,[10] Criegee,[13] Fritz,[27] Orgel,[49] and Zeiss.[67]

A. History

There is no early history of attempts to prepare cyclobutadiene–metal complexes. Philipps (1894)[51] and later Erdmann and Koethner (1898)[22] and Makowa (1908)[44] obtained a brownish-red precipitate of composition C_4H_5ClOPd by passing acetylene through an aqueous solution of palladous chloride. Many years later, Longuet-Higgins and Orgel suggested that the precipitate might be a cyclobutadiene-metal complex,[36] but a recent reinvestigation has uncovered evidence which indicates that the compound is a π-alkyl complex of acrolein.[61]

In 1940, Reppe made the surprising discovery that under the proper reaction conditions (high pressure, tetrahydrofuran or dioxane as solvent) and in the presence of certain nickel compounds [nickel cyanide or the nickel(II) halides], acetylene could be made to undergo polymerization to form a cyclopolyolefin mixture, the main constituent of which is cyclooctatetraene.[52, 53] Since the usual course of acetylene cyclopolymerization is trimerization to form benzene, it was inevitable that the suggestion should be made by subsequent workers (Bergmann[4] and Longuet-Higgins and Orgel[36]) that the tetrameric cyclopolymerization of acetylene is due to the formation and subsequent dimerization of a cyclobutadiene–metal complex. Much favorable (though inconclusive) evidence has been presented in support of the cyclobutadiene–metal complex mechanism, but Schrauzer has recently presented a disputandum which makes this mechanism no longer tenable[60] (see also Section B, 1, c.)

The stabilization of cyclobutadiene derivatives by coordination with transition metals was predicted in 1956 by Longuet-Higgins and Orgel from molecular orbital theory[36] and was confirmed three years later with the synthesis of tetramethylcyclobutadienenickel chloride,[17] tetraphenylcyclobutadieneiron tricarbonyl,[32] and tetraphenylcyclobutadienepalladium chloride.[45a]

B. Preparation of Cyclobutadiene–Metal Complexes*

Cyclobutadiene–metal complexes have been prepared in four ways: (1) by the reaction of acetylenes with transition metal compounds, (2) from cyclobutenyl metal complexes, (3) by the transformation of other cyclobutadiene–

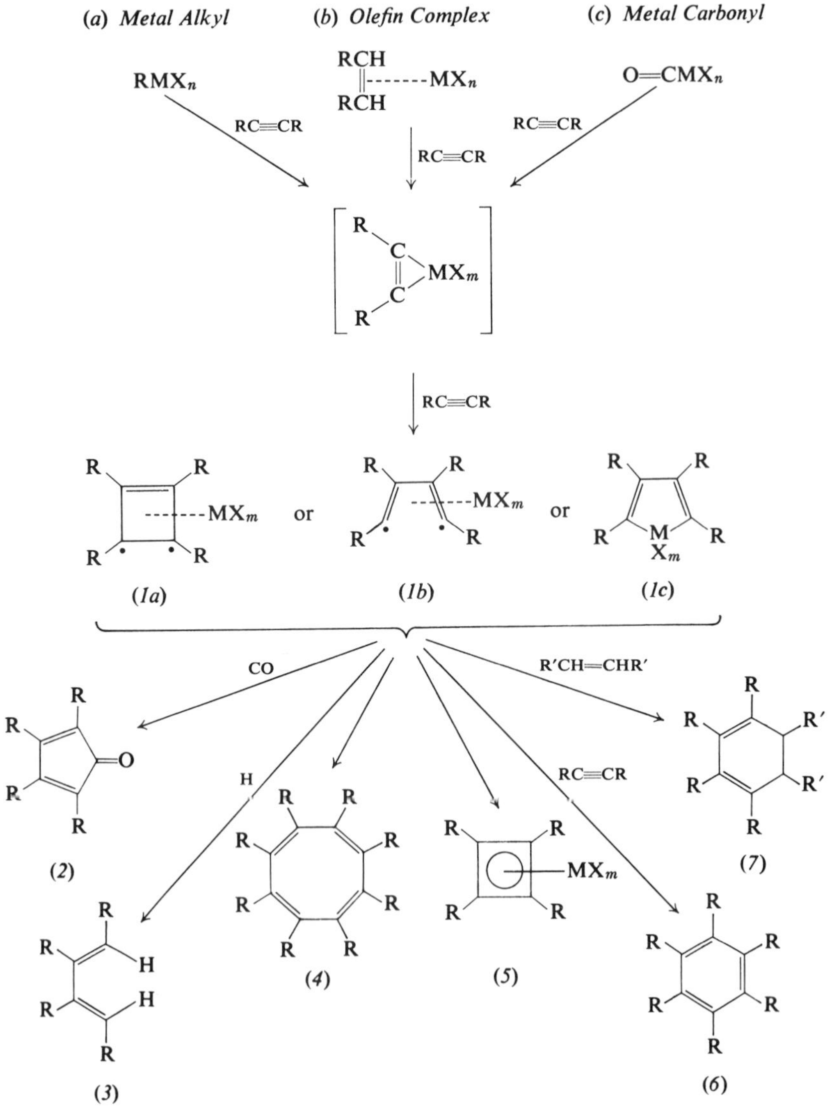

---

* Section B includes all examples of the formation of authentic cyclobutadiene–metal complexes, without regard to the question of their origin; in other words, whether by way of true cyclobutadienoid intermediates or not.

B. PREPARATION OF CYCLOBUTADIENE–METAL COMPLEXES

metal complexes, and (4), ostensibly, by trapping cyclobutadienes with transition metal compounds.

1. By the Reaction of Acetylenes with Transition Metal Compounds

Because of the highly reactive nature of the intermediates involved in the formation of cyclobutadiene–metal complexes from acetylenes, no generalization can be made that predicts all of the possible reaction pathways and, in fact, a slight variation in the reaction conditions often has a profound effect on the nature of the products.

The formation of cyclobutadiene–metal complexes from acetylenes is believed to proceed through intermediates such as (*1a*), (*1b*), or (*1c*). Depending on the nature of the organic substituents (R), the metal (M), and the ligands (mX), as well as on the conditions under which the reaction is carried out, the intermediate complex, (*1a*), (*1b*), or (*1c*), will give one or more of several possible reaction products (*2–7*), only one of which is a cyclobutadiene–metal complex.[47]

The following examples of the reactions of acetylenes with metal alkyls, olefin complexes, and metal carbonyls serve to demonstrate the versatility (and the complexity) of the method:

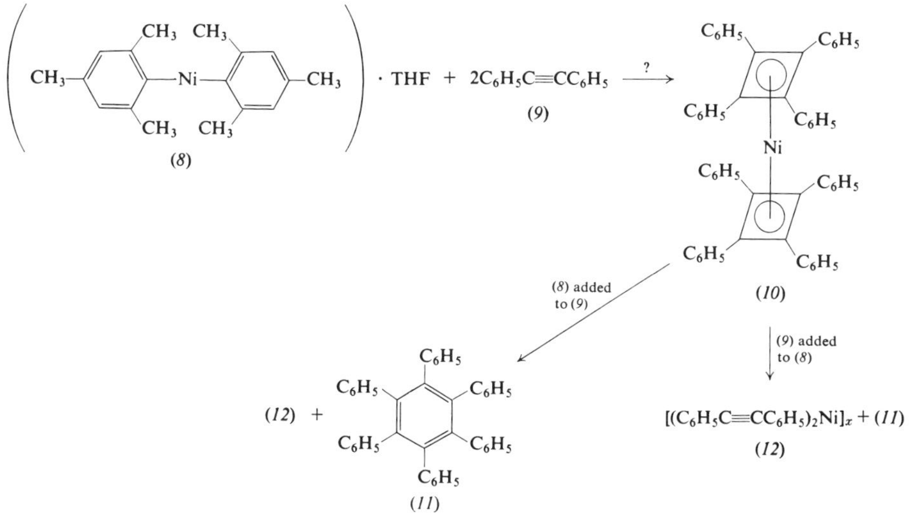

(a) When diphenylacetylene is added to dimesitylnickel (*8*) in tetrahydrofuran, a large amount of an orange polymeric material [(*12*), approximately of the empirical formula $(C_6H_5C{\equiv}CC_6H_5)_2Ni$] and a small amount of hexaphenylbenzene (*11*) are formed; when the order of addition is reversed (actually, dimesitylnickel is generated *in situ* by adding mesitylmagnesium bromide to a mixture of diphenylacetylene and nickelous bromide), the relative amounts of polymer and hexaphenylbenzene are also reversed, i.e., a small amount of polymer and a large amount of hexaphenylbenzene are produced.[63] These results have been ascribed to the formation of bis(tetramethylcyclobutadienyl)nickel(0) complex (*10*) as a transient intermediate.[63]

(b) 1,2,3,4-Tetraphenylcyclobutadiene cyclopentadienylcobalt(I) (*13*) is formed by the reaction of diphenylacetylene with cyclooctatetraene cyclopentadienylcobalt(I) (40% yield)[48]; by the reaction of diphenylacetylene with 1,5-cyclooctadiene cyclopentadienylcobalt(I) complex (47% yield)[48]; and by the reaction of diphenylacetylene with cobaltocene (30% yield).[7]

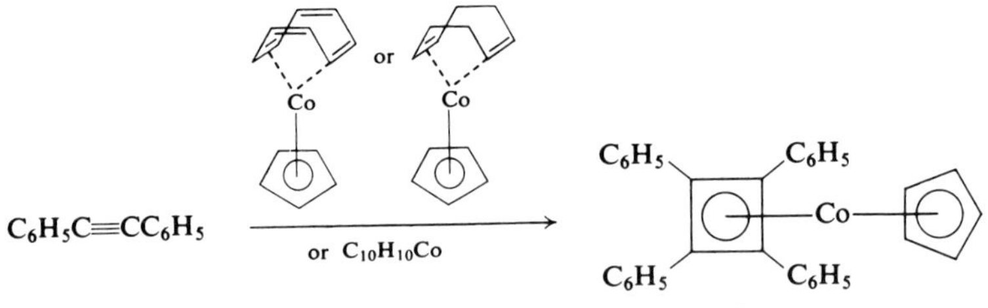

$$C_6H_5C{\equiv}CC_6H_5 \xrightarrow{\text{or } C_{10}H_{10}Co} (13)$$

(c) Acetylene reacts with nickel cyanide and the nickel halides to give a complex mixture of cyclopolyolefins, of which 40–50% is cyclooctatetraene (the Reppe reaction). Despite numerous statements to the contrary, it is unlikely that a cyclobutadiene–nickel complex is an intermediate in this reaction.[60] Rather, it is more likely that cyclooctatetraene is formed by the dimerization or ring enlargement of a heterocyclic intermediate such as (*1c*).[47] (See also Section E.)

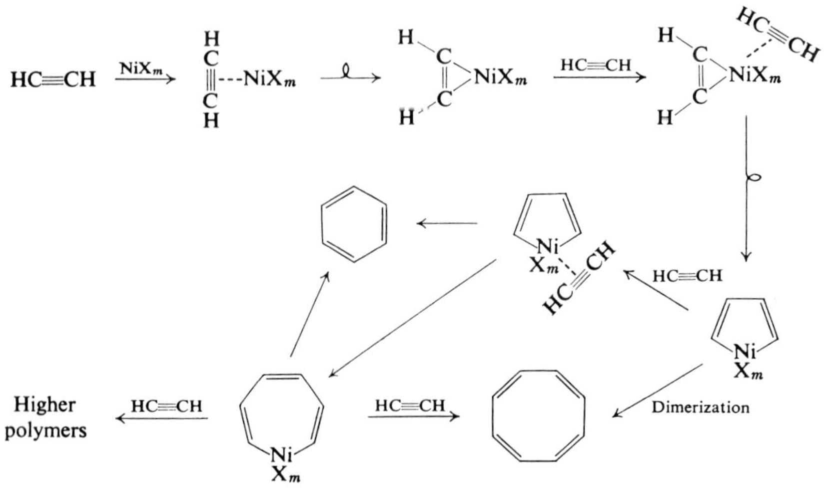

B. PREPARATION OF CYCLOBUTADIENE–METAL COMPLEXES

(d) More often than not, the reaction between an acetylene and a transition metal carbonyl affords several noncyclobutadienoid products along with the desired cyclobutadiene–metal complex (usually formed in small yield). Thus, the reaction of diphenylacetylene with iron dodecacarbonyl, [Fe(CO)$_4$]$_3$, in petroleum ether at 80°–90° affords a "small amount" of 1,2,3,4-tetraphenylcyclobutadieneiron tricarbonyl (*17*), together with three noncyclobutadienoid complexes, (*14*), (*15*), and (*16*).[31, 32, 47] The iron tricarbonyl complex (*17*) can be prepared in "somewhat larger yield" (16%)[30] by heating diphenylacetylene with iron pentacarbonyl, Fe(CO)$_5$, in an autoclave at 200°–240°.[32] It can also be prepared by heating diphenylacetylene with cyclooctatetraeneiron(0) tricarbonyl at 190°[47, 48a] and by treating diphenylacetylene with π-C$_5$H$_5$Fe(CO)$_2$NO.[11] Tetrakis(*p*-chlorophenyl)cyclobutadieneiron(0) tricarbonyl (*18*) is formed in 8% yield from *p*,*p*′-dichlorodiphenylacetylene and iron pentacarbonyl.[30]

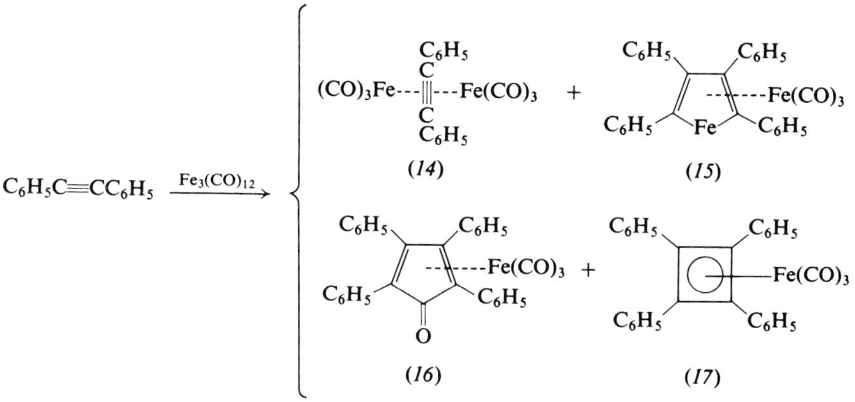

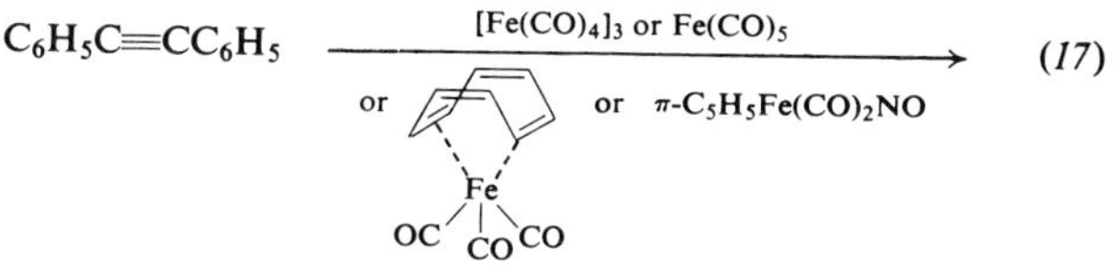

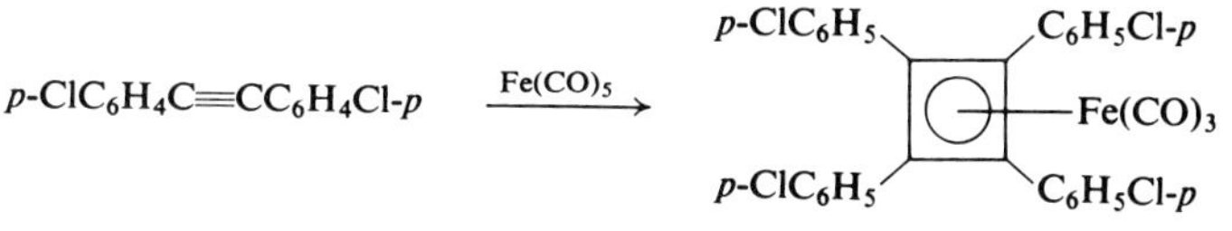

(e) The reaction of diphenylacetylene with bis(benzonitrile)palladium chloride in nonhydroxylic solvents yields mainly hexaphenylbenzene (11), together with smaller amounts of two palladium chloride complexes [(19a) and (19b)] of tetraphenylcyclobutadienepalladium chloride. Depending on the reaction conditions used, one or the other of the two complexes is formed. Thus, a benzene solution of 2 moles of diphenylacetylene and 1 mole of bis(benzonitrile)palladium chloride gave complex (19a) in 43% yield* after

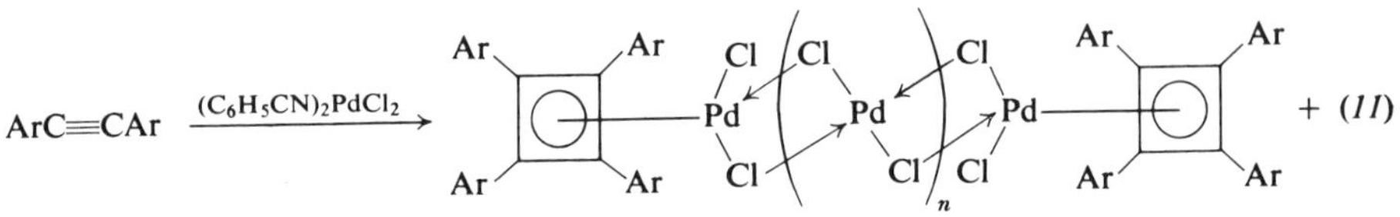

$ArC\equiv CAr \xrightarrow{(C_6H_5CN)_2PdCl_2}$ [structures] + (11)

$Ar = C_6H_5$ or $p\text{-}ClC_6H_4$

(19a): $Ar = C_6H_5$; $n = 3$
(19b): $Ar = C_6H_5$; $n = 4$
(20a): $Ar = p\text{-}ClC_6H_4$; $n = 2$
(20b): $Ar = p\text{-}ClC_6H_4$; $n = 4$

16 hours at room temperature.[42] In contrast, adding a dilute benzene solution of 1 mole of diphenylacetylene slowly to an identical solution of bis(benzonitrile)palladium chloride gave complex (19b) in 34% yield (based on the acetylene).[42] Two p-chlorophenyl analogs of the above complexes were prepared similarly and were shown to have the compositions $[C_{28}H_{16}Cl_4\text{-}(PdCl_2)_2]_2$ (20a) and $[C_{28}H_{16}Cl_4(PdCl_2)_3]_2$ (20b).[42]

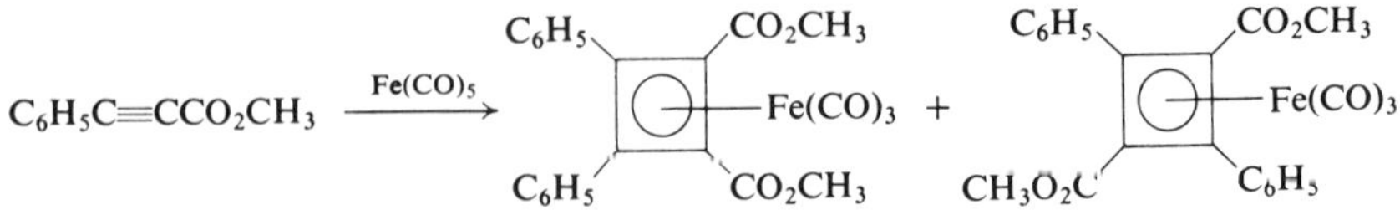

$C_6H_5C\equiv CCO_2CH_3 \xrightarrow{Fe(CO)_5}$ [structure (21a)] + [structure (21b)]

$RC\equiv CH \xrightarrow{[Co(CO)_4]_2Hg}$ [structure] or [structure]

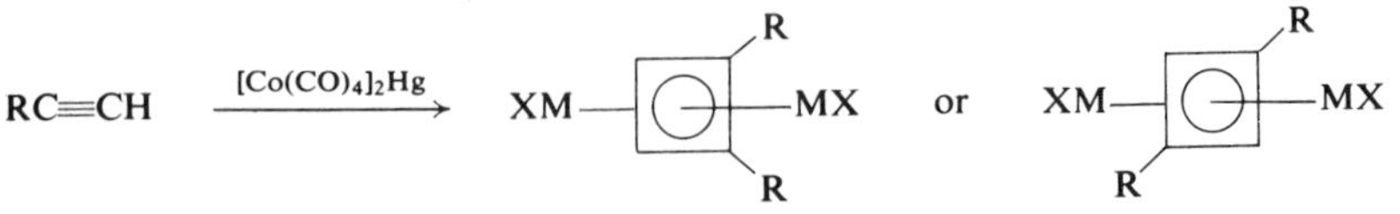

$R = C_6H_5$ or $(CH_3)_3C$

(22): $R = C_6H_5$; $MX = (CO)_4CoHg(CO)_2Co$
(23): $R = (CH_3)_3C$; $MX = (CO)_4CoHg(CO)_2Co$

* Based on bis(benzonitrile)palladium chloride. Percent yield calculated from weight data recorded in ref. 42.

(f) The method has also been used in the preparation of several compounds believed to be metal complexes of unsymmetrically substituted cyclobutadienes. Thus, methyl phenylpropiolate reacts with iron pentacarbonyl to give two complexes believed to be the syn and anti isomers of dimethyl diphenylcyclobutadienedicarboxylateiron tricarbonyl, (*21a*) and (*21b*).[30] Similarly, compounds believed to be cobalt–mercury carbonyl complexes of 1,2(or 3)-diphenyl- and 1,2(or 3)-di(*tert*-butyl)cyclobutadiene [(*22*) and (*23*) respectively] are formed when phenylacetylene and *tert*-butylacetylene are treated with Hg[Co(CO)$_4$]$_2$ at 85°.[30, 35] Degradation studies have shown that in the case of (*22*) and (*23*), only one of the two isomers possible for each of the compounds is formed.

2. From π-Cyclobutenyl Metal Complexes

Diphenylacetylene reacts with certain palladium salts and complexes to give hexaphenylbenzene and π-cyclobutenyl metal complexes [e.g., (*24a*)] which are stable and isolable but which can be converted readily into the corresponding cyclobutadiene–metal complexes [e.g., (*19c*)]. [The cyclobutenyl metal complexes were formerly believed to be butadienyl metal complexes, e.g., (*26*),[45a] but a recent X-ray structure determination has demonstrated the correctness of the cyclobutenyl formulation.[19]] Thus, (a) 3-ethoxy-1,2,3,4-tetraphenylcyclobutenylpalladium(II) chloride (*24a*),[39] bromide (*24b*), and iodide (*24c*) (prepared by treating solutions of diphenylacetylene in ethanol with palladium chloride and the appropriate hydrogen halides) are converted into the corresponding cyclobutadiene–metal complexes [(*19c*), (*19d*), and (*19e*), respectively] by treatment with excess hydrogen halide in nonhydroxylic solvents. The bromide (*19d*) can be produced in better yield (82%) by treating a chloroform solution of the cyclobutenylpalladium chloride (*24a*) with anhydrous hydrogen bromide.[39] [Compound (*24a*), 3-ethoxy-1,2,3,4-tetraphenylcyclobutenylpalladium chloride, can be prepared conveniently by treating diphenylacetylene with either bis(benzonitrile)palladium chloride in ethanol–chloroform[6] or with sodium or ammonium chloropalladite (2NH$_4^+$ + PdCl$_4^{2-}$) in aqueous ethanol.[39]] The tetraphenylcyclobutadienepalladium bromide complex (*19d*) can also be prepared in one operation from diphenylacetylene by treating the latter with palladium chloride and excess hydrogen bromide in ethanol, but the yields are always lower in the one-step method.[6, 39] [The cyclobutadienepalladium halides have a tendency to react with the ethanol used as solvent in the one-step method to give isomeric 3-ethoxy-1,2,3,4-tetraphenylcyclobutenylpalladium halides, e.g., (*24a'*).] Ethoxytetraphenylcyclobutenylpalladium chloride (*24a*) can be converted into tetraphenylcyclobutadienepalladium iodide (*19e*) in 50% yield with aqueous hydriodic acid.[39] Similarly, treating a methylene chloride solution of ethoxytetrakis(*p*-chlorophenyl)cyclobutenylpalladium chloride (*25*) with dry hydrogen bromide gave tetrakis(*p*-chlorophenyl)cyclobutadienepalladium bromide (*27*) in 82% yield.[42]

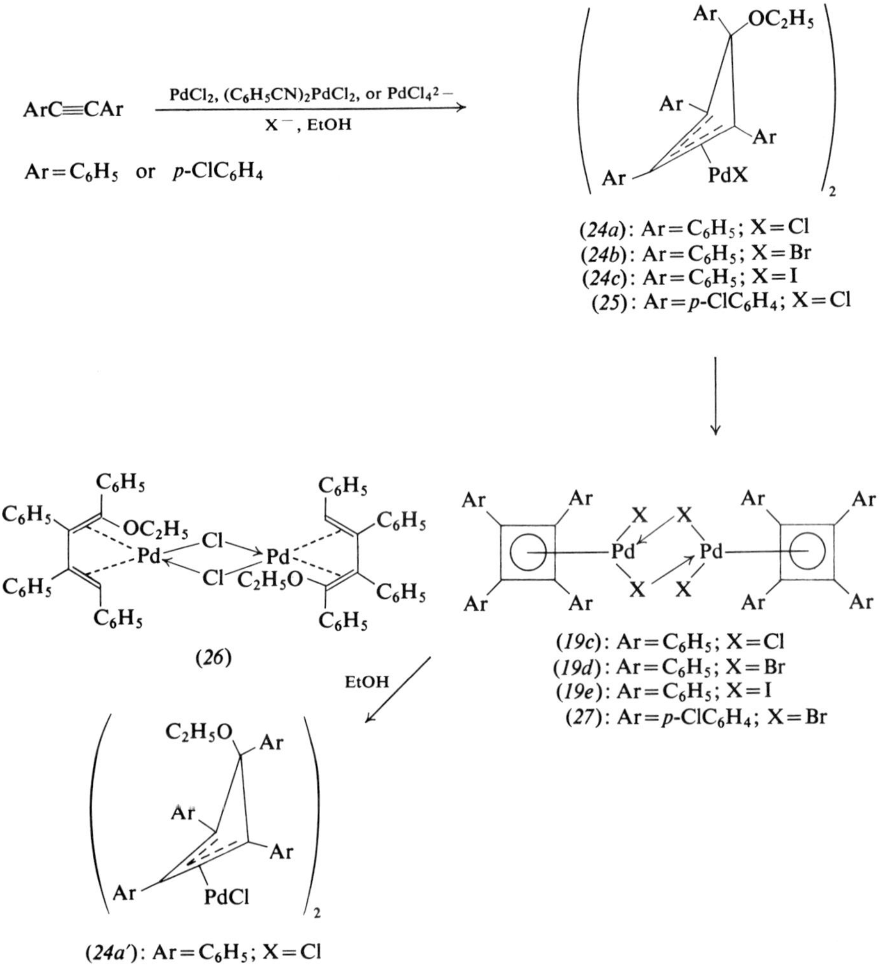

$$ArC\equiv CAr \xrightarrow[X^-, EtOH]{PdCl_2, (C_6H_5CN)_2PdCl_2, \text{ or } PdCl_4{}^{2-}}$$

Ar = C_6H_5 or *p*-ClC$_6$H$_4$

(*24a*): Ar = C_6H_5; X = Cl
(*24b*): Ar = C_6H_5; X = Br
(*24c*): Ar = C_6H_5; X = I
(*25*): Ar = *p*-ClC$_6$H$_4$; X = Cl

(*26*)

(*19c*): Ar = C_6H_5; X = Cl
(*19d*): Ar = C_6H_5; X = Br
(*19e*): Ar = C_6H_5; X = I
(*27*): Ar = *p*-ClC$_6$H$_4$; X = Br

(*24a′*): Ar = C_6H_5; X = Cl

3. By Transformation of Cyclobutadiene–Metal Complexes

The cyclobutadiene–metal complexes have been transformed into other cyclobutadiene–metal complexes in three ways, i.e., by displacement of the MX_m moiety with $M'X'_n$ (ligand transfer), by displacement of X by X', and by cleavage of certain metal complex oligomers.

B. PREPARATION OF CYCLOBUTADIENE–METAL COMPLEXES

a. Ligand Transfer

The preparation of cyclobutadiene–metal complexes by displacement of the entire MX_m moiety has been effected with the palladium halide complexes of tetraphenylcyclobutadiene only. Thus, treating tetraphenylcyclobutadienepalladium chloride (*19c*) or bromide (*19d*) with a large excess of iron pentacarbonyl in refluxing xylene affords tetraphenylcyclobutadieneiron(0) tricarbonyl (*17*) in good yield.[41] Similarly, treating the bromide (*19d*) with nickel tetracarbonyl gives tetraphenylcyclobutadienenickel(II) bromide (*28*).[41] The complexes formed by treating the palladium bromide complex (*19d*) and the palladium iodide complex (*19e*)* with molybdenum hexacarbonyl and tungsten hexacarbonyl have been assigned dimeric structures, viz., (*29a*), (*30a*), (*29b*), and (*30b*).[40] Cobaltocene reacts with the palladium bromide complex (*19d*) to give the sandwich complex, π-tetraphenylcyclobutadiene cyclobutadienylcobalt(I) (*13*).[41]

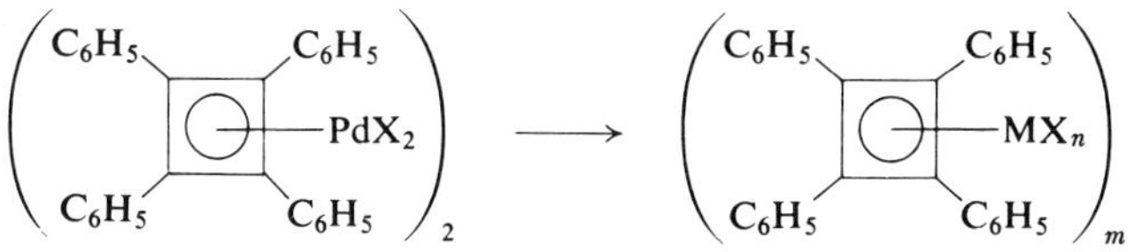

$$\left(\begin{array}{c} C_6H_5 \quad\quad C_6H_5 \\ \square\!\!-\!\!PdX_2 \\ C_6H_5 \quad\quad C_6H_5 \end{array} \right)_2 \longrightarrow \left(\begin{array}{c} C_6H_5 \quad\quad C_6H_5 \\ \square\!\!-\!\!MX_n \\ C_6H_5 \quad\quad C_6H_5 \end{array} \right)_m$$

(*19c*): X = Cl
(*19d*): X = Br
(*19e*): X = I

(*17*): $MX_n = Fe(CO)_3$; $m = 1$
(*28*): $MX_n = NiBr_2$; $m = 2$
(*29a*): $MX_n = Mo(CO)_3Br$; $m = 2$
(*30a*): $MX_n = W(CO)_3Br$; $m = 2$
(*29b*): $MX_n = Mo(CO)_3I$; $m = 2$
(*30b*): $MX_n = W(CO)_3I$; $m = 2$

(*19d*) $\longrightarrow$ [cobaltocene] → tetraphenylcyclobutadiene–Co–cyclopentadienyl complex

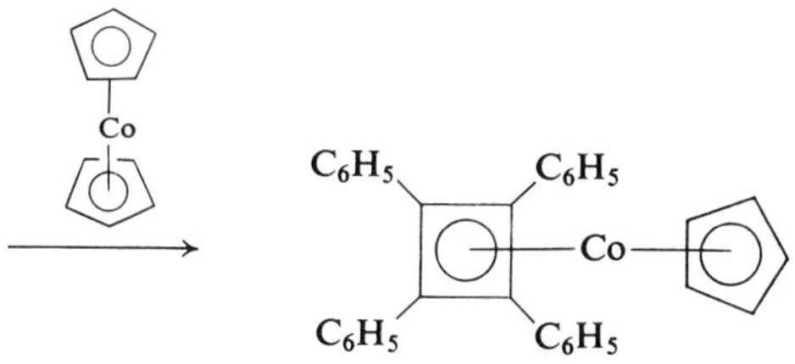

Tetrakis(*p*-chlorophenyl)cyclobutadieneiron tricarbonyl was prepared similarly by treating the corresponding palladium chloride complex or the palladium bromide complex with iron pentacarbonyl.[42]

The fact that the above reactions proceed in high yield and without the formation of tetraphenylcyclobutadiene dimer would seem to indicate that ligand transfer is a concerted reaction not involving a free cyclobutadienoid intermediate.

---
* Physical properties not described in the literature.

b. Anion Exchange

Numerous examples of the displacement of X by X' have been recorded, most of them involving tetramethylcyclobutadienenickel(II) chloride. Thus, treating an aqueous solution of the chloride (*31a*) with silver nitrate gives tetramethylcyclobutadienenickel(II) nitrate (*31b*).[50] Similarly, the sulfate, azide, acetate, and fluoride can be prepared by treating the chloride (*31a*) with an aqueous solution of two equivalents of the appropriate silver salt.[50] The bromide (*31c*) was prepared by repeatedly dissolving the chloride in aqueous hydrobromic acid and evaporating the resulting solution to dryness.[50]

Similar anion exchange reactions were used to prepare two complexes of tetraphenylcyclobutadiene. The nickel bromide complex (*28*) was treated with aqueous sodium acetate and with aqueous sodium hydroxide to give compounds believed to be tetraphenylcyclobutadienenickel(II) acetate and hydroxide, respectively.[25]

(*28*): R = C_6H_5; X = Br
(*31a*): R = CH_3; X = Cl
(*31b*): R = CH_3; X = NO_2
(*31c*): R = CH_3; X = Br

c. Cleavage of Metal Complex Oligomers

The metal atoms of such cyclobutadiene–metal halides as tetramethylcyclobutadienenickel chloride and tetraphenylcyclobutadienepalladium chloride lack two electrons of an inert gas configuration and tend to compensate for this shortage by forming dimers and oligomers which are held together by dative bonds.

Monomeric complexes can be prepared by treating the dimers and oligomers with electron-donating reagents: Thus, two monomeric complexes, (*32*) and (*33*), are produced when tetramethylcyclobutadienenickel chloride (*31a*) is treated with triphenylphosphine and phenanthroline, respectively.[13] Tetraphenylcyclobutadienepalladium chloride (*19c*) dissolves readily in chloroform saturated with hydrogen chloride, probably as a result of the formation of a monomeric hydrochloride (*34*).[6]

The dimeric tetrakis(*p*-chlorophenyl)cyclobutadienepalladium bromide complex (*27*) can be prepared in 23% yield by treating a solution of the oligomeric chloride (*20b*) in methylene chloride with hydrogen bromide.[42] [The phenyl analog of (*27*) cannot be prepared from the corresponding oligomers (see Section D).]

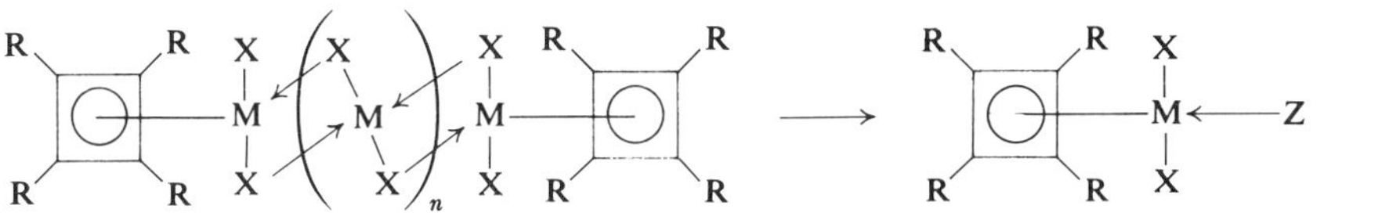

B. PREPARATION OF CYCLOBUTADIENE–METAL COMPLEXES

(19c): R = C_6H_5; MX_2 = $PdCl_2$; $n = 0$
(20b): R = p-ClC_6H_5; MX_2 = $PdCl_2$; $n = 4$
(27): R = p-ClC_6H_5; MX_2 = $PdBr_2$; $n = 0$
(31a): R = CH_3; MX_2 = $NiCl_2$; $n = 0$

(32): R = CH_3; MX_2Z = $NiCl_2P(C_6H_5)_3$
(33): R = CH_3; MX_2Z = $NiCl_2C_{12}H_{18}N_2$
(34): R = C_6H_5; MX_2Z = $PdCl_3H$

4. By Trapping Cyclobutadienes with Transition Metal Compounds

1,2,3,4-Tetramethylcyclobutadienenickel chloride (*31a*) can be prepared by the action of nickel tetracarbonyl on 3,4-dichloro-1,2,3,4-tetramethylcyclobutene (*35*).[17] Presumably, nickel tetracarbonyl acts as a dehalogenating agent to give tetramethylcyclobutadiene which subsequently reacts with the nickelous chloride produced in the dehalogenation step to give the final product, but a concerted dehalogenation–complex formation mechanism cannot be ruled out at present. Similarly, treating 3,4-dibromo-1,2,3,4-tetramethylcyclobutene (*36*) with nickel tetracarbonyl gives the nickel bromide complex (*31c*),[50] while treating 3,4-diiodo-1,2,3,4-tetramethylcyclobutene (*37*) with Raney nickel gives the nickel iodide complex (*38*).[13]

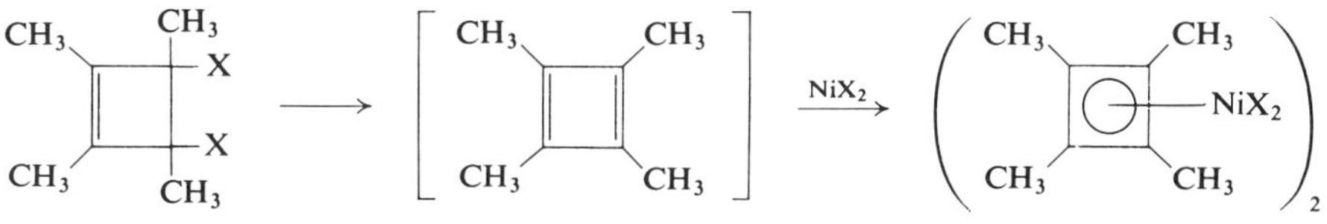

(35): X = Cl
(36): X = Br
(37): X = I

(31c): X = Br
(31a): X = Cl
(38): X = I

1,2,3,4-Tetraphenylcyclobutadienenickel chloride (*41*) and bromide (*28*) can be prepared (a) by the action of nickel tetracarbonyl on 1,4-bis(chloromercury)tetraphenylbutadiene (*39*)[30] and (b) by the thermolysis of (4-bromo-1,2,3,4-tetraphenyl-*cis,cis*-1,3-butadienyl)dimethyltin bromide (*40*) in the presence of nickelous bromide, respectively.[25]

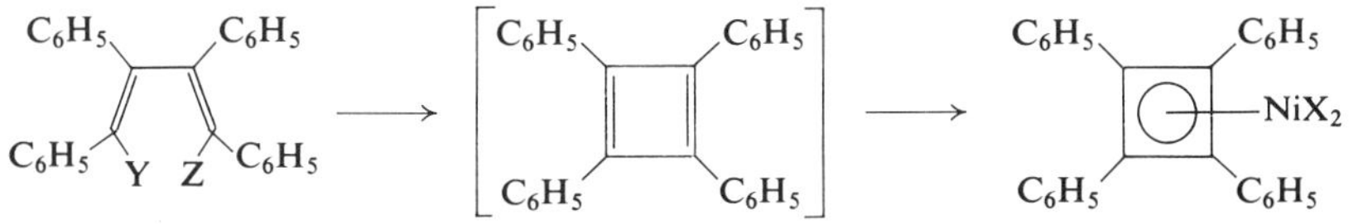

(39): Y = Z = HgCl
(40): Y = Sn(CH_3)$_2$Br; Z = Br

(28): X = Br
(41): X = Cl

1,4-Dilithio-1,2,3,4-tetraphenylbutadiene reacts with iron tetracarbonyl dibromide to give tetraphenylcyclobutadieneiron tricarbonyl in 1.5% yield.[8]

C. Cyclobutadiene–Metal Complexes as Transient Intermediates

A number of reactions are known which yield products attributable to transient cyclobutadiene–metal complexes, but the question of whether or not such complexes are actually generated in these reactions cannot be answered unambiguously at the present time.*

The reaction of diphenylacetylenecobalt hexacarbonyl (*1*) with bis(trimethylsilyl)acetylene (*2*) gave a small amount of 3,4-bis(trimethylsilyl)-2,5-diphenylcyclopentanone (*4*), probably by insertion of carbon monoxide between the two phenyl-bearing carbon atoms of a bis(trimethylsilyl)diphenylcyclobutadienecobalt carbonyl intermediate such as (*3*).[35] Tetraphenylcyclopentadienone (*5*) and several metal complexes of tetraphenylcyclopentadienone have been prepared by analogous reactions, but because all four substituents are identical, there is no direct evidence to implicate a tetraphenylcyclobutadiene–metal complex in these reactions.[11,31,32,35,59] The same mechanistic ambiguity exists in the conversion of phenylacetylene into 2,5-diphenylcyclopentadienone (*6*) by nickel tetracarbonyl[34,58]: The symmetrical arrangement of the two phenyl substituents in the complex does not constitute evidence in favor of a cyclobutadienoid intermediate, but neither does it constitute evidence against it. In this connection it is interesting to note that the stable palladium bromide complex of tetraphenylcyclobutadiene also

$(C_6H_5C{\equiv}CC_6H_5)_2Co(CO)_6 + (CH_3)_3SiC{\equiv}CSi(CH_3)_3 \longrightarrow$

(*1*)   (*2*)

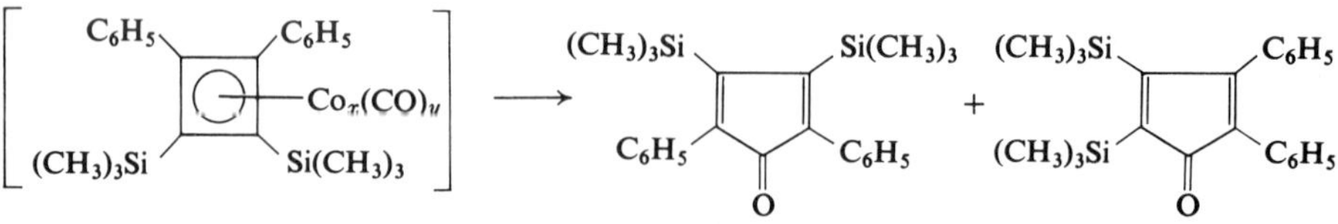

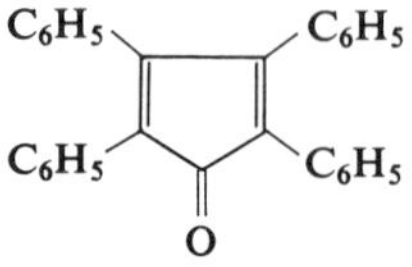

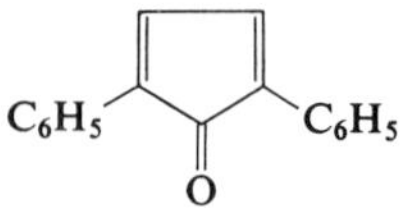

* The formation of cyclooctatetraene from acetylene in the presence of covalent nickel catalysts has been shown to proceed via noncyclobutadienoid intermediates. (See Section B, 1, c).

C. TRANSIENT INTERMEDIATES

gives a cyclopentadienone (5) when treated with nickel tetracarbonyl.[41]

Reppe and Vetter[54] and Wender and his co-workers[65] have described a number of acetylene–iron carbonyl complexes, many of them of unknown structure, which are too numerous to name here; it is not impossible that some of these substances are stable cyclobutadiene–iron carbonyl complexes or that they are compounds formed through transient complexes of this kind.

Octaphenylcyclooctatetraene (7) was formed, together with some hexaphenylbenzene, when diphenylacetylene was refluxed with phenylmagnesium bromide

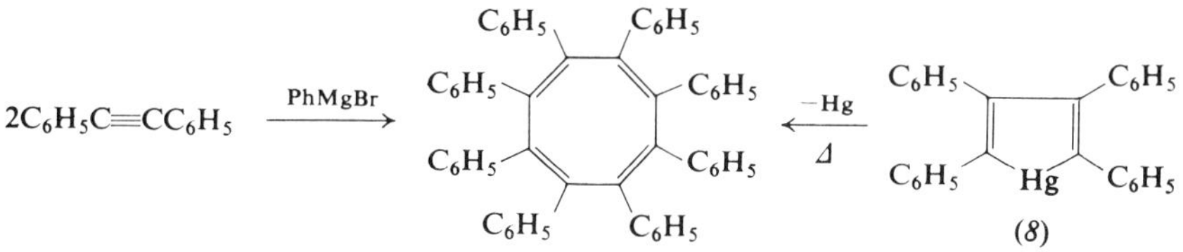

$$2C_6H_5C{\equiv}CC_6H_5 \xrightarrow{PhMgBr} (7) \xleftarrow{-Hg, \Delta} (8)$$

in xylene,[62] and when a compound believed to be tetraphenylmercurole was subjected to thermolysis (8).[8] (Octaphenylcyclooctatetraene was incorrectly assigned the structure of octaphenylcubane[24]; see also Chapter 1, Section E, 1, f.) It is possible (but not certain) that tetraphenylcyclobutadiene complexes of magnesium and mercury are involved as intermediates in these reactions.

A tetramethylcyclobutadienenickel(II) di(cyclopentadiene) complex (9) has been proposed as an intermediate in the reaction of tetramethylcyclobutadiene-

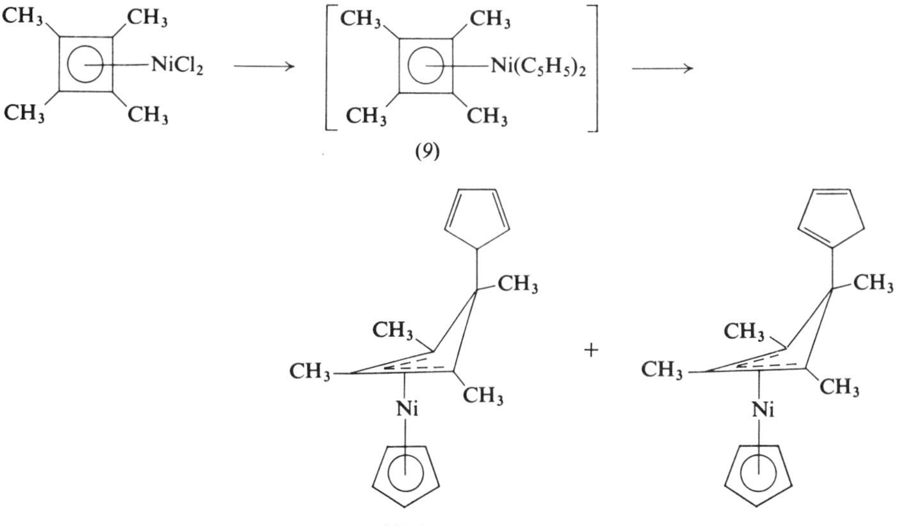

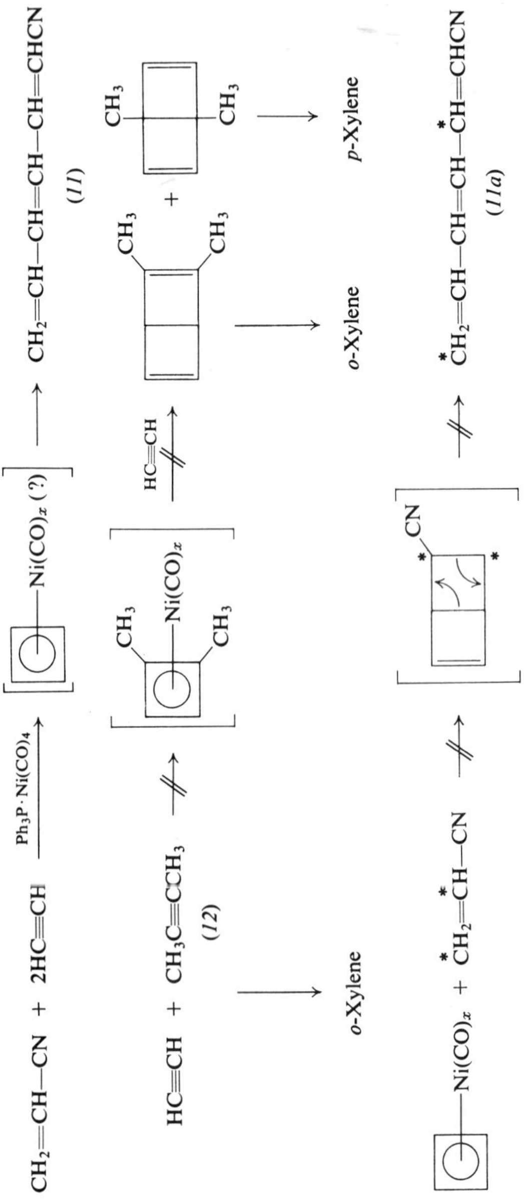

2. THE CYCLOBUTADIENE–METAL COMPLEXES

$$CH_2{=}CH{-}CN + 2HC{\equiv}CH \xrightarrow{Ph_3P\cdot Ni(CO)_4} [\text{cyclobutadiene-Ni(CO)}_x] \longrightarrow CH_2{=}CH{-}CH{=}CH{-}CH{=}CHCN$$
$$(11)$$

$$HC{\equiv}CH + CH_3C{\equiv}CCH_3 \xrightarrow{\;\;\not\to\;\;} [\text{dimethylcyclobutadiene-Ni(CO)}_x] \xrightarrow{HC{\equiv}CH}\;\not\to\; \text{(o-xylene + p-xylene products)} \longrightarrow \text{o-Xylene / p-Xylene}$$
$$(12)$$

$$[\text{cyclobutadiene-Ni(CO)}_x] + {}^*CH_2{=}{}^*CH{-}CN \;\not\to\; [\text{CN-substituted intermediate}] \;\not\to\; {}^*CH_2{=}CH{-}CH{=}CH{-}CH{=}{}^*CHCN$$
$$(11a)$$

nickel(II) chloride with cyclopentadienyl sodium,[16] the final product of which is a mixture of two isomeric π-cyclobutenylnickel cyclopentadiene complexes, (10a) and (10b).[14]

Acrylonitrile reacts with 2 moles of acetylene in the presence of nickel tetracarbonyl–triphenylphosphine complex to give 2,4,6-heptatrienonitrile (11) in high yield (74%).[9] A transient intermediate of the cyclobutadiene–metal-complex type has been considered and rejected for this reaction by Sauer and Cairns, largely because in a related reaction, i.e., the copolymerization of acetylene and dimethylacetylene (12), o-xylene was formed, rather than a mixture of o- and p-xylene, and because the copolymerization of acrylonitrile-2,3-C^{14} with acetylene did not give a product in which the C^{14}-labeled carbon atoms were separated (11a) as would be required if a cyclobutadienoid intermediate were involved.[55]

It has been pointed out by Bieber[5] and Schrauzer[57] that these are essentially negative arguments and that the results can also be interpreted to favor a cyclobutadiene–metal complex intermediate, but in a later paper Schrauzer also rejected the cyclobutadiene mechanism for one analogous to the simple trimerization of acetylenes.[56]

When diphenylacetylene was added to a solution of dimesitylnickel(II) in tetrahydrofuran, a large amount of an orange, insoluble material approximately of the empirical formula $[(C_6H_5C\equiv CC_6H_5)_2Ni]_x$ was formed, together with a small amount of hexaphenylbenzene. When the order of addition was reversed, the relative amounts of polymeric product and hexaphenylbenzene were also reversed, i.e., a small amount of polymer and a large amount of hexaphenylbenzene were formed.*,[63] These results have been attributed to the

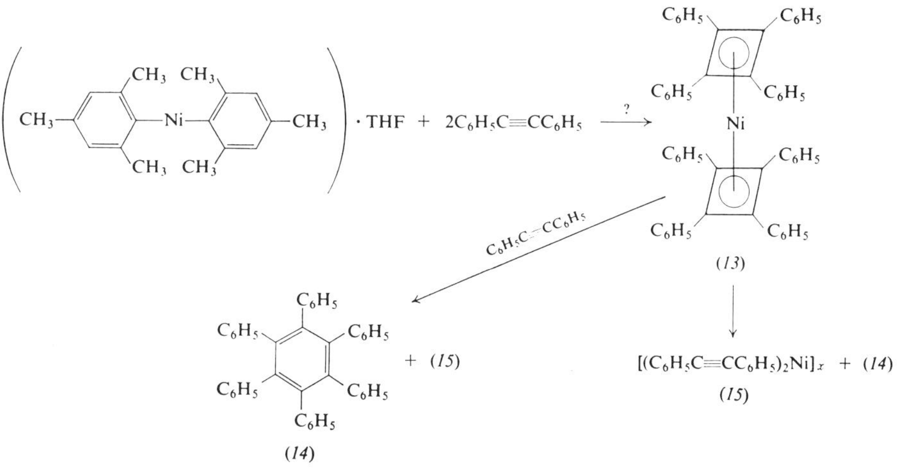

---

* In actuality, dimesitylnickel was generated *in situ* by slowly adding a solution of mesitylmagnesium bromide to a mixture of nickelous bromide and diphenylacetylene.

formation of a transient cyclobutadienylnickel(0) intermediate (*13*) which either reacts with excess diphenylacetylene to give hexaphenylbenzene [reverse addition, i.e., addition of dimesitylnickel(II) to diphenylacetylene] or undergoes polymerization to form $[(C_6H_5C\equiv CC_6H_5)_2Ni]_x$ [normal addition, i.e., addition of diphenylacetylene to dimesitylnickel(II)].

Similarly, the formation of hexaphenylbenzene from diphenylacetylene in the presence of dimesitylchromium or dimesityliron is also said to proceed via bis(tetraphenylcyclobutadiene)–metal complexes.[63]

D. Unsuccessful Approaches to the Cyclobutadiene–Metal Complexes

The factors which influence cyclobutadiene–metal complex formation are not well understood at present and in consequence, the methods used to prepare compounds of this type are not universally applicable and at times even appear to be quixotic. For example, it has been found that tetraphenylcyclobutadieneiron(0) tricarbonyl, which is readily prepared by heating diphenylacetylene with cyclooctatetraeneiron tricarbonyl at 190°, cannot be prepared from the same reactants if xylene is used as solvent [a compound of composition $(C_6H_5C\equiv CC_6H_5)_2Fe_2(CO)_6$ is formed].[47] Similarly, the formation of tetraphenylcyclobutadieneiron tricarbonyl from tetraphenylcyclobutadienepalladium bromide by ligand transfer could not at first be carried out in saturated hydrocarbons,[41] although subsequent work showed that under optimum conditions, the iron complex was formed in detectable yields in saturated hydrocarbon solvents.[38] (The reaction proceeds best when aromatic hydrocarbons are used as solvents.) An attempt to prepare tetraphenylcyclobutadienepalladium bromide by the thermolysis of (4-bromo-1,2,3,4-tetraphenyl-*cis*,*cis*-1,3-butadienyl)dimethyltin bromide in the presence of palladium bromide failed.[24] Since the expected product of the reaction, tetraphenylcyclobutadienepalladium bromide, is known to be stable and since the thermolysis of the butadienyltin compound in the presence of nickelous bromide readily affords tetraphenylcyclobutadienenickel(II) bromide, the cause of the failure with palladium bromide is not obvious. Similar attempts to prepare the iron(III) bromide and the copper(II) bromide complexes of tetraphenylcyclobutadiene were also unsuccessful,[24] as was an early attempt to prepare tetraphenylcyclobutadienepalladium chloride by the reaction of tolane with bis(benzonitrile)palladium chloride complex.*,[6]

Contrary to a published report,[7] tetrakis(trifluoromethyl)cyclobutadiene cyclopentadienecobalt(III) was not formed in the reaction of hexafluoro-2-butyne with π-cyclopentadienylcobalt dicarbonyl.[66] [The correct identity of the product formed in this reaction is tetrakis(trifluoromethyl)cyclopentadienone π-cyclopentadienylcobalt.][29]

* For a successful synthesis of tetraphenylcyclobutadienepalladium chloride from tolane and bis(benzonitrile)palladium chloride, see Section B, 1, e.

D. UNSUCCESSFUL APPROACHES TO THE COMPLEXES

An attempt to prepare a complex of cyclobutadiene with silver nitrate gave a substance of composition "$C_4H_4AgNO_3$." Earlier reports to the contrary notwithstanding,[1, 3, 28, 46] an NMR study of this complex has shown conclusively that it is not a cyclobutadiene complex but a *syn*-tricyclo[4.2.0.0^{2,5}]-octa-3,7-diene complex ($C_8H_8 \cdot 2AgNO_3$).[26] In the light of this fact, the preparation of "$C_4H_4AgNO_3$" by cleavage of tricyclo[4.2.0.0^{2,5}]octa-3,7-diene with silver nitrate[2] is seen to be incorrect and the alleged preparation of "$C_4H_4AgClO_3$" from tricyclo[4.2.0.0^{2,5}]octa-3,7-diene and from a "cyclobutadiene–mercury complex"[1] must be reclassified as attempted syntheses, as must the alleged preparation of a "cyclobutadiene–silver fluoborate complex" from tricyclo[4.2.0.0^{2,5}]octa-3,7-diene.[1]

The new structure assignments bring into question the structure of the alleged cyclobutadiene–mercury complex of unknown composition which is formed when 1,2,3,4-tetrabromocyclobutane is dehalogenated with lithium amalgam in ether and from which the $C_8H_8 \cdot 2AgNO_3$ complex can be prepared.[2] Attempts to convert the mercury complex into a nickel complex by treatment with nickelous chloride and with nickel tetracarbonyl failed.

An unsuccessful attempt was made to prepare a π-1,2,3,4-tetramethylcyclobutadiene π-cyclopentadienylnickel complex by treating 1,2,3,4-tetramethylcyclobutadienenickel chloride with cyclopentadienylsodium.[16] The product actually formed is a mixture of two isomeric π-cyclobutenylnickel–cyclopentadiene complexes (see Section C).

Treating methylphenylacetylene and ethylphenylacetylene with palladium chloride in ethanol gave products believed to be cyclobutadiene–metal complexes,[45] but a reinvestigation of the reaction has shown that the two alkylphenylacetylenes yield the 1,2,4-trialkyl-3,5,6-triphenylbenzenes and the complexes ($C_6H_5C\equiv CR$)$_4$PdCl and ($C_6H_5C\equiv CR$)$_4$Pd$_2$Cl$_2$,[68] which are probably not cyclobutadiene complexes.

All attempts to prepare dimeric tetraphenylcyclobutadienepalladium chloride (*1*) from its oligomers [(*2*) and (*3*)] by treatment with chloride ion, benzonitrile, and other solvents, were uniformly unsuccessful.*,[42]

(*1*): *n*=0
(*2*): *n*=3
(*3*): *n*=4

* But see also ref. 33 and the Appendix for a successful preparation of (*1*) from an oligomer by treatment of the latter with dimethylformamide.

E. Chemistry of the Cyclobutadiene Complexes

The most striking property of the cyclobutadiene–metal complexes is their great stability. For example, tetraphenylcyclobutadieneiron tricarbonyl melts sharply without decomposition at 234°, sublimes unchanged at 180° *in vacuo* and is unaltered when heated with carbon monoxide under high pressure.[31] Similarly, the metal ion of 1,2,3,4-tetramethylcyclobutadienenickel chloride is not abstracted by dimethylglyoxime,[17] and even more striking, π-tetraphenylcyclobutadiene π-cyclopentadiene cobalt(I) is inert to lithium aluminum hydride, alcoholic potassium hydroxide, hydrochloric acid, air, heat (360° under nitrogen), iodine, triphenylphosphine at 150°, carbon monoxide at 290° (100 atm pressure), and dimethyl acetylenedicarboxylate at 200°.[41, 48b] However, this is not to say that the cyclobutadiene–metal complexes are totally inert. On the contrary, a number of interesting transformations have been carried out which give promise of a highly diverse chemistry for these compounds.

1. Thermolysis

The thermal decomposition of the cyclobutadiene–metal complexes is of more than routine interest, because such decomposition has been postulated by Bergmann[4] and Longuet-Higgins[36] as being an integral step in the conversion of acetylene into cyclooctatetraene by nickel compounds. Thermolysis of a tetramethylcyclobutadienenickel complex (*1*) has in fact been effected by Criegee, but the results offer little in the way of support for Bergmann and Longuet-Higgins' theory. Thus, (a) the conditions necessary to effect thermal decomposition of tetramethylcyclobutadienenickel chloride (210°, no solvent, high vacuum) differ markedly from those used to prepare cyclooctatetraene from acetylene (60°, tetrahydrofuran, 5 atm pressure) and (b) the yield of octamethylcyclooctatetraene (*2*) obtained by thermolysis of (*1*) is small (7–21%)* in comparison with the yield of cyclooctatetraene obtained in the Reppe synthesis (70–90%).[53b] Two other products were obtained in the thermolysis of (*1*), viz., tetramethylbutadiene and a liquid hydrocarbon, $C_{16}H_{24}$ (*3*), previously obtained in the thermolysis of octamethyltricyclo-[4.2.0.0^{2,5}]octa-3,7-diene (*4*)[15] and believed to be either 4-methylene-1,2,3,5,6,7,8-heptamethylbicyclo[4.2.0]octa-2,7-diene (*3b*)[13b] or an almost inseparable mixture of (*3b*) with an isomeric hydrocarbon, 2-methylene-1,3,4,5,6,7,8-heptamethylbicyclo[4.2.0]octa-3,7-diene (*3a*).[13a]

Tetramethylcyclobutadienenickel chloride–triphenylphosphine complex $[C_8H_{12}NiCl_2 \cdot P(C_6H_5)_3]$ gave the same mixture of products on thermolysis [tetramethylbutadiene, (*2*), (*3a*), and (*3b*)], but the phenanthroline complex

---
* Calculated from weight data recorded in ref. 17a.

($C_8H_{12}NiCl_2 \cdot C_{12}H_8N_2$) gave the anti form of octamethyltricyclo[4.2.0.0^{2,5}]-octa-3,7-diene (4), 3-methylene-1,2,4-trimethylcyclobutene (5) and the $C_{16}H_{24}$ hydrocarbon (or mixture) (3a, b).[13, 37] Tetramethylcyclobutadienenickel azide complex decomposes on warming in benzene solution to give an insoluble precipitate of unknown composition.[50] It is not known whether free tetramethylcyclobutadiene is an intermediate in any of these transformations.

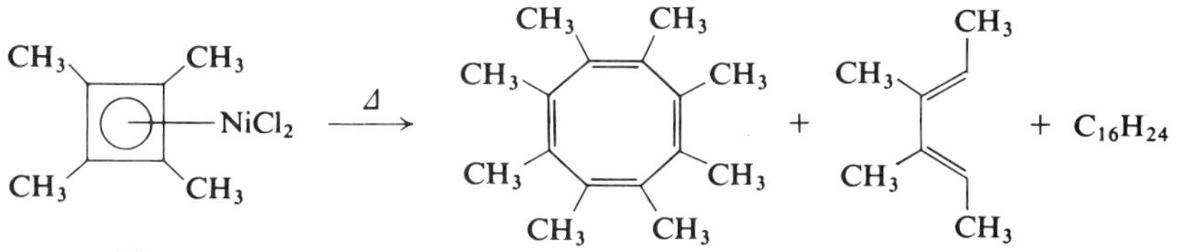

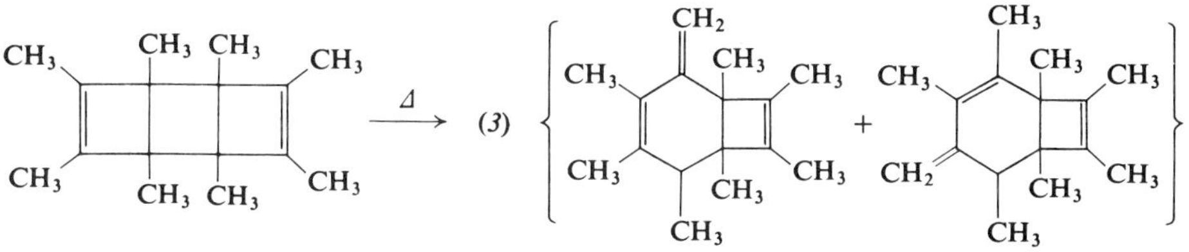

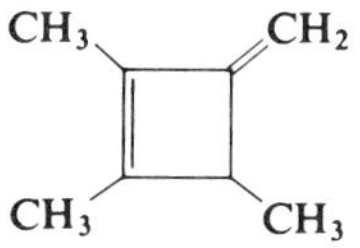

Tetraphenylcyclobutadienepalladium chloride gives mainly 1,4-dichloro-1,2,3,4-tetraphenylbutadiene on thermolysis, possibly via tetraphenylcyclobutadiene.[6]

2. SOLVOLYSIS

Tetramethylcyclobutadienenickel chloride (1) decomposes on heating in aqueous solution to give anti-1,2,3,4,5,6,7,8-octamethyltricyclo[4.2.0.0^{2,5}]-octa-3,7-diene (4a).[18] Inasmuch as the syn rather than the anti isomer of (4) is believed to be formed in the dimerization of tetramethylcyclobutadiene (Chapter 1, Section E, 1, b) some doubt exists as to whether the solvolytic transformation of the nickel complex proceeds via cyclobutadiene intermediate. At room temperature, tetramethylcyclobutadienenickel chloride dissolves in water to form a hydrate, possibly [$C_8H_{12}Ni(H_2O)_2$]Cl.[17]

2. THE CYCLOBUTADIENE–METAL COMPLEXES

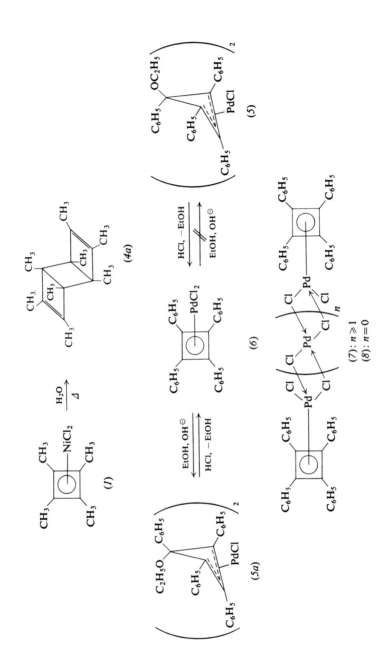

Blomquist and Maitlis[6] were unable to confirm the transformation of tetraphenylcyclobutadienepalladium chloride (6) into Malatesta's cyclobutenyl complex (5) by ethanolic alkali.[45] [Since the conversion of (5) into (6) by acid is known, transformation of (6) into (5) by base would constitute a reversible synthesis of a cyclobutadiene–metal complex and would therefore be of more than routine interest.] According to Blomquist and Maitlis, treatment of (6) with ethanol, either with or without alkali, gives a complex which is isomeric with (5) and from which (6) can be regenerated by treatment with acid.[6] The isomeric complex has recently been assigned structure (5a).[19]

Tetraphenylcyclobutadienepalladium chloride was at first believed to react with dimethylformamide to give a product of unknown structure,[6] but subsequent work has shown that dimethylformamide merely removes palladium chloride from oligomeric tetraphenylcyclobutadiene chloride (7) to give the dimeric species (8).[33]

Tetraphenylcyclobutadienepalladium chloride is insoluble in most organic solvents but it dissolves in complexing solvents with decomposition.[6] Similarly, tetraphenylcyclobutadienenickel chloride decomposes slowly in ethers, alcohols, and nitriles, but it can be recrystallized satisfactorily from methylene chloride and benzene.[30] Tetraphenylcyclobutadienenickel bromide is quite soluble in complexing solvents, in which it forms deep blue or violet solutions and in which it is readily decomposed by the application of heat or by the addition of water.[25] An attempt to remove the nickel atom from tetraphenylcyclobutadienenickel bromide by treating a solution of the compound in dimethylformamide with aqueous sodium acetate or with sodium hydroxide resulted only in exchange of the bromide by acetate or hydroxide ion.[25]

3. Oxidation

Tetramethylcyclobutadienenickel chloride reacts rapidly with aqueous sodium nitrite to give cis-3,4-dihydroxy-1,2,3,4-tetramethylcyclobutene [(1) → (9)] in yields greater than 50%.[17]

$$C_8H_{12}NiCl_2 + 2NaNO_2 + 2H_2O \rightarrow C_8H_{12}(OH)_2 + 2NO + 2NaCl + Ni(OH)_2$$
$$(1) \qquad\qquad\qquad\qquad\qquad (9)$$

Tetraphenylcyclobutadienenickel bromide is stable in air, but it is readily oxidized by sodium nitrite in aqueous dimethylformamide to give tetraphenylfuran and nickel hydroxide.[25] (Tetraphenylfuran is also formed when tetraphenylcyclobutadiene is generated in the presence of air by treatment of tetraphenylcyclobutadienepalladium chloride with either triphenylphosphine[12, 43] or trimethylphosphite.[43])

4. Reduction

An aqueous solution of tetramethylcyclobutadienenickel chloride absorbs exactly 2 moles of hydrogen in the presence of 5% palladium-on-charcoal catalyst but the identity of the product has not been reported.[17] Tetraphenylcyclobutadienenickel bromide undergoes hydrogenation, but only under drastic conditions [tetrachloroethane (solvent), platinum catalyst, 75°, 800 psi initial hydrogen pressure]; the hydrogenation product was assigned the structure of 1,2,3,4-tetraphenylcyclobutene,[25] but a more recent investigation indicates that the compound is 1,2,3,4-tetraphenyl-1-butene.[12a]

Tetraphenylcyclobutadieneiron tricarbonyl reacts with lithium aluminum hydride to give 1,2,3,4-tetraphenylbutadiene (m.p. 183°–184°).[31] Reduction of tetraphenylcyclobutadieneiron tricarbonyl with sodium in liquid ammonia gives tetraphenylbutane,[31] as does the reduction of tetraphenylcyclobutadiene cyclopentadienylcobalt(I) with sodium in liquid ammonia.[48b]

5. Ligand Transfer

The interconversion of cyclobutadiene–metal complexes by ligand-transfer reactions has been discussed in Section B, 3, a, under "Preparation of Cyclobutadiene–Metal Complexes." The fact that ligand transfer usually takes place in high yield and without the formation of the corresponding cyclobutadiene dimers would seem to indicate that the reaction is concerted and does not involve a free cyclobutadienoid intermediate.

6. Anion Exchange

Reactions of this type have been discussed in Section B, 3, b and merely involve the conversion of a cyclobutadiene–metal complex, $R_4C_4MX_n$, into its analog, $R_4C_4MY_m$.

7. Reaction with Complexing Agents

The metal atoms in such cyclobutadiene complexes as 1,2,3,4-tetramethylcyclobutadienenickel chloride (*1*) and 1,2,3,4-tetraphenylcyclobutadienepalladium chloride (*8*) lack two electrons of an inert gas configuration and tend to compensate for this shortage by forming dimeric or oligomeric species in which two pairs of dative electrons are shared by adjacent metal atoms as shown in formulas (*1*) and (*8*). A quick count of the valence electrons of the nickel atom in complex (*1*) shows that it has the 36-electron configuration of krypton. [In monomeric (*1*), 28 electrons are contributed by the nickel(II), the cyclobutadiene ring contributes another four, and the two halogens contribute one each, giving a total of 34 valence electrons; an additional two electrons are obtained by formation of a dative bond by the metal with an extraneous chlorine as shown in formula (*1*).] Similarly, palladium(II) in complex (*8*) has the 56-electron configuration of xenon.

The dimeric complexes react readily with electron-donating reagents. Thus, tetramethylcyclobutadienenickel chloride (*1*) reacts with triphenylphosphine and also with phenanthroline to give two stable monomeric complexes, (*12*) and (*13*), respectively.[13] In contrast, the monomeric species formed by treating tetraphenylcyclobutadienepalladium chloride with triphenylphosphine and with trimethyl phosphite appear to be unstable because the product isolated is tetraphenylcyclobutadiene dimer [octaphenylcyclooctatetraene, (*16*)], which is formed in 70% yield.[43] However, tetraphenylcyclobutadienepalladium chloride (*8*) can be recovered unchanged from its solutions in hydrogen chloride saturated chloroform and hence it appears that the hydrochloride complex (*14*) is stable, at least in solution.[6] The oligomer of tetrakis(*p*-chlorophenyl)cyclobutadienepalladium chloride (*10*) is readily converted into the dimeric bromide (*11*) by treatment with anhydrous hydrogen bromide, presumably by way of the monomeric species (*15*).[42]

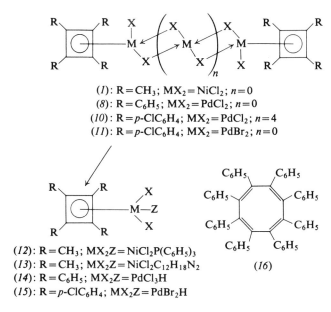

(*1*): R=CH$_3$; MX$_2$=NiCl$_2$; $n=0$
(*8*): R=C$_6$H$_5$; MX$_2$=PdCl$_2$; $n=0$
(*10*): R=*p*-ClC$_6$H$_4$; MX$_2$=PdCl$_2$; $n=4$
(*11*): R=*p*-ClC$_6$H$_4$; MX$_2$=PdBr$_2$; $n=0$

(*12*): R=CH$_3$; MX$_2$Z=NiCl$_2$P(C$_6$H$_5$)$_3$
(*13*): R=CH$_3$; MX$_2$Z=NiCl$_2$C$_{12}$H$_{18}$N$_2$
(*14*): R=C$_6$H$_5$; MX$_2$Z=PdCl$_3$H
(*15*): R=*p*-ClC$_6$H$_4$; MX$_2$Z=PdBr$_2$H

(*16*)

8. REACTION WITH ORGANOMETALLIC REAGENTS

Only one example of the reaction of a cyclobutadiene–metal complex with an organometallic reagent is known at the present time. 1,2,3,4-Tetramethylcyclobutadienenickel chloride reacts with cyclopentadienylsodium[16] to give a mixture of two isomeric π-cyclobutenylnickelcyclopentadiene complexes [(*17a*) and (*17b*)].[14]

9. Carbonylation

Tetraphenylcyclobutadienepalladium bromide reacts with excess nickel tetracarbonyl in refluxing benzene to give tetraphenylcyclopentadienone in 17% yield along with the expected ligand-transfer product, tetraphenylcyclobutadienenickel bromide.[41]

F. Physical Properties of the Cyclobutadiene–Metal Complexes

Authentic, stable, metal complexes of only three cyclobutadienes have been prepared, viz., the complexes of 1,2,3,4-tetramethylcyclobutadiene, 1,2,3,4-tetraphenylcyclobutadiene, and 1,2,3,4-tetrakis(p-chlorophenyl)cyclobutadiene, all of which are highly colored crystalline solids. (See Table 2.1).

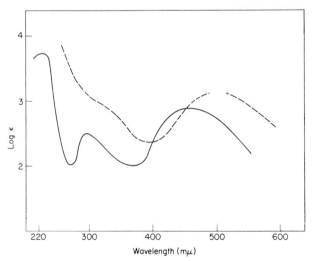

FIG. 2.1. Ultraviolet absorption spectrum of 1,2,3,4-tetramethylcyclobutadienenickel chloride in chloroform (----) and water (——).

1. Ultraviolet Absorption

The ultraviolet absorption spectra of the cyclobutadiene–metal complexes are largely characteristic of the organic ligand, and changing the metal atom or the anions merely produces band shifts in the absorption spectra. The ultraviolet spectra of 1,2,3,4-tetramethylcyclobutadienenickel chloride and 1,2,3,4-tetraphenylcyclobutadienenickel chloride are reproduced in Figs. 2.1 and 2.2, respectively. Absorption maxima and extinction coefficients of the cyclobutadiene–metal complexes are listed in Table 2.2.

2. Infrared Absorption

The infrared absorption spectra of the cyclobutadiene complexes are largely characteristic of the organic ligand (compare "Ultraviolet Absorption," above). Thus, the spectrum of the palladium bromide complex of tetraphenylcyclobutadiene is very similar to that of its nickel bromide complex.[39] The infrared spectra of tetramethylcyclobutadienenickel chloride[17a]

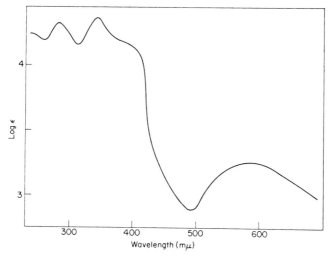

Fig. 2.2. Ultraviolet absorption spectrum of 1,2,3,4-tetraphenylcyclobutadienenickel chloride (in chloroform).

and tetraphenylcyclobutadienepalladium chloride[6] have been recorded in the literature and are reproduced in Figs. 2.3 and 2.4, respectively. The infrared spectrum of tetraphenylcyclobutadienenickel bromide is shown in Fig. 2.5.

The infrared spectrum of tetraphenylcyclobutadieneiron(0) tricarbonyl contains bands at 4.92, 5.09, and 5.18 μ (KBr disc), ascribed to the vibrational frequency of the carbonyl groups.[31] Similar bands have been observed in the

TABLE 2.1

Physical Properties of Substituted Cyclobutadiene–Metal Complexes[a]

Ligand	Method of preparation[b]	Physical properties and molecular formula[c]	Ref.
Tetramethylcyclobutadiene			
Nickel(II) chloride	4	M.p. 270°–280° (decomp.),[30] (pr) amorph, diam.; $(C_8H_{12}NiCl_2)_2 \cdot C_6H_6$ from benzene; $C_8H_{12}NiCl_2 \cdot CHCl_3$ from chloroform	17, 30
Tetraphenylcyclobutadiene			
Cobalt(I) π-cyclopentadiene	1	M.p. 256° (y-br); $[(C_6H_5)_2C_2]_2CoC_5H_5$	7, 48
Iron(0) tricarbonyl	1, 4	M.p. 234° (sharp), (y), monoclinic, d 1.39 gm/cc[31]; $[(C_6H_5)_2C_2]_2Fe(CO)_3$	11, 31, 32, 47, 48a
Molybdenum tricarbonyl bromide	3a	M.p. 280° (decomp.), (r); $[C_{28}H_{20}Mo(CO)_3Br]_2$	40
Molybdenum tricarbonyl iodide	3a	M.p. 290° (decomp.), (br); diam.; $[C_{28}H_{20}Mo(CO)_3I]_2$	40
Palladium(II) bromide	2	M.p. 347°–348° (decomp.); $(C_{28}H_{20}PdBr_2)_2$	39
Palladium(II) chloride	1, 2	M.p. 297°–298° (decomp.)[39]; (r), diam.[45]; $(C_{28}H_{20}PdCl_2)_2$ $[C_{28}H_{20}(PdCl_2)_{2.5}]_2$ $[C_{28}H_{20}(PdCl_2)_3]_2$; (r)	6, 39, 45 42 42
Tetrakis(p-chlorophenyl)cyclobutadiene			
Iron(0) tricarbonyl	1, 3a	M.p. 222°–226°,[30] 242°,[42] (y), $C_{31}H_{16}Cl_4FeO_3$	30, 42
Palladium(II) bromide	2, 3c	M.p. > 300° (decomp.), (r); diam.[42]; $[(C_{28}H_{16}Cl_4)PdBr_2]_2$	42, 45
Palladium(II) chloride	2	M.p. 297°–298° (decomp.)	39
	1	M.p. 244°–248° (decomp.), (r-br); $[C_{28}H_{16}Cl_4(PdCl_2)_2]_2$	42
	1	M.p. 232°–234° (decomp.), 256°–258° (rapid heating); $[C_{28}H_{16}Cl_4(PdCl_2)_3]_2$	42

[a] Complexes not characterized in the literature are not included in the Table. For the preparation of the following cyclobutadienenickel complexes see the appropriate passages in Section B: 1,2(or 3)di(*tert*-butyl)cyclobutadienecobalt mercury carbonyl, 1,2 (or 3)-diphenylcyclobutadienecobalt mercury carbonyl, dimethyl 1,2-diphenylcyclobutadiene-3,4-dicarboxylateiron tricarbonyl, and dimethyl 1,3-diphenylcyclobutadiene-2,4-dicarboxylateiron tricarbonyl (Section B, 1, f); tetramethylcyclobutadienenickel acetate, azide, bromide, fluoride, nitrate, sulfate (Section B, 3, b); iodide (Section B, 4); tetraphenylcyclobutadienenickel acetate and hydroxide (Section B, 3, b); tetraphenylcyclobutadienenickel bromide (Section B, 3, a); tetraphenylcyclobutadienepalladium iodide (Section B, 3, a); tetraphenylcyclobutadienetungsten tricarbonyl bromide and tricarbonyl iodide (Section B, 3, a).

[b] Method 1 is "By the Reaction of Acetylenes with Transition Metal Compounds," Section B, 1; Method 2 is "From π-Cyclobutenyl Metal Complexes," Section B, 4; Method 3 is "By Transformation of Cyclobutadiene–Metal Complexes," Section B, 3, and is subdivided into three parts, (a) "Ligand Transfer," (b) "Anion Exchange," and (c) "Cleavage of Metal Complex Oligomers," and Method 4 is "By Trapping Cyclobutadienes with Transition Metal Compounds," Section B, 4.

[c] Key: amorph. (amorphous), br (brown), diam. (diamagnetic), pr (purple), r (red), y (yellow).

TABLE 2.2

ULTRAVIOLET ABSORPTION CHARACTERISTICS OF TETRAMETHYL-, TETRAPHENYL-, AND TETRAKIS(p-CHLOROPHENYL)CYCLOBUTADIENE–METAL COMPLEXES

Ligand	Solvent	Absorption maxima, mμ, and extinction coefficients (log ϵ)	Ref.
Tetramethylcyclobutadiene			
Nickel(II) bromide	CHCl$_3$	510 (3.25)[a]	50
Nickel(II) chloride	CHCl$_3$	502 (3.15)	17a
	CHCl$_3$	245 (4.1); 315 (3.2); 503 (3.8)	26a
	H$_2$O	230 (3.75); 295 (2.50); 460 (2.88)	17a
Nickel(II) iodide	CHCl$_3$	530 (3.4)[a]	50
Tetraphenylcyclobutadiene			
Cobalt(I) π-cyclopentadiene	EtOH	241 (4.51); 257 (4.50); 278 (4.46); 287sh; 295sh; 400sh	48b
Iron(0) tricarbonyl	CHCl$_3$	258 (4.3); 308 (3.3)	26a
Nickel(II) chloride	CH$_2$Cl$_2$	238 (3.4); 281 (4.4); 335 (4.4); 388 (5.2); 577 (3.3)	26a
Palladium(II) chloride	CHCl$_3$	245 (3.0); 285 (4.0); 318 (4.1); 390 (3.8)	26a
Tetra(p-chlorophenyl)cyclobutadiene			
Iron(0) tricarbonyl	CHCl$_3$	268 (4.4); 320 (3.4)	26a

[a] Values interpolated from a spectral curve recorded in the literature.

F. PHYSICAL PROPERTIES

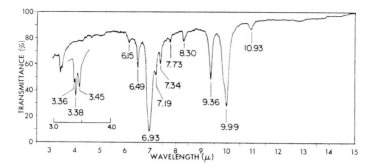

FIG. 2.3. Infrared spectrum of crystalline tetramethylcyclobutadienenickel chloride (potassium bromide disc) calibrated against polystyrene.

spectrum of tetraphenylcyclobutadienemolybdenum bromide tricarbonyl [ν_{max} 2026 cm^{-1} (4.93 μ) and 1976 cm^{-1} (5.06 μ)] and its iodide analog [ν_{max} 2021 cm^{-1} (4.95 μ), 1974 cm^{-1} (5.06 μ), and 2042 cm^{-1} (shoulder) (4.90 μ), carbon tetrachloride solution].[40]

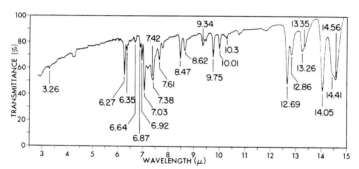

FIG. 2.4. Infrared spectrum of crystalline tetraphenylcyclobutadienepalladium chloride potassium bromide disc) calibrated against polystyrene.

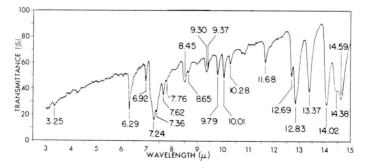

FIG. 2.5. Infrared spectrum of crystalline tetraphenylcyclobutadienenickel bromide (potassium bromide disc) calibrated against polystyrene.

3. Nuclear Magnetic Resonance

The NMR spectrum of tetramethylcyclobutadienenickel chloride in aqueous solution exhibits only one peak[7, 17a] (value not recorded in the literature), indicating that all four methyl groups are equivalent. Similarly, the NMR spectrum of tetraphenylcyclobutadienepalladium chloride shows the presence of aromatic hydrogens only[6] (value not recorded in the literature).

4. Structure

An X-ray crystallographic structure analysis of the benzene-containing solvate of tetramethylcyclobutadienenickel chloride $[(C_8H_{12}NiCl_2)_2 \cdot C_6H_6]$ has shown that the molecule is a dimer.[21] Each nickel atom has a cyclobutadiene carbocycle on one side and three chlorine atoms on the other, two of which are shared as shown in Fig. 2.6. The cyclobutadiene ring is planar and square (or very nearly so) with C—C bond distances of 1.40–1.45 Å. The methyl groups are displaced outwards (i.e., away from the nickel atom) by 0.12–0.19 Å,

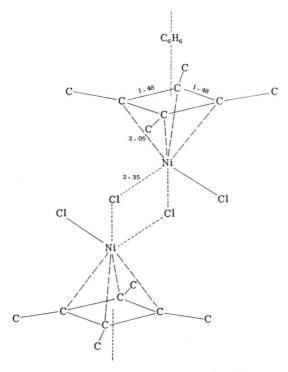

Fig. 2.6. Structure and bond distances of tetramethylcyclobutadienenickel chloride benzene solvate (bond distances in Ångstrom units) [after Dunitz et al.[21]].

probably as a result of steric interference with the chlorine atoms. The benzene rings lie between, and approximately parallel to, pairs of C_8H_{12} groups belonging to different molecules. The crystalline solvate loses benzene rapidly in air but is stable for several days in sealed glass tubes. Its unit cell is monoclinic, $a=8.311$, $b=11.944$, $c=12.715$ Å; $\beta=103°2'$; space group $P2_1/c$. Similarly, crystalline tetramethylcyclobutadienenickel chloride $[(C_8H_{12}NiCl_2)_2]$ is monoclinic, $a=16.01$, $b=8.01$, $c=16.45$ Å; $\beta=111°5'$; $Z=8$; space group $P2_1/n$.

An X-ray crystallographic analysis of tetraphenylcyclobutadieneiron tricarbonyl gave comparable results: The cyclobutadiene ring is planar and square within the error limitations of the method.[20]

References

1. Avram, M., Fritz, H. P., Keller, H., Mateescu, G., McOmie, J. F. W., Sheppard, N., and Nenitzescu, C. D., *Tetrahedron* **19**, 187 (1963).
2. Avram, M., Marica, E., and Nenitzescu, C. D., *Ber.* **92**, 1088 (1959).
3. Avram, M., Mateescu, G., Dinulescu, I. G., Marica, E., and Nenitzescu, C. D., *Tetrahedron Letters* **No. 1**, 21 (1961).
4. Bergmann, E. D., "The Chemistry of Acetylene and Related Compounds," Interscience, New York (1948), p. 93.
5. Bieber, T. I., *Chem. Ind. (London)* 1126 (1957).
6. Blomquist, A. T., and Maitlis, P. M., *J. Am. Chem. Soc.* **84**, 2329 (1962).
7. Boston, J. L., Sharp, D. W. A., and Wilkinson, G., *J. Chem. Soc.* 3488 (1962).
8. Braye, E. H., Hübel, W., Caplier, I., *J. Am. Chem. Soc.* **83**, 4406 (1961).
9. Cairns, T. L., Engelhardt, V. A., Jackson, H. L., Kalb, G. H., and Sauer, J. C., *J. Am. Chem. Soc.* **74**, 5636 (1952).
10. Coates, G. E., "Organo-Metallic Compounds," 2d ed., Wiley, New York (1960), p. 328.
11. Coffey, C. E., private communication.
12. (a) Cookson, R. C., and Jones, D. W., *J. Chem. Soc.* 1881 (1965); (b) *ibid.*, *Proc. Chem. Soc.* 115 (1963).
13. (a) Criegee, R., *Angew. Chem.* **74**, 703 (1962); *ibid.*, *Rev. Chim. Acad. Rep. Populaire Roumaine* **7**, 771 (1962); (b) *ibid.*, *Bull. Soc. Chim. (France)* 1 (1965).
14. Criegee, R., Förg, F., Brune, H.-A., and Schröder, G., *Ber.* **97**, 3461 (1964).
15. Criegee, R., and Louis, G., *Ber.* **90**, 417 (1957).
16. Criegee, R., and Ludwig, P., *Ber.* **94**, 2038 (1961).
17. (a) Criegee, R., and Schröder, G., *Ann.* **623**, 1 (1959); (b) *ibid.*, *Angew. Chem.* **71**, 70 (1959).
18. Criegee, R., Schröder, G., Maier, G., Fischer, H.-G., *Ber.* **93**, 1553 (1960).
19. (a) Dahl, L. F., and Oberhansli, W. E., *Proc. 8th Int. Conference on Coordination Chem.* Vienna (1964), p. 242; (b) Dahl, L. F., private communication cited in ref. 42.
20. Dodge, R. P., and Schomaker, V., *Nature* **186**, 798 (1960).
21. (a) Dunitz, J. D., Mez, H. C., Mills, O. S., and Shearer, H. M. M., *Helv. Chim. Acta* **45**, 647 (1962); (b) Dunitz, J. D., Mez, H. C., Mills, O. S., Pauling, P., and Shearer, H. M. M., *Angew. Chem.* **72**, 755 (1960).
22. Erdmann, H., and Köthner, P., *Z. anorg. Chem.* **18**, 48 (1898).
23. Fitzpatrick, J. D., Watts, L., Emerson, G. F., and Pettit, R., *J. Am. Chem. Soc.* **87**, 3255 (1965).

24. Freedman, H. H., *J. Am. Chem. Soc.* **83**, 2195 (1961).
25. Freedman, H. H., *J. Am. Chem. Soc.* **83**, 2194 (1961).
26. (a) Fritz, H.-P., *Z. Naturforsh.* **16B**, 415 (1961); (b) *ibid.*, private communication.
27. Fritz, H.-P., in Stone, F. G. A., and West, R. (eds.) "Advances in Organometallic Chemistry," Academic Press, New York (1964), Vol. 1, p. 260.
28. Fritz, H.-P., McOmie, J. F. W., and Sheppard, W., *Tetrahedron Letters* No. **26**, 35 (1960).
29. Gerloch, M., and Mason, R., *Proc. Royal Soc.* **A279**, 170 (1964).
30. Hübel, W., private communication.
31. Hübel, W., and Braye, E. H., *J. Inorg. Nucl. Chem.* **10**, 250 (1959).
32. Hübel, W., Braye, E. H., and Claus, A., Weiss, E., Kruerke, U., Brown, D. A., King, G. S. D., and Hoogzand, C., *J. Inorg. Nucl. Chem.* **9**, 204 (1959).
33. Hüttel, R., and Neuegebauer, H. J., *Tetrahedron Letters* 3541 (1964).
34. Jones, E. R. H., Wailes, P. C., and Whiting, M. C., *J. Chem. Soc.* 4021 (1955).
35. Kruerke, U., and Hübel, W., *Ber.* **94**, 2829 (1961).
36. Longuet-Higgins, H. C., and Orgel, L. E., *J. Chem. Soc.* 1969 (1956).
37. Ludwig, P., and Noll, K., unpublished work, cited in ref. 13.
38. Maitlis, P. M., private communication.
39. Maitlis, P. M., and Games, M. L., *Can. J. Chem.* **42**, 182 (1964).
40. Maitlis, P. M., and Games, M. L., *Chem. Ind. (London)* 1624 (1963).
41. Maitlis, P. M., and Games, M. L., *J. Am. Chem. Soc.* **85**, 1887 (1963).
42. Maitlis, P. M., Pollock, D., Games, M. L., and Pryde, W. J., *Can. J. Chem.* **43**, 470 (1965).
43. Maitlis, P. M., and Stone, F. G. A., *Proc. Chem. Soc.* 330 (1962).
44. Makowa, O., *Ber.* **41**, 824 (1908).
45. (a) Malatesta, L., Santarella, G., Vallarino, L., and Zingales, F., *Accad. Naz. Lincei Rend. Classe Fis. Matemat. Naturali* **27**, 230 (1959); (b) *ibid.*, *Angew. Chem.* **72**, 34 (1960).
46. Martologu, N., and Mumuianu, D., *Acad. Rep. Populaire Roumaine Rev. Chim.* **6**, 303 (1961).
47. Nakamura, A., Address, presented before the Symposium of Organic Synthetic Chemistry, Tokyo, Nov. 1961.
48. (a) Nakamura, A., and Hagihara, N., *Nippon Kagaku Zasshi* **84**, 339 (1963); (b) *ibid.*, *Bull. Chem. Soc. Japan* **34**, 452 (1961).
49. Orgel, L. E. "An Introduction to Transition-Metal Chemistry," Wiley, New York (1960), p. 150.
50. Pfrommer, J. F., Ph.D. diss., Karlsruhe (1961).
51. Philipps, F. C., *Z. anorg. Chem.* **6**, 229 (1894).
52. Reppe, W., *Publications Board of the U.S. Dept. of Commerce*, 46996 and 62593, cited in Copenhauer, J. W., and Bigelow, M. H., "Acetylene and Carbon Monoxide Chemistry," Reinhold, New York (1949).
53. (a) Reppe, W., Schlichting, O., Klager, K., and Toepel, T., *Ann.* **560**, 1 (1948); (b) Reppe, W., "Neue Entwicklungen auf dem Gebiete der Chemie des Acetylens und Kohlenoxyds," Springer, Berlin (1949), p. 68.
54. Reppe, W., and Vetter, H., *Ann.* **582**, 133 (1953).
55. Sauer, J. C., and Cairns, T. L., *J. Am. Chem. Soc.* **79**, 2659 (1957).
56. Schrauzer, G. N., *Ber.* **94**, 1403 (1961).
57. Schrauzer, G. N., *J. Am. Chem. Soc.* **81**, 5310 (1959).
58. Schrauzer, G. N., *Chem. Ind. (London)* 1404 (1958).
59. Schrauzer, G. N., *Chem. Ind. (London)* 1403 (1958).
60. (a) Schrauzer, G. N., and Eichler, S., *Ber.* **95**, 550 (1962); (b) *ibid.*, *Angew. Chem.* **73**, 546 (1961).

61. Temkin, O. N., Brailovskii, S. M., Flid, R. M., Sturkova, M. P., Belyanin, V. B., and Zaitseva, M. G., *Kinetika i Kataliz* **5**, 192 (1964).
62. Tsutsui, M., *Chem. Ind. (London)* 780 (1962).
63. Tsutsui, M., and Zeiss, H., *J. Am. Chem. Soc.* **82**, 6255 (1960).
64. Watts, L., Fitzpatrick, J. D., and Pettit, R., *J. Am. Chem. Soc.* **87**, 3253 (1965).
65. (a) Wender, I., Friedel, R. A., Markby, R., Sternberg, H. W., *J. Am. Chem. Soc.* **77**, 4946 (1955); (b) Sternberg, H. W., Friedel, R. A., Markby, R., and Wender, I., *J. Am. Chem. Soc.* **78**, 3621 (1956); (c) Sternberg, H. W., Markby, R., and Wender, I., *J. Am. Chem. Soc.* **80**, 1009 (1958).
66. Wilkinson, G., private communication.
67. Zeiss, H., in Zeiss, H. (ed.), "Organometallic Chemistry," Reinhold, New York (1960), p. 380 *ff*.
68. Zingales, F., *Ann. Chim. (Rome)* **52**, 1174 (1962).

CHAPTER 3

Cyclobutadiene Divalent Ions

Cyclobutadiene   Cyclobutadiene
   dication   dianion

The removal of two π electrons from the cyclobutadiene molecule results in the formation of the cyclobutadiene dication, $C_4H_4^{2+}$, a delocalized 2π-electron system which is electronically related to the stable cyclopropenium cation, $C_3H_3^+$. Similarly, addition of two π electrons to the cyclobutadiene molecule should afford the cyclobutadiene dianion, $C_4H_4^{2-}$, a delocalized 6π-electron system which is isoelectronic with the stable cyclopentadiene anion C_5H_5. Although the parent cyclobutadiene dication is unknown, strong experimental evidence has been obtained for the existence of several of its derivatives. In contrast, no positive evidence has yet been found for the existence of a cyclobutadiene dianion.

The Hückel orbitals for cyclobutadiene predict equal resonance energy for the dianion and the neutral molecule. However, since the double negative charge of the dianion will reduce the effective electronegativity of the carbocyclic system, the energy of the π orbitals will be raised and the highest filled orbitals will become nonbonding.[10] In other words, the dianion must be expected to be *less* stable than cyclobutadiene itself. In contrast, the dication may be expected to be *more* stable than the neutral molecule because its double positive charge will increase the effective electronegativity of the carbocyclic system and hence will lower the energy of the bonding orbitals.

In addition, the highest π orbitals of the dication are not degenerate as they are in the neutral molecule and consequently would not be subject to distortion by Jahn–Teller forces. (See also Chapter 12.)

A. History

There is no early history of attempts to prepare cyclobutadiene divalent ions. Hart and Fish (1960) appear to have been the first to point out that a cyclobutene bearing two positive charges might have some aromaticity.[7]

B. Cyclobutadiene Dication

Only two examples of the cyclobutadiene dication are known at the present time, viz., 1,3-dihydroxy-2,4-diphenylcyclobutadiene dication and tetraphenylcyclobutadiene dication, both of which are known only in solution.

1. Synthesis of Cyclobutadiene Dications

a. By Acid Treatment of a Halocyclobutenone

1,3-Dihydroxy-2,4-diphenylcyclobutadiene dication (2) is formed when 4-bromo-3-hydroxy-2,4-diphenylcyclobutenone (1) is dissolved in 96% sulfuric acid.[4] The reaction probably proceeds by protonation of the carbonyl oxygen, followed by solvolytic loss of bromide ion.

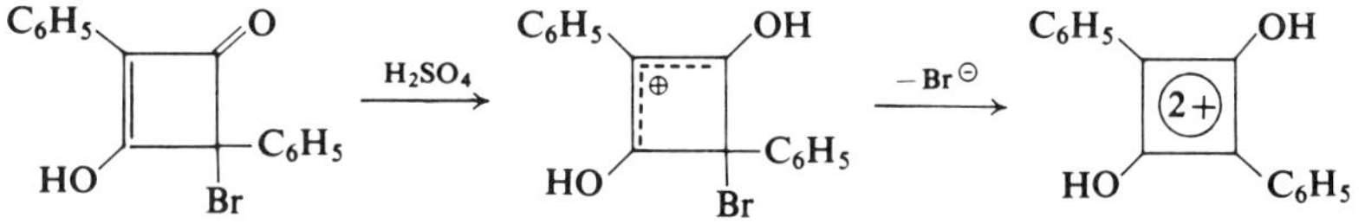

b. By Reaction of Silver Ion with a 3,4-Dihalocyclobutene

The reaction of 3,4-dibromo-1,2,3,4-tetraphenylcyclobutene (3) with two equivalents of silver fluoborate in methylene chloride solution affords tetraphenylcyclobutadiene dication (4) as its fluoborate salt. An almost quantitative precipitation of two equivalents of silver bromide is produced.[6]

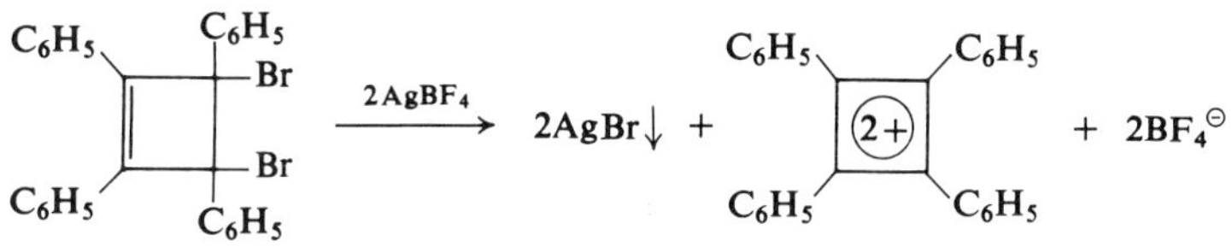

2. Unsuccessful Attempts to Prepare Cyclobutadiene Dications

Even though cyclobutadiene dication is undoubtedly an electronically favorable cyclic 2π-electron system, an appreciable destabilizing effect can be attributed to the two positive charges, distributed as they are over only four carbon atoms. Indeed, the fact that methods which are applicable to the synthesis of phenyl-substituted cyclobutadiene dications are unsuccessful when applied to the analogous methyl-substituted dications (see below), confirms the destabilizing effect of the two positive charges, since the phenyl substituents are more effective in delocalizing the charges.

The reaction of 3,4-dichloro-1,2,3,4-tetramethylcyclobutene (6) with silver hexafluoroantimonate in sulfur dioxide solution affords 4-chloro-1,2,3,4-tetramethylcyclobutenyl cation (8), rather than tetramethylcyclobutadiene dication (5).[9] Similarly, reaction of dichloride (6) or the corresponding dibromide (7) with aluminum chloride in methylene chloride solution gives only the corresponding monocations (8) and (9).[8] The latter [(8) and (9)] have been characterized unambiguously by their NMR spectra, which show clearly that each contains methyl groups in three different environments.[8, 9]

An NMR study has shown that 1,3-dihydroxy-2,4-dimethylcyclobutadiene dication (12) is not formed when either 4-bromo-3-hydroxy-2,4-dimethylcyclobutenone (10) or 3,4-dihydroxy-2,4-dimethylcyclobutenone (11) is treated with 96% sulfuric acid. Indeed, bromoketone (10) is completely resistant to solvolysis in 96% sulfuric acid over a period of several days at room temperature.[3] Bromoketone (10), its chloro analog (13), and the methyl ether (14) of the bromoketone all failed to react with stannic chloride or antimony pentachloride to give an α-ketocarbonium ion [(15) or (16)], which might have been stabilized through a cyclobutadiene dication resonance contributor [(15a) or (16a)].[3]

3,4-Dibromo-1,2,3,4-tetraphenylcyclobutene (3) reacts with stannic chloride in benzene solution to give a brick-red crystalline product having the composition $C_{28}H_{20}Cl_6Sn$. The product was at first believed to be the hexachlorostannate salt ($SnCl_6^{2-}$) of the tetraphenylcyclobutadiene dication (4),[5] but a subsequent X-ray structure analysis demonstrated that it is actually the pentachlorostannate salt ($SnCl_5^-$) of 4-chlorotetraphenylcyclobutenyl cation

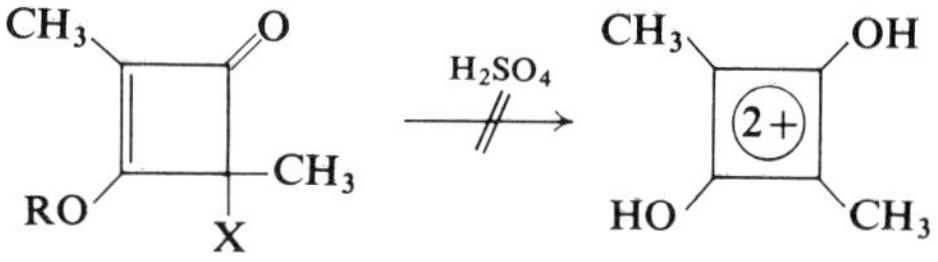

B. CYCLOBUTADIENE DICATION

(10): X=Br; R=H
(11): X=OH; R=H
(13): X=Cl; R=H
(14): X=Br; R=CH₃

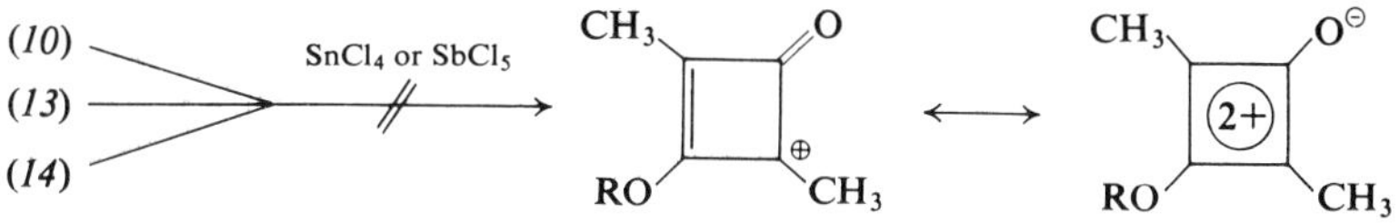

(15): R=H
(16): R=CH₃

(15a): R=H
(16a): R=CH₃

(19).[2] It is not possible to state with certainty, however, that solutions of monocation (19) are not in equilibrium with dication (4). Thus, a solution of the red salt in methylene chloride shows only low-field aromatic protons

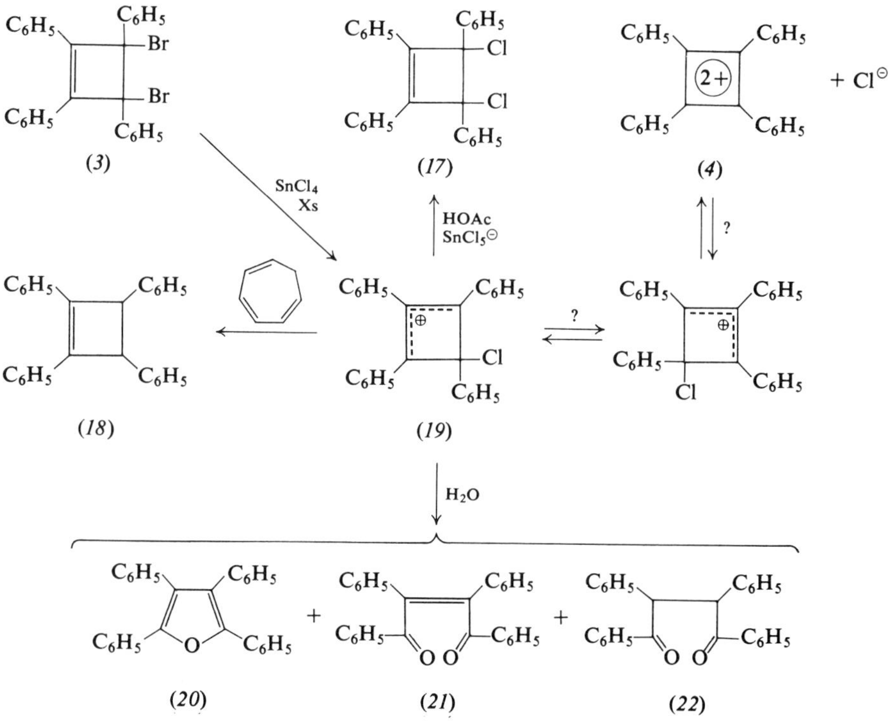

($\delta > 7.5$). This observation appears to support the existence of the totally symmetrical dication (*4*) in solution; the observed spectrum, however, could result from rapid tautomerism of monocation (*19*).[6] A methylene chloride solution of the red salt also undergoes several chemical reactions which are attributable to either dication (*4*) or monocation (*19*): Cycloheptatriene reacts with the red salt to give, by hydride exchange, 1,2,3,4-tetraphenylcyclobutene (*18*); acetic acid gives a compound believed to be 3,4-dichloro-1,2,3,4-tetraphenylcyclobutene (*17*); and hydrolysis affords a mixture of tetraphenylfuran (*20*), *cis*-dibenzoylstilbene (*21*), and didesyl (*22*).[5]

3,4-Dibromo-1,2,3,4-tetraphenylcyclobutene (*3*) dissolves in 96% sulfuric acid to give a deep red solution (λ_{max} 482 mμ, $\epsilon = 50,000$).[5] It is not possible to state unequivocably that the major ionic species present in the solution is the dication (*4*) and not the monocation (*23*). Dilution of the acid solution with water affords "bromine-free products characteristic of the dication."[5]

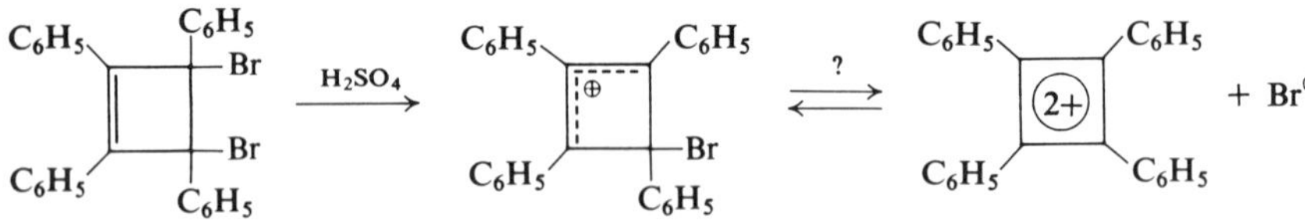

3. Chemical and Physical Properties of a Cyclobutadiene Dication

A solution of 1,3-dihydroxy-2,4-diphenylcyclobutadiene dication (*2*) in 96% sulfuric acid is colored deep red and shows a strong absorption maximum at 487 mμ ($\epsilon = 144,000$). The NMR spectrum of dication (*2*) shows only bands typical of benzylic carbonium ions, the ortho protons appearing at $\delta = 8.5$ and the other aromatic protons appearing at $\delta = 8.0$. Dilution of the sulfuric acid solution of (*2*) with cold water causes the solvolysis of (*2*) to 3,4-dihydroxy-2,4-diphenylcyclobutenone (*24*).[4]

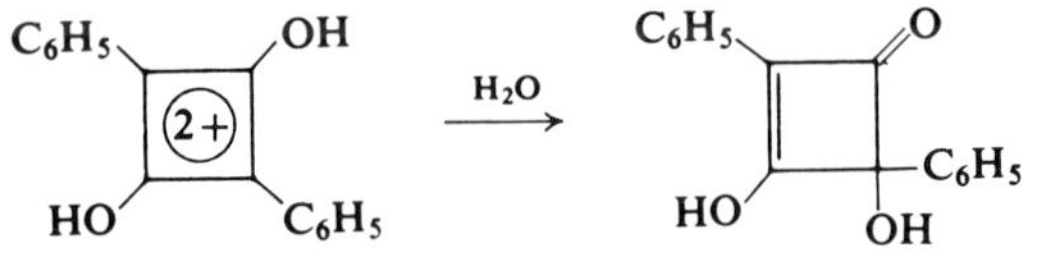

C. Cyclobutadiene Dianion

Simple molecular orbital theory predicts that cyclobutadiene dianion (*1*) should be a stable cyclic 6π-electron system but the theory does not take into account the electronic repulsion terms, which can be the source of considerable destabilization.[1]

A number of unsuccessful attempts have been made to generate tetramethylcyclobutadiene dianion (5) from 3,4-dichloro-1,2,3,4-tetramethylcyclobutene (2) or the corresponding dibromide (3) and diiodide (4).[1] The reaction of dihalides (2), (3), and (4) with *n*-butyllithium under a variety of conditions, followed by decomposition with methanol, does *not* give 1,2,3,4-tetramethylcyclobutene (6), which is the expected protonation product of dianion (5). On the other hand, the reaction of dichloride (2) with lithium in liquid ammonia, followed by treatment with deuterium oxide, gave tetramethylcyclobutene (6) among other products, but none of the expected deuterated tetramethylcyclobutene (7).[1] At best, the latter experiment indicates that dianion (5), if it is produced in the initial reaction, is unstable and is strong enough as a base to abstract two protons from the liquid ammonia solvent.

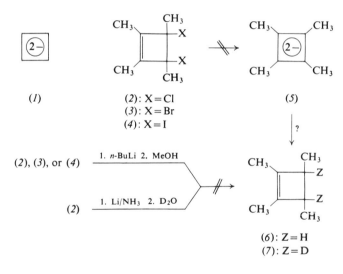

References

1. Adam, W., *Tetrahedron Letters* 1387 (1963).
2. Bryan, R. F., *J. Am. Chem. Soc.* **86**, 733 (1964).
3. (a) Farnum, D. G., Heybey, M. A. T., and Webster, B., *J. Am. Chem. Soc.* **86**, 673 (1964); (b) *ibid.*, *Tetrahedron Letters* 307 (1963).
4. Farnum, D. G., and Webster, B., *J. Am. Chem. Soc.* **85**, 3502 (1963).
5. Freedman, H. H., and Frantz, A. M., Jr., *J. Am. Chem. Soc.* **84**, 4165 (1962).
6. Freedman, H. H., and Young, A. E., *J. Am. Chem. Soc.* **86**, 734 (1964).
7. Hart, H., and Fish, R. W., *J. Am. Chem. Soc.* **82**, 5419 (1960) and ref. 24 cited therein.
8. Katz, T. J., and Gold, E. H., *J. Am. Chem. Soc.* **86**, 1600 (1964).
9. Katz, T. J., Hall, J. R., and Neikam, W. C., *J. Am. Chem. Soc.* **84**, 3199 (1962).
10. Waack, R., private communication.

CHAPTER 4

Cyclobutadienequinone*

The parent compound, cyclobutadienequinone, is not known but a number of substituted cyclobutadienequinones have been prepared and have been found to possess considerable stability. In a formal sense, these compounds are the stable four-membered-ring quinones of the corresponding highly unstable cyclobutadienes, but in actuality they have none of the chemical properties of true quinones. Since the cyclobutadienes and the cyclobutadienequinones have approximately the same ring strain, the remarkable difference in stability between the two classes must be attributed to electronic factors.†

The chemistry of phenylcyclobutadienequinone has been reviewed by Roberts.[14]

A. History

There is no early history of attempts to prepare cyclobutadienequinone or substituted cyclobutadienequinones. The first such compound, phenylcyclobutadienequinone (2), was prepared only recently, in 1955, by Smutny and Roberts by the hydrolysis of 4-chloro-1-phenyl-3,3,4-trifluorocyclobutene (1) in 92% sulfuric acid.[18]

* Cyclobutadienequinone is listed under "cyclobutenedione" in *Chemical Abstracts*.
† See Section D, and Chapter 12.

B. Synthesis of Cyclobutadienequinones

Cyclobutadienequinones have been prepared by two methods: (a) by the acid-catalyzed hydrolysis of tetrahalocyclobutenes, dihalocyclobutenones, and similar compounds and (b) by the transformation of substituted cyclobutadienequinones prepared by the first method.

1. Acid Hydrolysis of Halogenated Cyclobutenes and Related Compounds

The preparation of substituted cyclobutadienequinones by the acid-catalyzed hydrolysis of appropriately substituted 3,3,4,4-tetrahalocyclobutenes, 4,4-dihalocyclobutenones, and similar compounds appears to be a method of general applicability. Strong mineral acid (usually 50–90% sulfuric acid) is the most widely used hydrolytic medium, but in two instances, (a) the hydrolysis of 2-chloro-4,4-difluoro-1,3,3-trimethoxycyclobutene [13a] and (b) the hydrolysis of perchlorocyclobutenone,[10b] only water and autogenous hydrogen halide were used.*

The acid-hydrolysis method is useful as a synthetic tool for preparing cyclobutadienequinones because the tetrahalocyclobutenes, dihalocyclobutenones, etc., which are required as starting materials are available in good yield through the cycloaddition reaction.[15] Thus, in all but one of the examples described below, the synthetic precursor of the cyclobutadienequinone was obtained through cycloaddition of a fluorinated alkene to a substituted alkene or alkyne.

Phenylcyclobutadienequinone (2) can be prepared in good yield (75%) by the sulfuric acid hydrolysis of 4-chloro-1-phenyl-3,3,4-trifluorocyclobutene

---

* In the hydrolysis of polyfluorocyclobutenes, the strength of the sulfuric acid used, the temperature to which it is heated, and the period of time that the reaction is allowed to continue are all critical factors and can greatly affect the yield of product. For this reason, it is sometimes difficult to scale up a reaction using conditions that have been worked out with small quantities. But this is not a serious drawback, because the hydrolysis is usually over in a few minutes and so a large number of small runs can easily be made, one after the other. The serial technique sometimes confers an unexpected benefit, for it has been observed that the use of an etched flask (one that has been used in a previous hydrolysis) can improve the yield of product or at least make the reaction less erratic (refs. 2 and 4).

[(*1*), prepared by the cycloaddition of chlorotrifluoroethylene to phenylacetylene].[18] The quinone (*2*) can also be obtained in a similar manner, but in lower yield, from 1-phenyl-3,3,4,4-tetrafluorocyclobutene (*5*). The intermediates in these two hydrolysis reactions are 4-chloro-4-fluoro-3-phenylcyclobutenone (*3*) and 4,4-difluoro-3-phenylcyclobutenone (*4*), respectively.

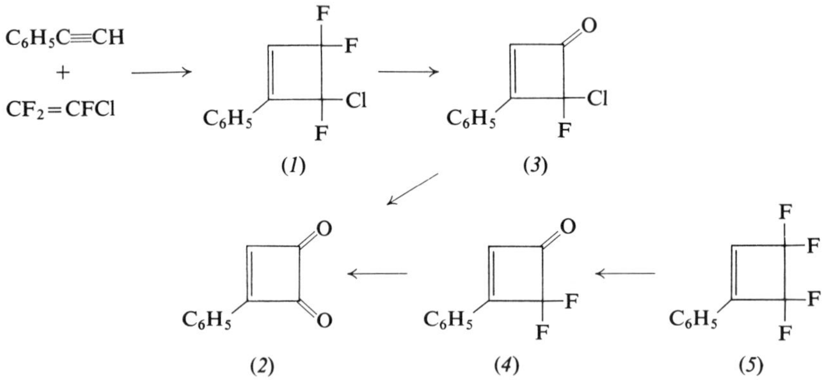

Intermediates (*3*) and (*4*) can be isolated when the tetrahalides are subjected to shorter periods of hydrolysis, and they can be converted in turn into phenylcyclobutadienequinone (*2*) by further treatment with sulfuric acid.[18] (See also Table 4.1.)

Phenylcyclobutadienequinone is also available by the sulfuric acid hydrolysis

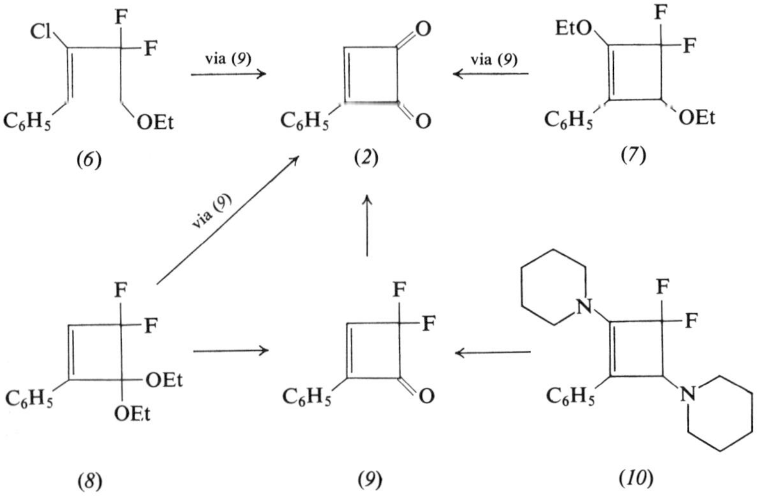

B. SYNTHESIS OF CYCLOBUTADIENEQUINONES

of the following ethoxycyclobutenes: 2-chloro-3,3-difluoro-4-ethoxy-1-phenyl-cyclobutene (6), 2,4-diethoxy-3,3-difluoro-1-phenylcyclobutene (7), and 3,3-difluoro-4,4-diethoxy-1-phenylcyclobutene (8).[8] It is likely that a single intermediate, 4,4-difluoro-2-phenylcyclobutenone (9), is common to all of these transformations; this view is supported by the fact that mild acid hydrolysis of ketal (8) affords the difluorocyclobutenone (9),[8] which is readily hydrolyzable to the cyclobutadienequinone. 4,4-Difluoro-2-phenylcyclobutenone was also prepared by mild acid hydrolysis of 3,3-difluoro-2,4-dipiperidino-1-phenyl-cyclobutene (10) and was converted in 79% yield into phenylcyclobutadienequinone (2) when heated with sulfuric acid.[18a]

3,4-Diphenylcyclobutadienequinone (13) was obtained by the sulfuric acid hydrolysis of 1,2-diphenyl-3,3,4,4-tetrafluorocyclobutene (12) [prepared by the reaction of phenyllithium with perfluorocyclobutene (11)].[2]

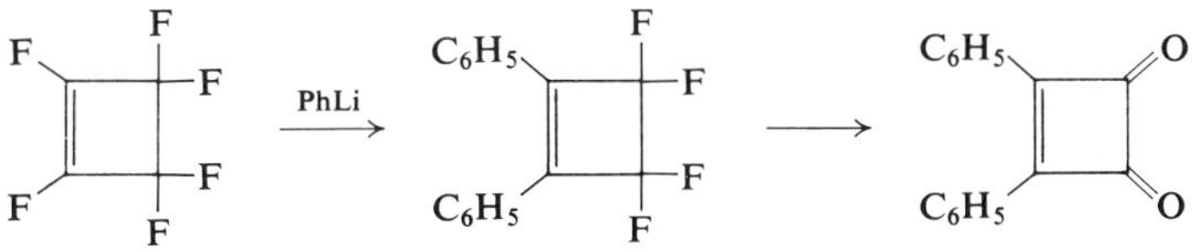

Similarly, acid hydrolysis of 4-chloro-1-(1'-cyclohexenyl)-3,3,4-trifluoro-cyclobutene (15) [prepared by the cycloaddition of 1-ethynylcyclohexene (14) to chlorotrifluoroethylene] gave 3-(1'-cyclohexenyl)cyclobutadienequinone (16).[16]

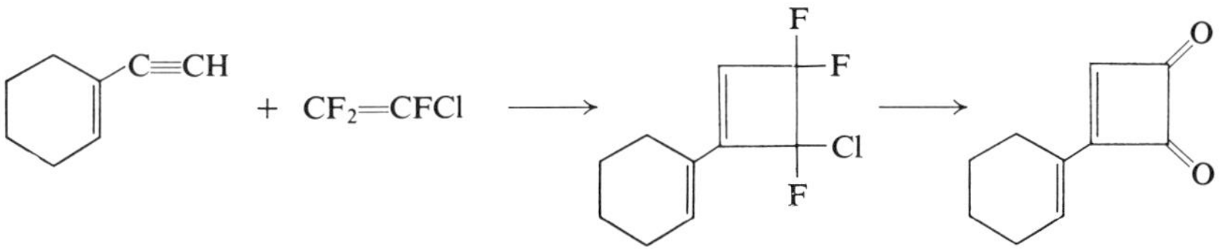

3,4-Dimethylcyclobutadienequinone (18) was obtained by the acid hydrolysis of 1,2-dimethyl-3,3,4,4-tetrafluorocyclobutene (17) [prepared by the reaction of methyllithium with perfluorocyclobutene (13)].[3]

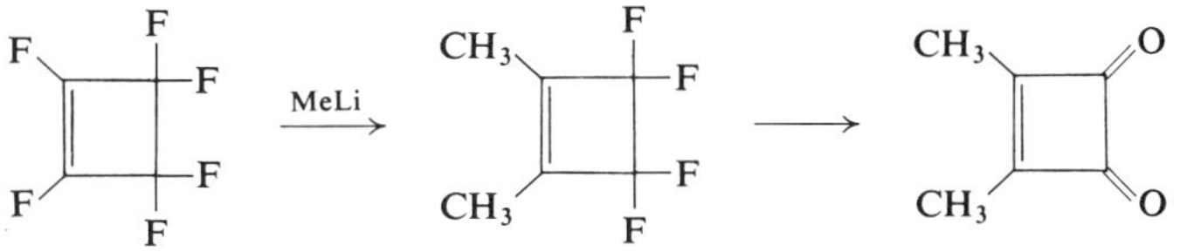

132   4. CYCLOBUTADIENEQUINONE

TABLE 4.1

SUBSTITUTED CYCLOBUTADIENEQUINONES PREPARED FROM 3,3,4,4-TETRAHALOCYCLOBUTENES, 4,4-DIHALOCYCLOBUTENONES AND SIMILAR COMPOUNDS

Cyclobutadienequinone	Reactants and reaction conditions	Yield %	Ref.
3-(1'-Cyclohexenyl)-	4-Chloro-1-(1'-cyclohexenyl)-3,3,4-trifluorocyclobutene and 97% sulfuric acid at 40°–45° for 10–15 minutes and 60°–65° for 20–30 minutes	23–41	16
3,4-Dibutoxy-	Perchlorocyclobutenone and more than 4 moles of butyl alcohol	—	10a
3,4-Dihydroxy-	1-Chloro-3,3-difluoro-2,4,4-triethoxycyclobutene and water at 101° for 5 hours	20[a]	13a
	1-Chloro-3,3-difluoro-2,4,4-trimethoxycyclobutene and hot hydrochloric acid	85	20
	1-Chloro-3,3-difluoro-2,2,4,4-tetramethoxycyclobutane and 70% sulfuric acid at 100° for 12 hours	70	20
	1,2-Diethoxy-3,3,4,4-tetrafluorocyclobutene and 50% sulfuric acid at 100° for 12 hours	94	13
	Perchlorocyclobutenone and		
	(1) a large volume of water at 100°	35	10a
	(2) 20% sulfuric acid at 100° without stirring	57	10a
	(3) 20% sulfuric acid at reflux temp. with rapid stirring	84–87	10b
	(4) 98% sulfuric acid at 110°, with dropwise addition of a calculated amount of water	90	10a
3,4-Dimethyl-	1,2-Dimethyl-3 3,4,4-tetrafluorocyclobutene and		
	(1) orthophosphoric acid	Poor	3
	(2) polyphosphoric acid at 100°–105° for 1.5 hours	43	3
	(3) 96–97% sulfuric acid at 65°–70° for 70 minutes	60	3

Product	Starting material and conditions	Yield (%)	Ref.
3,4-Diphenyl-	1,2-Diphenyl-3,3,4,4-tetrafluorocyclobutene and 98% sulfuric acid at 98° for 35 minutes	83.5	2
3-Phenyl-	4-Chloro-1-phenyl-3,3,4-trifluorocyclobutene and 92% sulfuric acid at 100°	75	18
	2-Chloro-3,3-difluoro-4-ethoxy-1-phenylcyclobutene and 1:4 water–sulfuric acid at 100° for 15 minutes	—	8
	4-Chloro-4-fluoro-3-phenylcyclobutenone and 90–92% sulfuric acid for 20 minutes on the steam bath	86	18
	1,3-Diethoxy-4,4-difluoro-2-phenylcyclobutene and 1:6 water–sulfuric acid at 100° for 3 minutes	—	8
	4,4-Diethoxy-3,3-difluoro-1-phenylcyclobutene and 1:4 water–sulfuric acid at 100° for 15 minutes	—	8
	4,4-Difluoro-2-phenylcyclobutenone and sulfuric acid at 100° for 3 minutes	—	7
	4,4-Difluoro-3-phenylcyclobutenone and sulfuric acid at 95° for 30 minutes	79[a]	18
	1-Phenyl-3,3,4,4-tetrafluorocyclobutene and sulfuric acid at 98° for 15 minutes	1[a]	8

[a] Calculated from weight data recorded in the literature.

Dihydroxycyclobutadienequinone (22), which is also known by the trivial name "squaric acid," was prepared by the hydrolysis of (a) 1,2-diethoxy-3,3,4,4-tetrafluorocyclobutene (19) in 50% sulfuric acid and (b) 2-chloro-4,4-difluoro-1,3,3-triethoxycyclobutene (21) in water.[13] Subsequently, (22) was also prepared by the oxidative acid hydrolysis of 2-chloro-4,4-difluoro-1,1,3,3-tetramethoxycyclobutane (20) and by the acid hydrolysis of 2-chloro-4,4-difluoro-1,3,3-trimethoxycyclobutene (23).[20] The latter is probably the most convenient synthesis of squaric acid because (23) is easily prepared by the action of methanolic potassium hydroxide on 1,2-dichloro-3,3,4,4-tetrafluorocyclobutene (24).[20]

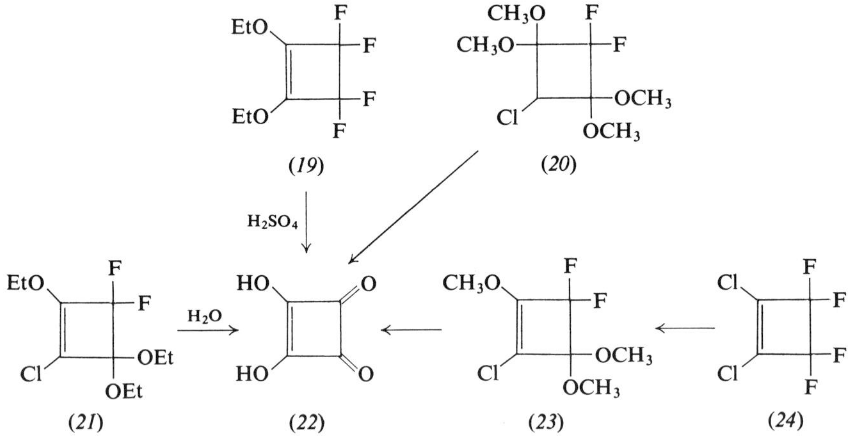

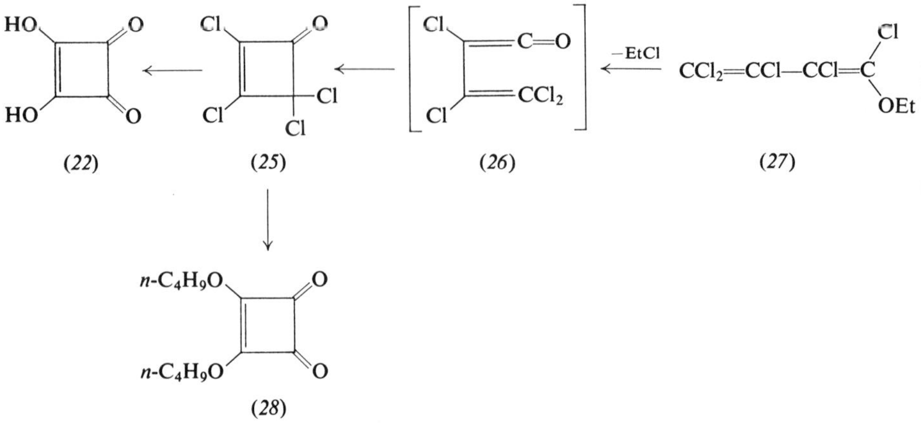

B. SYNTHESIS OF CYCLOBUTADIENEQUINONES 135

Squaric acid (22) has also been obtained by the acid hydrolysis of perchlorocyclobutenone (25).[10] The dibutyl "ester" of squaric acid, dibutoxycyclobutadienequinone (28), was prepared by treating perchlorocyclobutenone with butanol.[10] Perchlorocyclobutenone is unique as a cyclobutadienequinone precursor in that it is not obtained by the bimolecular cycloaddition reaction of a fluoroolefin, but rather by the pyrolysis of 1-ethoxypentachloro-1,3-butadiene (27) at 200°.[10] It appears quite likely, however, that perchlorocyclobutenone is formed by the intramolecular cycloaddition of an unsaturated ketene intermediate [(27) → (26) → (25)].

2. TRANSFORMATION OF CYCLOBUTADIENEQUINONES

A number of substituted phenylcyclobutadienequinones have been prepared from phenylcyclobutadienequinone by transformation reactions (substitution, displacement, ether formation, etc.) (Table 4.2.) Thus, the halogenation of phenylcyclobutadienequinone (2) with chlorine or bromine in acetic acid gives 4-chloro- (29) and 4-bromophenylcyclobutadienequinone (31) respectively.[18] (No reaction takes place when the solvent is carbon tetrachloride.)

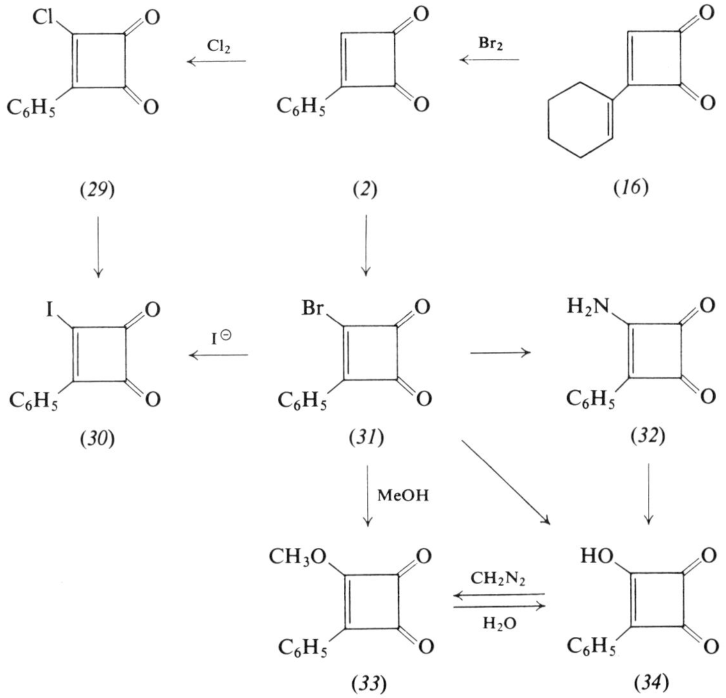

TABLE 4.2

SUBSTITUTED CYCLOBUTADIENEQUINONES PREPARED BY TRANSFORMATION REACTIONS

Cyclobutadienequinone	Reactants and reaction conditions	Yield (%)	Ref.
3-Amino-4-phenyl-	1-Bromo-2-phenylcyclobutadienequinone and ammonia in dry benzene	82	18
3-Bromo-4-phenyl-	3-Phenylcyclobutadienequinone and bromine in acetic acid–acetic anhydride	53	18
3-Chloro-4-phenyl-	3-Phenylcyclobutadienequinone and chlorine in acetic acid	70	18
3,4-Dibutoxy-	3,4-Dihydroxycyclobutadienequinone and more than 2 moles of butyl alcohol	—	10
3-Hydroxy-4-phenyl-	3-Amino-4-phenylcyclobutadienequinone and 10% aqueous sodium hydroxide[a]	—	18
	3-Methoxy-4-phenylcyclobutadienequinone and hot water	82	18
	3-Phenylcyclobutadienequinone and bromine in acetic acid followed by water and heat	82	18
3-Iodo-4-phenyl-	3-Bromo-4-phenylcyclobutadienequinone and sodium iodide in acetone	—	18
3-Methoxy-4-phenyl-	3-Bromo-4-phenylcyclobutadienequinone and hot methanol	71	18
	3-Hydroxy-4-phenylcyclobutadienequinone and diazomethane	83[b]	18

[a] The 3-hydroxy-4-phenylcyclobutadienequinone was not isolated. Its presence in the reaction product was inferred from a positive ferric chloride test.
[b] Calculated from weight data recorded in the literature.

The halogen atoms of quinones (29) and (31) are highly reactive and are readily replaced by other atoms or groups. Thus, iodide ion converts both the chloroquinone and the bromoquinone into 4-iodophenylcyclobutadienequinone (30). Similarly, the bromoquinone reacts with ammonia to give 4-aminophenylcyclobutadienequinone (32), with methanol to give 4-methoxyphenylcyclobutadienequinone (33), and with aqueous acetic acid to give 4-hydroxyphenylcyclobutadienequinone (34).[18] The hydroxyquinone (34) was also obtained by hydrolysis of the methoxyquinone (33) with water and by hydrolysis of the aminoquinone (32) with aqueous base. The methoxyquinone (33) was also prepared by methylation of the hydroxyquinone (34) with diazomethane.[18]

In a different kind of transformation reaction, i.e., one not involving the four-membered ring, cyclohexenylcyclobutadienequinone (16) was aromatized by the action of bromine to give phenylcyclobutadienequinone (2) in low yield (1%).[16] [Bromoquinone (31) is the expected product from this reaction, but its formation was not reported.] (See also Table 4.2.)

C. Unsuccessful Attempts to Prepare Cyclobutadienequinones

A not unreasonable approach to the synthesis of cyclobutadienequinones involves the dehydrochlorination of substituted succinoyl chlorides (1) to the corresponding bis(ketenes) (2), which would be expected to undergo ring-closure to form the tautomeric cyclobutadienequinones (3). However, two attempts to effect such a synthesis have been unsuccessful: diphenylcyclobutadienequinone (5) was not obtained when α,α'-diphenylsuccinoyl chloride (4) was treated with triethylamine.[2] Similarly, an expected bicyclic quinone (6) was not formed when *cis*-hexahydrophthaloyl chloride (7) was treated with triethylamine, and only a bis(ketene) dimer (8) could be isolated from the reaction mixture (6% yield).[9]

An attempt to prepare 3-cyclohexylcyclobutadienequinone (9) by the catalytic reduction of 3-(1'-cyclohexenyl)cyclobutadienequinone (10) gave

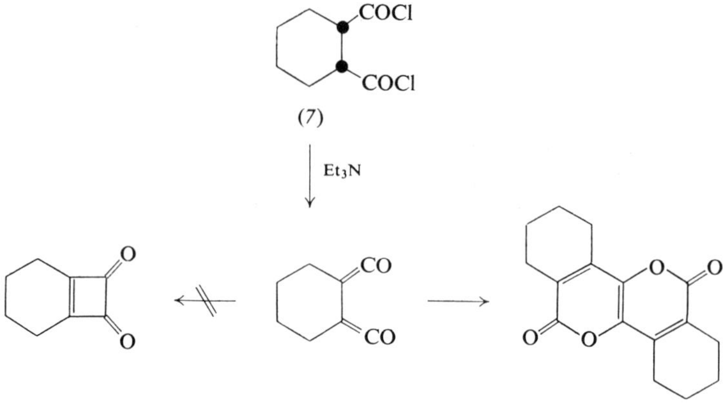

2-hydroxy-3-cyclohexenylcyclobutenone (*11*), instead. The mechanism of this unusual reduction has not been elucidated. The bromination of phenylcyclobutadienequinone, which is successful when carried out in acetic acid [(*13*) → (*12*)], could not be effected in carbon tetrachloride.[18] The bromination of 3-(1'-cyclohexenyl)cyclobutadienequinone (*10*) failed to give 3-bromo-4-(1'-cyclohexenyl)cyclobutadienequinone (*14*) and gave phenylcyclobutadienequinone (*13*), instead.

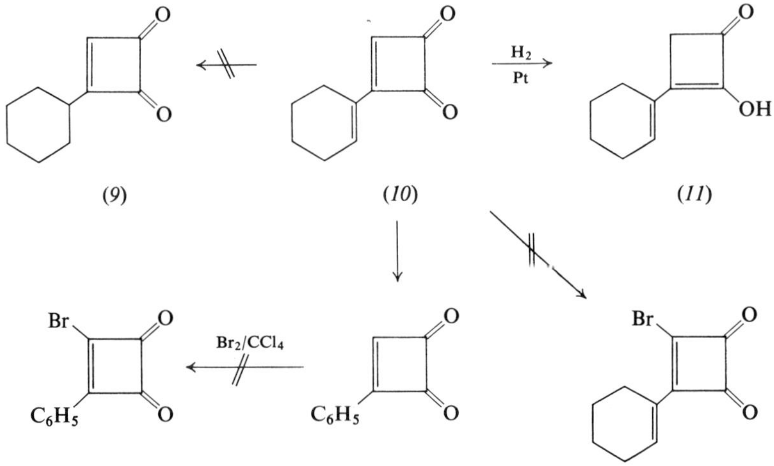

D. Chemistry of the Cyclobutadienequinones

Molecular orbital calculations predict that cyclobutadienequinone should have an appreciable delocalization energy (1.24 β or ~25 kcal/mole),[12] and that conjugative substituents at the 3- and the 4-positions should exert

D. CHEMISTRY OF THE CYCLOBUTADIENEQUINONES 139

an additional stabilizing influence on the system. The calculated delocalization energies (DE) of vinylcyclobutadienequinone, phenylcyclobutadienequinone, and diphenylcyclobutadienequinone are appreciably greater than the corresponding values ($\overline{DE}$) obtained by simply adding together the DE values of cyclobutadienequinone (1.24 β) and of the substituent group or groups.[12]

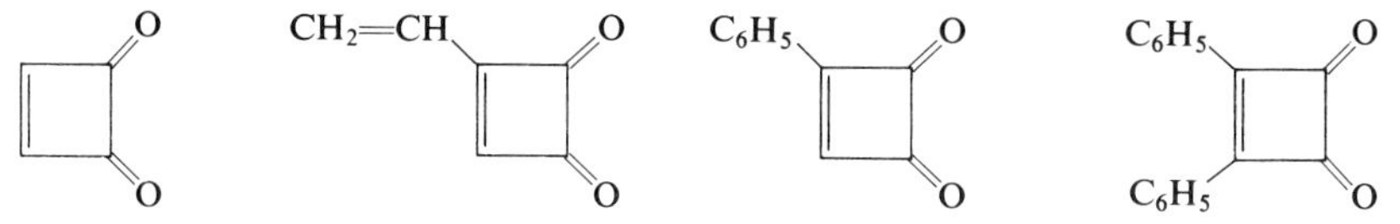

DE = 1.24 β

DE = 1.74 β
$\overline{DE}$ = 1.24 β + 0 = 1.24 β

DE = 3.68 β
$\overline{DE}$ = 1.24 β + 2 β
= 3.24 β

DE = 6.03 β
$\overline{DE}$ = 1.24 β + 4 β
= 5.24 β

CHART 4.1. Effect of conjugative substituents on the calculated delocalization energy of cyclobutadienequinone.

Nine canonical resonance forms can be written for cyclobutadienequinone. In addition to one uncharged structure (*1a*), two structures may be written which are of the electronically favored cyclobutadiene dication type [(*1b*) and

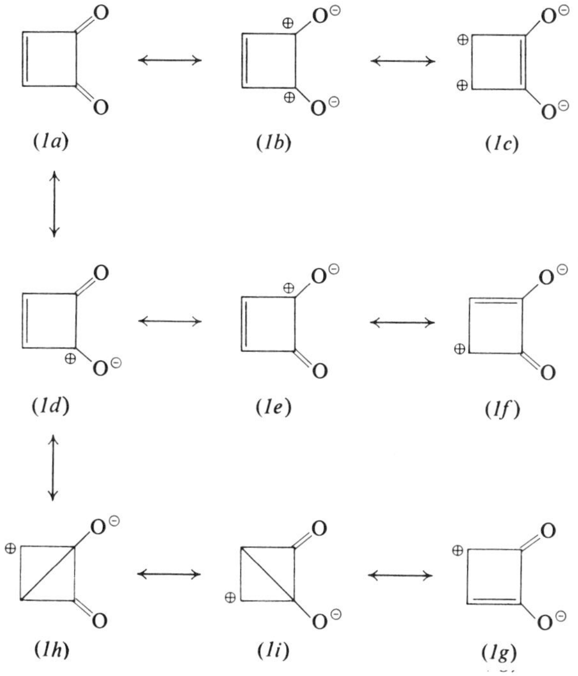

(*1a*)   (*1b*)   (*1c*)

(*1d*)   (*1e*)   (*1f*)

(*1h*)   (*1i*)   (*1g*)

(*1c*)].* The remaining major contributors to the hybrid [(*1d*), (*1e*), (*1f*), and (*1g*)] are all α-ketocationic forms, which are of lower energy than might otherwise be expected for a cyclobutenone ring.[5]

Two additional contributors to the ground state of cyclobutadienequinone can be envisaged [(*1h*) and (*1i*)]. In these, additional stabilization is achieved by electronic cross-ring interaction between the orbitals of the carbon atoms in the 1,3- and the 2,4-positions. When interactions of this type are invoked, the calculated delocalization energy of cyclobutadienequinone is increased from the value cited above (1.24 β) to a much higher value (1.99 β).[1b]

The chemistry of the cyclobutadienequinones may be conveniently divided into two categories: (a) cyclobutadienequinones substituted with unreactive groups such as alkyl and aryl groups and (b) cyclobutadienequinones substituted with labile or reactive groups such as the halogens, etc. In general, the alkyl- and aryl-substituted cyclobutadienequinones serve to illustrate the basic "quinone chemistry" of the ring system while the functionally substituted compounds illustrate its "acid chemistry." (In many instances, functionally substituted cyclobutadienequinones behave like vinylogous carboxylic acids and the derivatives thereof, i.e., like acid halides, amides, esters, and anions.)

1. REACTIONS OF ALKYL- AND ARYL-SUBSTITUTED CYCLOBUTADIENEQUINONES

a. Stability to Heat

Dimethyl- (*2*),[3] cyclohexenyl- (*3*),[16] phenyl- (*4*),[18] and diphenylcyclobutadienequinone (*5*)[2] are all stable at room temperature and can be purified by distillation or sublimation at moderate temperatures. Diphenylcyclobutadienequinone (*5*) decomposes at its melting point (152°–153°) with the evolution of carbon monoxide and the formation of polymeric material.[18]

b. Photolysis

It is quite possible that susceptibility to photochemical alteration is a general characteristic of the cyclobutadienequinones. Photolysis of dimethylcyclobutadienequinone (*2*), for example, yields a yellow dimer (m.p. 222°–223°) of unknown structure[3]; similarly, photolysis of diphenylcyclobutadienequinone (*5*) yields a high-melting, tan solid (m.p. 318°–320°), also of unknown structure.†,[7]

---

* For a discussion of the cyclobutadiene dication, see Chapter 3 and Chapter 12.

† See also Section D, 1, g for the photolytic ring cleavage of phenylcyclobutadienequinone in alcohol.

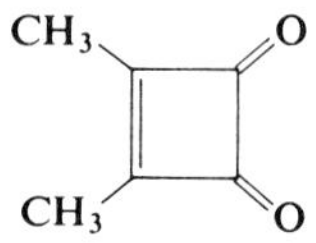

D. CHEMISTRY OF THE CYCLOBUTADIENEQUINONES

(2), 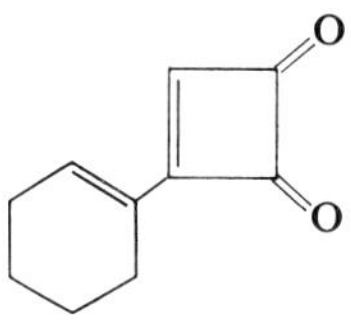 (3), 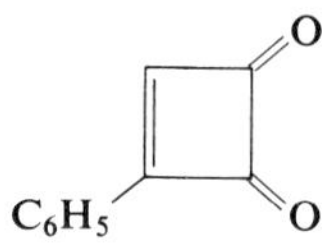 (4), 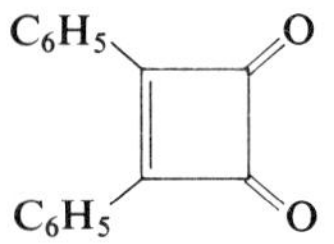 (5)

c. Oxidation

Vigorous permanganate oxidation of both phenylcyclobutadienequinone[18] and diphenylcyclobutadienequinone[2] yields benzoic acid. Dimethyl-,[3] phenyl-,[18] and diphenylcyclobutadienequinone[2] all undergo a Bayer–Villiger type oxidation with neutral hydrogen peroxide to give the corresponding substituted maleic anhydrides (6), (7), and (8), respectively.

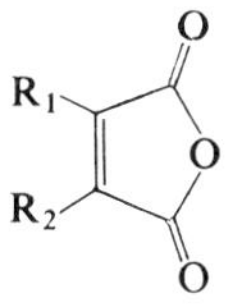

(6): $R_1 = R_2 = CH_3$
(7): $R_1 = H; R_2 = C_6H_5$
(8): $R_1 = R_2 = C_6H_5$

d. Reduction

Several attempts to reduce phenylcyclobutadienequinone (4) to the corresponding cyclobutadienoid hydroquinone (10) were unsuccessful. The quinone (4) was completely resistant to catalytic reduction in the presence of various catalysts and, in fact, was found to act as a catalyst poison. Neither could reduction be effected with catechol at 110°, for no indication of quinhydrone formation could be detected. The results of a preliminary polarographic study suggest that reduction of phenylcyclobutadienequinone is an irreversible process. Vigorous reduction of the quinone (4) under the conditions of the Clemmensen reduction afforded phenylcyclobutane (9).[18]

Cyclohexenylcyclobutadienequinone (3) readily absorbs one equivalent of hydrogen in the presence of a platinum catalyst. However, the product is neither cyclohexylcyclobutadienequinone (11) nor the hydroquinone of cyclohexenylcyclobutadiene (12), but rather 2-hydroxy-3-(1'-cyclohexenyl)-cyclobutenone (13), formed in quantitative yield.[16] The mechanism of this transformation is obscure.

e. Reaction with Halogens

Phenylcyclobutadienequinone (4) reacts with chlorine or bromine in acetic acid (but not in carbon tetrachloride) to give 4-chloro- (14a) and 4-bromo-phenylcyclobutadienequinone (14b), respectively. While an addition–elimina-

(9) (4) (10)

(11) (3) (12)

(13)

tion mechanism for this reaction cannot be excluded, attempts to isolate an intermediate addition product have been unsuccessful, and it appears likely that substitution occurs by direct electrophilic attack of halogen on the quinone.[18]

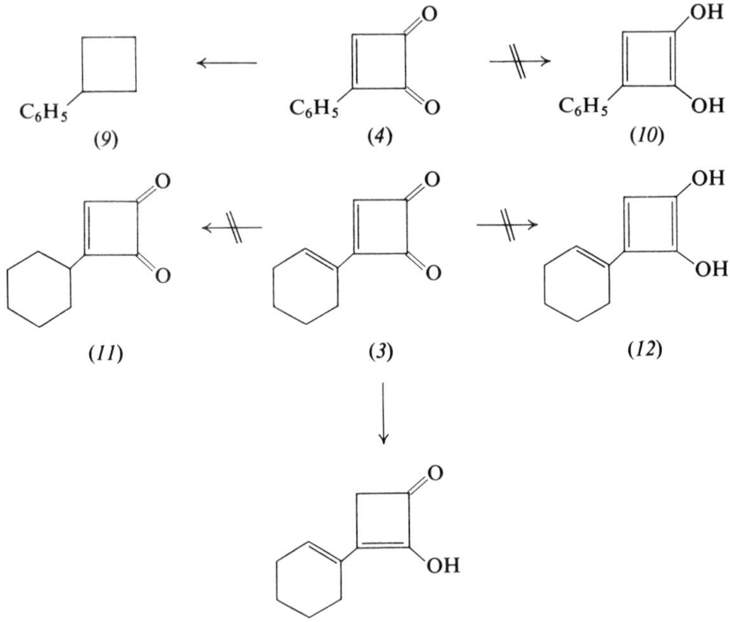

On the other hand, the reaction of cyclohexenylcyclobutadienequinone (*3*)

(4) (*14a*): X=Cl
(*14b*): X=Br

(15) (3) (16)

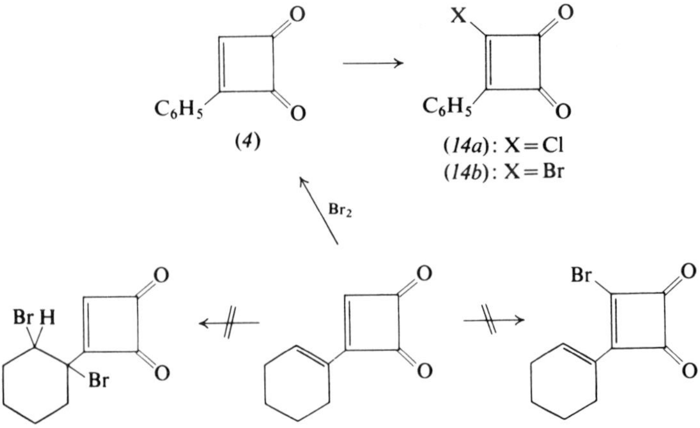

with one equivalent of bromine does not lead to the formation of the analogous 4-bromo derivative (16) or to the dibromocyclohexyl derivative (15) but to phenylcyclobutadienequinone (4). The latter was obtained in this reaction in 1% yield, ostensibly by direct aromatization of the cyclohexyl substituent.

f. Formation of Carbonyl Derivatives

Unlike benzocyclobutadienequinone, which sometimes gives cleavage products (Chapter 7, Section E, 6), the cyclobutadienequinones react with the usual reagents (2,4-dinitrophenylhydrazine, tosylhydrazine and hydroxylamine) to give normal carbonyl derivatives.* Thus, 3-(1'-cyclohexenyl)-,[16] 3,4-dimethyl-,[3] 3-phenyl-,[18] and 3,4-diphenylcyclobutadienequinone[2] react normally with 2,4-dinitrophenylhydrazine to give the corresponding mono-2,4-dinitrophenylhydrazones. Similarly, 3-phenylcyclobutadienequinone reacts with hydroxylamine to give a monooxime,[18] and 3,4-diphenylcyclobutadienequinone reacts with tosylhydrazine to give the monotosylhydrazone.[2] Dimethyl-[3] and diphenylcyclobutadienequinone[2] are the only cyclobutadienequinones reported thus far to react with *two* moles of a carbonyl reagent: The two compounds react normally with tosylhydrazine to give the corresponding ditosylhydrazones.

Theoretically, mono-substituted cyclobutadienequinones which react with one mole of carbonyl reagent can give either of two possible isomeric derivatives, depending upon which of the two carbonyl groups is attacked. Up to the present time, structure assignments have not been made for any of the mono derivatives of mono-substituted cyclobutadienequinones described above, viz., 3-(1'-cyclohexenyl)- and 3-phenylcyclobutadienequinone mono-2,4-dinitrophenylhydrazone and 3-phenylcyclobutadienequinone monooxime.

g. Ring-Cleavage Reactions

Condensation of a substituted cyclobutadienequinone with *o*-phenylenediamine should lead to the formation of a substituted quinoxaline having a fused cyclobutadiene ring, i.e., a substituted 3,8-diazanaphtho[*b*]cyclobutadiene. However, in each of the examples reported thus far, condensation was found to follow a different course, involving rupture of the four-membered ring of the quinone, and resulting in the formation of a simple quinoxaline. Thus, 2-phenylquinoxaline (18) and not 1-phenyl-3,8-diazanaphtho[*b*]cyclobutadiene (17) was formed when 3-phenylcyclobutadienequinone (4) was

---

* However, the possibility that some of the cyclobutadienequinone carbonyl derivatives might be ring cleavage or rearrangement products cannot be ruled out at the present time. (See Section D, 1, g, below.) Physical constants of the carbonyl derivatives described above are listed in Table 4.2.

heated with *o*-phenylenediamine,[18] possibly in accord with the following three-step mechanism: (*i*) formation of a Schiff base by the condensation of one of the carbonyl groups of the quinone with one of the amine groups of *o*-phenylenediamine, (*ii*) 1,4-addition of the other amine group of the diamine to the enone system of the resulting Schiff base, and (*iii*) ring-cleavage of the four-membered ring to give the observed product (*18*).[18b] Similarly, diphenylcyclobutadienequinone reacts with *o*-phenylenediamine in the presence of sodium acetate (ethanol, 1 hour at 78°) to give the analogous ring-cleaved product (*20*, $C_{22}H_{16}N_2O$), and not 1,2-diphenyl-3,8-diazanaphtho[*b*]cyclobutadiene (*19*).[2] Under slightly different reaction conditions, 3,4-diphenylcyclobutadienequinone reacts with *o*-phenylenediamine to give products of unknown structure [$C_{28}H_{20}N_4$ (ethanol, 1 hour at 78°, no sodium acetate) and $C_{28}H_{18}N_4$ (acetic acid, 45 minutes at 118°)].[7]

Dimethylcyclobutadienequinone also failed to give a diazanaphtho[*b*]cyclobutadiene, but the product in this case was a polymeric substance from which no crystalline material could be isolated.[3b]

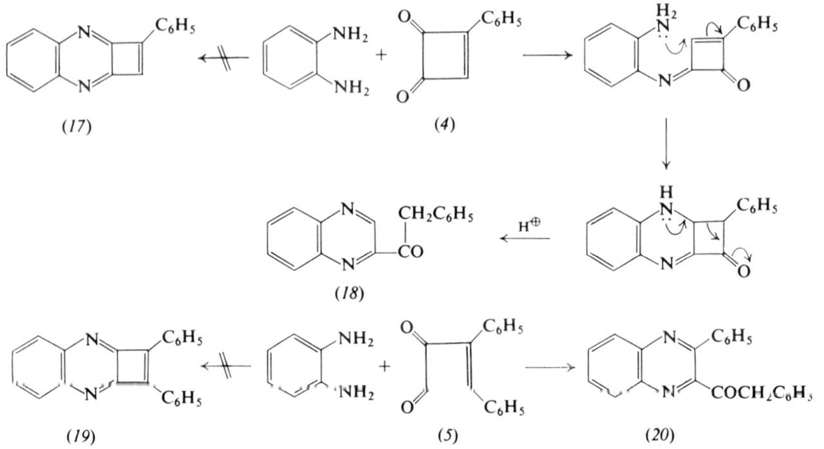

When it is heated with methanolic sodium hydroxide, phenylcyclobutadienequinone (*4*) undergoes ring-cleavage to give a mixture of benzaldehyde and benzylidenepyruvic acid (*21*). The latter is believed to be the primary cleavage product, from which benzaldehyde is no doubt formed by a reverse aldol condensation.[17] Diphenylcyclobutadienequinone (*5*) is cleaved in a similar manner by hot methanolic sodium hydroxide to give a mixture of benzaldehyde and benzylidenephenylpyruvic acid (*22*), isolated as the γ-lactone, α-keto-β,γ-diphenylbutyrolactone (*23*).[1]

Two mechanisms have been suggested for the alkaline cleavage of the

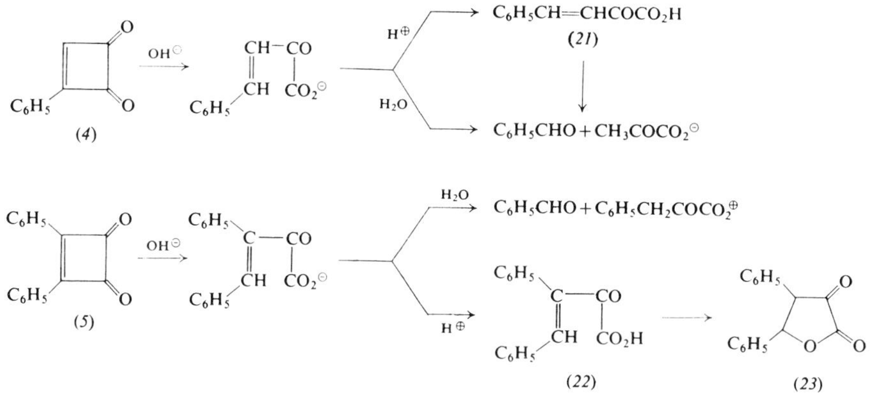

D. CHEMISTRY OF THE CYCLOBUTADIENEQUINONES

cyclobutadienequinones,[1,17] both of which are applied below to 3-phenyl-cyclobutadienequinone by way of illustration. In the first mechanism, the quinone (4) undergoes a benzilic acid rearrangement to give a hydroxycyclopropenecarboxylic acid intermediate (24), which undergoes base cleavage to give the observed product, benzylidenepyruvic acid (25).[17] According to the second mechanism, the quinone first undergoes 1,4-addition of hydroxide ion to give a hydroxy-1,2-dione (26) which then undergoes further attack by

Cyclopropenol mechanism[17]

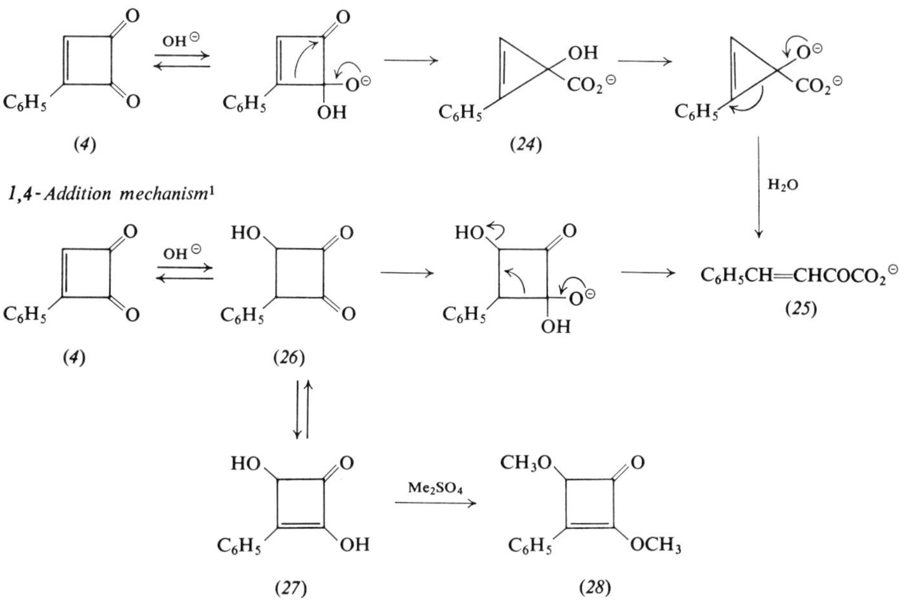

1,4-Addition mechanism[1]

hydroxide ion in a concerted cleavage–elimination transformation, thereby giving rise to the observed product (25).[1]

In support of the second mechanism, it may be noted that (26), the initial intermediate resulting from the 1,4-addition of hydroxide ion, can be trapped in the enol form (27) as the dimethyl ether (28), which is formed in appreciable yield (about 10%) when the reaction mixture is treated with dimethyl sulfate.[17] The alkaline cleavage of quinone (4) in deuterium oxide yields benzaldehyde-d_1,[17] but this observation does not serve to distinguish between the two possible mechanisms.

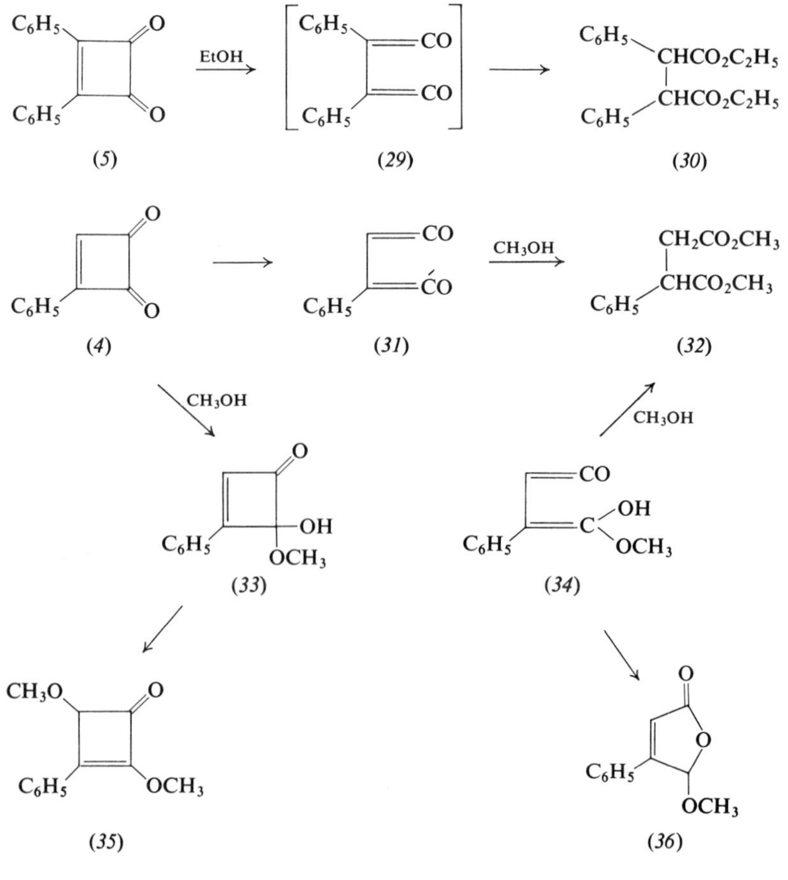

Diphenylcyclobutadienequinone (5) is completely destroyed in neutral ethanolic solution during 12 hours. The product is a mixture of the meso and racemic forms of diethyl α,α'-diphenylsuccinate (30) which is probably formed by way of a bis(phenylketene) intermediate (*29*).[2] In contrast, dimethylcyclobutadienequinone is only one-third decomposed under similar conditions after 16 days (the decomposition products were not isolated).[3]

Phenylcyclobutadienequinone (*4*) is destroyed when heated in either neutral or acidic methanol solution at temperatures of 90°–150°. Three products, viz., 2,4-dimethoxy-3-phenylcyclobutenone (*35*), 3-phenyl-4-hydroxy-4-methoxy-2-butenoic acid lactone (*36*), and dimethyl phenylsuccinate (*32*), are obtained in varying yields, depending upon the reaction conditions used. The cyclobutenone (*35*) is obtained in higher yield when an acidic methanol solution is used, while the diester (*32*) is obtained best from a neutral solution.[11]

The hemiketal (*33*) may be a common intermediate in the formation of all three of these products. Thus, acid-catalyzed formation of a carbonium ion derived from (*33*) could lead to the cyclobutenone (*35*). Similarly, thermal cleavage of (*33*) could give a mixture of stereoisomeric hydroxyketenes (*34*), from which both the butenolide (*36*) and the diester (*32*) could arise. However, it is also possible that diester (*32*) is formed through direct thermal cleavage of the quinone (*4*) by way of an intermediate bis(ketene) (*31*). Furthermore, it is possible that the bis(ketene) (*31*) is an intermediate in the photolysis of quinone (*4*) in methanol at 60°; the photolysis reaction proceeds rapidly to give a mixture of diester (*32*) and a crystalline compound of unknown structure $(C_{21}H_{16}O_5)$.[11]

2. REACTIONS OF FUNCTIONALLY SUBSTITUTED CYCLOBUTADIENEQUINONES

a. *3-Substituted 4-Phenylcyclobutadienequinones*

The halogen atoms in 3-chloro- (*14a*) and 3-bromo-4-phenylcyclobutadienequinone (*14b*) are unusually labile, and both compounds behave as vinylogous acid halides. Thus, both (*14a*) and (*14b*) give an immediate precipitate of silver halide with alcoholic silver nitrate solution, both react readily with sodium iodide in acetone to give 3-iodo-4-phenylcyclobutadienequinone (*37*) and both give the dehalogenated compound phenylcyclobutane (*9*) in the Clemmensen reduction.[18]

3-Bromo-4-phenylcyclobutadienequinone (*14b*) reacts with refluxing methanol to give the vinylogous ester, 3-methoxy-4-phenylcyclobutadienequinone (*39*) and with dry ammonia to give 3-amino-4-phenylcyclobutadienequinone (*38*).[18] The latter is practically neutral, is insoluble in concentrated hydrochloric acid (though not in concentrated sulfuric acid), does not form a benzylidene derivative, and hence is more like a vinylogous amide than an amine. Hydrolysis of 3-bromo- (*14b*), 3-amino- (*38*), and 3-methoxy-4-phenylcyclobutadienequinone (*39*) affords 3-hydroxy-4-phenylcyclobutadienequinone, which may be viewed as the parent vinylogous acid of the series. The

3-hydroxy compound reacts readily with diazomethane to give the corresponding methyl ether (39), yet it gives an intense magenta color like a typical enol when treated with ferric chloride.[*,1]

3-Hydroxy-4-phenylcyclobutadienequinone (40) is a remarkably strong acid; its pK_a value (0.35 ± 0.04) is even smaller than that of picric acid (0.8). In order to account for the remarkable stability of the anion of the hydroxyquinone, it has been suggested that cross-ring resonance structures such as (41d) and (41e) must contribute significantly to the hybrid (41a)–(41e). Indeed, according to molecular orbital theory, inclusion of cross-ring interactions raises the calculated delocalization energy of the anion from 6.11 β to the considerably higher value of 6.94 β.[18]

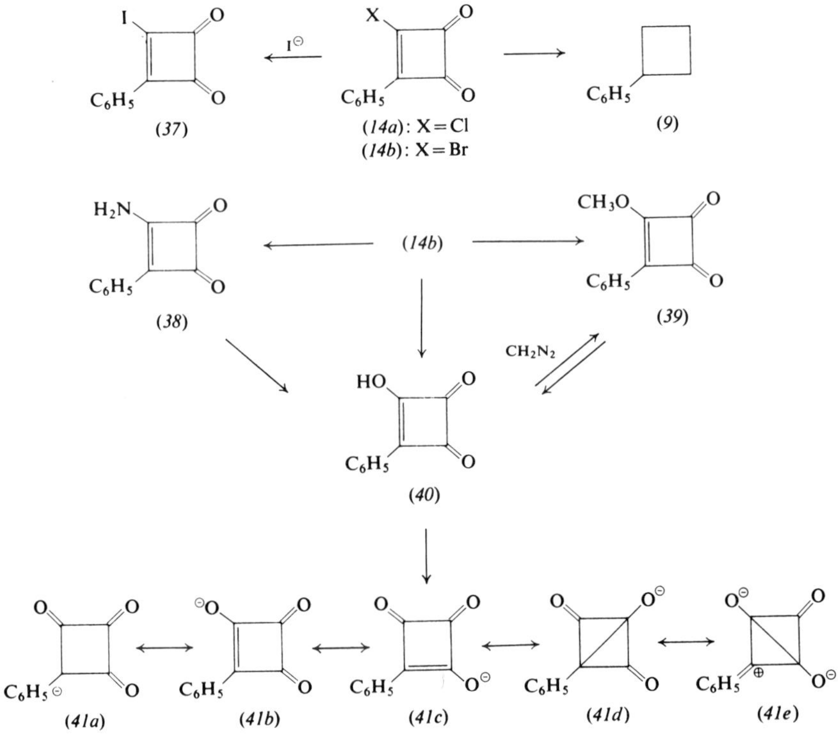

3-Hydroxy-4-phenylcyclobutadienequinone (40) is destroyed by prolonged boiling with 10% aqueous sodium hydroxide solution. The degradation products, viz., oxalic acid, 1,3-diphenylpropene (43) and phenylpyruvic acid (44) may be viewed as rational transformation products of a postulated

* Compare squaric acid, Section D, 2, b.

primary cleavage product, β-formylphenylpyruvic acid (42). A mechanism has been proposed for the formation of the aldehydic acid (42) which involves a cyclopropenol intermediate.*,22

b. Squaric Acid

Dihydroxycyclobutadienequinone (46), which is also known by the trivial

$$\underset{(40)}{\text{HO}\diagup\diagdown\text{O}\atop\text{C}_6\text{H}_5\diagdown\diagup\text{O}}\longrightarrow\underset{(42)}{\overset{\text{COCO}_2\text{H}}{\underset{\text{C}_6\text{H}_5\text{CHCHO}}{|}}}\longrightarrow \text{HO}_2\text{CCO}_2\text{H}+\text{C}_6\text{H}_5\text{CH}_2\text{CHO}$$

$$\downarrow$$

$$\text{C}_6\text{H}_5\text{CH}=\text{CHCH}_2\text{C}_6\text{H}_5$$
$$(43)$$

$$\underset{(44)}{\overset{\text{COCO}_2\text{H}}{\underset{\text{C}_6\text{H}_5\text{CH}_2}{|}}}$$

name "squaric acid," behaves as a strong dibasic acid.[3] The ionization constants of this compound ($pK_1 = 1$ and $pK_2 = 3.0$) indicate that it is an acid comparable in strength to sulfuric acid ($pK_2 = 1.5$).[13b] The squarate dianion (49) may be expected to be greatly stabilized by resonance structures (49a)–(49d), such that all four oxygens of the anion will be essentially equivalent.[13] Furthermore, it has been pointed out that the squarate ion is only one member of a series of cyclic, symmetrical, electron-delocalized anions of composition $C_nH_n^{2-}$ which have considerable aromatic character.[21] Simple molecular orbital calculations predict a delocalization energy of 1.92 β for the squarate dianion[21b] and this figure is considerably greater than that calculated in a similar manner for the parent cyclobutadienequinone system (1.24 β).[12, 18b]

The two carbonyl groups of squaric acid are unreactive to phenylhydrazine[3] and thus resemble the carbonyl group of a carboxylic acid rather than that of a ketone. Indeed, the hydroxyl groups of squaric acid are readily esterified with butyl alcohol to give a dibutyl "ester," dibutoxycyclobutadienequinone (45),[10] but they also behave as enolic hydroxyls, for the compound reacts with ferric chloride to give an intense purple color.[13]

Squaric acid is readily attacked by relatively weak oxidizing agents such as ferricyanide and bromine water, as well as by nitric acid, periodic acid, permanganate, and ceric ion.[13] (The latter oxidizes the acid quantitatively to carbon dioxide.) At room temperature, the oxidation of squaric acid with

---

* However, see Chapter 7, Section E, 4, for a related alkaline cleavage reaction of benzocyclobutadienequinone wherein a benzocyclopropenol intermediate is excluded.

nitric acid or with bromine gives a mixture of carbon dioxide and oxalic acid, but at ice-bath temperature these reagents afford octahydroxycyclobutane (*47*), which may be viewed as the tetrahydrated form of the unknown cyclobutadienebis(quinone) (*48*). Infrared studies show that the octahydroxy compound is not in equilibrium with any carbonyl-containing species, yet in being reducible to squaric acid by sulfur dioxide [(*47*) → (*46*)] it behaves as though it were the tetraketone (*48*).[20]

A number of metal salts of squaric acid have been prepared, viz., (a) the potassium salt, $K_2C_4O_4 \cdot H_2O$[13]; (b) the calcium, cobaltous, cupric, ferrous, magnesium, manganous, nickelous, and zinc salts (all expressible by the formula $MC_4O_4 \cdot 2H_2O$)[19]; and (c) the aluminum, chromous, and ferric salts [all expressible by the formula $MC_4O_4(OH) \cdot 2H_2O$].[19] With the exception of the calcium and the cupric salts, all of the divalent metal salts are isostructural, as are the three trivalent metal salts. A study of the magnetic moments of the divalent salts indicates that they are high-spin complexes; a polymeric structure (*50*) has been proposed for the divalent metal complexes.[19]

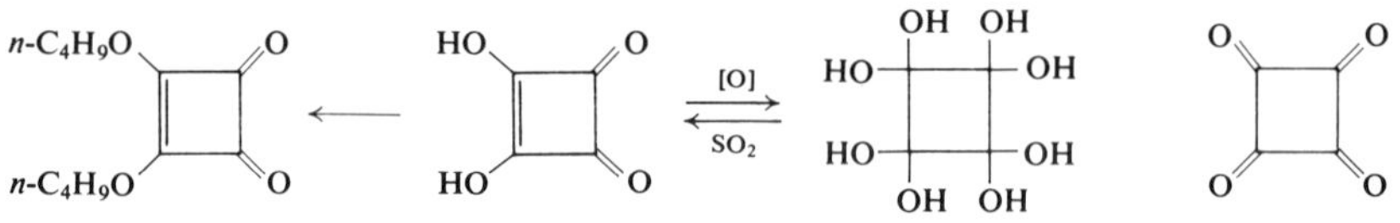

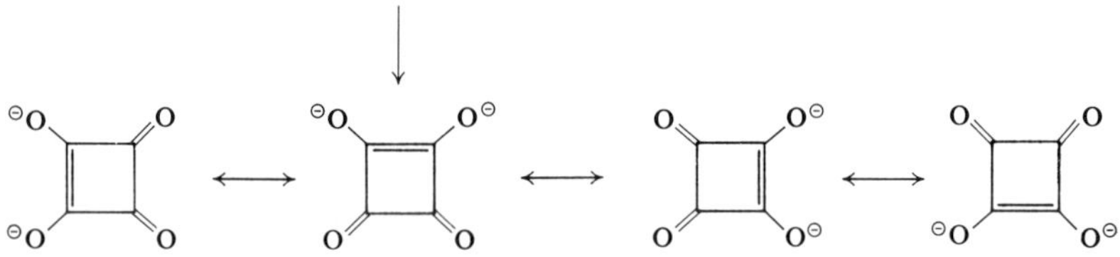

E. Physical Properties of the Cyclobutadienequinones

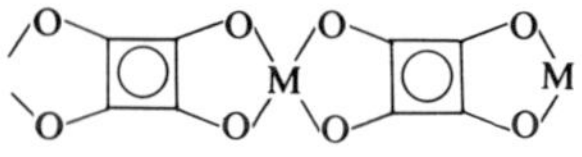

With the exception of four cyclobutadienequinones substituted with polar groups, viz., 3,4-dihydroxycyclobutadienequinone, 3-amino-4-phenyl-, 3-hydroxy-4-phenyl-, and 3-methoxy-4-phenylcyclobutadienequinone, all of the cyclobutadienequinones prepared up to the present time are colored

E. PHYSICAL PROPERTIES OF THE CYCLOBUTADIENEQUINONES   151

TABLE 4.3

ULTRAVIOLET ABSORPTION CHARACTERISTICS OF CYCLOBUTADIENEQUINONES

Cyclobutadienequinone	Solvent	Absorption maxima, mμ, and extinction coefficients (log ϵ)[a]	Ref.
3-Amino-4-phenyl-	EtOH	315 (4.18); 328 (4.42)	18a
3-Bromo-4-phenyl-	CHCl$_3$	304 (4.38)	18a
3-Chloro-4-phenyl-	Isooctane	296 (4.47); 309 (4.31)	18a
	CHCl$_3$	299 (4.42)	18a
	Isooctane	292 (4.43); 306 (4.32)	18a
3-(1'-Cyclohexenyl)-	Cyclohexane	272 (4.31)	16
3,4-Dihydroxy-	Water	269.5 (4.57)	13
3,4-Dimethyl-	EtOH	216 (4.27); 340 (1.41); 355sh (1.36)	3
3,4-Diphenyl-	MeCN	223sh (4.20); 265 (4.23); 318 (4.31)	2
	EtOH	223sh (4.16); 266 (4.12); 322 (4.21); 410 (2.21)	2
3-Hydroxy-4-phenyl-	EtOH	287 (4.30)	18a
	Isooctane	281 (4.36); 295 (4.15)	18a
3-Iodo-4-phenyl-	CHCl$_3$	315 (4.25)	18a
	Isooctane	308 (4.16); 320 (4.07)	18a
3-Methoxy-4-phenyl-	CHCl$_3$	302 (4.45); 316 (4.42)	18a
	EtOH	301 (4.28); 314 (4.35)	18a
3-Phenyl-	EtOH	287 (4.30)[b]	18a
	Isooctane	281 (4.36); 295sh (4.15)	18a
	CHCl$_3$	288[b] (4.40)[b]	18a

[a] Log ϵ values were calculated from ϵ values recorded in the literature.
[b] Value interpolated from a spectral curve recorded in the literature.

yellow. Furthermore, with the exception of the 3,4-dibutoxy and the 3,4-dimethyl derivatives, all of the cyclobutadienequinones are solids at room temperature.

1. Ultraviolet Absorption Spectra

Data have been recorded for the ultraviolet absorption maxima of 11 substituted cyclobutadienequinones (Table 4.3). In general, cyclobutadienequinones having alkyl, aryl, or halogen substituents in the 3- and 4-positions have a primary absorption band in the 250–300 mμ region and a weaker absorption band (or a long absorption tail) extending from the ultraviolet into the visible. (The latter accounts for the yellow color of these compounds.) In some cases, the major absorption band in the ultraviolet can be resolved into several bands by using a nonpolar solvent such as isooctane.

The ultraviolet spectra of 3-phenyl-, and 3-hydroxy-4-phenylcyclobutadienequinone are recorded in Figs. 4.1 and 4.2. (See also Table 4.3.)

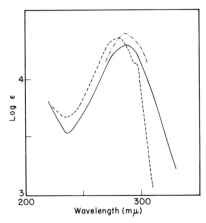

FIG. 4.1. Ultraviolet absorption spectrum of phenylcyclobutadienequinone in ethanol (——), isooctane (----), and chloroform (-.-.-).[18a]

2. Infrared Absorption Spectra

The two carbonyl groups of the cyclobutadienequinones exhibit strong absorption in the rather low range 5.5–5.6 mμ, which may be ascribed to the stretching frequency of a strained carbonyl group attached to a four-membered ring.[2] Infrared absorption bands have been recorded for the following cyclobutadienequinones: 3-(1′-cyclohexenyl)- (state not specified), 3.24, 3.30, 5.62,

6.17 μ*,[16]; 3,4-dihydroxy- (solid state), 4.3 (strong H bonding), 5.5, 6.1 μ†,[13]; 3,4-dimethyl- (neat), 2.83 (w), 5.49sh, 5.60, 5.67 (unresolved doublet),

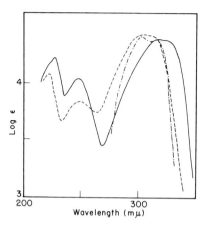

FIG. 4.2. Ultraviolet absorption spectrum of 3-hydroxy-4-phenylcyclobutadienequinone in water (———), 10N sulfuric acid (----), and chloroform (-·-·-).[18a]

5.84sh, 6.21 μ^3; 3,4-diphenyl- (potassium bromide disc), 5.60, 5.63 (doublet), 5.68sh, 6.37 μ^4; 3-phenyl- (state not specified), 5.58, 6.21, 6.42, 6.72, 6.87 μ.[18]

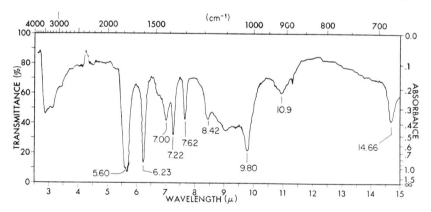

FIG. 4.3. Infrared absorption spectrum of 3,4-dimethylcyclobutadienequinone (neat), calibrated against polystyrene.

* Calculated from frequency values recorded in ref. 16.
† When squaric acid is converted into the dianion, the bands at 5.5 (carbonyl) and 6.1 μ (ethylene) disappear and are replaced by a single, intense band at 6.5–6.75 μ (ref. 13a) [1400–1700 cm^{-1} (5.88–7.14), ref. 19] due to C—C and C—O stretching vibrations in the delocalized dianion.

The infrared spectrum of solid $K_2C_4O_4$ and the Raman spectrum of an aqueous solution indicate a planar symmetrical structure for the ion. Analysis of the spectral data indicates that the ion is aromatic.[6, 21]

The infrared spectrum of 3,4-dimethylcyclobutadienequinone is recorded in Fig. 4.3.

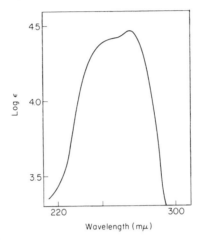

Fig. 4.4. Ultraviolet absorption spectrum of 3,4-dihydroxycyclobutadiene quinone (in water).

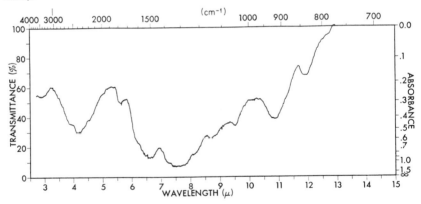

Fig. 4.5. Infrared absorption spectrum of 3,4-dihydroxycyclobutadienequinone (potassium bromide disc) calibrated against polystyrene.

3. Other Physical Properties

Melting points, boiling points, and refractive indices of the cyclobutadienequinones are recorded in Table 4.4. Dipole moments, NMR values, and ultraviolet and infrared spectra are given below for a select group of representative substituted cyclobutadienequinones.

E. PHYSICAL PROPERTIES OF THE CYCLOBUTADIENEQUINONES   155

TABLE 4.4

PHYSICAL PROPERTIES OF SUBSTITUTED CYCLOBUTADIENEQUINONES

Cyclobutadienequinone	Method of preparation[a]	Physical properties[b]	Properties of derivatives[b]	Ref.
3-Amino-4-phenyl-	2	M.p. 282°–283° (decomp.) (w)	—	18
3-Bromo-4-phenyl-	2	M.p. 128°–129° (from benzene–pet. ether)	DNP	18a
3-Chloro-4-phenyl-	2	M.p. 113.8°–114.8° (y) (from carbon tetrachloride)	DNP	18a
3-(1'-Cyclohexenyl)-	1	M.p. 77.8°–78.4° (y) subl. 50° (1 mm)	MonoDNP, (r) m.p. 260°	16
3,4-Dibutoxy-	1, 2	B.p. 139° (0.5 mm), n_D^{20} 1.4932	—	10
3,4-Dihydroxy-	1	Decomp. 292° (from water)	Metal salts	10,13,15,21
3,4-Dimethyl-	1	B.p. 74°–76° (2 mm) (y) n_D^{24} 1.4908	DiTsH, m.p. 190°–191° (decomp.); monoDNP, m.p. 205°–208° (decomp.)	3
3,4-Diphenyl-	1	M.p. 97°–97.2° (y) (from CHCl$_3$–C$_5$H$_{12}$)	MonoDNP, m.p. 253° (from dioxane); monoTsH, m.p. 188°–189° (from EtOH); diTsH (?) m.p. 239°–240°	2
3-Hydroxy-4-phenyl-	2	M.p. 208°–211° (cl) (from Et$_2$O–benzene)	DNP, Cu(II) salt	18a
3-Iodo-4-phenyl-	2	M.p. 162.5°–165.5° (decomp.)	—	18
3-Methoxy-4-phenyl-	2	M.p. 151°–152.2°	—	18
3-Phenyl-	1	M.p. 152°–153° (decomp.) (y) (from Me$_2$CO), subl., $\mu = 5.3\ D$	Monooxime, m.p. 165°–166.8°; monoDNP, m.p. 212°–213.4° (o) (from dioxane)[c]	18

[a] Method 1 is "Acid Hydrolysis of Halogenated Cyclobutenes and Related Compounds," Section B, 1; Method 2 is "Transformation of Cyclobutadienequinones," Section B, 2.
[b] Key to abbreviations: cl (colorless), DNP (dinitrophenylhydrazone), o (orange), r (red), subl. (sublimes), TsH (tosylhydrazone), w (white), y (yellow).
[c] For other derivatives of 3-phenylcyclobutadienequinone, see 3-bromo-, 3-chloro-, and 3-iodo-4-phenylcyclobutadienequinone in this table.

3-(1'-Cyclohexenyl)cyclobutadienequinone has a dipole moment of 5.62 D. The ultraviolet and infrared spectra of squaric acid are recorded in Figs. 4.4 and 4.5, respectively. The magnetic moments of a number of divalent and trivalent metal cation salts of squaric acid have been determined and the results show that the divalent cation salts are high-spin complexes and that the trivalent cation complexes exhibit somewhat reduced paramagnetism.[19]

The NMR spectrum of 3,4-dimethylcyclobutadienequinone contains one unresolved peak only, at $\delta = 7.60$ (compare $\delta = 7.75$ for biacetyl).

3,4-Diphenylcyclobutadienequinone shows only one NMR absorption peak (unresolved) due to 10 phenyl protons.[2]

REFERENCES

1. (a) Blomquist, A. T., and LaLancette, E. A., *J. Am. Chem. Soc.* **84**, 220 (1962); (b) LaLancette, E. A., *Dissertation Abstr.* **21**, 1760 (1961).
2. Blomquist, A. T., and LaLancette, E. A., *J. Am. Chem. Soc.* **83**, 1387 (1961).
3. (a) Blomquist, A. T., and Vierling, R. A., *Tetrahedron Letters* No. **19**, 655 (1961); (b) Vierling, R. A., *Dissertation Abstr.* **23**, 79 (1962).
4. Breslow, R., and Mitchell, M. J., unpublished work.
5. Farnum, D. G., Heybey, M. A. T., and Webster, B., *J. Am. Chem. Soc.* **86**, 673 (1964).
6. Ito, M., and West, R., *J. Am. Chem. Soc.* **85**, 2580 (1963).
7. Jenny, E. F., and Druey, J., *J. Am. Chem. Soc.* **82**, 3111 (1960).
8. Kitahara, Y., Caserio, M. C., Scardiglia, F., and Roberts, J. D., *J. Am. Chem. Soc.* **82**, 3106 (1960).
9. LeGoff, E., *Dissertation Abstr.* **21**, 2112 (1961).
10. (a) Maahs, G., *Angew. Chem.* **75**, 982 (1963); (b) *ibid.*, private communication.
11. Mallory, F. B., and Roberts, J. D., *J. Am. Chem. Soc.* **83**, 393 (1961).
12. Manatt, S. L., and Roberts, J. D., *J. Org. Chem.* **24**, 1336 (1959).
13. (a) Park, J. D., Cohen, S., and Lacher, J. R., *J. Am. Chem. Soc.* **84**, 2919 (1963); (b) Cohen, S., Lacher, J. R., and Park, J. D., *J. Am. Chem. Soc.* **81**, 3480 (1959).
14. Roberts, J. D., *Rec. Chem. Progress* **17**, 95 (1956).
15. Roberts, J. D., and Sharts, C. M., "Organic Reactions," Wiley, New York (1962), Vol. 12, p. 8.
16. Sharts, C. M., and Roberts, J. D., *J. Am. Chem. Soc.* **83**, 871 (1961).
17. Skattebøl, L., and Roberts, J. D., *J. Am. Chem. Soc.* **80**, 4085 (1958).
18. (a) Smutny, E. J., Caserio, M. C., and Roberts, J. D., *J. Am. Chem. Soc.* **82**, 1793 (1960); (b) Smutny, E. J., and Roberts, J. D., *J. Am. Chem. Soc.* **77**, 3420 (1955).
19. West, R., Niu, H. Y., *J. Am. Chem. Soc.* **85**, 2589 (1963).
20. West, R., Niu, H. Y., and Ito, M., *J. Am. Chem. Soc.* **85**, 2584 (1963).
21. (a) West, R., and Powell, D. L., *J. Am. Chem. Soc.* **85**, 2577 (1963); (b) West, R., Niu, H. Y., Powell, D. L., and Evans, M. V., *J. Am. Chem. Soc.* **82**, 6204 (1960).

CHAPTER 5

Methylene Analogs of Cyclobutadienequinone

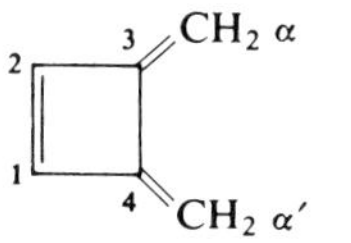

3,4-Dimethylenecyclobutene

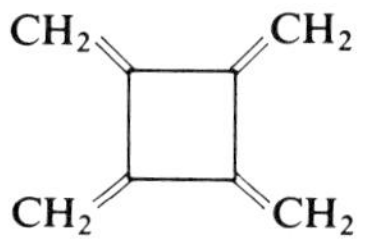

1,2,3,4-Tetramethylenecyclobutane

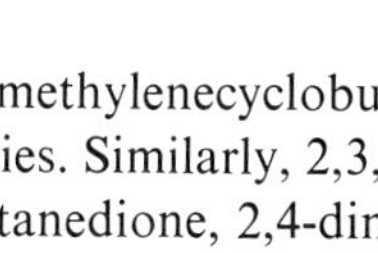

2,3,4-Trimethylene-
cyclobutanone

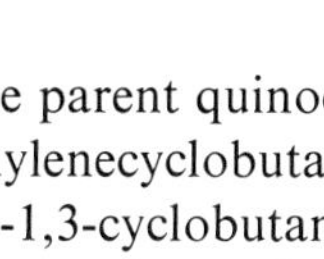

3,4-Dimethylene-
1,2-cyclobutanedione

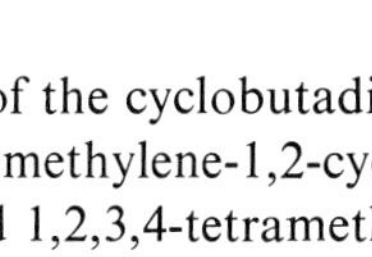

2,4-Dimethylene-
1,3-cyclobutanedione

Dimethylenecyclobutene is the parent quinodimethane of the cyclobutadiene series. Similarly, 2,3,4-trimethylenecyclobutanone, 3,4-dimethylene-1,2-cyclobutanedione, 2,4-dimethylene-1,3-cyclobutanedione, and 1,2,3,4-tetramethylenecyclobutane are methylene analogs of the unknown cyclobutadienebis-(quinone). All five systems have a four-membered ring composed exclusively of trigonal carbon atoms and all five are known either in the form of the parent compound (e.g., 1,2,3,4-tetramethylenecyclobutane) or as substituted derivatives thereof (e.g., 1,2-dimethyl-3,4-dimethylenecyclobutene, $\alpha,\alpha,\alpha',\alpha',\alpha'',\alpha''$-hexaphenyl-2,3,4-trimethylenecyclobutanone, etc.). A monomethylene analog may be written for both cyclobutadienequinone and cyclobutadienebis-(quinone) (Chart 5.1), but no example of either is known as yet.

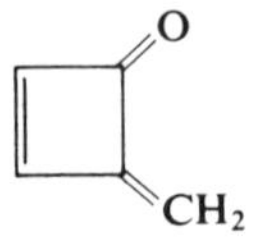

158   5. METHYLENE ANALOGS OF CYCLOBUTADIENEQUINONE

4-Methylenecyclobutenone 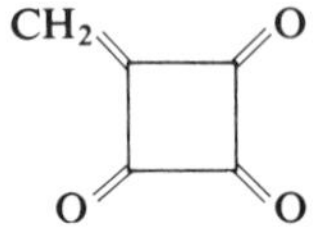   4-Methylene-1,2,3-cyclobutanetrione

CHART 5.1. Unknown methylene analogs of cyclobutadienequinone and cyclobutadienebis(quinone).

A. History

The first methylene analog of a cyclobutadienequinone was prepared unknowingly by Brand in 1921 by the photodimerization of tetraphenylbutatriene (*1*).[6] The identity of Brand's product was not recognized until 1962 when Uhler, Shechter, and Tiers demonstrated that it is octaphenyltetramethylenecyclobutane (*2*).[23, 24]

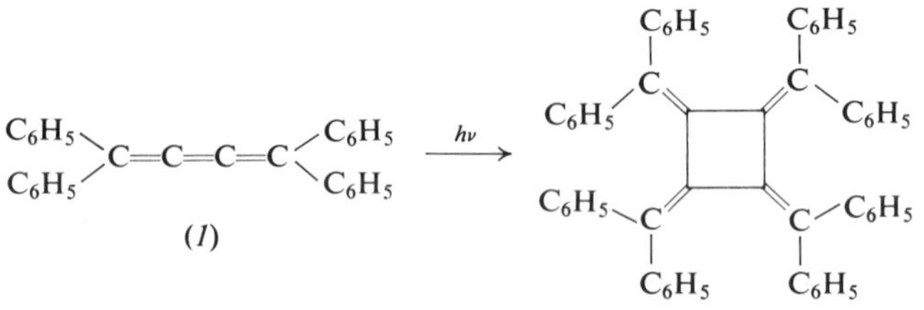

B. Dimethylenecyclobutenes

Molecular orbital calculations indicate that dimethylenecyclobutene has a delocalization energy of 1.21 β, which is almost equal to that of cyclobutadienequinone (1.24 β).[15, 18] It was inferred correctly from the high free-valence index calculated for the α positions of dimethylenecyclobutene (0.86) that the hydrocarbon should polymerize readily.[18] The molecular orbital calculations predict that vinyl and phenyl substituents at the 1- and 2-positions of dimethylenecyclobutene should exert a stabilizing effect on the system. Thus, the calculated values for the delocalization energy (DE) of 1-vinyldimethylenecyclobutene, 1-phenyldimethylenecyclobutene, and 1,2-diphenyldimethylenecyclobutene are greater than the corresponding values ($\overline{DE}$) obtained by simply adding together the DE values of dimethylenecyclobutene (1.21 β) and of the substituent group or groups (Chart 5.2).[15]

In addition to the parent hydrocarbon, a number of substituted dimethylenecyclobutenes have been synthesized and isolated in a state of purity. With the

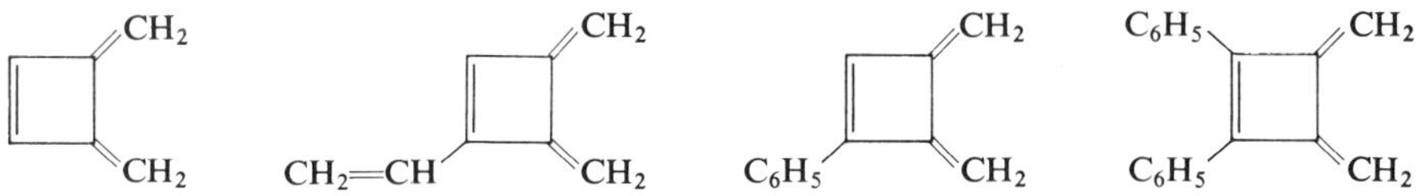

CHART 5.2. Effect of conjugative substituents on the calculated delocalization energy of dimethylenecyclobutene.

DE = 1.21 β

DE = 1.69 β
$\overline{DE}$ = 1.21 β + 0
= 1.21

DE = 3.47 β
$\overline{DE}$ = 1.21 β + 2 β
= 3.21 β

DE = 6.07 β
$\overline{DE}$ = 1.21 β + 4 β
= 5.21 β

exception of the perchloro derivative, all of the known members of the dimethylenecyclobutene series contain unsubstituted methylene groups and are generally characterized by a marked tendency to polymerize.

1. SYNTHESIS OF DIMETHYLENECYCLOBUTENES

a. *By Base-Catalyzed Elimination Reactions*

Base-catalyzed elimination reactions have been employed in the synthesis of a number of dimethylenecyclobutenes from appropriately substituted cyclobutenes. Thus, 1,2-dimethyl-3,4-dimethylenecyclobutene (*2*) was prepared by treating 3,4-dichloro-1,2,3,4-tetramethylcyclobutene (*1*) with a potassium *tert*-butoxide (yield not reported)[11a] or with quinoline (33 % yield).[8] All other known examples of the synthesis of dimethylenecyclobutenes by elimination methods were carried out by the Hofmann degradation of quaternary ammonium hydroxides. For example, the pyrolysis of 3,4-bis(dimethylaminomethyl)-cyclobutene bis(methohydroxide) (*3*) gave the parent dimethylenecyclobutene (*4*).[3]

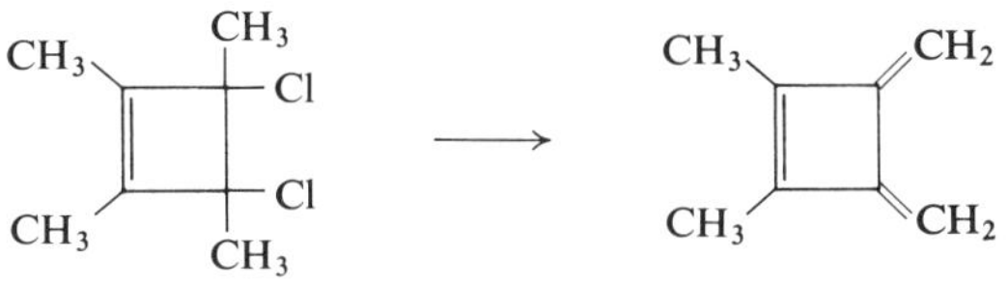

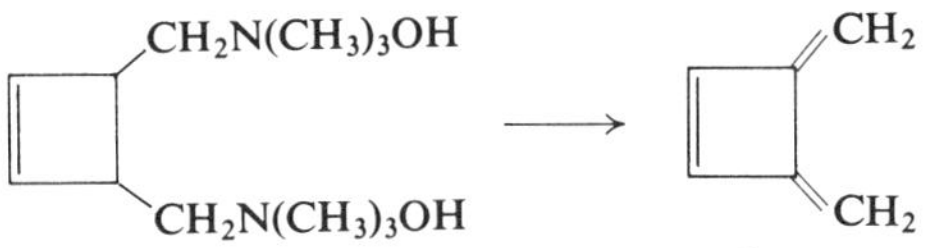

Similarly, the synthesis of 1,2-diphenyldimethylenecyclobutene (*8*) was carried out by pyrolyzing 1,2-diphenyl-3,4-bis(dimethylaminomethyl)cyclobutene bis(methohydroxide) (6b) or its isomer, 1,4-diphenyl-2,3-bis(dimethylaminomethyl)cyclobutene bis(methohydroxide) (7b), at 120°–140° under reduced pressure. The isomeric bis(methohydroxides) were prepared by treating 1-bromo-2,3-bis(bromomethyl)-1,4-diphenylcyclobutane (5) with excess trimethylamine to give a separable mixture of two bis(methobromides), (6a) and (7a) which were treated in turn with silver hydroxide; in practical preparations of triene (8), the two isomeric bis(methobromides) were not separated, but were converted as a mixture into the bis(methohydroxides), (6b) and (7b), and thence into the triene (8).[4]

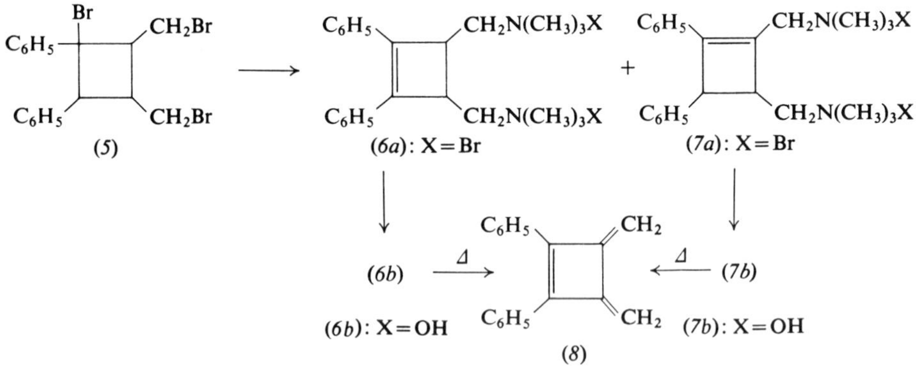

1-Methyl-3,4-dimethylenecyclobutene (12) was obtained as a by-product in the attempted synthesis of the highly unstable trimethylene compound (11) by the thermal decomposition of 2,3-bis(dimethylaminomethyl)methylenecyclobutane bis(methohydroxide) (9).[26] It is possible that the 1-methyl-3,4-dimethylenecyclobutene (12) arises in this reaction by the base-catalyzed isomerization of trimethylenecyclobutane [(11), not isolated)] but it is equally possible that at least some of compound (12) arises by the double-bond isomerization of a water-soluble intermediate, e.g., (10) → (13) → (12). (See the following section.)

b. *By Double-Bond Migration*

When the thermal decomposition products of 2,3-bis(dimethylaminomethyl)methylenecyclobutane bis(methohydroxide) (9) were trapped at a low temperature and purified in a special way, a product other than (12) was found to have been formed (11?). On warming, the product underwent an exothermic rearrangement (double-bond migration?) and gave 1-methyl-3,4-dimethylenecyclobutene (12).[26]

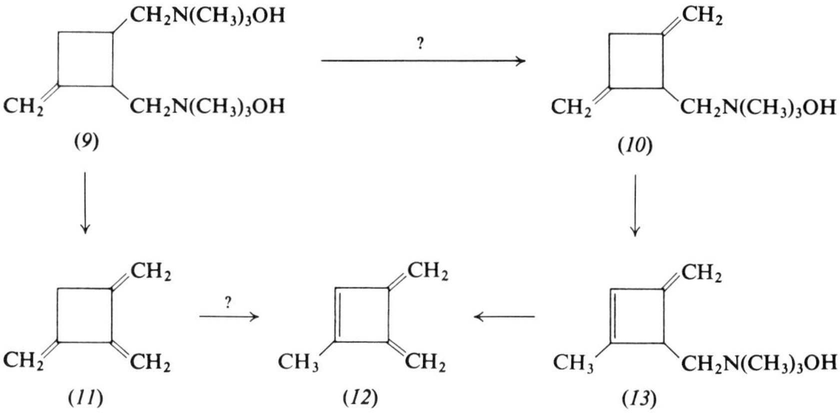

c. From Tetramethylenecyclobutane

The dimerization and Diels–Alder reactions of tetramethylenecyclobutane (15)* have afforded several polycyclic compounds containing the dimethylenecyclobutene system. Thus, spontaneous dimerization of tetramethylenecyclobutane at room temperature gave a tricyclic cyclobutene hydrocarbon (17), while the reaction of tetramethylenecyclobutane with tetracyanoethylene on the one hand and with N-phenylmaleimide on the other gave a bicyclic nitrile (14) and a tricyclic imide (16), respectively.[11a]

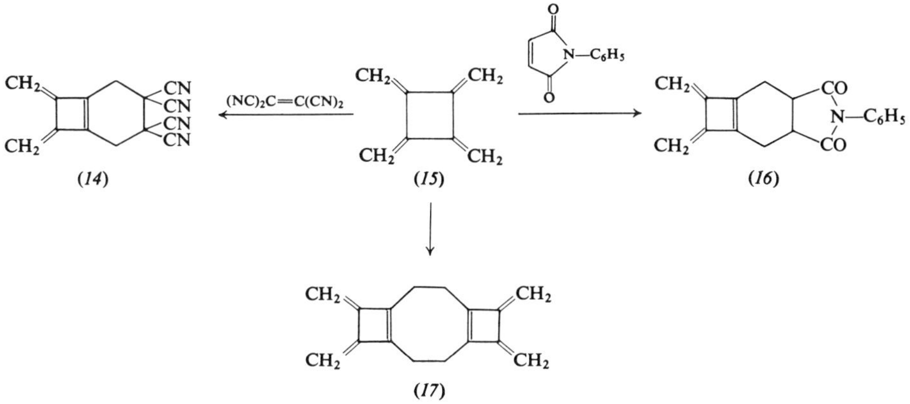

d. From Diacetylenes

The formation of dimethylenecyclobutene by valence isomerization of a diacetylene has been observed in three cases. Pyrolysis of 1,5-hexadiyne (18) at 350° under carefully controlled conditions gave dimethylenecyclobutene (4)

* For the preparation and properties of tetramethylenecyclobutane, see Section C.

in remarkably high yield (85%).[14] Similarly, 1,5-heptadiyne gave 1-methyl-3,4-dimethylenecyclobutene, and 2,6-octadiyne gave 1,2-dimethyl-3,4-dimethylenecyclobutene.[13] The reaction, which probably proceeds through an intermediate diallene (*19*), gives promise of being by far the most direct and most practical method of preparing the 3,4-dimethylenecyclobutenes.

$$HC\equiv C-CH_2 \atop HC\equiv C-CH_2 \quad (18) \longrightarrow \begin{bmatrix} CH=C=CH_2 \\ CH=C=CH_2 \end{bmatrix} \quad (19) \longrightarrow \underset{(4)}{\text{cyclobutene with exocyclic }CH_2\text{ groups}}$$

(*18*)   (*19*)   (*4*)

e. By a Dehalogenation Reaction

Perchloro-3,4-dimethylenecyclobutene (*23*) was prepared by dehalogenation of perchloro-1,2-dimethylenecyclobutane (*22*) with either Raney nickel–potassium hydroxide in aqueous dioxane (50% yield) or with aluminum metal–aluminum chloride in ether (94% yield). [The perchlorodimethylenecyclobutane (*22*) was prepared by the spontaneous dimerization of perchloroallene (*21*), which was itself prepared by the dehydrochlorination in the gaseous phase of pentachloropropene (*20*) with solid potassium hydroxide[17]; or by treating pentachloropropene (*20*) with potassium hydroxide in refluxing toluene or with sodium amide in liquid ammonia.[19, 20]]

$$Cl_2C=CClCHCl_2 \longrightarrow (Cl_2C=C=CCl_2) \longrightarrow$$

(*20*)   (*21*)

(*22*) $\xrightarrow{Zn}$ (*23*)

2. Unsuccessful Attempt to Prepare a Dimethylenecyclobutene

Only one unsuccessful synthesis has been recorded: An attempt to dehalogenate 1,1,2,2-tetrachloro-$\alpha,\alpha,\alpha',\alpha'$-tetraphenyl-3,4-dimethylenecyclobutane with zinc dust gave benzophenone [*sic*] rather than 1,2-dichloro-$\alpha,\alpha,\alpha',\alpha'$-tetraphenyl-3,4-dimethylenecyclobutene.[21]

3. Physical Properties of Dimethylenecyclobutenes

Dimethylenecyclobutene is a mobile, colorless liquid, b.p. 51°[3] (b.p. 72°[14]) having a strong olefinic odor. In general, dimethylenecyclobutenes that have

B. DIMETHYLENECYLOBUTENES

no conjugative substituents exhibit two strong absorption bands in the ultraviolet, one at 209–213 and the other at 245–249 mμ. Since the ultraviolet absorption spectrum of 1,2-dimethylenecyclobutane shows maxima in the same region [237, 246, and 255 mμ (shoulder)], it has been suggested that dimethylenecyclobutene is better viewed as a cross-conjugated diene rather than as a conjugated triene.[3,5] On the other hand, the ultraviolet maxima reported for 1,2-diphenyldimethylenecyclobutene, i.e., 237, 262, and 328 mμ, are not derivable by the simple addition of the spectra of dimethylenecyclobutane and cis-stilbene (about 285 mμ),[4a,c] and it must be concluded that some direct conjugation exists over the entire π-electron system of the dimethylenecyclobutenes. A comparison of the NMR spectra of 3,4-dimethylenecyclobutene and 1,2-dimethylenecyclobutane would no doubt prove instructive.

3,4-Dimethylenecyclobutene and the 1,2-disubstituted 3,4-dimethylenecyclobutenes show several bands in the infrared which are characteristic of conjugated methylene groups. The values for these and other bands are listed in Table 5.1. In general, (a) bands due to C—H stretching vibrations are found in the range 3.23–3.29 μ; (b) bands due to C=C stretching vibrations are found as doublets in the ranges 5.84–5.89 μ and 6.05–6.21 μ; and (c) bands due to C—H out-of-plane bending vibrations are found in the range 11.4–11.8 μ. The infrared spectrum of 3,4-dimethylenecyclobutene has been recorded in the literature.[3] (See Fig. 5.1.)

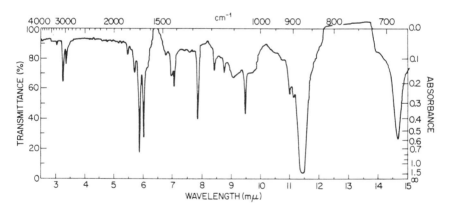

FIG. 5.1. Infrared absorption spectrum of 3,4-dimethylenecyclobutene (carbon tetrachloride solution).[13]

The nuclear magnetic resonance spectrum of 1,2-dimethyl-3,4-dimethylenecyclobutene shows a peak due to methyl protons at $\delta = 1.85$ and a symmetrical doublet due to methylene protons at $\delta = 4.56$ and 4.44.[11a] The methylene doublet is shifted to lower field in compounds (14) and (16) (p. 161) because of

TABLE 5.1
Spectral Characteristics of the 3,4-Dimethylenecyclobutenes

Compound	Ultraviolet absorption bands: absorption maxima, mμ, extinction coefficients (log ε), and [solvent]	Infrared absorption bands: absorption, maxima, μ, and [solvent]	NMR Peaks, ppm (δ), and [solvent][a]	Ref.
1,2-Dimethyl-3,4-dimethylene-cyclobutene	213 (4.21, min), 249 (4.55, min) [isooctane]	3.24, 5.90(m); 6.09(s); 6.24(m); 11.7(vs)[b] [CCl$_4$]	4.56, 4.44, 1.83 [CHCl$_3$]	11a
3,4-Dimethylenecyclobutene	211.5 (5.0), 248 (4.3) [isooctane]	5.88, 6.08, 6.16, 6.21, 6.30, 11.8 3.29 (triplet), 3.39, 11.5, 12.5	—	8
7,8-Dimethylene-3,3,4,4-tetracyano-bicyclo[4.2.0]oct-1(6)-ene	209 (4.42),[c] 247.5sh (3.61)[c]	3.23, 3.34, 3.37, 3.41, 4.44, 5.84, 6.05, 7.01, 11.4 [KBr]	4.87, 4.76, 3.55 [MeCN]	3, 14 11a
1,2-Diphenyl-3,4-dimethylene-cyclobutene	237 (4.42), 262 (4.50), 328 (4.20) [isooctane]	3.29, 3.34, 5.89, 6.10, 11.56(vs) [CHCl$_3$]	—	4a, c
1-Methyl-3,4-dimethylenecyclo-butene	210 (4.50),[c] 245 (4.76) [isooctane]	5.8, 6.0, 11.5–12.1[d]	5.69, 4.74,[e] 1.91[d]	26
Perchloro-3,4-dimethylenecyclo-butene	238s-1 (4.79), 241 (4.56), 249sh (4.51), 278 (3.91), 288 (3.91)[d]	—	—	20a
N-Phenyl-7,8-dimethylenebicyclo-[4.2.0]oct-1(6)-ene-3,4-dicarb-oximide	212.5 (4.58),[c] 248sh (3.83)	3.23, 3.43, 3.51(w); 5.62, 5.72(m); 5.85(vs); 6.21, 6.99, 11.7(m) [CHCl$_3$]	7.38, 4.70, 4.59, 3.32, 2.79 [CDCl$_3$]	11a
5,6,11,12-Tetramethylenetricyclo-[8.2.0.0^{4,7}]dodeca-1(10),4(7)-diene	214, 245sh [EtOH]	3.24, 5.89(m); 6.10(s); 6.35(m); 11.7(vs)	4.55, 4.45, 2.48 [CDCl$_3$]	11a

[a] δ Values calculated from τ values recorded in the literature.
[b] Wavelength values (microns, μ) calculated from wavenumber values (cm^{-1}).
[c] Log ε values calculated from ε values recorded in the literature.
[d] Values interpolated from a spectrum recorded in the literature.
[e] An unresolved multiplet?

long-range interactions between the methylene protons and the cyano or carbonyl groups.

The mass spectrum of 1,2-dimethyl-3,4-dimethylenecyclobutene contains a band at $m/e = 106$ which is almost as intense as the base peak at $m/e = 91$, due to the fragment formed by loss of a methyl group.[11a]

4. CHEMISTRY OF THE DIMETHYLENECYCLOBUTENES

a. *Oxidation, Polymerization, Reduction, and Addition of Halogen and Hydrogen Halides*

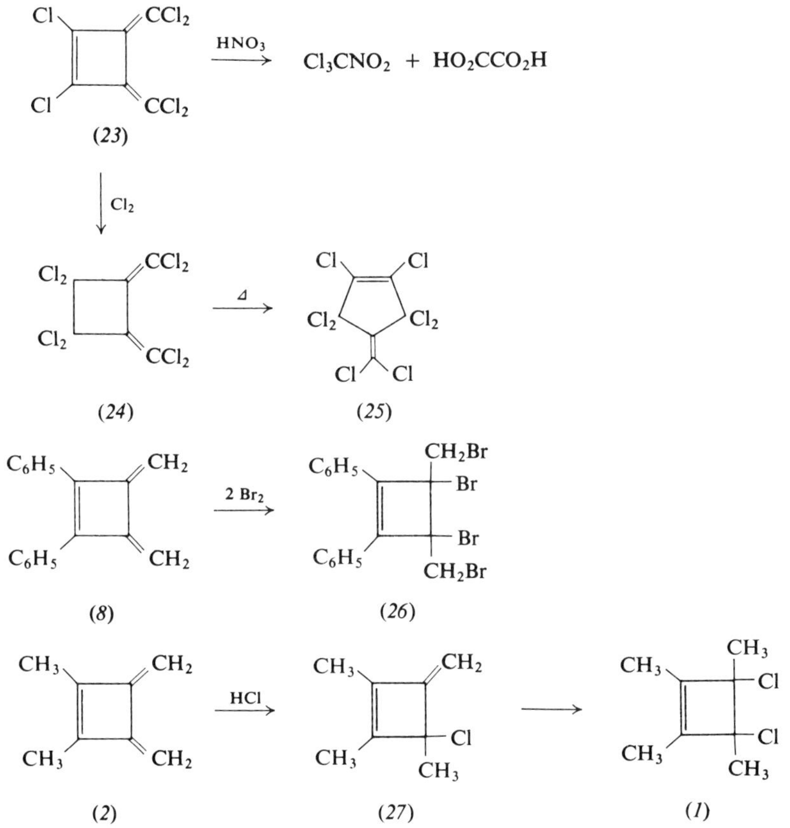

The reductive ozonolysis of 1,2-diphenyldimethylenecyclobutene affords formaldehyde as expected but no other identifiable oxidation products.[4a,c] The oxidation of perchloro-3,4-dimethylenecyclobutene (23) with nitric acid affords chloropicrin and oxalic acid dihydrate.[20a] Dimethylenecyclobutene may be kept for several days under nitrogen in a refrigerator without undergoing appreciable change,[3, 14] but it polymerizes rapidly when exposed to air, giving an oxygen-containing polymer which may deflagrate or detonate on being rubbed.[14] The known dimethylenecyclobutene derivatives, all of which are unsubstituted at the methylene functions, are also prone to polymerization.[4a,c, 11a] 7,8-Dimethylene-3,3,4,4-tetracyanobicyclo[4.2.0]oct-1(6)-ene [(*14*), p. 161], perhaps the most stable dimethylenecyclobutene prepared up to the present time, reacts slowly with oxygen at room temperature to give an intractable tar[9]; in contrast, *N*-phenyl-7,8-dimethylenebicyclo[4.2.0]oct-1(6)-ene-3,4-dicarboximide and 5,6,11,12-tetramethylenetricyclo[8.2.0.0^{4,7}]dodeca-1(10),4(7)-diene [(*16*) and (*17*), p. 161] were found to be extremely sensitive to oxygen in the air.[11a]

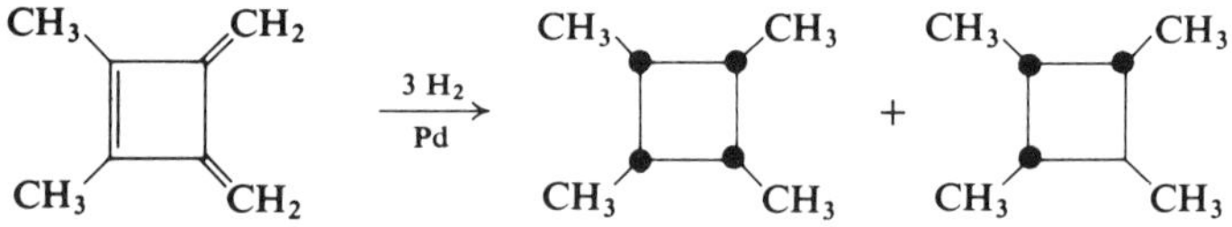

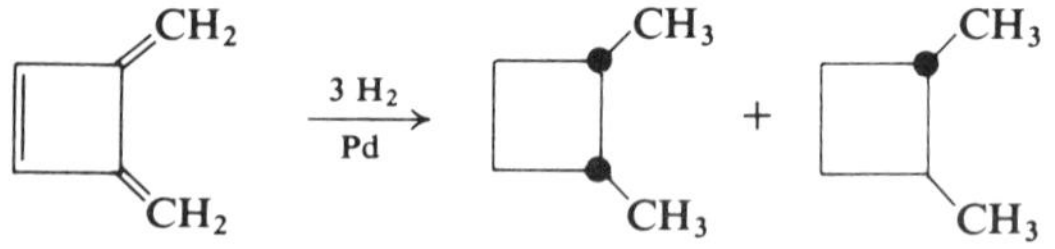

1,2-Diphenyldimethylenecyclobutene (*8*) readily adds two equivalents of bromine at room temperature; the addition of halogen occurs only at the exocyclic double bonds, giving 1,2-diphenyl-3,4-dibromo-3,4-bis(bromomethyl)cyclobutene (*26*).[4a,c] Hydrogen chloride behaves similarly towards 1,2-dimethyl-3,4-dimethylenecyclobutene (*2*) and addition takes place at the methylene double bonds; the reaction can be carried out stepwise to give, first, 4-chloro-3-methylene-1,2,4-trimethylcyclobutene (*27*) and, finally, 3,4-dichloro-1,2,3,4-tetramethylcyclobutene (*1*).[8] In contrast, the addition of chlorine to perchloro-3,4-dimethylenecyclobutene (*23*) is believed to take place at the 1,2-positions, because the product of the reaction is perchloromethylenecyclopentene (*25*), which is known to be formed when perchloro-1,2-dimethylenecyclobutane (*24*) is heated to 290° or treated with chlorine at 200°.[20a]

The dimethylenecyclobutenes are easily reduced under catalytic hydrogenation conditions. Thus, 1,2-dimethyl-3,4-dimethylenecyclobutene (2) absorbs exactly 3 moles of hydrogen and affords primarily *all-cis*-1,2,3,4-tetramethylcyclobutane (28a),[8,9,11a] together with a small amount of *cis,cis,trans*-1,2,3,4-tetramethylcyclobutane (28b).[8] The parent compound, dimethylenecyclobutene (4), gives a mixture of two hydrogenation products, presumably *cis*-1,2-dimethylcyclobutane (29a) and *trans*-1,2-dimethylcyclobutane (29b).[3] The reduction of 1,2-diphenyl-3,4-dimethylenecyclobutene (8) also requires 3 moles of hydrogen, but the configuration of the reduction product, 1,2-diphenyl-3,4-dimethylcyclobutane (30) has not been established.[4a,c]

b. *Reaction with Dienophiles*

The dimethylenecyclobutenes do not react normally in the Diels–Alder reactions. An unusually large energy barrier to normal addition undoubtedly exists because the expected Diels–Alder adducts would be cyclobutadiene derivatives. Thus, 1,2-diphenyl-3,4-dimethylenecyclobutene (8) reacts with tetracyanoethylene to give a crystalline cycloaddition product (32) rather than the expected Diels–Alder product (31).[4a,b] 1,2-Diphenyl-3,4-dimethylenecyclobutene does not react with maleic anhydride, with *N*-phenylmaleimide, or with dimethyl acetylenedicarboxylate at moderate temperatures (25°–75°); under forcing reaction conditions (150°), the triene reacts with dimethyl acetylenedicarboxylate to give an amorphous copolymer.[4a,b] 1,2-Dimethyl-3,4-dimethylenecyclobutene (2) reacts in a similar manner with tetracyanoethylene and gives compound (34) rather than the cyclobutadiene (33).[8,11a] On the other hand, the parent compound, dimethylenecyclobutene reacts with tetracyanoethylene to give a complex mixture of uncharacterized products.[3]

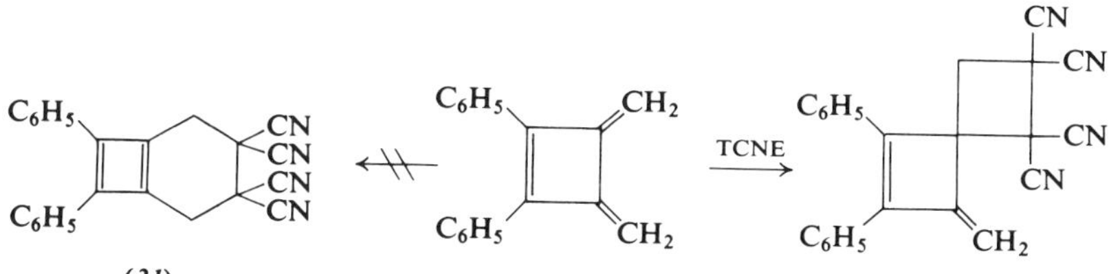

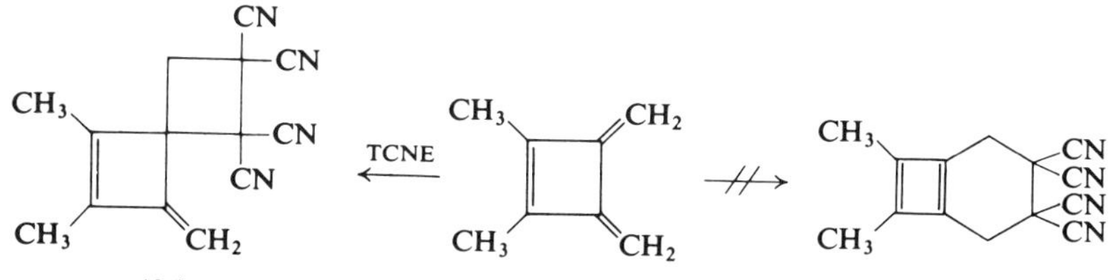

C. Tetramethylenecyclobutanes

Only two tetramethylenecyclobutanes are known at the present time, viz., the parent hydrocarbon, tetramethylenecyclobutane (*1*), and a derivative, octaphenyltetramethylenecyclobutane (*2*).

Molecular orbital calculations predict that tetramethylenecyclobutane (*1*) should have a delocalization energy of 1.66 β, which is substantially greater than that of either dimethylenecyclobutene (1.21 β) or cyclobutadienequinone (1.24 β).[18] Tetramethylenecyclobutane is also predicted to exist in a singlet ground state, but the high free-valence index at the exocyclic methylene carbon atoms indicates quite correctly that the hydrocarbon will nevertheless be prone to undergo polymerization.[11,18] The presence of conjugative substituents at the methylene carbons should greatly reduce polymerization and, as a matter of fact, octaphenyltetramethylenecyclobutane (*2*) has been found to be quite stable.[22]

1. Synthesis of Tetramethylenecyclobutanes

a. By Elimination Reactions

Tetramethylenecyclobutane (*1*) has been prepared from a number of α-substituted 1,2,3,4-tetramethylcyclobutanes by use of common elimination reactions. Thus, the dehydrobromination of *all-trans*-1,2,3,4-tetra(bromomethyl)cyclobutane (*3*) at 0° gave tetramethylenecyclobutane in good yield (about 50%).[11] Comparable results were obtained under the same conditions in the dehydrohalogenation of *all-trans*-1,2,3,4-tetra(iodomethyl)cyclobutane (*4*), *cis,trans,cis*-1,2,3,4-tetra(bromomethyl)cyclobutane (*5*), and *cis,trans,cis*-1,2,3,4-tetra(iodomethyl)cyclobutane (*6*).[11]

Tetramethylenecyclobutane is also formed, albeit in poor yield (about 1%), by subjecting *cis,trans,cis*-1,2,3,4-tetra(dimethylaminomethyl)cyclobutane tetroxide (*7*) to the conditions of the Cope elimination and by subjecting *cis,trans,cis*-1,2,3,4-tetra(dimethylaminomethyl)cyclobutane tetramethohydroxide (*8*) to the conditions of the Hofmann elimination.[11] The low yields obtained in these reactions may be attributed at least in part to the relatively high reaction temperatures required (100°–250°).

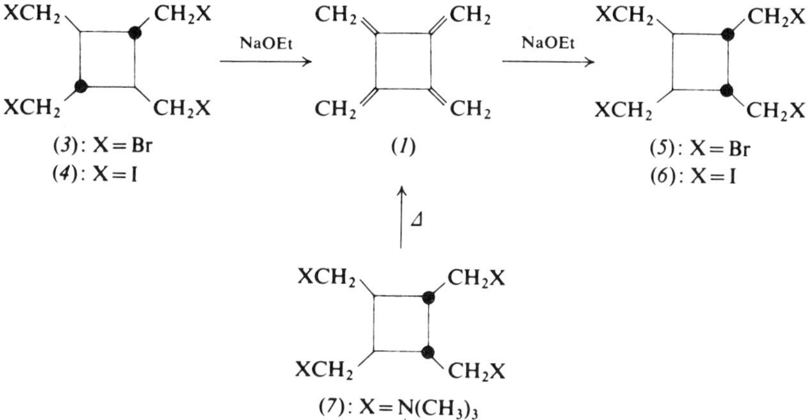

C. TETRAMETHYLENECYCLOBUTANES

[Structures: compounds (3) X=Br, (4) X=I converted by NaOEt to (1), then by NaOEt to (5) X=Br, (6) X=I; compound (7) X=N(CH₃)₃→O converts by Δ to (1); (8): X=N(CH₃)₃OH]

(3): X = Br
(4): X = I

(1)

(5): X = Br
(6): X = I

(7): X = N(CH$_3$)$_3$
              ↓
              O

(8): X = N(CH$_3$)$_3$OH

b. By Photochemical Dimerization of a Butatriene

Irradiation of crystalline tetraphenylbutatriene (9) affords octaphenyltetramethylenecyclobutane (2).[24] The scope of the reaction in the synthesis of tetramethylenecyclobutanes is not known.*

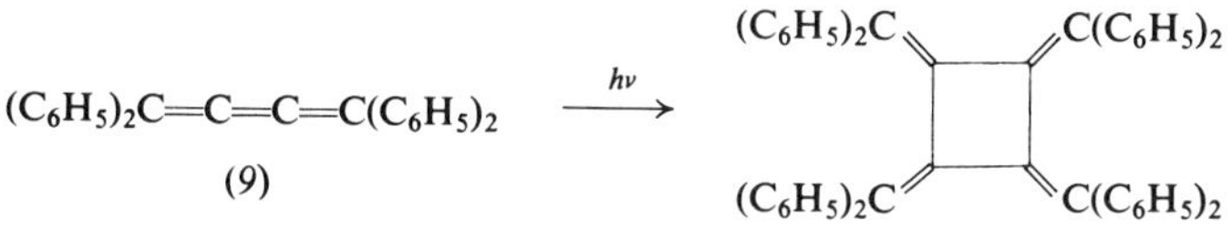

$(C_6H_5)_2C=C=C=C(C_6H_5)_2$ $\xrightarrow{h\nu}$ [cyclobutane structure with four $=C(C_6H_5)_2$ groups]

(9)

(2)

2. Unsuccessful Attempts to Prepare Tetramethylenecyclobutanes

Several unsuccessful attempts to prepare tetramethylenecyclobutanes have been recorded. (a) Tetramethylenecyclobutane (1) was not obtained, even in small yield, by the pyrolysis of 1,2,3-butatriene (10) at temperatures in the range 150°–550°.[11a] Similarly, (b) the pyrolysis of cis,trans,cis-1,2,3,4-tetraacetoxymethylcyclobutane (11) at 540° gave no tetramethylenecyclobutane, but rather a mixture of six aromatic hydrocarbons (benzene, toluene, ethylbenzene, o-xylene, styrene, and anthracene).[11a, 25] (c) Dehydration of all-trans-1,2,3,4-tetra(diphenylhydroxymethyl)cyclobutane (13)[12] and of all-trans-1,2,3,4-tetra(phenylhydroxymethyl)butane (14)[11] failed to give the expected

* Irradiation of 1,1,4,4-tetra(p-anisyl)butatriene gives *two* photo products of unknown structure (ref. 7).

products, octaphenyltetramethylenecyclobutane (*12*) and tetrabenzylidine-cyclobutane (*15*), respectively, even when a variety of reaction conditions was used.

CH$_2$=C=C=CH$_2$ $\xrightarrow{\Delta}$ [H$_2$C, CH$_2$ / H$_2$C, CH$_2$ cyclobutane] (*1*) $\xleftarrow{540°}$ [CH$_3$CO$_2$CH$_2$, CH$_2$O$_2$CCH$_3$ / CH$_3$CO$_2$CH$_2$, CH$_2$O$_2$CCH$_3$ cyclobutane] (*11*)

(*10*)

(C$_6$H$_5$)$_2$C, C(C$_6$H$_5$)$_2$ / (C$_6$H$_5$)$_2$C, C(C$_6$H$_5$)$_2$ (*12*)

R'R(HO)C, C(OH)RR' / R'R(HO)C, C(OH)RR'
(*13*): R = R' = C$_6$H$_5$
(*14*): R = H; R' = C$_6$H$_5$

C$_6$H$_5$CH, CHC$_6$H$_5$ / C$_6$H$_5$CH, CHC$_6$H$_5$ (*15*)

3. Physical Properties of Tetramethylenecyclobutanes

Because of its marked tendency to undergo polymerization, tetramethylene-cyclobutane has been characterized only in dilute solution by spectroscopic methods.[11] Octaphenyltetramethylenecyclobutane, on the other hand, has been found to be a stable, crystalline compound (m.p. 290°–293°) and has been thoroughly characterized.[24]

Dilute solutions of tetramethylenecyclobutane can be stored for prolonged periods at −78° and the hydrocarbon can be freed from polymeric material and other nonvolatile impurities by codistillation with hexane at 0° under reduced pressure (0.02 mm).[11a] The ultraviolet spectrum of tetramethylene-cyclobutane (Fig. 5.2) is complex: In addition to a strong (apparent) maximum at 208 mμ, six longer-wavelength maxima appear at 263, 271, 281.5, 295.5, 312, and 325 mμ (solvent, chloroform).[11a] Only the relative intensities of these bands are known accurately. In contrast, the ultraviolet spectrum of octa-phenyltetramethylenecyclobutane shows only two maxima, at 274 (ϵ = 18,300) and at 307–308 mμ (ϵ = 39,500) (solvent, chloroform).*,[11a]

The infrared spectrum of tetramethylenecyclobutane has bands (a) at 3.23 μ, due to C—H stretching vibrations; (b) at 5.68 and 5.85 μ (doublet) due to C=C stretching vibrations; (c) at 7.14 μ, presumably due to C—H in-plane bending vibrations; and (d) at 11.37 μ, due to C—H out-of-plane bending vibrations.[11a]

The NMR spectrum of tetramethylenecyclobutane consists of a single unsplit peak at δ = 5.19.[11a] As would be expected, the NMR spectrum of octaphenyltetramethylenecyclobutane shows energy absorption due to aromatic protons only, in the vicinity of δ = 7.0. The observed shielding of these

* See also Table 5.2.

C. TETRAMETHYLENECYCLOBUTANES 171

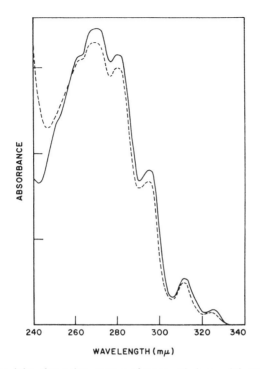

FIG. 5.2. Ultraviolet absorption spectra of tetramethylenecyclobutane generated from *cis, trans, cis*-1,2,3,4-tetra(dimethylaminomethyl)cyclobutane tetroxide (—) (in hexane) and from *all-trans*-tetra(iodomethyl)cyclobutane (- - - -) (in ethanol). (The ordinate scale is omitted because exact concentrations were unknown.)[11a]

protons apparently results from interactions between adjacent phenyl groups, which are prevented sterically from becoming totally coplanar.[24]

The mass spectrum of tetramethylenecyclobutane, determined in hexane solution, exhibits the expected parent peak at $m/e = 104$. The cracking pattern of tetramethylenecyclobutane has not been determined, however, because of the unavoidable presence of solvent in the sample.[11a]

4. CHEMISTRY OF THE TETRAMETHYLENECYCLOBUTANES

Octaphenyltetramethylenecyclobutane (2) reverts to 1,1,3,3-tetraphenylbutatriene (9) at its melting point (290°–293°).[24] Partial ozonolysis of hydrocarbon (2) in chloroform results in the formation of hexaphenyltrimethylenecyclobutanone (16) and benzophenone.[23, 24] More extensive ozonolysis of (2) in chloroform gives 2,4-bis(diphenylmethylene)-1,3-cyclobutanedione [(17), 10% yield] and benzophenone.[24] In methanol–chloroform, the product of

5. METHYLENE ANALOGS OF CYCLOBUTADIENEQUINONE

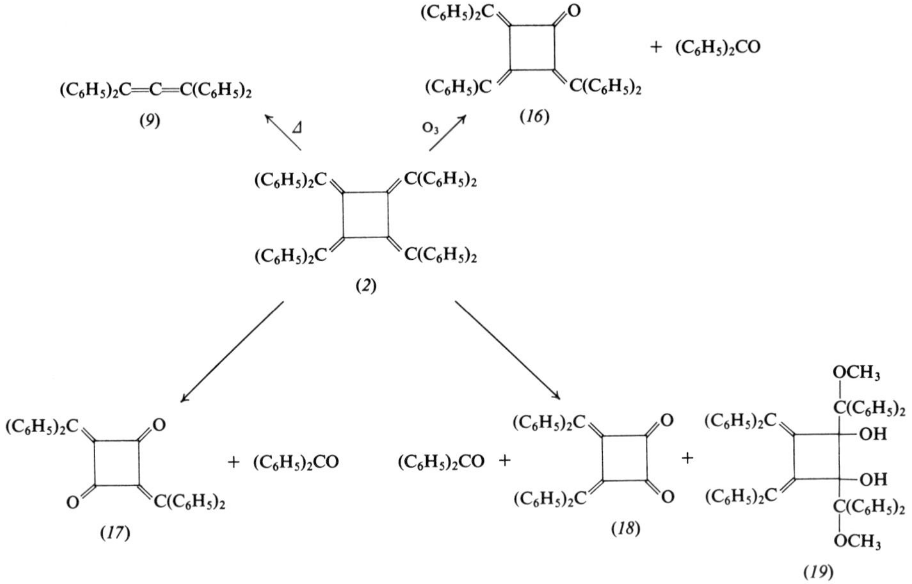

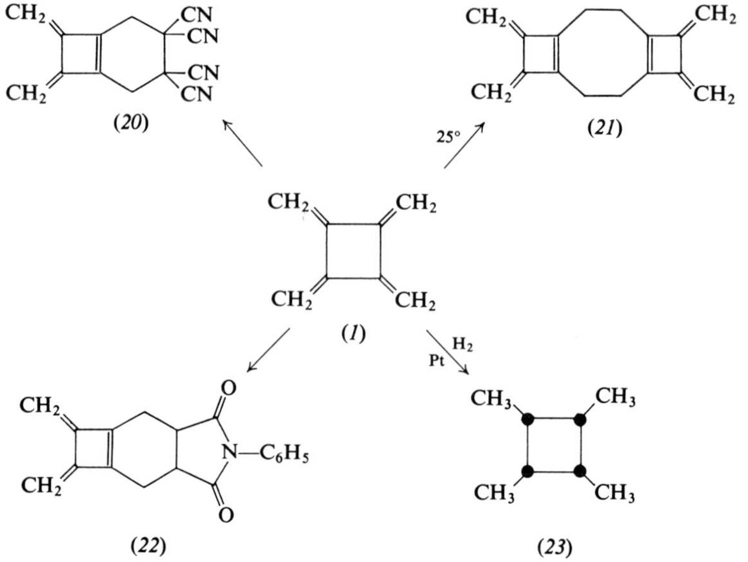

extensive ozonolysis is a mixture of the 1,3-dione (*17*), 3,4-bis(diphenyl-methylene)-1,2-cyclobutanedione (*18*), benzophenone, and 3,4-bis(diphenyl-methylene)-1,2-bis(diphenylmethoxymethyl)-1,2-cyclobutanediol (*19*), which is the major product, formed in 35% yield.[24] The formation of compound (*19*) may proceed by epoxidation of two contiguous benzhydrylene residues, followed by methanolysis of the resulting ozonide intermediate.[24]

Octaphenyltetramethylenecyclobutane is not prone to polymerization under ordinary conditions. It is not sensitive to air and it does not react to form adducts with maleic anhydride or with tetracyanoethylene. In contrast, the parent hydrocarbon, tetramethylenecyclobutane (1), behaves as a typical, highly reactive polyolefin. On brief exposure to air, it forms an intractable oxygen-containing polymer; it reacts readily with tetracyanoethylene at 0° and with N-phenylmaleimide at 25° to give the normal 1:1 Diels–Alder adducts (20) and (22), respectively; and it undergoes dimerization readily at room temperature to give 5,6,11,12-tetramethylenetricyclo[8.2.0.0^{4,7}]dodeca-1(10),-4(7)-diene (21).[11] (But it is indefinitely stable when stored in dilute solution at −78°). Catalytic reduction of tetramethylenecyclobutane gives all-cis-1,2,3,4-tetramethylcyclobutane (23).[11]

D. Trimethylenecyclobutanone and the Dimethylenecyclobutanediones

Trimethylenecyclobutanone (1), the dimethylenecyclobutanediones (2) and (3), and methylenecyclobutanetrione (4) all have four-membered rings made up entirely of trigonal carbon atoms and, like tetramethylenecyclobutane, are methylene analogs of the unknown bis(quinone) of cyclobutadiene.

None of the parent compounds [(1), (2), (3) and (4)] are known at the present time and only one derivative of ketones (1), (2), and (3) has been described.

1. Synthesis of the Trimethylene and Dimethylenecyclobutanones

The three known methylene analogs of cyclobutadienebisquinone, viz., hexaphenyltrimethylenecyclobutanone (6), 3,4-bis(diphenylmethylene)-1,2-cyclobutanedione (7), and 2,4-bis(diphenylmethylene)-1,3-cyclobutanedione (8), were all prepared by partial ozonolysis of octaphenyltetramethylenecyclobutanedione (5).[23,24] The parent methylenecyclobutanones corresponding to (6), (7), and (8) are not known. Monoketone (6) was prepared in 9% yield[23] by limited ozonolysis of (5) in chloroform solution.[24] Similarly, 3,4-bis(diphenylmethylene)-1,2-cyclobutanedione (7) was prepared in 1.9% yield by limited ozonolysis of (5) in methanol–chloroform solution.[23] Extensive ozonolysis of (5) in chloroform afforded the isomeric diketone, 2,4-bis-(diphenylmethylene)-1,3-cyclobutanedione (8) in 10% yield.[23] In methanol–chloroform, extensive ozonolysis of (5) gave diketone (7) in 2% yield and diketone (8) in 10–12% yield.[24]

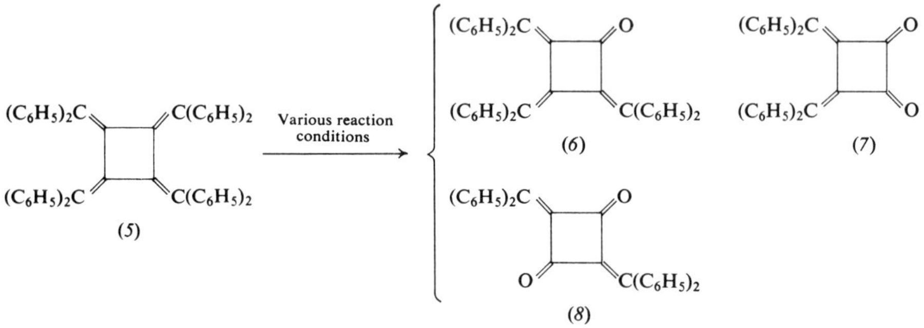

2. Chemistry of the Methylenecyclobutanones

Hexaphenyltrimethylenecyclobutanone (6) does not form a semicarbazone or a 2,4-dinitrophenylhydrazone, even under forcing conditions, but it reacts readily with lithium aluminum hydride to give the corresponding hydroxy compound (9).[24] The 1,3-diketone is similarly reduced by lithium aluminum hydride [(8) → (10)] in tetrahydrofuran[24] (but not in ether).[23] In contrast to the monoketone (6), which is stable under ordinary conditions, 3,4-bis(diphenylmethylene)-1,2-cyclobutanedione decomposes readily in ethanol at 25°–30° and also during recrystallization from hexane or ethyl acetate (decomposition products not specified).[24]

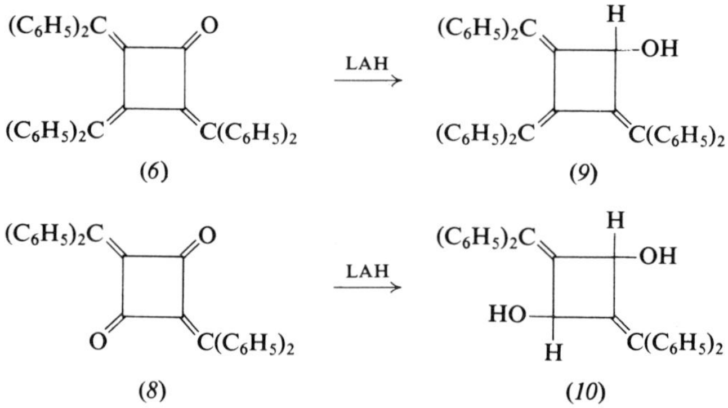

TABLE 5.2

SPECTRAL CHARACTERISTICS OF OCTAPHENYLTETRAMETHYLENECYCLOBUTANE AND THE ANALOGOUS METHYLENEKETONES[a]

Compound	Ultraviolet absorption bands: absorption maxima, mμ, extinction coefficients (log ε), and [solvent][b]	Infrared absorption bands: absorption maxima, μ	NMR peaks, ppm (δ)
2,4-Bis(diphenylmethylene)-1,3-cyclobutanedione	224 (4.51); 256 (3.72); 308 (2.36); 324 (2.48); 362 (2.45) [cyclohexane]	5.73 (unsplit), 13.9 (s)	7.32
3,4-Bis(diphenylmethylene)-1,2-cyclobutanedione	258sh (4.04); 307 (3.74); 376 (3.86); 392 (3.83) [cyclohexane]	5.78, 5.88 (doublet)	
Hexaphenyltrimethylenecyclobutanone	279 (4.21); 287 (4.21); 295 (4.16) [EtOH][c]	5.6	7.15
Octaphenyltetramethylenecyclobutane	274 (4.26); 307–308 (4.60) [CHCl$_3$]	13.9	~7.0

[a] All data in Table 5.2 are taken from ref. 24.
[b] Extinction coefficients (log ε) calculated from extinction values (ε) recorded in ref. 24.
[c] Which read for "95% methanol" in ref. 24 (ref. 22).

TABLE 5.3

PREPARATION, PHYSICAL PROPERTIES, AND DERIVATIVES OF THE METHYLENE ANALOGS OF CYCLOBUTADIENEQUINONE AND CYCLOBUTADIENEBIS(QUINONE)[a]

Compound	Method[b] and yield (%)	Physical properties[c]	Derivatives	Ref.
3,4-Dimethylenecyclobutenes				
3,4-Dimethylenecyclobutene	a	B.p. 51°	cis- and trans-Dimethylcyclobutane	3
	d, 85	B.p. 72°	—	14
1,2-Dimethyl-3,4-dimethylenecyclobutene	a	—	—	11a
	a, 33[d]	B.p. 40°–43° (35 mm)	all-cis-Tetramethylcyclobutane	8
	d	—	—	13
7,8-Dimethylene-3,3,4-tetracyanobicyclo[4.2.0]oct-1(6)-ene	c, 51[d]	(w)	—	11a
1,2-Diphenyl-3,4-dimethylenecyclobutene	a, 46–58	M.p. 42°–43°, subl. 40° (0.3 mm), (w)	TCNE adduct, m.p. 175°–176° (decomp.)	4a, c
1-Methyl-3,4-dimethylenecyclobutene	a or b, 52[d]	B.p. 44°–45° (100 mm)	—	26
Perchloro-3,4-dimethylenecyclobutene	e, 50 and 94	M.p. 147°–148°	Perchloro-4-methylenecyclopentene, m.p. 179°–180°	20a
N-Phenyl-7,8-dimethylenebicyclo[4.2.0]oct-1(6)-ene-3,4-dicarboximide	c	(w)	—	11a
5,6,11,12-Tetramethylenetricyclo[8.2.0.0^{4,7}]dodeca-1(10),4(7)-diene	c, 19	Amorphous solid (w)	—	11a

D. TRIMETHYLENECYCLOBUTANONE

Tetramethylenecyclobutanes and Analogous Methyleneketones

Tetramethylenecyclobutane	a, 50	—	all-cis-Tetramethylcyclobutane	11a
2,4-Bis(diphenylmethylene)-1,3-cyclobutanedione	c, 10[e]	M.p. 252.5°–253° (w)	—	23, 24
3,4-Bis(diphenylmethylene)-1,2-cyclobutanedione	c, 1.9[e]	M.p. 182°–183° (y, fl)	—	23, 24
Hexaphenyltrimethylenecyclobutanone	c, 9	M.p. 171.5°–172° (w, pl); 184°–185° (w, nd)[f]	Corresponding alcohol, m.p. 205°–206°	24
Octaphenyltetramethylenecyclobutane	b, 75[e]	M.p. 290°–293° (y-gr, fl)	—	23, 24

[a] See also Table 5.1 and Table 5.2.
[b] For the 3,4-dimethylenecyclobutenes, Method a is "Base-Catalyzed Elimination Reactions," Section B, 1, a; Method b is "By Rearrangement of a Double Bond," Section B, 1, b; Method c is "From Tetramethylenecyclobutane," Section B, 1, c; Method d is "From Diacetylenes," Section B, 1, d; and Method e is "By a Dehalogenation Reaction," Section B, 1, c. For the tetramethylenecyclobutanes etc., Method a is "By Elimination Reactions," Section C, 1, a; Method b is "By Photochemical Dimerization of a Butatriene," Section C, 1, b; and Method c is ozonolysis of octaphenyltetramethylenecyclobutane (see "Synthesis of Trimethylene and Dimethylenecyclobutanones," Section D, 1).
[c] Key: fl (fluorescent), gr (green), nd (needles), pl (plates), subl. (sublimes), w (colorless), y (yellow).
[d] Calculated from weight data recorded in the literature.
[e] From ref. 23.
[f] Polymorphic.

Like the monoketone, the 1,3-diketone is unaffected by semicarbazide and by 2,4-dinitrophenylhydrazine; furthermore, the 1,3-diketone does not react with hydrogen peroxide or with *o*-phenylenediamine.

3. Physical Properties of the Methylenecyclobutanones

The three known quinomethides of cyclobutadienebis(quinone) are all crystalline solids: Hexaphenyltrimethylenecyclobutanone forms polymorphic crystals (white plates, m.p. 171.5°–172.0°, and white needles, m.p. 184.0°–185°)[24]; 2,4-bis(diphenylmethylene)-1,3-cyclobutanedione is also colorless (white needles, m.p. 252.5°–253°)[24]; but the vicinal diketone, 3,4-bis(diphenylmethylene)-1,2-cyclobutanedione, is yellow and fluorescent (m.p. 182°–183°).[24]

The spectral characteristics of the three methyleneketones are compared with those of octaphenyltetramethylenecyclobutane in Table 5.2.

E. Condensed Methylenecyclobutene Aromatic Systems

Several unsuccessful attempts have been made to prepare condensed methylenecyclobutene aromatic systems. Thus, attempts to prepare the benzologs (2) and (3) of tricyclo[5.3.0.0^{2,6}]deca-1,3,5,7,9-pentaene (*1*) were unsuccessful,[1, 9] as was an attempt to prepare 1,3,4,6-tetraphenylcyclobuta-[1,2-*c*:3,4-*c'*]difuran (*4*).[10]

The positively charged ion $C_{10}H_6^+$ (*m/e* = 126) found in the mass spectra of the dihydroxyanthraquinones has been ascribed to a positively charged tricyclo[5.3.0.0^{2,6}]deca-1,3,5,7,9-pentaenyl ion.[2]

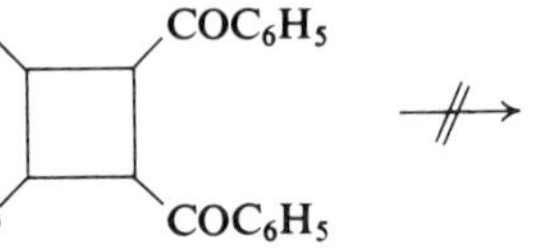

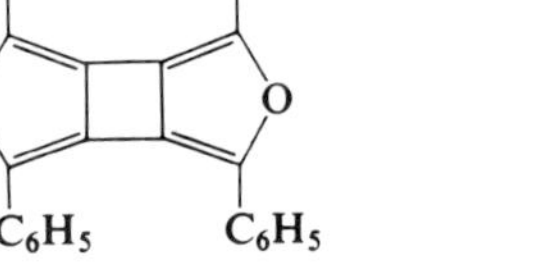

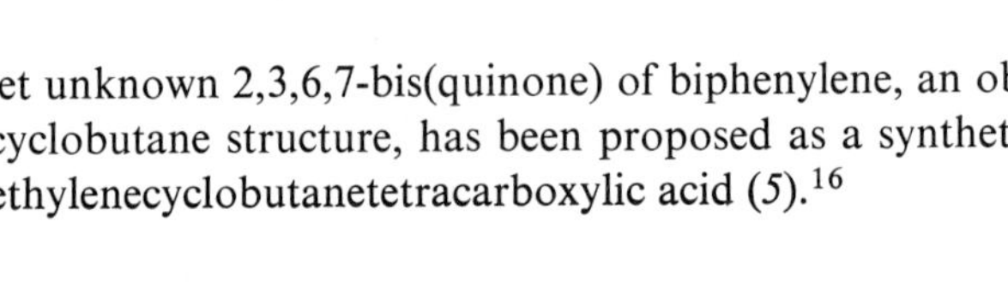

The as yet unknown 2,3,6,7-bis(quinone) of biphenylene, an obvious tetramethylenecyclobutane structure, has been proposed as a synthetic precursor for tetramethylenecyclobutanetetracarboxylic acid (*5*).[16]

References

1. Anastassiou, A. G., Ph.D. diss., Yale (1963).
2. Beynon, J. H., and Williams, A. E., *Appl. Spectry.* **14**, 156 (1960).
3. Blomquist, A. T., and Maitlis, P. M., *Proc. Chem. Soc.* 332 (1961).
4. (a) Blomquist, A. T., and Meinwald, Y. C., *J. Am. Chem. Soc.* **81**, 667 (1959); (b) *ibid.*, *J. Am. Chem. Soc.* **79**, 5316 (1957); (c) *ibid.*, *J. Am. Chem. Soc.* **79**, 5317 (1957).
5. Blomquist, A. T., and Verdol, J. A., *J. Am. Chem. Soc.* **77**, 1806 (1955).
6. Brand, K., *Ber.* **54**, 1987 (1921).
7. Brand, K., and Kercher, F., *Ber.* **54**, 2007 (1921).
8. Criegee, R., Dekker, J., Engel, W., Ludwig, P., and Noll, K., *Ber.* **96**, 2362 (1963).
9. Griffin, G. W., private communication.
10. Griffin, G. W., Hager, R. B., and Veber, D. F., *J. Am. Chem. Soc.* **84**, 1008 (1962).
11. (a) Griffin, G. W., and Peterson, L. I., *J. Am. Chem. Soc.* **85**, 2268 (1963); (b) *ibid.*, *J. Am. Chem. Soc.* **84**, 3398 (1962).
12. Griffin, G. W., and Vellturo, A. F., *J. Org. Chem.* **26**, 5183 (1961).
13. Huntsman, W. D., private communication.
14. Huntsman, W. D., and Wristers, H. J., *J. Am. Chem. Soc.* **85**, 3308 (1963).
15. Manatt, S. L., and Roberts, J. D., *J. Org. Chem.* **24**, 1336 (1959).
16. McOmie, J. F. W., *Rev. Chim. Acad. Rep. Populaire Roumaine* **7**, 1071 (1962).
17. Pilgram, K., and Korte, F., *Tetrahedron Letters* **No. 19**, 883 (1962).
18. Roberts, J. D., Streitwieser, A., Jr., and Regan, C. M., *J. Am. Chem. Soc.* **74**, 4579 (1952).
19. Roedig, A., and Bernemann, P., *Ann.* **600**, 1 (1956), footnote 9 (p. 5) and ref. cited therein. See also ref. 20.
20. (a) Roedig, A., Bischoff, F., Heinrich, B., and Märkl, G., *Ann.* **670**, 8 (1963); (b) Roedig, A., and Bischoff, F., *Naturwissen.* **49**, 448 (1962); (c) Roedig, A., Märkl, G., and Heinrich, B., *Angew. Chem.* **75**, 88 (1963).
21. Roedig, A., and Niedenbrück, H., *Ber.* **90**, 673 (1957).
22. Shechter, H. S., private communication.
23. Uhler, R. O., *Dissertation Abstr.* **21**, 765 (1960).
24. Uhler, R. O., Shechter, H., Tiers, G. V. D., *J. Am. Chem. Soc.* **84**, 3397 (1962).
25. Vellturo, A. F., Ph.D. diss., Yale (1962).
26. Williams, J. K., and Sharkey, W. H., *J. Am. Chem. Soc.* **81**, 4269 (1959).

CHAPTER 6

Benzocyclobutadiene

Benzocyclobutadiene, the monobenzo derivative of cyclobutadiene, is intermediate in structure between the unstable parent compound, cyclobutadiene, and its stable dibenzo analog, biphenylene. Benzocyclobutadiene and its derivatives resemble cyclobutadiene rather than biphenylene in being known only as transient reaction intermediates. In contrast to cyclobutadiene, however, no isolable metal complexes of benzocyclobutadiene have been reported thus far.*

Benzocyclobutadiene is listed in *Chemical Abstracts* under "bicyclo[4.2.0]-octa-1,3,5,7-tetraene." An "Acebenzylen" nomenclature was proposed for benzocyclobutadiene by Wreden in 1876 but was never put to use.[56]

Benzocyclobutadiene has been reviewed by Cava[15] and, in part, by Baker and McOmie.[6]

A. History

The earliest experimental work in the chemical literature dealing with benzocyclobutadiene is described in a note published in 1906 by Bakunin and Parlati.[9] The authors claimed that a minor side-product formed in the reaction of *o*-nitrobenzaldehyde with sodium phenylacetate in acetic anhydride at

* After the manuscript of this chapter had been prepared, a paper appeared in which the preparation of benzocyclobutadieneiron tricarbonyl was described (see ref. 27 and the Appendix).

A. HISTORY   181

120° was 3-nitro-1-phenylbenzocyclobutadiene (1). Their claim, based only on the results of an elemental analysis and a cryoscopic molecular weight determination, was withdrawn in 1927 when it was discovered that the compound was in fact 2-phenylisatogen (2).[10]

         C$_6$H$_5$

  NO$_2$
   (1)   (2)

                      O
                      N—C$_6$H$_5$
                      O

The first purposeful attempt to prepare a benzocyclobutadiene was reported in 1907 by Willstätter and Veraguth, who heated α,α′-dibromo-o-xylene with (a) solid potassium hydroxide and (b) quinoline; the only products isolated were naphthalene,* in the first case, and 1,3-dihydroisobenzofuran, in the second.[55] Roman Malachowski, another of Willstätter's co-workers, made several attempts to prepare benzocyclobutene derivatives suitable for conversion into benzocyclobutadiene, but his work was also unsuccessful and was recorded only in dissertation form.[45] According to Malachowski, Willstätter and Veraguth also attempted to prepare benzocyclobutadiene by the Hofmann degradation of α,α′-bis(dimethylamino)-o-xylene dimethohydroxide.

A few years later, Hans Finkelstein, working in Johannes Thiele's laboratory at the University of Strasbourg, carried out what is now believed to be the first reaction involving a genuine benzocyclobutadiene intermediate. He found that 1,2-dibromobenzocyclobutene (3), obtained from the reaction of α,α,α′,α′-tetrabromo-o-xylene with sodium iodide, could be dehalogenated with zinc suspended in acetone or acetic acid, but he was unable to isolate benzocyclobutadiene (4). Similarly, the reaction of the dibromide (3) with ethanolic potassium hydroxide proceeded readily but gave only a crystalline yellow compound of undetermined structure, C$_{16}$H$_9$Br, rather than 1-bromobenzocyclobutadiene (5).[30] Many years later, both of these reactions were reinvestigated and were shown to proceed via benzocyclobutadiene intermediates.[20]

Thiele and Finkelstein's attempts to synthesize benzocyclobutadiene were recorded in the latter's dissertation (1909)[30] but were not published during the lifetime of either investigator.† Indeed, in a 1910 publication on the reaction of vicinal dihalides with sodium iodide, Finkelstein referred to the synthesis of

---
* See also Section C, 2.

† Finkelstein's dissertation on the synthesis and properties of 1,2-dibromobenzocyclobutene was published posthumously in 1959 (ref. 29) along with a brief biography of the author (ref. 7). In this connection, it is interesting to note that Malachowski, and therefore Willstätter, was aware of Thiele and Finkelstein's work (see p. 48 of ref. 45).

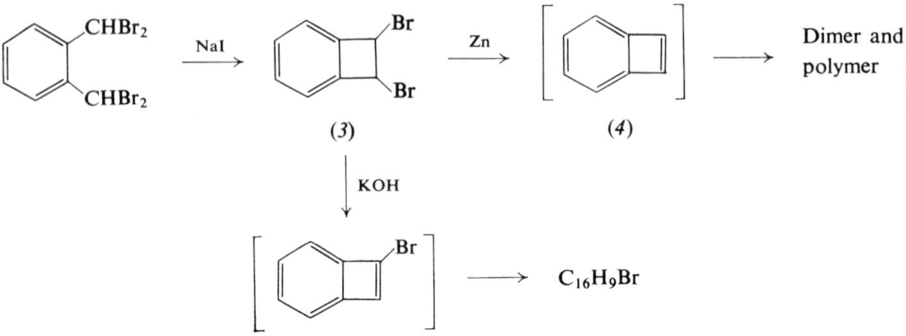

1,2-dibromobenzocyclobutene only briefly and without experimental details.[31] It is not surprising, therefore, that Finkelstein's work lay unnoticed for many years. In fact, it was only through happenstance that is was rediscovered: In 1953, Cava noticed the title of Finkelstein's dissertation in Thiele's obituary in the *Berichte* and, finding the dissertation itself to be of unusual interest, decided to reinvestigate the work in detail. As a result, the structure of 1,2-dibromobenzocyclobutene and the existence of benzocyclobutadiene and 1-bromobenzocyclobutadiene were confirmed and the study of the chemistry of these substances was resumed after a lapse of more than four decades.

B. Generation of Benzocyclobutadiene and Substituted Benzocyclobutadienes

The formation of benzocyclobutadiene, a high-energy species, can be effected only through reactions which have considerable driving force and which are not easily diverted into side-paths. Benzocyclobutadiene and substituted benzocyclobutadienes have been generated for the most part by the dehalogenation of 1,2-dihalobenzocyclobutenes and by the dehydrohalogenation of 1-halobenzocyclobutenes. In this way, one of the most readily available of all the benzocyclobutenes, *trans*-1,2-dibromobenzocyclobutene (2), has served as a convenient starting material for the generation of both benzocyclobutadiene (*1*) and 1-bromobenzocyclobutadiene (*3*).

The only other method by which benzocyclobutadienes have been generated unambiguously consists in the cycloaddition of benzyne to an acetylene.[54]

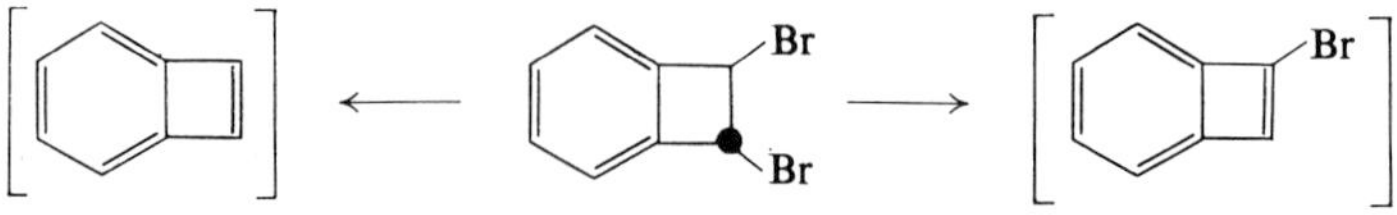

B. GENERATION OF BENZOCYCLOBUTADIENE

This novel method [(4) → (5)], which is probably applicable in only a few special cases, is nevertheless of great interest from the theoretical point of view.

$$\left[\begin{array}{c}\bigcirc\end{array}\right] + \begin{array}{c}R\\|\\C\\|||\\C\\|\\R'\end{array} \longrightarrow \left[\begin{array}{c}\bigcirc\!\!\square\!\!\begin{array}{c}R\\R'\end{array}\end{array}\right]$$

(4)   (5)

In general, the question of whether or not a benzocyclobutadiene intermediate is formed during the course of a reaction must be decided on the basis of secondary evidence, i.e., through the isolation of a polymer, a Diels–Alder adduct, a dimer, or the transformation product of a dimer (a dimeride). The most rigorously established cases of benzocyclobutadiene formation are those in which the diene was isolated in the form of both a dimer (or a dimeride) and a Diels–Alder adduct.

A benzocyclobutadiene structure (6) has been proposed for a fragment of mass 139 observed in the mass spectrum of 2-hydroxyanthraquinone.[11]

(6)

1. By Dehalogenation of 1,2-Dihalobenzocyclobutenes

Benzocyclobutadienes have been generated by the dehalogenation of 1,2-dihalobenzocyclobutenes with (a) zinc dust, (b) lithium amalgam, and, in a few instances, with (c) sodium amalgam, (d) sodium iodide, and (e) nickel tetracarbonyl. (See Table 6.1 for specific examples.) Zinc dust is the most convenient dehalogenating agent to use, but it has one pronounced disadvantage. The zinc halides which are formed in the reaction appear to act as Lewis acids and may promote the polymerization of benzocyclobutadiene. However, if certain precautionary measures are taken, i.e., adding hydroquinone to the reaction mixture and carrying out the reaction under high-dilution conditions, polymer formation is minimized. These precautionary measures appear to be unnecessary when lithium amalgam is the dehalogenation agent because the resulting lithium halides are not Lewis acids and do not promote polymerization.

Benzocyclobutadiene (1) is readily generated by the action of zinc dust (preferably zinc dust activated with aqueous ammonium chloride solution) on either *trans*-1,2-dibromo- or *trans*-1,2-diiodobenzocyclobutene, (2) or (7a), in ethanol.[20, 23, 53] Under these reaction conditions, dehalogenation of the diiodide proceeds more readily than dehalogenation of the dibromide. The fact that the diiodide (7a) is readily dehalogenated was used to advantage in preparing the Diels–Alder adduct of benzocyclobutadiene and cyclopentadiene.[16] (Cyclopentadiene dimerizes rapidly at temperatures much above 25°.) Benzocyclobutadiene is also generated smoothly and under very mild conditions by the action of 0.5% lithium amalgam on an ether solution of the dibromide (2).[48] The lithium amalgam method has been used in the preparation of benzocyclobutadiene dimer and a variety of Diels–Alder adducts.[1, 3, 4]

Benzocyclobutadiene is also believed to be generated during the conversion of dibromide (2) into the corresponding diiodides (7a) and (7b).[20a]

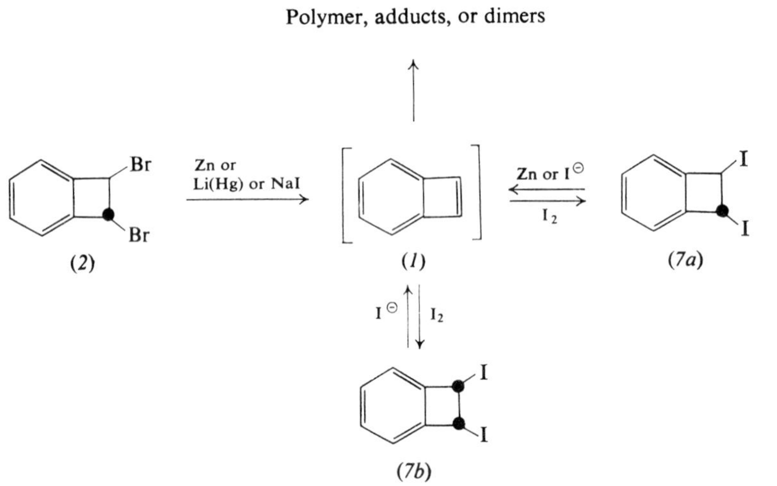

1,2-Dibromobenzocyclobutadiene (9) was generated by the action of sodium iodide on 1,1,2,2-tetrabromobenzocyclobutene (8) in warm dimethylformamide. The diene was isolated in the form of a Diels–Alder adduct with diphenylisobenzofuran and in the form of a dimeride, $C_{16}H_8Br_2$.[17a]

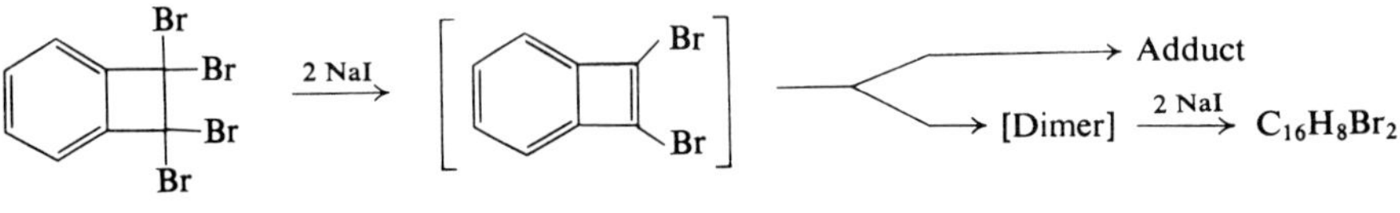

Formal dimers of 1-methyl-2-phenylbenzocyclobutadiene (11) were obtained when 1,2-dibromo-1-methyl-2-phenylbenzocyclobutene (10) was treated with zinc dust, lithium amalgam, sodium amalgam, sodium iodide, or nickel tetracarbonyl.[13] Some doubt exists as to whether or not 1-methyl-2-phenylbenzocyclobutadiene was in fact an intermediate in the dehalogenation of (10) because all attempts to trap the benzocyclobutadiene as a Diels–Alder adduct failed.

$$\underset{(10)}{\text{benzocyclobutene with CH}_3, \text{Br, Br, C}_6\text{H}_5} \xrightarrow[\text{NaI or Ni(CO)}_4]{\text{Zn, Li(Hg)} \atop \text{Na(Hg)}} \left[\underset{(11)}{\text{benzocyclobutadiene with CH}_3, \text{C}_6\text{H}_5}\right] \longrightarrow \text{Dimers}$$

2. By Dehydrohalogenation of 1-Halobenzocyclobutenes

Benzocyclobutadiene and several substituted benzocyclobutadienes have been generated in good yield by the dehydrohalogenation of the appropriate 1-halobenzocyclobutenes with potassium *tert*-butoxide in *tert*-butyl alcohol. Polymer formation is negligible in the dehydrohalogenation reaction and high-dilution conditions are not required. (Compare Section B, 1, above.) Furthermore, since 1-halobenzocyclobutenes are resistant to both solvolysis and to halogen displacement,[20b] ether formation is not a competitive process. In fact, only one exception to this generalization is known, that of 1-methyl-2-bromo-2-phenylbenzocyclobutene, which affords exclusively the methyl, ethyl, and *tert*-butyl ethers [(12a), (12b), and (12c)] and none of the expected 1-methyl-2-phenylbenzocyclobutadiene dimer when it is treated with the corresponding alkoxides.[13] No doubt, in this case solvolysis is facilitated by the α-phenyl substituent.

$$\text{benzocyclobutene (CH}_3\text{, H, Br, C}_6\text{H}_5) \xrightarrow{\text{RO}^\ominus/\text{ROH}} \text{benzocyclobutene (CH}_3\text{, H, OR, C}_6\text{H}_5)$$

(12a): R = *tert*-Bu
(12b): R = Et
(12c): R = Me

Benzocyclobutadiene was generated by the dehydrohalogenation of 1-bromobenzocyclobutene (13) and was isolated in 84% yield in the form of a dimer.[19]

1-Bromobenzocyclobutadiene and 1-iodobenzocyclobutadiene, (3) and (15), were generated by the action of potassium *tert*-butoxide on the corresponding *cis*- and *trans*-1,2-dihalobenzocyclobutenes [(14), (7b), (2), (7a)] and were isolated in the form of dimerides or Diels–Alder adducts.[5, 18, 23, 25] Dehydrohalogenation of the dihalides with potassium *tert*-butoxide and *tert*-butyl alcohol proceeds rapidly at room temperature and gives the dimerides in high yield. Special reaction conditions (high-dilution, etc.) are not necessary and polymer formation is unappreciable.

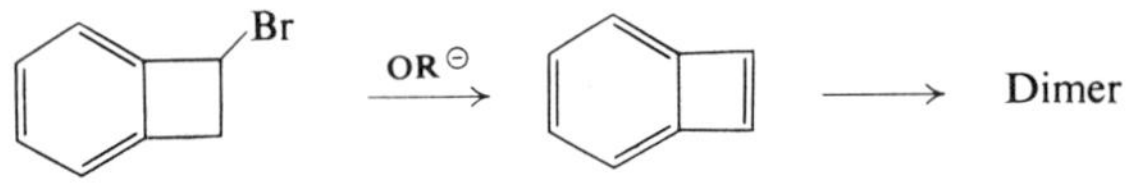

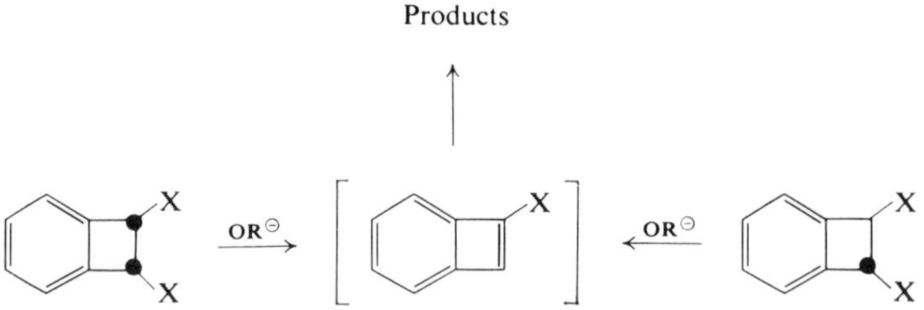

3. By Cycloaddition of Benzyne to Acetylenes

Benzyne, generated by the thermolysis of benzenediazonium-2-carboxylate, reacts with phenylacetylene to give 5,6-diphenyldibenzo[*a,e*]cyclooctatetraene (*17*) in 29% yield and phenanthrene in 8% yield.[54] The origin of the latter is obscure, but the cyclooctatetraene derivative almost certainly arises via 1-phenylbenzocyclobutadiene (*16*), itself formed by the cycloaddition of benzyne to phenylacetylene. Under similar reaction conditions, 1-phenylpropyne reacts with benzyne to give 5,6-dimethyl-11,12-diphenyldibenzo[*a,e*]-cyclooctatetraene (*18*) in 0.5% yield, presumably via 1-methyl-2-phenylbenzocyclobutadiene (*11*), and 9-methylphenanthrene in 3.6% yield.[54]

The benzyne–acetylene cycloaddition reaction will probably prove to be very limited in scope; neither diphenylacetylene nor ethoxyacetylene gave a product attributable to a benzocyclobutadiene intermediate.[54]

4. By Free-Radical Halogenation

The free-radical bromination of benzocyclobutene and of 1-bromobenzocyclobutene is accompanied by polymer formation, attributed to the generation of benzocyclobutadiene and 1-bromobenzocyclobutadiene, respectively.[18, 19]

(*16*): R=H
(*11*): $R=CH_3$

(*17*): R=H
(*18*): $R=CH_3$

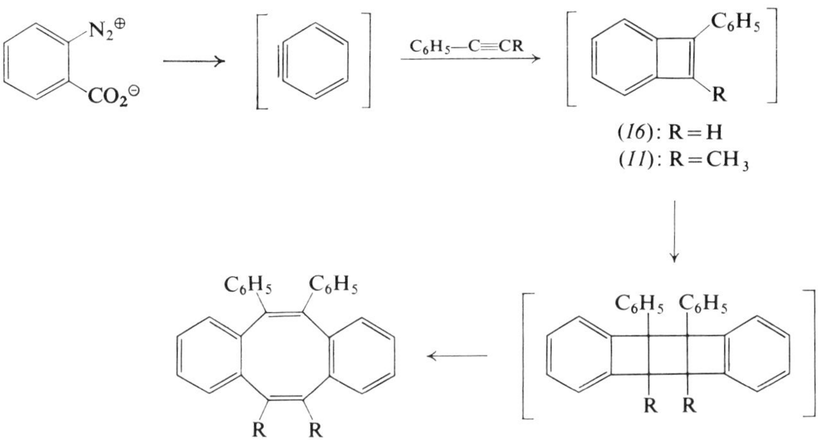

C. Unsuccessful Approaches to the Benzocyclobutadienes

1. By Repositioning an Unsaturated Linkage

Three attempts, all of them unsuccessful, have been made to generate a benzocyclobutadiene by repositioning (shifting) an unsaturated linkage. In the most straightforward of these, 3,4,5,6-tetramethylbenzocyclobutenone (*1*) was treated with heavy water in an effort to determine whether the ketone is capable of undergoing enolization.[38] Contrary to the published report, the ketone does not undergo deuterium exchange at C-2, and hence the benzocyclobutadienoid enol form (*2*) is not in equilibrium with the keto form.[37]

(*1*) (*2*)

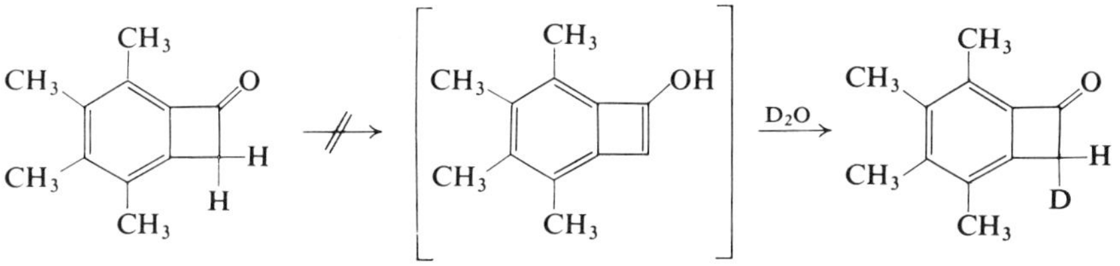

In a variant of this approach, biphenylene-2,3-quinone (*3*) was subjected to the conditions of the Thiele acetylation reaction (acetic anhydride and an acid catalyst).[12] The "normal" Thiele acetylation product, a benzocyclobutadiene derivative (*4*), was not formed.

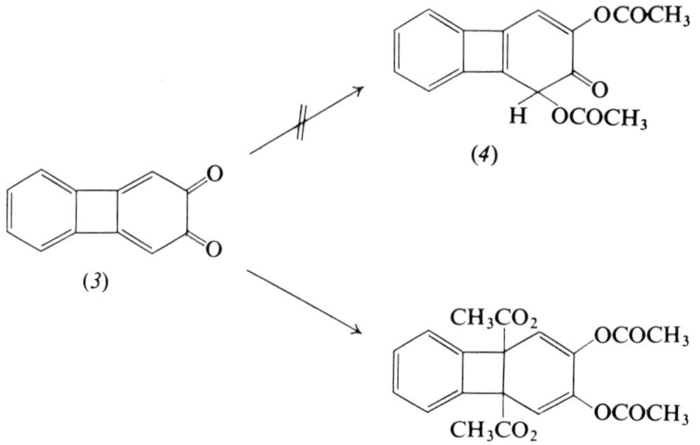

In still another variant, biphenylene-2,3-quinone was treated with maleic anhydride in the hope that a Diels–Alder reaction would take place, giving a benzocyclobutadiene derivative (5).[12] However, the quinone proved to be quite inert to maleic anhydride. In a similar attempt, 1,2-dimethylenebenzocyclobutene (6) was treated with tetracyanoethylene but it also failed to give a benzocyclobutadienoid Diels–Alder adduct (7) and gave an indeterminate polymer instead.[21]

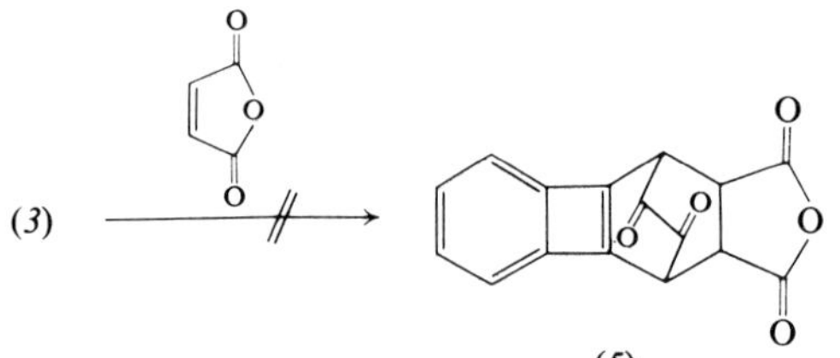

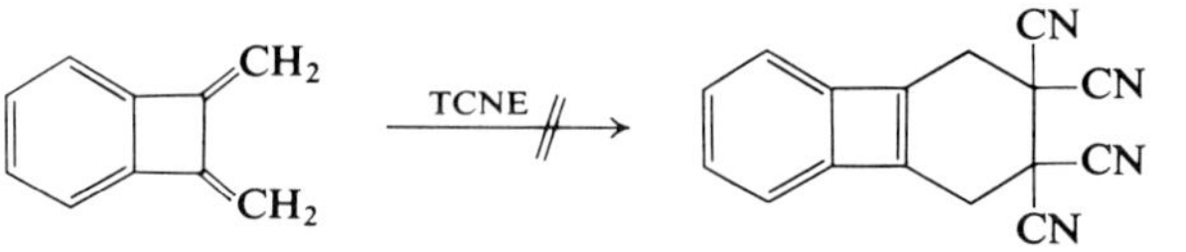

2. From α,α'-Disubstituted o-Xylenes

The double dehydrohalogenation of α,α'-dibromo-o-xylene (8) with quinoline gave a resinous reaction product from which only naphthalene could be isolated, rather than the expected product, benzocyclobutadiene (9).[55] (The

C. UNSUCCESSFUL APPROACHES TO THE BENZOCYLOBUTADIENES

origin of the naphthalene is obscure and it is not impossible that it was present in the quinoline as an impurity.) When the dibromide was heated with powdered potassium hydroxide, the only product that could be obtained was 1,3-dihydroisobenzofuran (*10*).[55] Similarly, an attempted double Hofmann reaction with α,α′-bis(dimethylamino)-*o*-xylene dimethohydroxide (*11*) gave only a resin of composition $(C_8H_8O)_x$ and not the expected product, benzocyclobutadiene.*,[45]

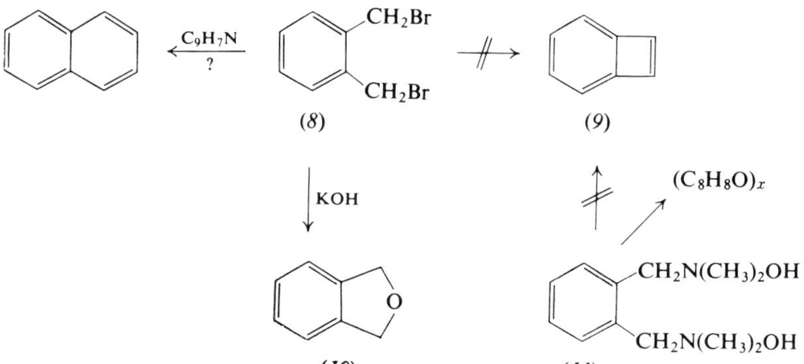

3. BY DEHYDROHALOGENATION OF THE ADDUCT OF 1-IODOBENZOCYCLO-
BUTADIENE AND 1,3-DIPHENYLISOBENZOFURAN

5a-Iodo-5,10-diphenyl-5,5a,9b,10-tetrahydro-5,10-oxidobenzo[*b*]biphenylene [(*12*), prepared by generating 1-iodobenzocyclobutadiene in the presence of 1,3-diphenylisobenzofuran] was found to be extraordinarily resistant to the usual dehydrohalogenating agents.[23] Under vigorous reaction conditions, i.e., (a) potassium *tert*-butoxide in refluxing *p*-xylene and (b) sodium amide in refluxing piperidine, the adduct was reduced to (*14*). It has been suggested that a benzocyclobutadienoid species (*13*) might be produced as a transient intermediate in this transformation; if so, (*13*) must abstract two hydrogen atoms from the

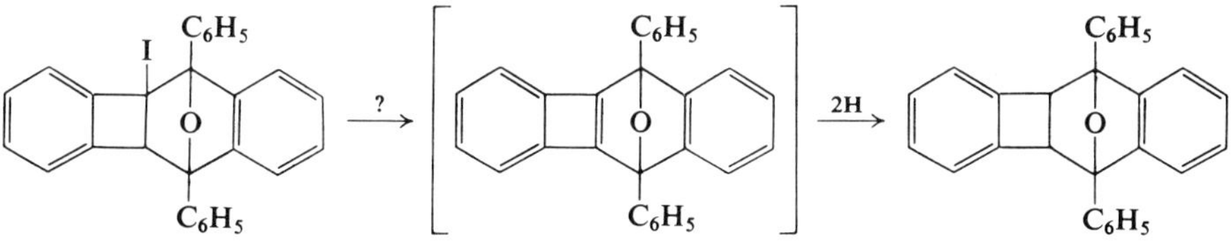

---

* This unpublished experiment is ascribed by Malachowski to Willstätter and Veraguth. See p. 48 of ref. 45.

solvent to give (*14*). The reluctance of iodide (*12*) to undergo dehydrohalogenation has been ascribed to an unusual degree of strain in the benzocyclobutadiene intermediate (*13*).[23]

4. By Dehalogenation of the Adduct of 1,2-Dibromobenzocyclobutadiene and 1,3-Diphenylisobenzofuran

5a,9b-Dibromo-5,10-diphenyl-5,5a,9b,10-tetrahydro-5,10-oxidobenzo[*b*]biphenylene [(*15*), prepared by generating 1,2-dibromobenzocyclobutadiene in the presence of 1,3-diphenylisobenzofuran] was found to be resistant to dehalogenation.[23] Under vigorous reaction conditions, i.e., (a) zinc dust in refluxing ethanol and (b) lithium amalgam in ether, the adduct was converted into 5,10-diphenylbenzo[*b*]biphenylene (*17*) by elimination of both the oxide bridge and the bromine atoms [(*15*) → (*16*) → (*17*)]. It is unlikely that a benzocyclobutadienoid species is produced as a transient intermediate in this transformation.[23]

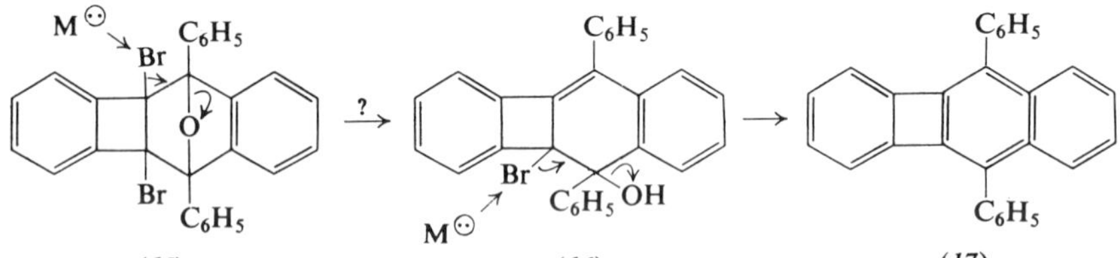

5. By Reaction of Benzyne with Acetylenes

Benzyne failed to react with diphenylacetylene to give the expected dimer (*19*) of 1,2-diphenylbenzocyclobutadiene (*18*). With ethoxyacetylene, benzyne gave 2-ethoxyphenylacetylene (*20*) and not a dimer such as would be expected to be formed via the benzocyclobutadiene intermediate (*21*).[54]

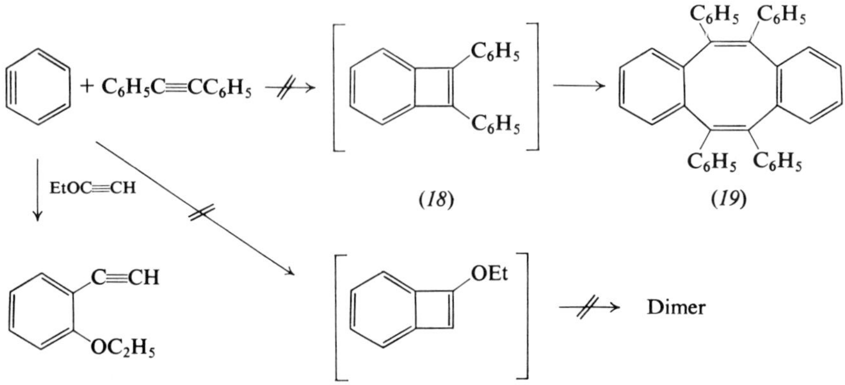

C. UNSUCCESSFUL APPROACHES TO THE BENZOCYCLOBUTADIENES

6. BY DEHYDROHALOGENATION OF 1,2-DIIODOBENZOCYCLOBUTENE WITH DIMETHYLAMINE

Treating *trans*-1,2-diiodobenzocyclobutene with dimethylamine gave an unstable compound believed to be 1,2-(dimethylamino)benzocyclobutene (*24*).[36b] A mechanism involving benzocyclobutadiene intermediates (*22*) and (*23*) was considered for this transformation[36b] but was rejected when the product was found to have structure (*25*) and not (*24*).[36a]

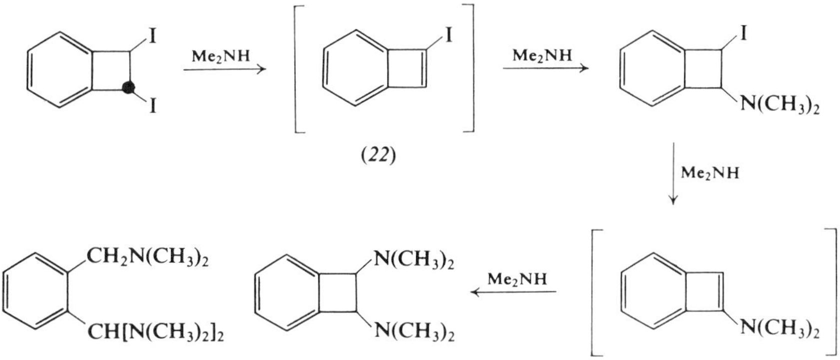

7. BY THE FRIEDEL–CRAFTS REACTION

Lagidze and Petrov assigned a benzocyclobutadiene structure (*26*) to a hydrocarbon, "$C_{14}H_{10}$," m.p. 102°, which they obtained by the reaction of benzene with 1,4-diacetoxy-2-butyne and aluminum chloride.[43a] The authors considered a tetramethylenebenzodicyclobutene structure for the hydrocarbon (*27*) but rejected it in favor of structure (*26*), because the hydrocarbon was found to be stable to permanganate. Structure (*26*) was subsequently withdrawn in favor of (*28*), a valence tautomer of anthracene.[43b,43c] Lagidze's "$C_{14}H_{10}$" hydrocarbon has been shown to be 2-phenylnaphthalene ($C_{14}H_{12}$).[28,34,44]

Lagidze proposed benzocyclobutadiene structure (*29*) for the product resulting from the reaction of 2,5-diacetoxy-2,5-dimethyl-3-hexyne with benzene and aluminum chloride.[43d] The correct structure of the product is that of 5,5,10,10-tetramethyl-5,5a,10,10a-tetrahydroindeno[2,1-*a*]indene (*30*).[35] Similarly, the product resulting from the reaction of 1,4-diacetoxy-1-cyclohexyl-4-methyl-2-pentyne with benzene and aluminium chloride was said to be either (*31*) or (*32*).[43e] The structure of the product is not yet known,[33] but it is almost certain not to be (*31*) or (*32*).

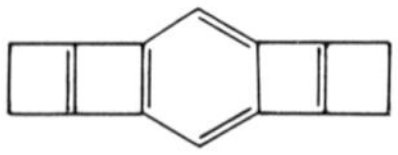

(26) 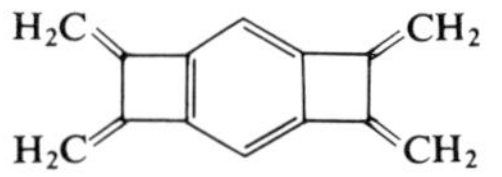   (27) 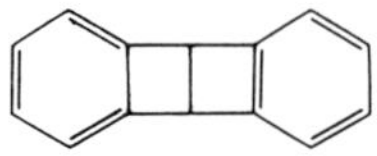   (28)
(29) 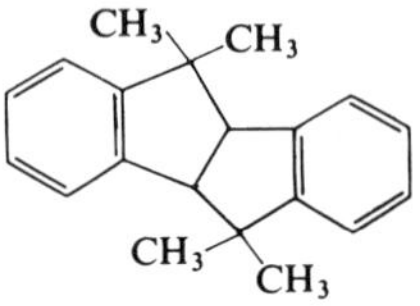   (30) 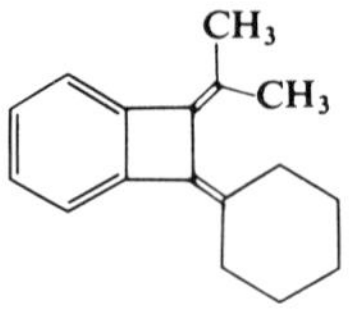   (31) 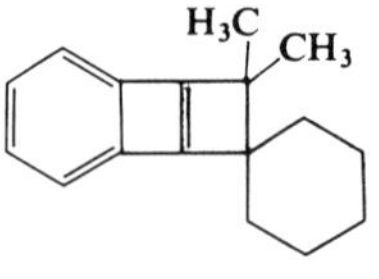   (32) 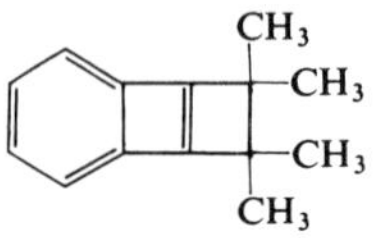

8. From a Reduced Biphenylene

When a reduced biphenylene derivative (*34*) having an *N*-oxide substituent[42] was subjected to the Cope elimination, a methylenecyclobutene compound (*35*) was formed rather than the corresponding benzocyclobutadiene compound (*33*).[41]

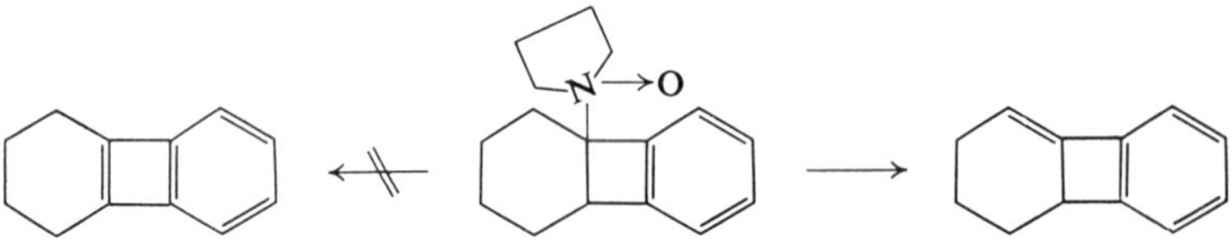

(33)   (34)   (35)

9. By Dehydrohalogenation of a 1-Halobenzocyclobutene

Treatment of 1-bromo-2-methyl-1-phenylbenzocyclobutene (*37*) with sodium ethoxide, sodium methoxide, or potassium *tert*-butoxide failed to give 1-methyl-2-phenylbenzocyclobutadiene (*36*) and gave the corresponding ethers (*38*) instead.[13]

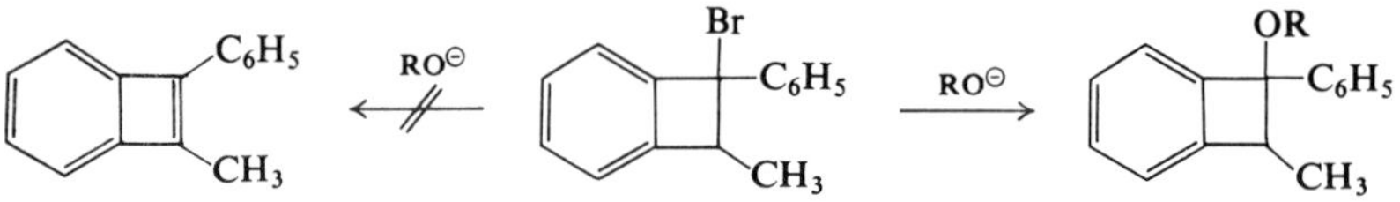

(36)   (37)   (38)

R = Me, Et, or *tert*-Bu

D. Chemistry of Benzocyclobutadiene

Molecular orbital calculations predict that, in contrast to cyclobutadiene, benzocyclobutadiene should have a singlet ground state and substantial delocalization energy (2.38 β or ~48 kcal/mole[51]) and that phenyl substituents

D. CHEMISTRY OF BENZOCYCLOBUTADIENE   193

at the 1- and 2-positions should exert a substantial stabilizing effect on the system. The calculated values for the delocalization energy (DE) of 1-phenylbenzocyclobutadiene and of 1,2-diphenylbenzocyclobutadiene are appreciably larger than the corresponding values ($\overline{DE}$) obtained by simply adding together the DE values of benzocyclobutadiene (2.38 β) and of one or two phenyl groups (2 β per phenyl group).[46] (See Chart 6.1). The highest free-valence index for benzocyclobutadiene (0.62 at the 1- and 2-positions) is less than that accepted for ethylene (0.73).[49, 51]

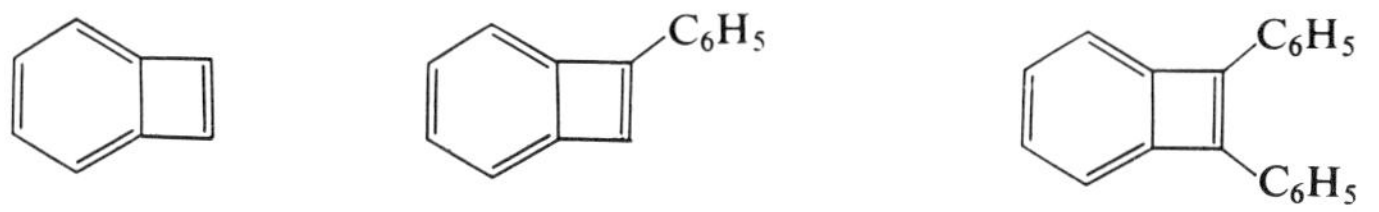

DE = 2.38 β

DE = 4.84 β
$\overline{DE}$ = 2.38 β + 2 β = 4.38 β

DE = 7.32 β
$\overline{DE}$ = 2.38 β + 4 β = 6.38 β

CHART 6.1. Effect of phenyl substituents on the calculated delocalization energy of benzocyclobutadiene.

Three canonical resonance forms may be written for benzocyclobutadiene, (1a), (1b), and (1c). Of these, the two Kekulé structures (1a) and (1b) may be expected to be much more important as contributors to the hybrid than the o-quinonoid structure (1c). Furthermore, of the two Kekulé structures, the second (1b), which is not a cyclobutadienoid form, should be of lower energy and should therefore be the major contributor. These qualitative results are in agreement with the bond orders in benzocyclobutadiene as calculated by molecular orbital theory.[51]

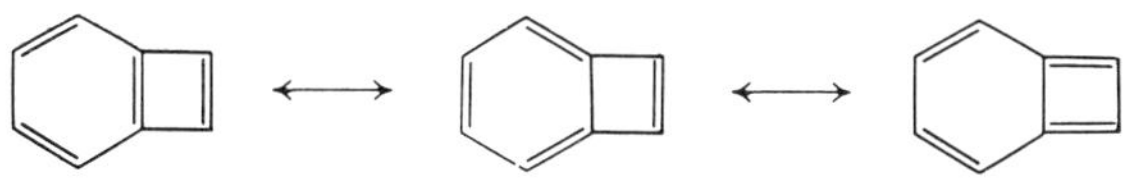

(1a)   (1b)   (1c)

Despite the favorable properties predicted for benzocyclobutadiene, experience has shown that it is a compound of very low stability and, in fact, benzocyclobutadiene is known only as a transient intermediate. Nevertheless, it is not impossible that a suitably substituted benzocyclobutadiene having moderate stability (perhaps 1,2-diphenylbenzocyclobutadiene) will be synthesized eventually.

1. POLYMER FORMATION

When benzocyclobutadiene is generated by the dehalogenation of 1,2-dibromo- and 1,2-diiodobenzocyclobutene with zinc dust, a substantial amount of polymer (*3*) is always formed along with the expected dimer, even when precautionary measures (high-dilution conditions, hydroquinone, etc.) are taken to guard against it. It has been suggested that polymer formation is promoted by the zinc halides which are formed in the dehalogenation reaction.[19] This hypothesis is supported by the fact that benzocyclobutadiene and substituted benzocyclobutadienes are not prone to polymerization when generated under basic reaction conditions. Thus, benzocyclobutadiene (*1*)[19] and 1-bromobenzocyclobutadiene (*6*)[25] gave the dimeric products, (*5*) and (*7*), respectively, in good yield (84% and 86.1%) and without the formation of polymeric material when generated by the dehydrohalogenation of 1-bromo- (*4*) and 1,2-dibromobenzocyclobutene (*2a*) with potassium *tert*-butoxide in *tert*-butyl alcohol. The high-dilution technique was not required under these conditions.

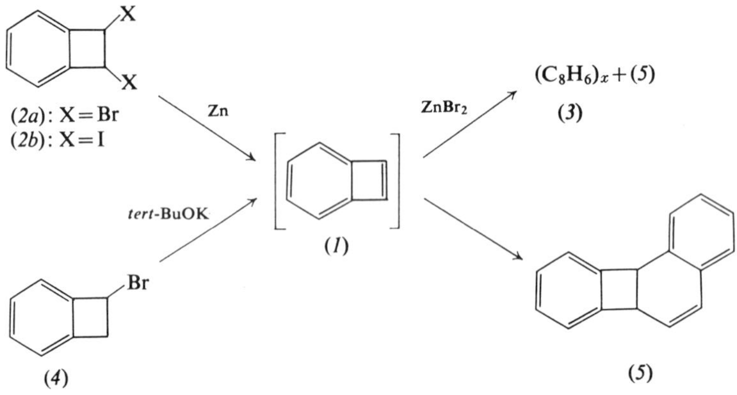

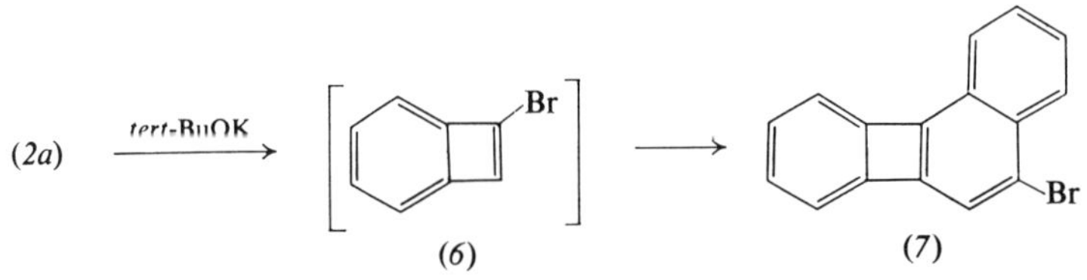

In contrast, the free-radical bromination of benzocyclobutene[19] and of 1-bromobenzocyclobutene[18] is accompanied by much polymer formation and this has been attributed to the generation of benzocyclobutadiene and 1-bromobenzocyclobutadiene, respectively, during the reaction.

2. Dimer Formation

a. By a Diels–Alder Reaction Path

Benzocyclobutadiene and the substituted benzocyclobutadienes can form two kinds of dimers, the so-called "angular" dimers and "linear" dimers. The angular dimers are formed through a Diels–Alder reaction in which one molecule of benzocyclobutadiene, acting as a diene, adds to another molecule of the same species, acting as a dienophile. The initial adduct rearranges as it is formed to give 6a,10b-dihydrobenzo[a]biphenylene (5). It is interesting to note that in this reaction one of the bonds of the fused benzene ring (the 2a,6a-bond) acts as part of a diene system. The angular dimerization of benzocyclobutadiene can take place at room temperature or below and is closely related mechanistically to the thermolysis of thianaphthene dioxide (9), which gives a tetracyclic sulfone (11), presumably via intermediate (10).[14]

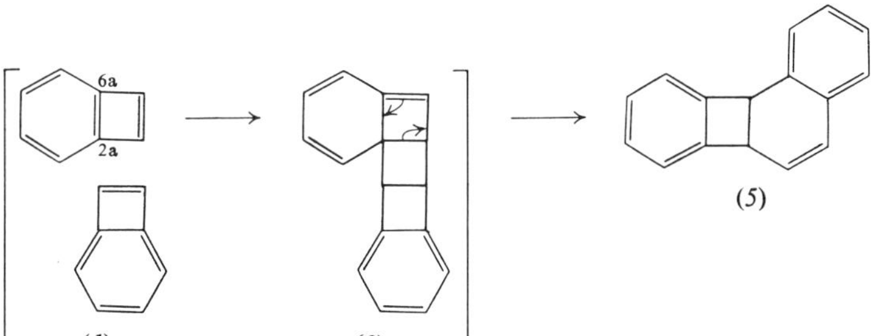

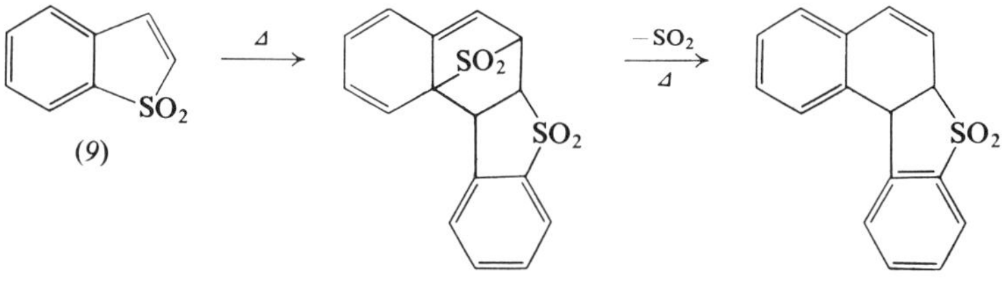

1-Halobenzocyclobutadienes, generated by the dehydrohalogenation of 1,2-dihalobenzocyclobutenes, also undergo the Diels–Alder dimerization reaction. However, the resulting halogen-containing dimers of the 6a,10b-dihydrobenzo[a]biphenylene type [(13a) and (13b)] are not isolable and are dehydrohalogenated in situ to give the corresponding fully aromatic benzo[a]-biphenylenes. Thus, 1-bromobenzocyclobutadiene (6), generated from 1,2-dibromobenzocyclobutene (2a) by the action of excess base, gives the angular dimer 5,6a(or 10b)-dibromo-6a,10b-dihydrobenzo[a]biphenylene (13a) initially, which is dehydrohalogenated in situ to give the final product, 5-bromobenzo[a]biphenylene (7).*,[25] Similarly, the dehydrohalogenation of

* The exact structure of intermediate (13a) is not known.

1,2-diiodobenzocyclobutene (*2b*) gives as the final product, 5-iodobenzo[*a*]-biphenylene (*14*),[24] and the dehydrohalogenation of 1,1,2-tribromobenzocyclobutene (*15*) affords 5,6-dibromobenzo[*a*]biphenylene (*18*).[17] In the latter case, 1,2-dibromobenzocyclobutadiene (*16*) is formed, and the intermediate dimer, 5,6,6a,10b-tetrabromo-6a,10b-dihydrobenzo[*a*]biphenylene (*17*), undergoes *debromination* (rather than dehydrobromination) to give the final product (*18*). The same end product is also formed when 1,2-dibromobenzocyclobutadiene is generated by the sodium iodide dehalogenation of 1,1,2,2-tetrabromobenzocyclobutene [(*19*) → (*16*) → (*17*) → (*18*)].[17a]

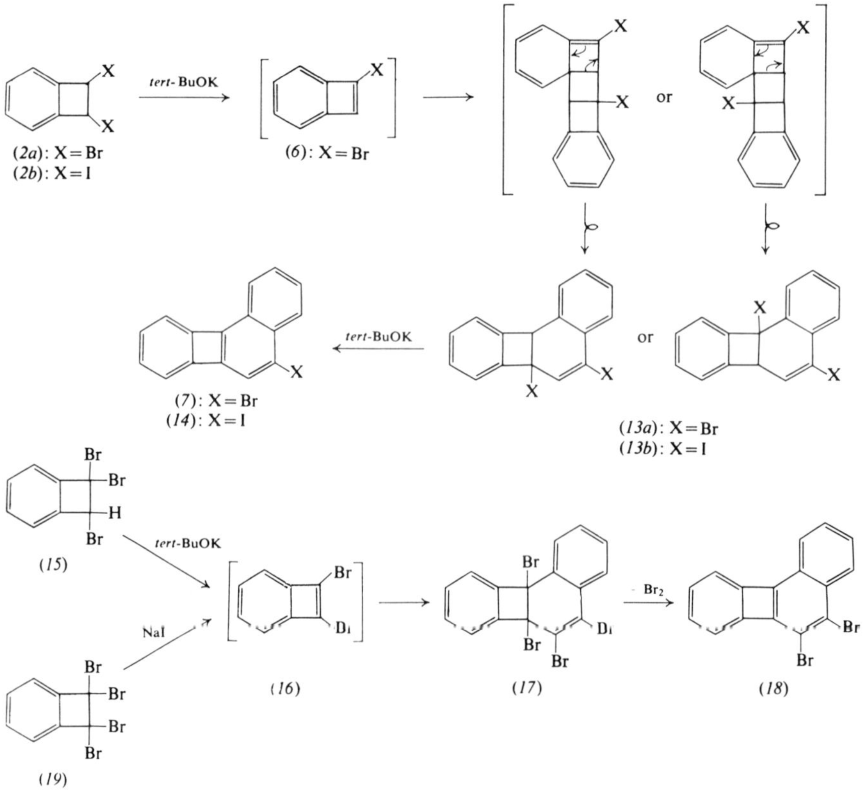

The dehydrohalogenation of α,α,α',α'-tetrabromo-*o*-xylene with potassium *tert*-butoxide is of considerable interest because all three of the products of the reaction, viz., 5,6,11,12-tetrabromodibenzo[*a*,*e*]cyclooctatetraene (*26*), 5,10-dibromobenzo[*b*]biphenylene (*30*), and 5,6-dibromobenzo[*a*]biphenylene (*18*),[5] appear at first sight to be dimers or dimerides of 1,2-dibromobenzocyclobutadiene (*16*).[39] However, when 1,2-dibromobenzocyclobutadiene is

D. CHEMISTRY OF BENZOCYCLOBUTADEINE

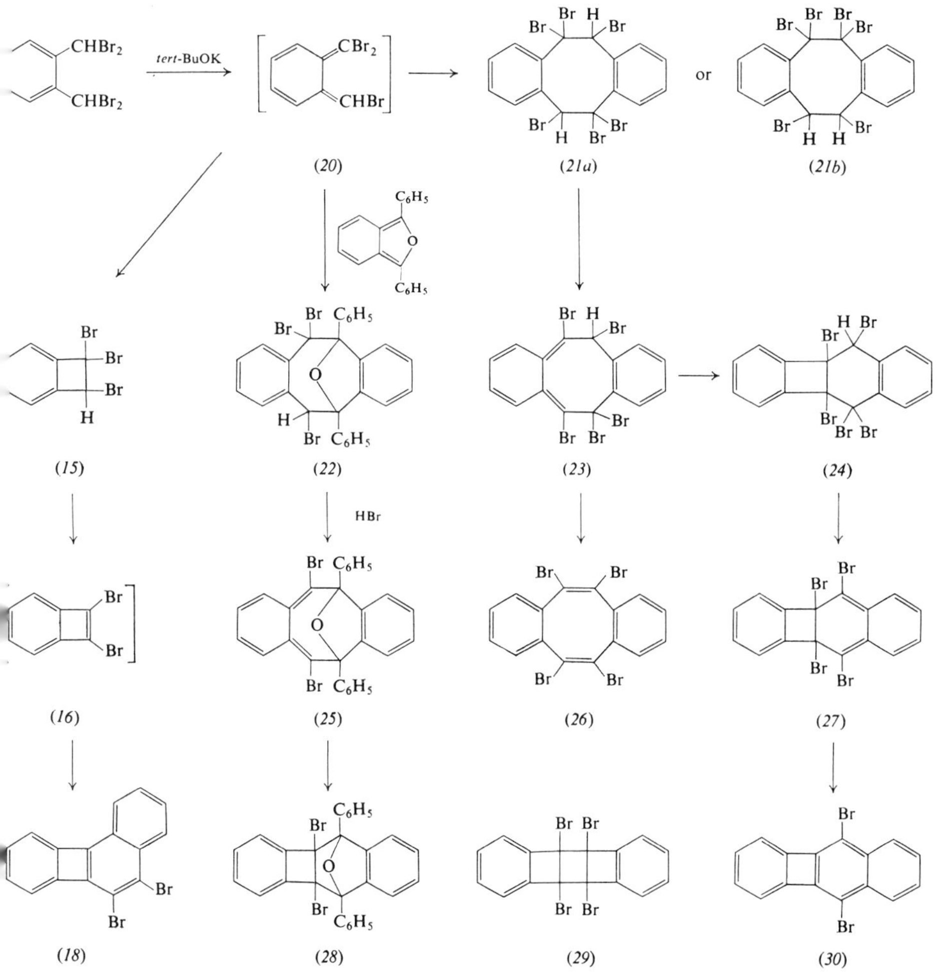

generated by unambiguous methods, i.e., by the dehydrohalogenation of 1,1,2-tribromobenzocyclobutene or by the dehalogenation of tetrabromobenzocyclobutene, only 5,6-dibromobenzo[a]biphenylene (18) is formed. Consequently, while it is almost certain that the small amount of 5,6-dibromobenzo[a]biphenylene which is obtained in the tetrabromo-o-xylene transformation stems from 1,2-dibromobenzocyclobutadiene, it is unlikely that the other two products of this reaction, (26) and (30), are formed in this way. Instead, it has been suggested that products (26) and (30) are formed via a dibenzocyclooctadiene intermediate (21a) or (21b).[17] This hypothesis is supported by the fact that when 1,3-diphenylisobenzofuran is added to the tetrabromo-o-xylene reaction mixture, an adduct (28) derived from o-quinodimethane intermediate (20) is obtained in high yield. Adduct (28) is probably formed by the same kind of reaction path as that leading to compounds (26) and (30), i.e., by the 1,4-coupling of two o-quinonoid species.[23]

Of course, it might at first appear that adduct (28) is formed by a straightforward Diels–Alder reaction between 1,2-dibromobenzocyclobutadiene and 1,3-diphenylisobenzofuran; undoubtedly a small portion of the adduct is formed in this way. However, the entire 86% yield of adduct (28) cannot be ascribed to 1,2-dibromobenzocyclobutadiene because its dimeride (18) is formed only in small yield (4%) when the dehydrobromination of tetrabromo-o-xylene is carried out in the absence of 1,3-diphenylisobenzofuran.

Finally, it is necessary to note that the product to which we have assigned structure (26) is described in the literature as 1,2,9,10-tetrabromopentacyclo-[$8.6.0.0^{2,9}.0^{3,8}.0^{11,16}$]hexadeca-3,5,7,11,13,15-hexaene (29).[5, 17, 39] The correctness of structure (26) has been demonstrated by an X-ray crystallographic analysis.[47]

b. *By Cycloadditive Dimerization*

When 1-methyl-2-phenylbenzocyclobutadiene (32) was generated by the reaction of benzyne with methylphenylacetylene[54] or by the reaction of 1,2-dibromo-1-methyl-2-phenylbenzocyclobutene(31)with zinc dust,[13] it was found that the end product was not the usual angular dimer of the dihydrobenzo[a]-biphenylene type (34), but rather a linear dimer which is a dibenzo[a,e]cyclooctatetraene derivative. Oxidative degradation of the dibenzocyclooctatetraene to tetraphenylethylene [(35) → (37) → (40)] and to o-benzoylbenzoic acid (36) showed that it is the 5,6-dimethyl-11,12-diphenyl isomer (35), undoubtedly formed by valence isomerization of a dibenzotricyclooctadiene intermediate (33).[13, 54] Since this intermediate is precisely the one to be expected from a head-to-head, cis-oriented cycloadditive dimerization,[50] it is not unreasonable to assume that it has the syn configuration as well (33a). The facile transformation of dimer (33) (syn configuration) into a dibenzo[a,e]cyclooctatetraene ("tub" conformation) probably reflects the close configurational similarity between the two compounds.

When 1-phenylbenzocyclobutadiene (38) was generated by the reaction of benzyne with phenylacetylene, the end product of the reaction was 5,6-diphenyldibenzo[a,e]cyclooctatetraene (39),[54] which also probably results from a head-to-head, cis-oriented dimerization.

c. *By Cycloadditive Dimerization via a Metal-Complex Intermediate*

When benzocyclobutadiene is generated from 1,2-dibromobenzocyclobutene by debromination in the presence of nickel tetracarbonyl, the usual "angular" benzocyclobutadiene dimer is not formed. Instead, a linear benzocyclobutadiene dimer, 3,4,7,8-dibenzotricyclo[4.2.0.0^{2,5}]octadiene (42),[1,2] is obtained which has a moderate degree of thermal stability,* and has the anti configuration of the central rings, as was demonstrated by ozonolysis to cis,trans,cis-1,2,3,4-cyclobutanetetracarboxylic acid (43).[32] The formation of the linear dimer of benzocyclobutadiene undoubtedly proceeds by way of an unstable nickel complex,[1] which may be observed as a transient red color during the course of the reaction.[2] A number of possible structures, e.g., (41a)–(41d), may be written for the nickel complex; structures (41a) and

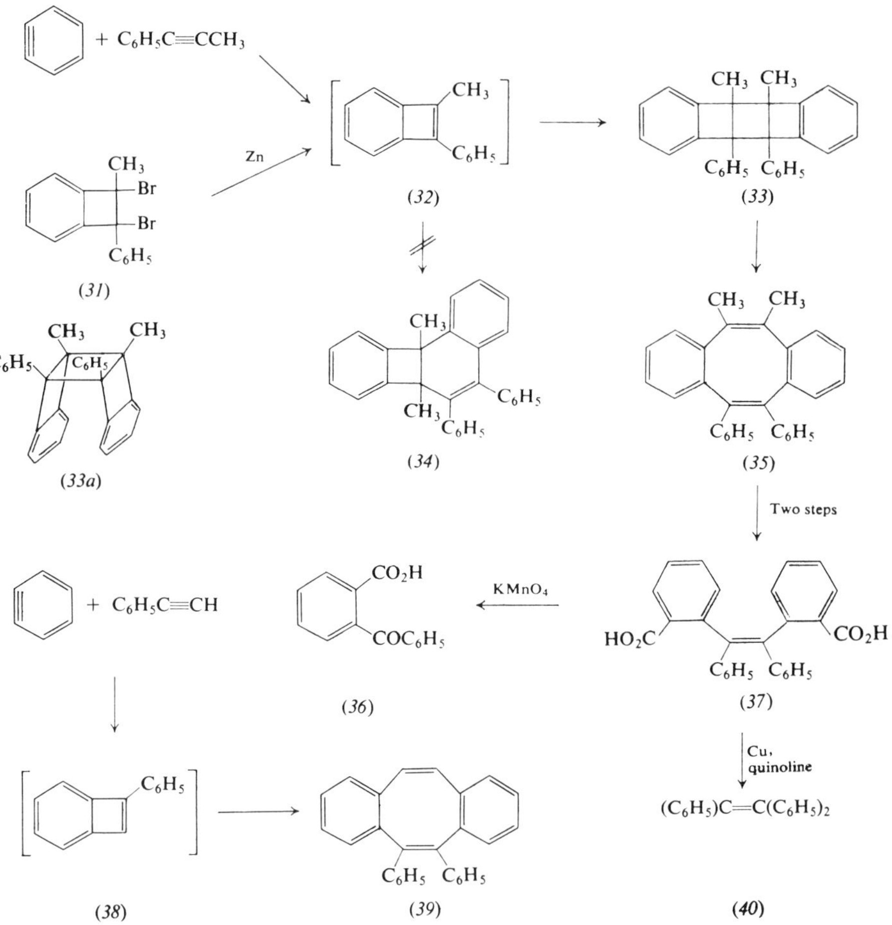

* On being heated to its melting point (131°) or above, the compound rearranges to dibenzo[a,e]cyclooctatetraene (ref. 1). The mechanism of this rearrangement has been studied (ref. 2, 22).

(*41d*) seem more probable than (*41b*) and (*41c*) since they are more consistent with the anti stereochemistry of the final product (*42*).

When 1-methyl-2-phenylbenzocyclobutadiene was generated in the presence of nickel tetracarbonyl, a stable dimer which is believed to be 1,2(or 5)-dimethyl-5(or 2),6-diphenyl-3,4,7,8-dibenzotricyclo[4.2.0.0^{2,5}]octadiene (*33*) [or (*44*)] was formed.[13] The exact skeletal configuration and substitution pattern of the dimer are not known, but it is probable that it will be found to have the trans substitution pattern and that, like (*42*), it will have the anti configuration (*44a*).

(*41a*) (*41b*)

(*44a*)

or

(*41c*)
X = Br or CO

(*41d*)

(*41*)

(*42*) (*43*)

(*32*) (*33*) (*44*)

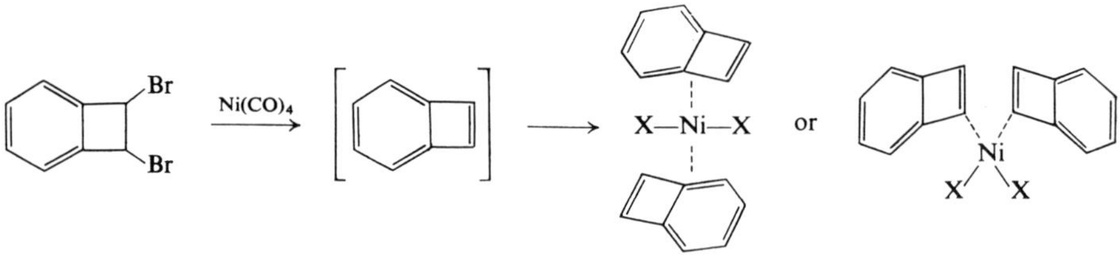

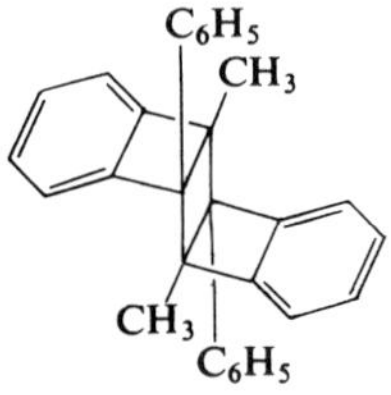

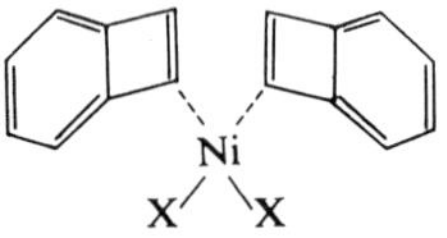

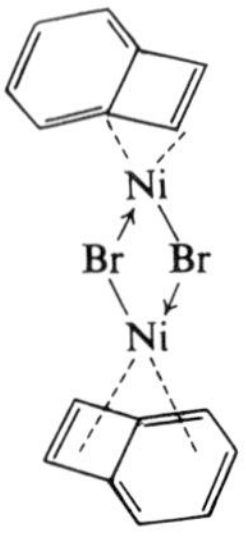

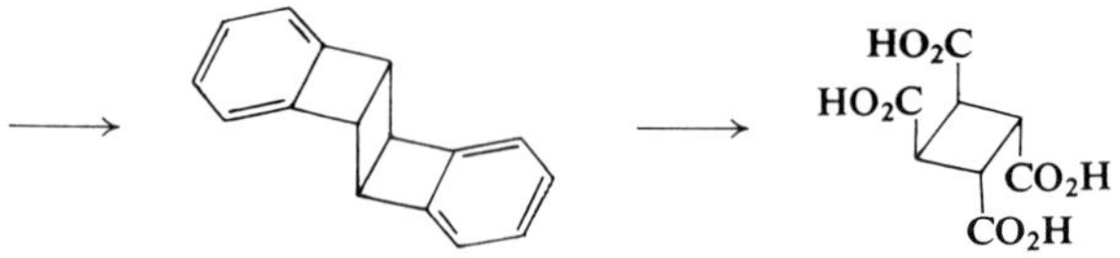

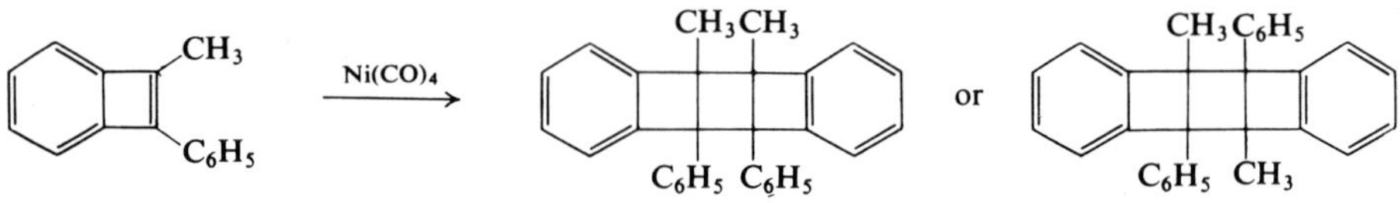

D. CHEMISTRY OF BENZOCYCLOBUTADIENE

3. Diels–Alder Adduct Formation

a. With Dienes

The existence of benzocyclobutadienes as distinct, though transient, substances has been demonstrated convincingly by trapping experiments with reactive dienes.

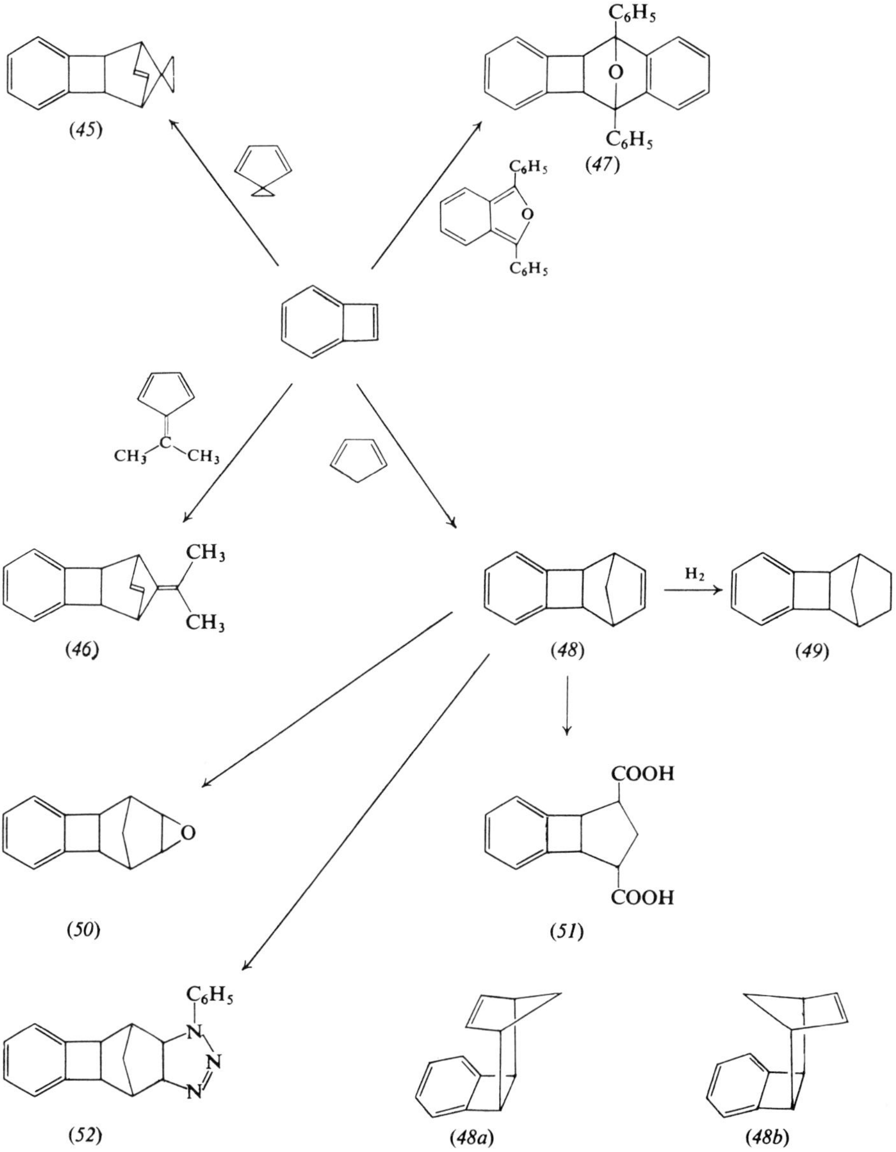

Benzocyclobutadiene itself (as generated from 1,2-dihalobenzocyclobutenes and metals) is a highly reactive dienophile and reacts readily with spiroheptadiene,[4] dimethylfulvene,[4] 1,3-diphenylisobenzofuran,[4,23] and cyclopentadiene[16,48] to give the corresponding adducts (45), (46), (47), and (48), respectively. The chemistry of the cyclopentadiene adduct has been studied in some detail and confirms the assigned structure (48). Thus, adduct (48) reacts with phenyl azide to give a triazole derivative (52)[16,48] and with monoperphthalic acid to give a monoepoxide (50)[16]; it is oxidized by permanganate to give a dicarboxylic acid (51)[16] and reduced catalytically to give a dihydro derivative (49).[16b,52] The adduct was found to have the endo configuration (48a) by a comparative NMR study with its exo isomer (48b), obtained by the cycloaddition of benzyne to norbornadiene.[52] Hence, the rule of maximum accumulation of double bonds is applicable in the Diels–Alder addition of benzocyclobutadiene to cyclopentadiene and, presumably, in its addition to other dienes as well.

When dienes of only moderate reactivity are employed as trapping agents, the rate of dimerization of benzocyclobutadiene becomes comparable with, or greater than, the rate of adduct formation. Thus, butadiene gave none of the expected adduct (53) and only the usual "angular" benzocyclobutadiene dimer (5) could be isolated.[16] The furan adduct (54) was obtained in low yield (about 5%) and then only by generating benzocyclobutadiene in the presence of a large excess of furan. Here again, the major reaction product was the angular dimer (5).*,[16] The furan adduct gave a crystalline derivative (55) with phenyl azide, but it could not be converted into biphenylene (56) by treatment with mineral acid (see also Chapter 10, Section B, 5).[16]

The fate of benzocyclobutadiene in the presence of anthracene appears to be dependent on the temperature. Thus, at room temperature only the dimer of benzocyclobutadiene (5) is obtained,[22] while at a higher temperature (refluxing ethanol) both the dimer (5) and the expected adduct (57) are obtained.[53] Further, at the temperature of refluxing dimethylformamide the yield of adduct (33%)[22] is more than twice that (15%) obtained in refluxing ethanol. The structure of the anthracene adduct (57) has been confirmed by an independent synthesis from p-benzoquinone.[22]

Two reactions have been recorded in which benzocyclobutadiene formed adducts with unstable o-quinonoid hydrocarbons: (a) benzocyclobutadiene and o-quinodimethane gave a 1:1 adduct (58) when generated simultaneously by debromination of a mixture of trans-1,2-dibromobenzocyclobutene and o-xylylenedibromide.[2,3] (b) In a similar manner, benzocyclobutadiene and 2,3-dihydronaphthalene gave an adduct (59) when generated together.[2,3] These

---

* In two earlier reports, furan and benzocyclobutadiene were said to give as reaction products an "unstable oxygen-containing polymer" (ref. 8) and an unstable product of m.p. 155–160° (ref. 48), respectively.

reactions are remarkable in that they proceed in good yield (over 40%) even though they involve the addition of two transient species. It is therefore likely that the lifetimes and rates of formation of benzocyclobutadiene and the *o*-quinonoid hydrocarbons are fortuitously of the same order of magnitude.

An attempt to prepare the furan adduct (*60*) of 1-iodobenzocyclobutadiene was unsuccessful and only the usual dimeride (*14*) could be obtained.[16b] However, 1-bromo-, 1-iodo-, and 1,2-dibromobenzocyclobutadiene were all trapped successfully with the more reactive diene, 1,3-diphenylisobenzofuran. Thus, 1-bromo- and 1-iodobenzocyclobutadiene, generated from 1,2-dibromo- and 1,2-diiodobenzocyclobutene by dehydrohalogenation with potassium

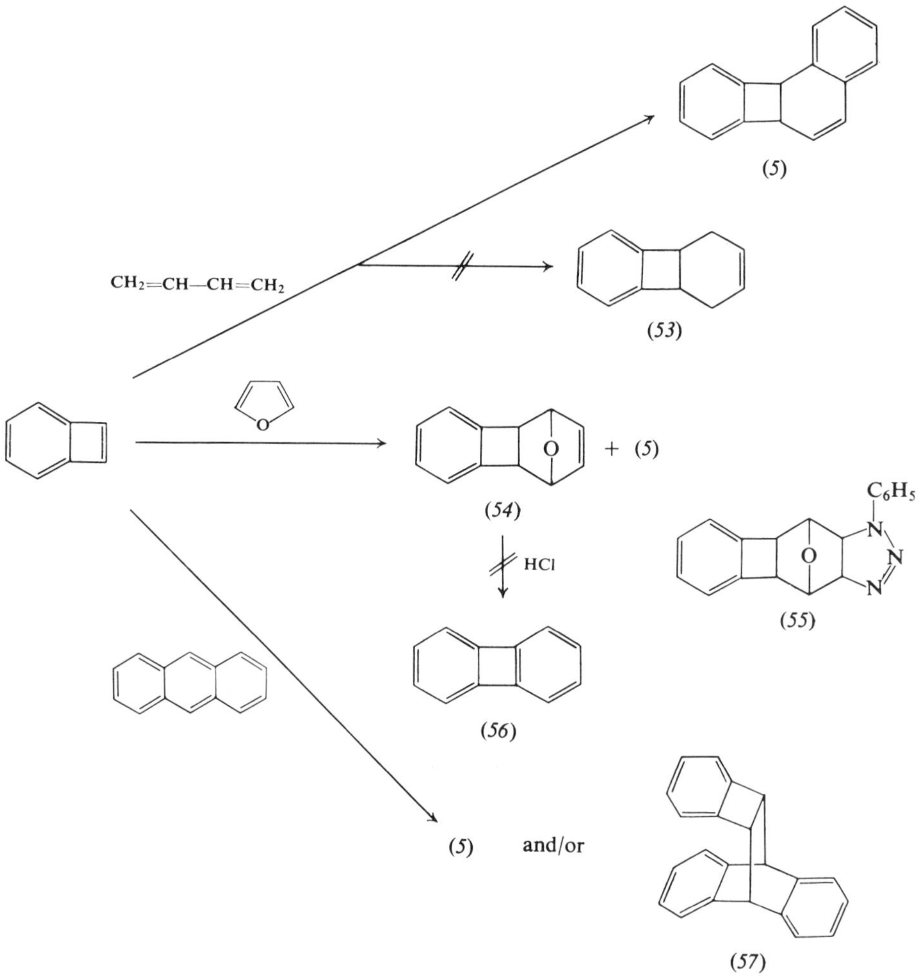

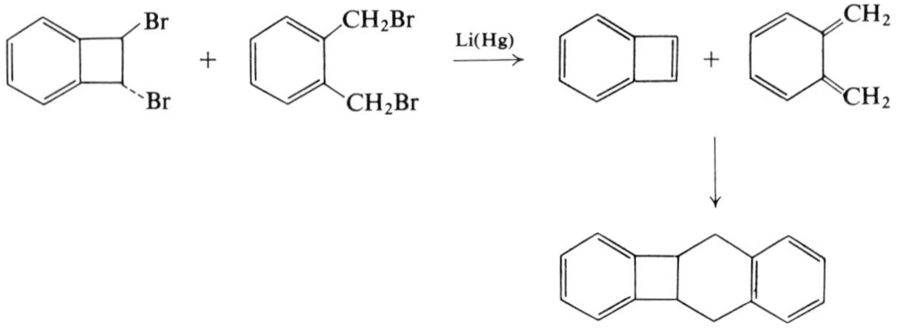

6. BENZOCYCLOBUTADIENE

(58)

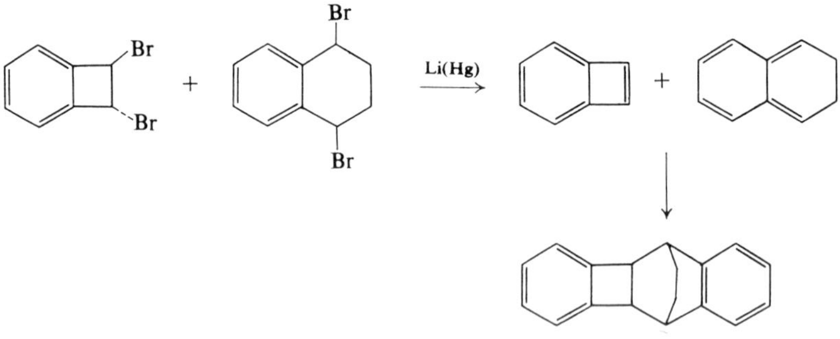

(59)

tert-butoxide, reacted with 1,3-diphenylisobenzofuran to give the expected adducts (61) and (62), respectively, in good yield.[23] Similarly, 1,2-dibromobenzocyclobutadiene, generated either by the dehydrohalogenation of 1,1,2-tribromobenzocyclobutene (with potassium tert-butoxide) or by the dehalogenation of 1,1,2,2-tetrabromobenzocyclobutene (with sodium iodide), gave the 1,3-diphenylisobenzofuran adduct (28).[23] 1-Bromobenzocyclobutadiene was also trapped by the extremely reactive diene, 1,3-diphenylnaphtho[2,3-c]-furan, to give the expected adduct (63).[26]

b. With a Dienophile

Although benzocyclobutadiene behaves as both a diene and a dienophile during dimerization, it appears to possess only limited reactivity as a diene. Thus, it forms no adduct with either dimethyl maleate or dimethyl acetylenedicarboxylate,[16b] but forms the angular benzocyclobutadiene dimer instead. The dimer (5) is the major product isolated, even when benzocyclobutadiene is generated in the presence of the highly reactive dienophile, N-phenylmaleimide. However, in the N-phenylmaleimide reaction, the expected Diels–

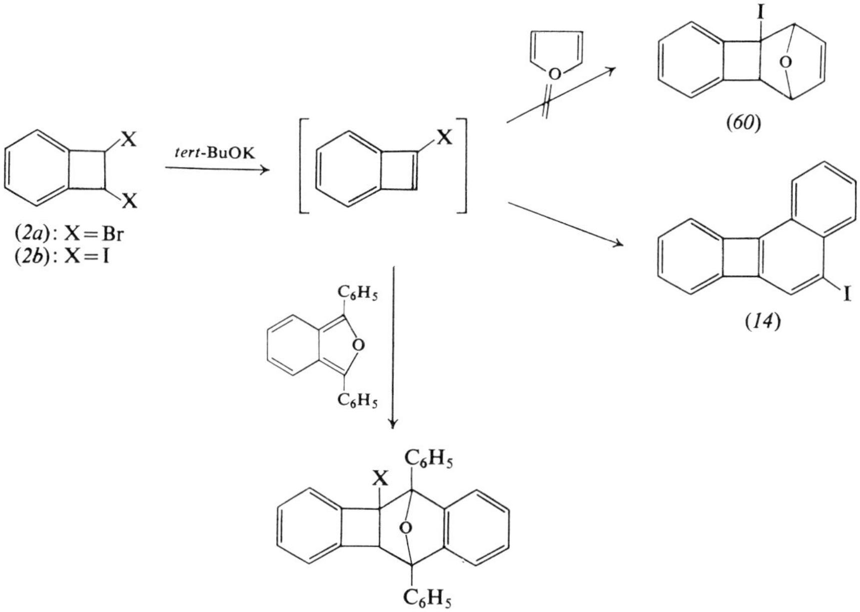

D. CHEMISTRY OF BENZOCYCLOBUTADIENE

(2a): X=Br
(2b): X=I

Alder adduct, N-phenyl-1,2-dihydronaphthalene-1,2-dicarboximide (64) is also formed, although in low yield, together with its dehydrogenation product, N-phenylnaphthalene-1,2-dicarboximide (65).[16] The formation of the expected Diels–Alder adduct in this reaction is of more than passing interest, since it

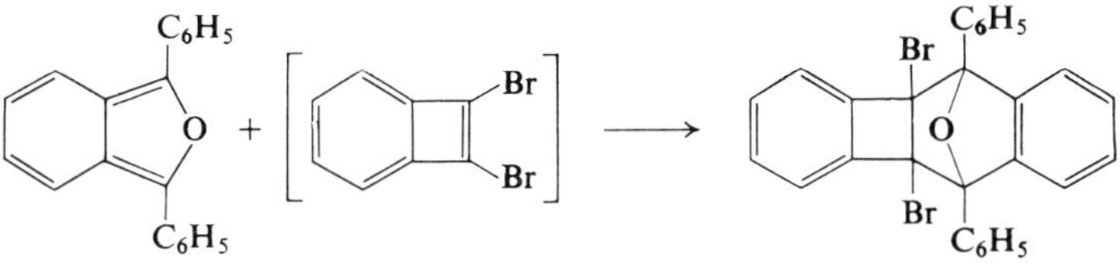

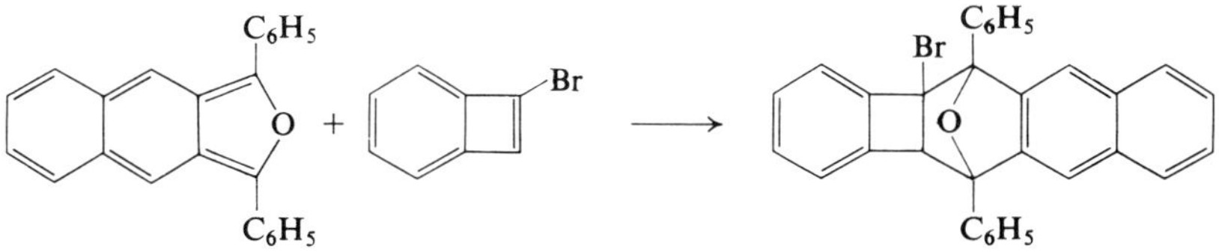

TABLE 6.1

BENZOCYCLOBUTADIENES

Part A. Generation of benzocyclobutadienes Part B. Formation of dimers, adducts, etc., of benzocyclobutadienes

Method of generation[a]	Reactants	Trapping agent and/or reaction conditions[b]	Product[c]	Yield (%)	Ref.
Benzocyclobutadiene					
1	*trans*-1,2-Dibromobenzocyclobutene and lithium amalgam	Et_2O (solvent), room temp.	I	80	48
		Et_2O (solvent), nickel tetracarbonyl	II	67	1
		Cyclopentadiene, Et_2O (solvent), room temp.	III	52	48
		2,3-Dihydronaphthalene, Et_2O, room temp.	IV	65	3
		Dimethylfulvene, Et_2O, room temp.	V	53	4
		DPIBF, Et_2O, room temp.	VI	57	4
		Furan, Et_2O, room temp.	"Compd, m.p. 155°–160°"	—	48
		o-Quinodimethane, Et_2O, room temp.	VII	44	3
		Spiro[2.4]hepta-1,3-diene	VIII	41	4
	1,2-Dibromobenzocyclobutene and (1) sodium iodide (2) zinc dust	EtOH (solvent, reflux), I_2	"IXa or IXb"	78	20a
		Acetone (solvent)	—	—	29, 30
		Acetic acid (solvent)	—	—	29, 30

D. CHEMISTRY OF BENZOCYCLOBUTADIENE 207

Reactants	Conditions	Products	Yield	Ref.
1,2-Dibromobenzocyclobutene and activated zinc dust	EtOH (solvent, reflux), h-d, HQ, N_2	Polymer and I	83	20
	Anthracene, room temp.	I	—	22
	Anthracene, DMF (reflux)	X	33	22
	Anthracene, EtOH (reflux), h-d, HQ, N_2	I and X	65 and 15	53
trans-1,2-Dibromobenzocyclobutene and activated zinc dust	DPIBF, EtOH (reflux)	VI	55	23
cis-1,2-Diiodobenzocyclobutene and sodium iodide	EtOH (solvent, reflux)	IXa and IXb	91 and 9	40
trans-1,2-Diiodobenzocyclobutene and				
(1) sodium iodide	EtOH (solvent, reflux)	IXa and IXb	91 and 9	40
(2) activated zinc dust	EtOH (solvent, reflux)	Polymer and I	70	20
	Butadiene, EtOH, h-d	Polymer and I	—	16
	Cyclopentadiene, EtOH, h-d, N_2, 30°	III	64.9	16
	Dimethylacetylenedicarboxylate	I	—	16b
	Dimethyl maleate	I	—	16b
	Furan, EtOH, h-d, N_2, 30°	I and XI	64 and 5.6	16
	NPMI, EtOH (reflux), h-d, HQ, N_2	XII and XIII	5.2 and 1.3	16
2 1-Bromobenzocyclobutene and potassium tert-butoxide	tert-Butyl alc. (solvent), room temp.	I	84	19
4 Benzocyclobutene and NBS	Carbon tetrachloride, light	1-Bromobenzocyclobutene and polymer	—	19
1-Bromobenzocyclobutadiene				
2 1,1-Dibromobenzocyclobutene and potassium tert-butoxide	tert-Butyl alc. (solvent)	XIV	75	18
cis-1,2-Dibromobenzocyclobutene and potassium tert-butoxide	tert-Butyl alc. (solvent)	XIV	86.1	25
trans-1,2-Dibromobenzocyclobutene and potassium tert-butoxide	tert-Butyl alc., THF (solvent), DPIBF	XV	60	23
	tert-Butyl alc., THF (solvent), DPIBF	XV	73	23
1,2-Dibromobenzocyclobutene and potassium hydroxide	MeOH (solvent, reflux)	XIV	31.4[a]	29, 30

Table 6.1 (continued)

Method of generation[a]	Reactants	Trapping agent and/or reaction conditions	Product[c]	Yield (%)	Ref.
4	1-Bromobenzocyclobutene and NBS	Carbon tetrachloride, light	1,2-Dibromobenzocyclobutene and polymer	—	18
1,2-Dibromobenzocyclobutadiene					
1	1,1,2,2-Tetrabromobenzocyclobutene and sodium iodide	DMF (solvent)	XVI	41	17a
	α,α',α'-Tetrabromo-o-xylene and potassium *tert*-butoxide	EtOH–THF, DPIBF, 80°–90°	XVII	35	23
		tert-Butyl alc. (solvent)	XVI and other products	4	5
2	1,1,2-Tribromobenzocyclobutene and potassium *tert*-butoxide	*tert*-Butyl alc. (solvent)	XVI	85	17
		tert-Butyl alc., DPIBF	XVII	88	23
1-Iodobenzocyclobutadiene					
2	*cis*-1,2-Diiodobenzocyclobutene and potassium *tert*-butoxide	*tert*-Butyl alc., DPIBF	XVIII	73	23
	trans-1,2-Diiodobenzocyclobutene and potassium *tert*-butoxide	*tert*-Butyl alc. (solvent)	XIX	71	24
		tert-Butyl alc., DPIBF	XVIII	92	23
1-Methyl-2-phenylbenzocyclobutadiene					
1	1,2-Dibromo-1-methyl-2-phenylbenzocyclobutene and				
	(1) lithium amalgam	—	XX	45	13
	(2) sodium amalgam	—	XX	56	13

D. CHEMISTRY OF BENZOCYCLOBUTADIENE

(3) nickel tetracarbonyl	Benzene (solvent, reflux)	XXIa (or b)	—	13
(4) activated zinc dust	EtOH (solvent)	XX	10	13
	EtOH (solvent)	XX	85	13
3 Benzyne (from benzenediazonium-2-carboxylate) and methylphenylacetylene	—	XX	0·5	54
1-Phenylbenzocyclobutadiene				
3 Benzyne (from benzenediazonium-2-carboxylate) and phenylacetylene	—	XXII	29	54

Part C. Structural formulas and physical properties of the dimers, adducts, etc. of benzocyclobutadiene (I–XIII), 1-bromobenzocyclobutadiene (XIV–XV), 1,2-dibromobenzocyclobutadiene (XVI–XVII), 1-iodobenzocyclobutadiene (XVIII–XIX), 1-methyl-2-phenylbenzocyclobutadiene (XX–XXI), and phenylbenzocyclobutadiene (XXII).[b]

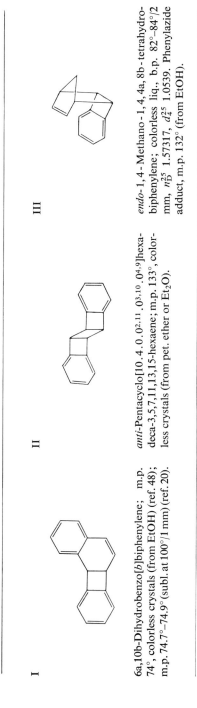

I

II

III

6a,10b-Dihydrobenzo[b]biphenylene; m.p. 74°, colorless crystals (from EtOH) (ref. 48); m.p. 74.7°–74.9° (subl. at 100°/1 mm) (ref. 20).

anti-Pentacyclo[10.4.0.0^{2,11}.0^{3,10}.0^{4,9}]hexadeca-3,5,7,11,13,15-hexaene; m.p. 133°, colorless crystals (from pet. ether or Et$_2$O).

endo-1,4-Methano-1,4,4a,8b-tetrahydrobiphenylene; colorless liq., b.p. 82°–84°/2 mm, n_D^{25} 1.57317, d_4^{25} 1.0539. Phenylazide adduct, m.p. 132° (from EtOH).

Table 6.1 (*continued*)

IV	V	VI
5,10-Ethano-5, 5a, 9b, 10-tetrahydrobenzo-[*b*]biphenylene; m.p. 181°.	Spiro[cyclopropane-1, 9′-(1, 4-methano-1,4,4a,8b-tetrahydrobiphenylene)]; b.p. 115°–117°/1 mm. Phenylazide adduct, m.p. 63° (from MeOH).	5,10-Diphenyl-5,10-oxido-5, 5a, 9b, 10-tetrahydrobenzo[*b*]biphenylene; m.p. 194° (from Me₂CO–MeOH) (ref. 4).
VII	VIII	IX
5, 5a, 9b, 10-Tetrahydrobenzo[*b*]bipheryl-ene; m.p. 124°.	9-Isopropylidenyl-1, 4-methano-1,4,4a,8b-tetrahydrobiphenylene; oily liq., b.p. 130°/5 mm. Phenylazide adduct, m.p. 208° (from EtOH).	1,2-Diiodobenzocyclobutene: IXa (*trans*-), colorless rhombs, m.p. 62.7°–62.8° (from pet. ether) (ref. 20a). IXb (*cis*-), colorless needles, m.p. 150.1°–150.8° (decomp).

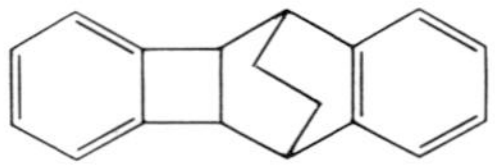

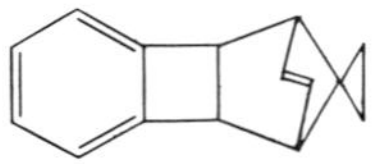

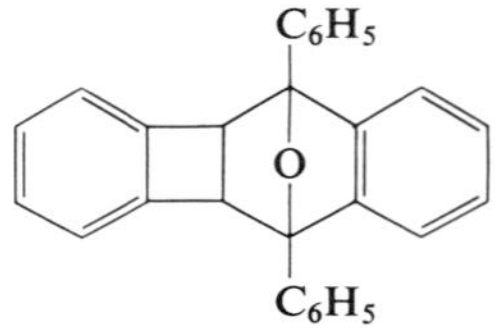

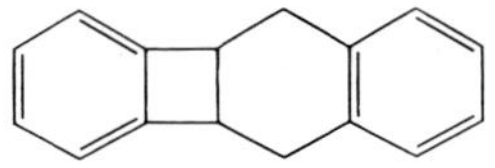

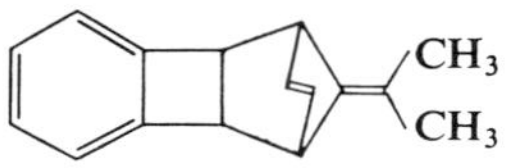

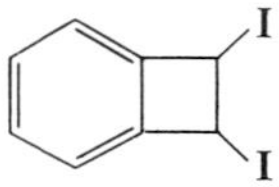

D. CHEMISTRY OF BENZOCYCLOBUTADIENE

X

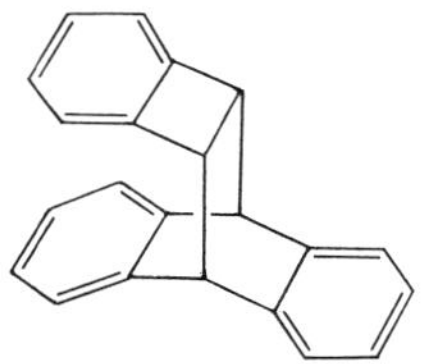

4b, 5, 10, 10a-Tetrahydro-5,10-*o*-benzeno-benzo[*b*]biphenylene; m.p. 164°–165° (from EtOH); λ_{max} 745, 733 cm^{-1} (Nujol) (ref. 53).

XI

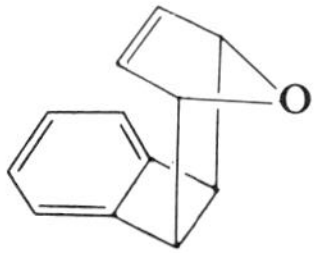

1,4-Oxido-1,4,4a,8b-tetrahydrobiphenyl-ene; white crystals, m.p. 51.5°–53.5° (subl. at 38°/1 mm).

XII

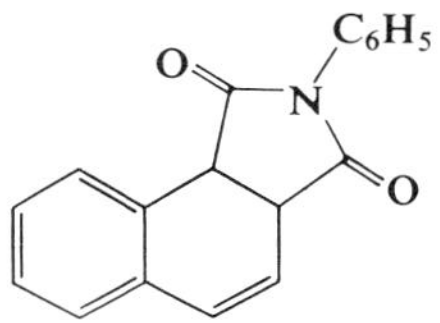

N-Phenyl-1, 2-dihydronaphthalene-1, 2-dicarboximide; colorless crystals, m.p. 137°–138°, $\lambda_{max}^{n\text{-hex}}$ 261 mμ (log ϵ, 3.94), 272 (3.92), 296 (2.56, shoulder).

XIII

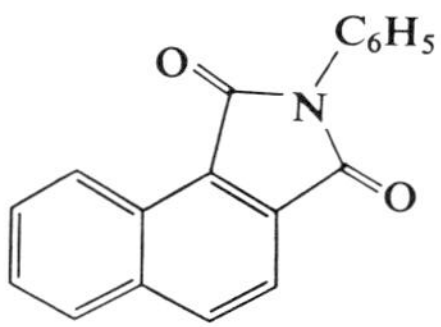

N-Phenylnaphthalene-1, 2-dicarboximide; yellow needles, m.p. 162°–163°.

XIV

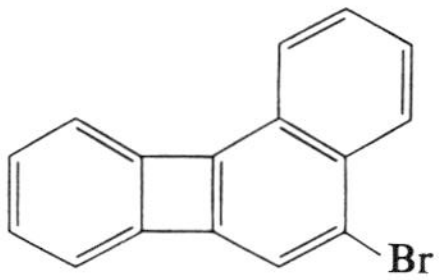

5-Bromobenzo[*a*]biphenylene; orange needles, m.p. 126°–127.5° (from EtOH) (subl. *in vacuo*) (ref. 25).

XV

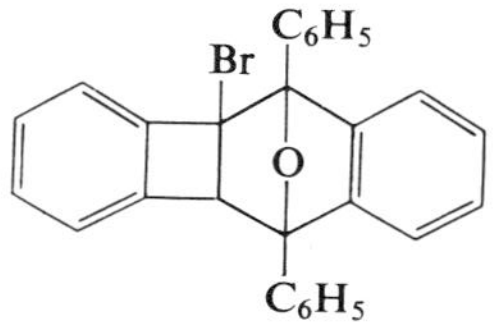

5a-Bromo-5, 10-diphenyl-5, 10-oxido-5, 5a, 9b, 10-tetrahydrobenzo[*b*]biphenylene m.p. 170°–173° (from MeOH–CHCl$_3$).

Table 6.1 (*continued*)

XVI	XVII	XVIII
5,6-Dibromobenzo[b]biphenylene; orange needles, m.p. 149°–150°.	5a, 9b - Dibromo - 5, 10 - diphenyl - 5, 10 - oxido - 5, 5a, 9b, 10 - tetrahydrobenzo[b]-biphenylene; m.p. 200° (from EtOH–CHCl$_3$).	5a - Iodo - 5, 10 - diphenyl - 5, 10 - oxido-5, 5a, 9b, 10 - tetrahydrobenzo[b]biphenylene m.p. 208°–210° (from HOAc).

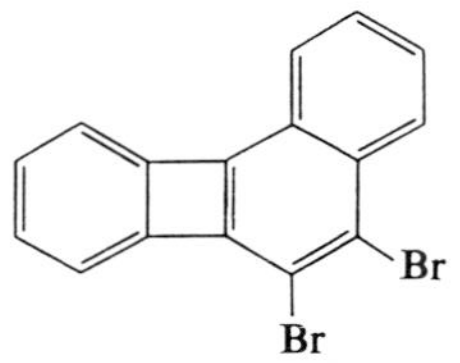

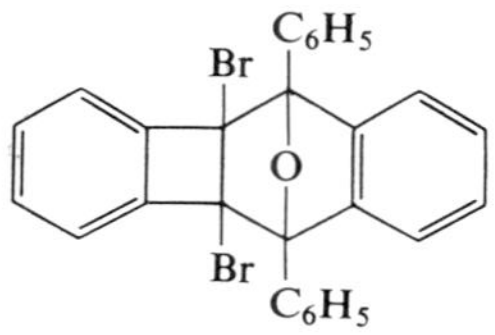

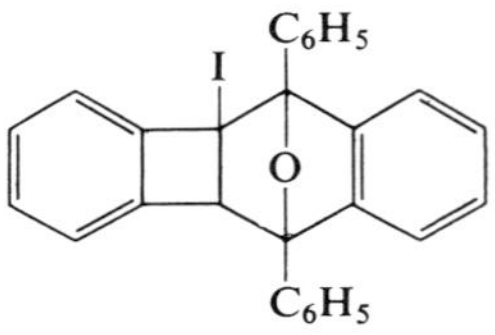

XIX	XX
5-Iodobenzo[a]biphenylene; m.p. 131°–132° (from EtOH–H$_2$O).	5,6-Diphenyl-11,12-dimethyldibenzo[a,e]cyclooctatetraene; m.p. 215°–216.0°, λ_{max}^{EtOH} 285 mμ (log ϵ, 3.50).

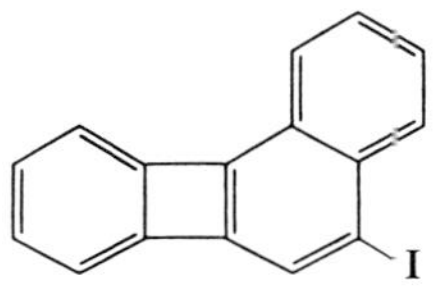

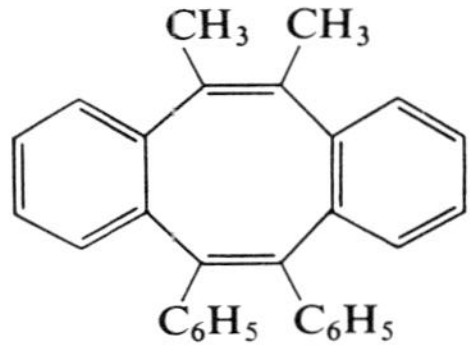

D. CHEMISTRY OF CYCLOBUTADIENE   213

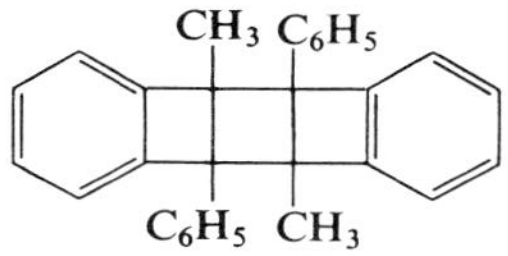

XXIa (or b)

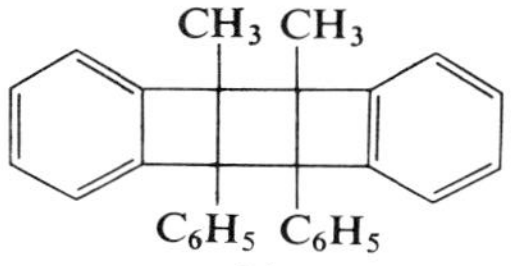

XXII

1,2(or 9)-Dimethyl-9(or 2), 10-diphenylpentacyclo[8.6.0.0^{2,9}.0^{3,8}.0^{11,16}]-hexadeca-3,5,7,11,13,15-hexaene; m.p. 165°, λ_{max} 269.7 mμ (log ϵ, 3.77), 275.5 (3.72), 299.0 (2.39).

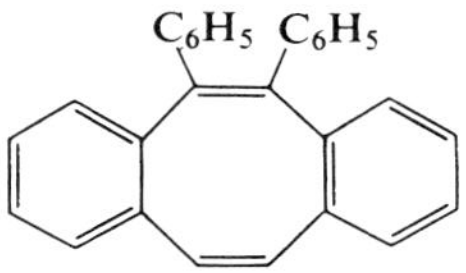

5,6-Diphenyldibenzo[a,e]cyclooctatetraene; m.p. 195.5°–196.0° (from Et$_2$O); λ_{max} 288 mμ (log ϵ, 3.95), 240 (4.49).

[a] Method 1 is "By Dehalogenation of 1,2-Dihalobenzocyclobutenes," Section B, 1; Method 2 is "By Dehydrohalogenation of 1-Halobenzocyclobutenes," Section B, 2; Method 3 is "By Cycloaddition of Benzyne to Acetylenes," Section B, 3; and Method 4 is "By Free-Radical Halogenation," Section B, 4.
[b] Key to abbreviations: DPIBF (1,3-diphenylisobenzofuran), h-d (high-dilution technique), HQ (hydroquinone), NPMI (N-phenylmaleimide), subl. (sublimes), yel. (yellow).
[c] Products represented by Roman numerals are listed in Part C of this table.
[d] Calculated from weight data recorded in the literature.

confirms the expectation implicit in the benzocyclobutadiene dimerization mechanism that benzocyclobutadiene can react as both a diene and a dienophile. (See p. 215.)

4. Addition of Halogen

It has been suggested that several transformations of 1,2-dihalobenzocyclobutenes proceed by reaction paths that have as a common feature the addition of iodine to benzocyclobutadiene. Thus, (a) the conversion of trans-1,2-dibromobenzocyclobutene [trans-(2a)] into trans-1,2-diiodobenzocyclobutene [trans-(2b)] by sodium iodide,[20a] (b) the iodide-ion-catalyzed interconversion of cis- and trans-1,2-diiodobenzocyclobutene [cis-(2b) ⇌ trans-(2b)][40] and (c) the light-catalyzed interconversion of the two diiodides[40] may involve the generation of and subsequent addition of iodine to benzocyclobutadiene. The lack of any tendency of diiodides cis-(2b) and trans-(2b) to form benzocyclobutadiene dimer during these transformations is consistent with the postulate that benzocyclobutadiene adds iodine very readily at the 1,2-positions.

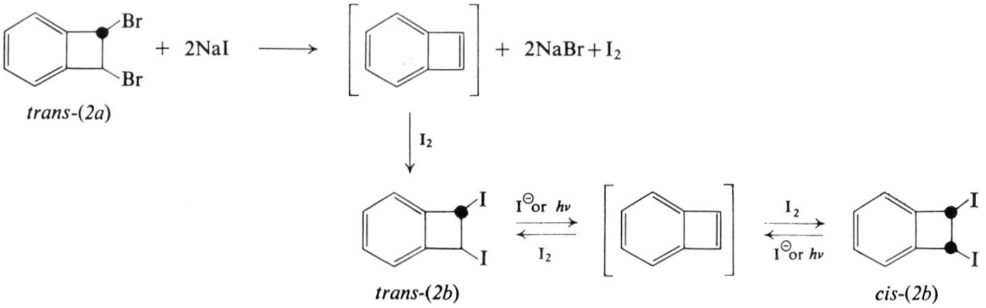

5. Possible Rearrangement of a Benzocyclobutadiene

Benzyne reacts with phenylacetylene to give 5,6-diphenyldibenzo[a,e]cyclooctatetraene (39) and phenanthrene (66).[54] Similarly, 1-phenylpropyne gives 5,6-dimethyl-11,12-diphenyldibenzo[a,e]cyclooctatetraene (35) and 9-methylphenanthrene (67).[54] The two dibenzocyclooctatetraenes (35) and (39) are undoubtedly formed through dimerization of the corresponding substituted benzocyclobutadienes (32) and (38), respectively, but the origins of phenanthrene (66) and 9-methylphenanthrene (67) are obscure. It has been suggested that the phenanthrenes are formed either by rearrangement of the corresponding substituted benzocyclobutadienes or by a 1,4-cycloaddition mechanism.[54]

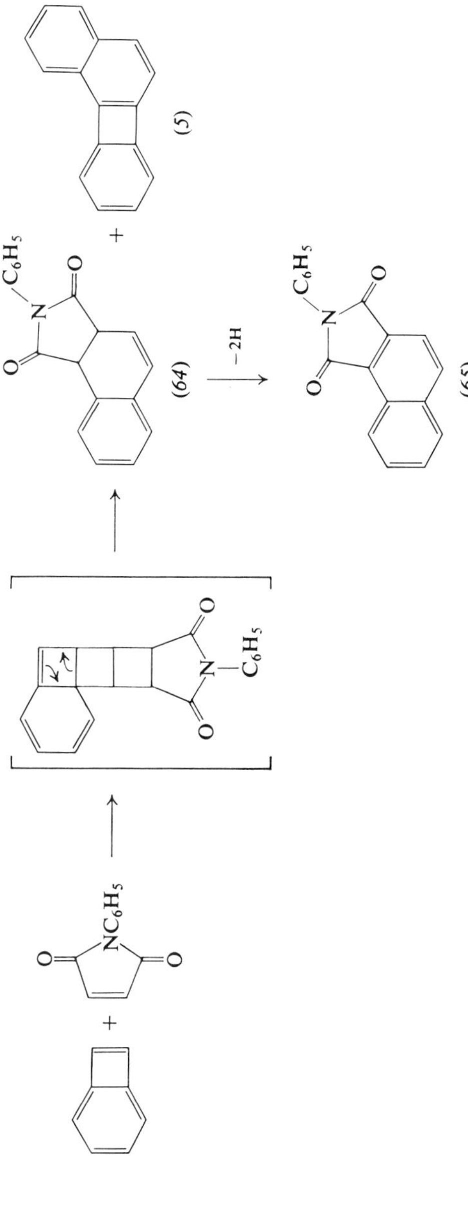

D. CHEMISTRY OF BENZOCYCLOBUTADIENE

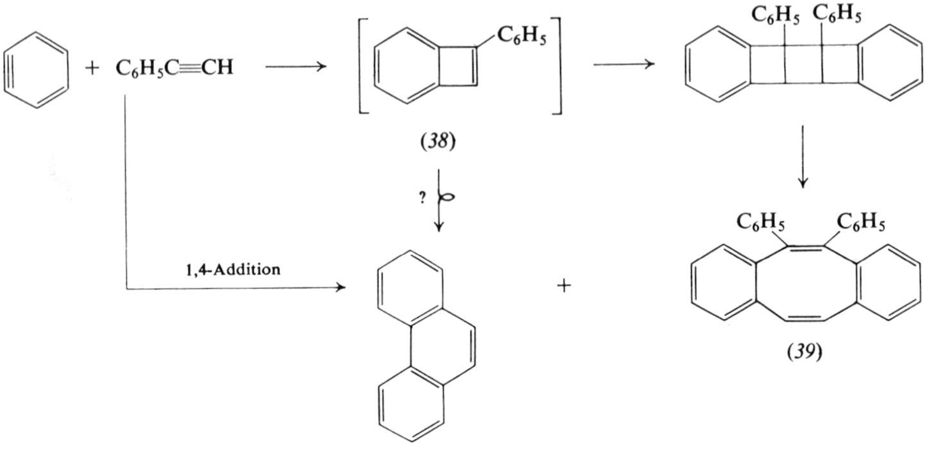

6. BENZOCYCLOBUTADIENE

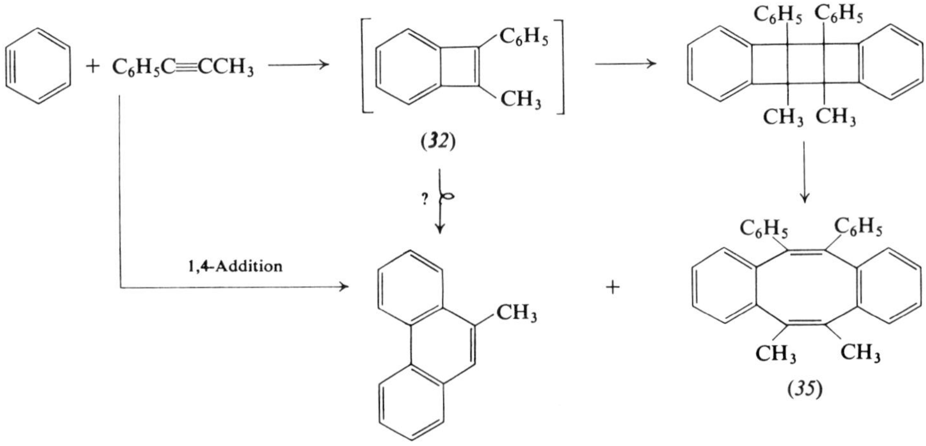

Since 1-methyl-2-phenylbenzocyclobutadiene has been found to give only the dibenzocyclooctatetraene (*35*) and no 9-methylphenanthrene, when generated by another method [(*31*) → (*32*)],[13] rearrangement appears less likely than 1,4-addition.

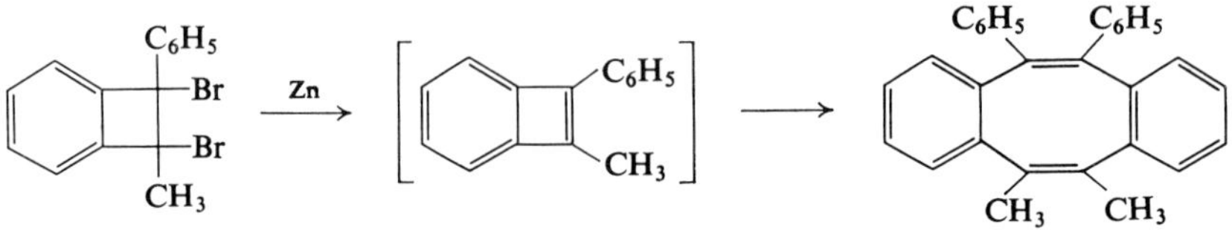

References

1. (a) Avram, M., Dinu, D., Mateescu, G., and Nenitzescu, C. D., *Ber.* **93**, 1789 (1960); (b) Avram, M., Dinu, D., and Nenitzescu, C. D., *Chem. Ind. (London)* 257 (1959).
2. Avram, M., Dinulescu, I. G., Dinu, D., Mateescu, G., and Nenitzescu, C. D., *Tetrahedron* **19**, 309 (1963).
3. Avram, M., Dinulescu, I. G., Dinu, D., and Nenitzescu, C. D., *Chem. Ind. (London)* 555 (1962).
4. Avram, M., Mateescu, G. D., Dinulescu, I. G., and Nenitzescu, C. D., *Accad. Rep. Populare Romîne, Studii Cercetari Chim.* **9**, 435 (1961). See also *ibid., Rev. Chim. Roumaine* **8**, 77 (1963).
5. Baker, W., Barton, J. W., McOmie, J. F. W., and Searle, R. J. G., *J. Chem. Soc.* 2633 (1962).
6. Baker, W., and McOmie, J. F. W., in Ginsburg, D., "Non-Benzenoid Aromatic Compounds," Interscience, New York (1959), p. 43.
7. Baker, W., and McOmie, J. F. W., *Ber.* **92**, xxxvii (1959).
8. Baker, W., and McOmie, J. F. W., *Chem. Soc. Special Publication No.* **12**, 49 (1958).
9. Bakunin, M., and Parlati, L., *Gazz. Chim. Ital.* **36B**, 264 (1906).
10. Bakunin, M., and Vitale, T., *Rend. Accad. Sci. Napoli* [3] **33**, 270 (1927); [*Chem. Abstr.* **23**, 2969 (1929)].
11. Beynon, J. H., Lester, G. R., and Williams, A. E., *J. Phys. Chem.* **63**, 1861 (1959).
12. Blatchly, J. M., McOmie, J. F. W., and Thatte, S. D., *J. Chem. Soc.* 5090 (1962).
13. (a) Blomquist, A. T., and Bottomley, C. G., *Trans. N.Y. Acad. Sci.* [2] **24**, 823 (1962); (b) Bottomley, C. G., *Dissertation Abstr.* **22**, 2188 (1962).
14. Bordwell, F. G., McKellin, W. H., and Babcock, D., *J. Am. Chem. Soc.* **73**, 5566 (1951).
15. Cava, M. P., *Bull. Soc. Chim. (France)* [5] 1744 (1959).
16. (a) Cava, M. P., and Mitchell, M. J., *J. Am. Chem. Soc.* **81**, 5409 (1959); (b) Mitchell, M. J., Ph.D. diss., The Ohio State Univ. (1960).
17. (a) Cava, M. P., and Muth, K., *J. Org. Chem.* **27**, 1561 (1962); (b) *ibid., Tetrahedron Letters* **No. 4**, 140 (1961).
18. Cava, M. P., and Muth, K., *J. Org. Chem.* **27**, 757 (1962).
19. Cava, M. P., and Napier, D. R., *J. Am. Chem. Soc.* **80**, 2255 (1958).
20. (a) Cava, M. P., and Napier, D. R., *J. Am. Chem. Soc.* **79**, 1701 (1957); (b) *ibid., J. Am. Chem. Soc.* **78**, 500 (1956).
21. Cava, M. P., Pohl, R. J., and Mitchell, M. J., *J. Am. Chem. Soc.* **85**, 2080 (1963); (b) Cava, M. P., Mitchell, M. J., and Pohl, R. J., *Tetrahedron Letters* **No. 18**, 825 (1962).
22. Cava, M. P., and Pohlke, R., *J. Org. Chem.* **28**, 1012 (1963).
23. Cava, M. P., and Pohlke, R., *J. Org. Chem.* **27**, 1564 (1962).
24. Cava, M. P., Ratts, K. W., and Stucker, J. F., *J. Org. Chem.* **25**, 1101 (1960).
25. Cava, M. P., and Stucker, J. F., *J. Am. Chem. Soc.* **79**, 1706 (1957).
26. Cava, M. P., and Van Meter, J. A., unpublished work.
27. Emerson, G. F., Watts, L., and Pettit, R., *J. Am. Chem. Soc.* **87**, 131 (1965).
28. Fenton, S. W., Lipscomb, W. N., Rossman, M. G., and Osborn, J. H., *J. Org. Chem.* **23**, 994 (1958).
29. Finkelstein, H., *Ber.* **92**, xi (1959).
30. Finkelstein, H., Ph.D. diss., Strasbourg (1909).
31. Finkelstein, H., *Ber.* **43**, 1528 (1910).
32. Griffin, G. W., and Weber, D. F., *Chem. Ind. (London)* 1162 (1961).
33. Hancock, J. E. H., private communication.

34. (a) Hancock, J. E. H., and Taber, H. W., *Tetrahedron* **3**, 132 (1958); (b) Hancock, J. E. H., Taber, H. W., and Scheuchenpflug, D. R., *Chem. Ind.* (*London*) 437 (1958).
35. (a) Hancock, J. E. H., and Scheuchenpflug, D. R., *J. Am. Chem. Soc.* **80**, 3621 (1958); (b) *ibid.*, *Abstr. American Chemical Society 133d Meeting*, 75N (1958).
36. (a) Hanna, M. W., and Fenton, S. W., *J. Org. Chem.* **26**, 1371 (1961); (b) Fenton, S. W., and Hanna, M. W., *J. Org. Chem.* **24**, 579 (1959).
37. Hart, H., and Hartlage, J., private communication.
38. Hart, H., and Fish, R. W., *J. Am. Chem. Soc.* **82**, 749 (1960).
39. Jensen, F. R., and Coleman, W. E., *Tetrahedron Letters* **No. 20**, 7 (1959).
40. Jensen, F. R., and Coleman, W. E., *J. Org. Chem.* **23**, 869 (1958).
41. Kuehne, M. E., private communication.
42. Kuehne, M. E., *J. Am. Chem. Soc.* **84**, 837 (1962).
43. (a) Lagidze, R. M., and Petrov, A. D., *Doklady Akad. Nauk S.S.S.R.* **83**, 235 (1952); *Chem. Abstr.* **47**, 4321^g (1953). (b) Lagidze, R. M., *Soobshchenya Akad. Nauk Gruzin. S.S.R.* **19**, 279 (1957). (c) *ibid.*, *Trud. Inst. Khim. Akad. Nauk Gruzin. S.S.R.* **12**, 157 (1956). (d) Lagidze, R. M., and Loladze, N. R., *Soobshchenya Akad. Nauk Gruzin. S.S.R.* **16**, 607 (1955); *Chem. Abstr.* **50**, 11960^b (1956); (e) Lagidze, R. M., and Kuprava, Sh. D., *Doklady Akad. Nauk S.S.S.R.* **110**, 795 (1956); *Chem. Abstr.* **51**, 8059^d (1957).
44. Maier, G., *Ber.* **90**, 2949 (1957).
45. Malachowski, R., Ph.D. diss., ETH (1911).
46. Manatt, S. L., and Roberts, J. D., *J. Org. Chem.* **24**, 1336 (1959).
47. McDonald, S. G. G., private communication.
48. Nenitzescu, C. D., Avram, M., Dinu, D., *Ber.* **90**, 2541 (1957).
49. Roberts, J. D., "Notes on Molecular Orbital Calculations," Benjamin, New York (1961), p. 58.
50. Roberts, J. D., and Sharts, C. M., "Organic Reactions," Wiley, New York (1962).
51. Roberts, J. D., Streitwieser, A., and Regan, C. M., *J. Am. Chem. Soc.* **74**, 4579 (1952).
52. Simmons, H. E., *J. Am. Chem. Soc.* **83**, 1657 (1961).
53. Sisido, K., Noyori, R., Kozaki, N., and Nozaki, H., *Tetrahedron* **19**, 1185 (1963).
54. Stiles, M., Burckhardt, U., and Haag, A., *J. Org. Chem.* **27**, 4715 (1962).
55. Willstätter, R., and Veraguth, H., *Ber.* **40**, 957 (1907).
56. Wreden, F., *Ber.* **9**, 590 (1876).

CHAPTER 7

1,2-Benzocyclobutadienequinone

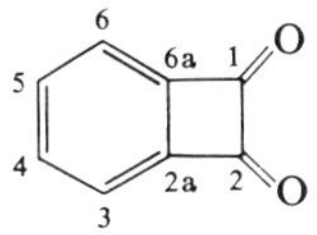

1,2-Benzocyclobutadienequinone, or benzocyclobutene-1,2-dione, is both the simplest and the only known example of a fused aromatic analog of cyclobutadienequinone. In a formal sense at least, it is the stable four-membered ring quinone of the highly unstable benzocyclobutadiene. The stability of benzocyclobutadienequinone, which contrasts markedly with the lack of stability of benzocyclobutadiene, thus parallels the stability of cyclobutadienequinone and the instability of cyclobutadiene.

Four additional benzocyclobutadienequinone structures may be written, but neither attempts to prepare these compounds nor molecular orbital calculations predicting their properties have been recorded. On the basis of the simplest theoretical considerations, the 4,5-isomer (a 1,2-dimethylenecyclobutene structure) may be expected to be the most stable member of this group of four isomers, and the 3,6-isomer (a cyclobutadienoid structure) may be expected to be the least stable.

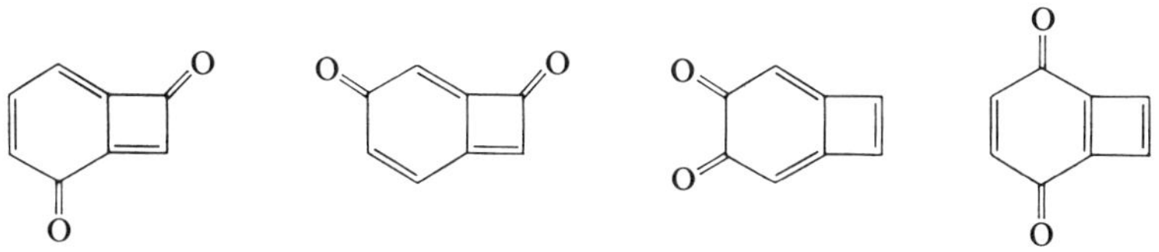

CHART 7.1. Unknown benzocyclobutadienequinones.

The chemistry of benzocyclobutadienequinone has been reviewed briefly by Cava.[1]

In the remainder of this chapter, 1,2-benzocyclobutadienequinone will be referred to simply as benzocyclobutadienequinone and no further mention of the 1,3-, 1,5-, 3,6-, and 4,5-isomers will be made.

A. History

There is no early history of attempts to prepare benzocyclobutadienequinone. The quinone (2) was first prepared only recently (1957) by Cava and Napier by the action of triethylamine on *cis-* or *trans-*benzocyclobutenediol dinitrate, (*1a*) or (*1b*).[5b]

(*1a*): *cis-*
(*1b*): *trans-*   (2)

B. Synthesis of Benzocyclobutadienequinone

Benzocyclobutadienequinone has been prepared in three ways: (1) by the elimination of nitrous acid from *cis-* and *trans-*1,2-benzocyclobutenediol dinitrate, (2) by silver-ion-catalyzed hydrolysis of 1,1,2,2-tetrabromobenzocyclobutene, and (3) by hydrolysis of a compound believed to be 2,2-bis(trifluoroacetoxy)benzocyclobutenone.

1. From Benzocyclobutenediol Dinitrate

The cis and trans dinitrates (*1a*) and (*1b*) of benzocyclobutenediol react with triethylamine to give benzocyclobutadienequinone (2) and nitrous acid (in the form of the triethylamine salt).[5] In the case of the trans dinitrate, yields as high as 87% were obtained, but for the most part the yields were not larger than 80% (or smaller than 60%). The cis dinitrate proved to be far more resistant in the elimination reaction and the yield of quinone did not exceed 18% under conditions that were quite effective with the trans dinitrate.* However, when a modified procedure was used, the yield of quinone was much improved (35%).†,[5a]

* In a preliminary report (ref. 5b), Cava and Napier stated that benzocyclobutadienequinone was obtained in 75% yield from either the cis or the trans dinitrate under the same reaction conditions (triethylamine, methylene chloride, 1 hour refluxing). In subsequent work, the yield of quinone obtained from the cis isomer never exceeded 18% (ref. 5a).

† A solution of cis dinitrate in dimethylacetamide and a solution of triethylamine in dimethylacetamide were added slowly and simultaneously to a solution of urea in the same solvent at 70°.

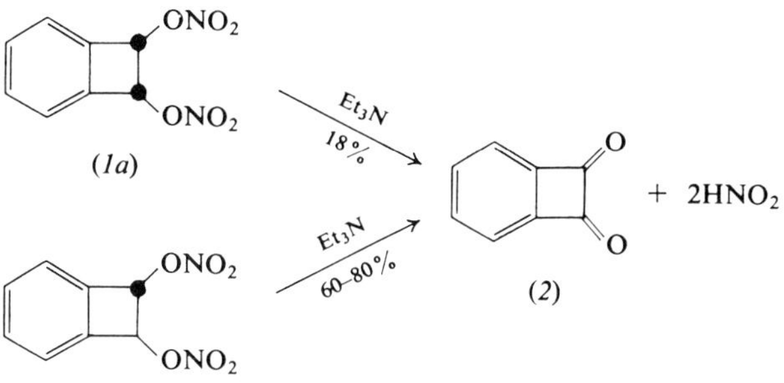

B. SYNTHESIS OF BENZOCYCLOBUTADIENEQUINONE

The conversion of either *cis-* or *trans-*1,2-benzocyclobutenediol dinitrate into benzocyclobutadienequinone was accompanied by the formation of a small amount (less than 5%) of a dimer which was shown to be (*3*) and not 5,6,11,12-tetraketodibenzocyclooctadiene (*4*) as might have been expected.[5a] The mechanistic path by which dimer (*3*) is formed remains obscure, but benzocyclobutadienequinone itself is apparently not a precursor, since the quinone is unaffected by refluxing triethylamine–methylene chloride.[3]

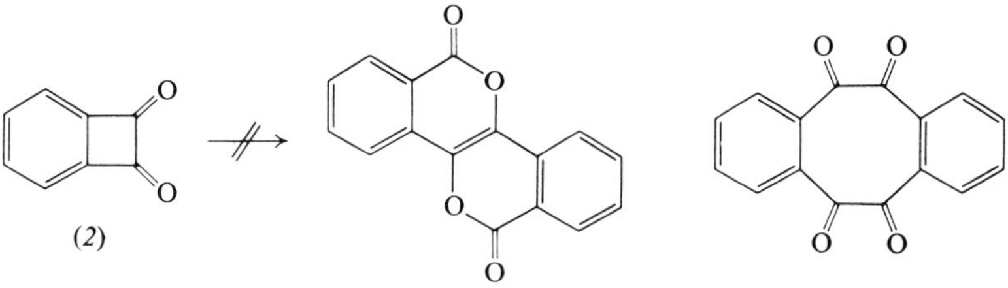

2. FROM 1,1,2,2-TETRABROMOBENZOCYCLOBUTENE

The conversion of 1,1,2,2-tetrabromobenzocyclobutene (*5*) into benzocyclobutadienequinone by hydrolysis was first used as a means of proving the structure of the tetrabromide.[4] Presumably, treatment of the tetrabromide with silver trifluoroacetate in anhydrous benzene first gives the tetrakis(trifluoroacetate) (*6*) which, on treatment with water, affords the quinone (*2*). Under these conditions the yield of quinone (*2*) is 41% (based on the tetrabromide).[4]

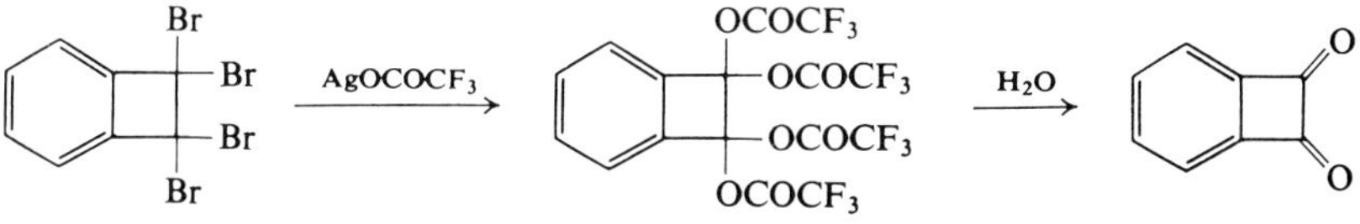

A closer investigation of the reaction revealed the formation of a colorless by-product to which the structure of 2,2-bis(trifluoroacetoxy)benzocyclobutenone was assigned (7).[5a] It was also found that formation of the by-product was suppressed and a better yield of benzocyclobutadienequinone (89%) was obtained when the reaction was carried out with a mixture of silver trifluoroacetate–aqueous acetonitrile.[5a]

The tetrabromobenzocyclobutene method is a more practical and more convenient method for preparing benzocyclobutadienequinone than the benzocyclobutenediol dinitrate method.[5a] Thus, the overall yield of quinone obtained from *trans*-1,2-dibromobenzocyclobutene via the tetrabromide [(8) → (5) → (2)] is 82%, while that obtained via the *trans*-dinitrate [(8) → (1b) → (2)] is only 31%.

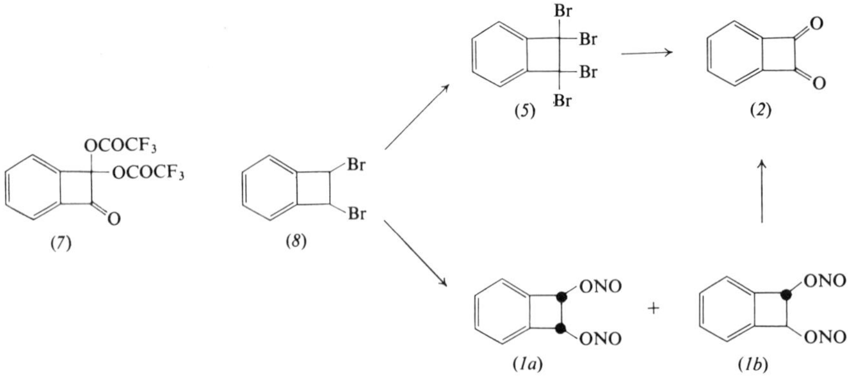

3. FROM 2,2-BIS(TRIFLUOROACETOXY)BENZOCYCLOBUTENE

The method has no practical value since the starting material, 2,2-bis(trifluoroacetoxy)benzocyclobutenone (7), m.p. 80°–85°, is a by-product obtained in small yield (not specified) in the conversion of tetrabromobenzocyclobutene into 1,1,2,2-tetrakis(trifluoracetoxy)benzocyclobutene with silver trifluoroacetate in anhydrous benzene[5a] (see above). Treatment of the diacyloxyketone (7) with boiling aqueous methanol gives benzocyclobutadienequinone (2).[5a]

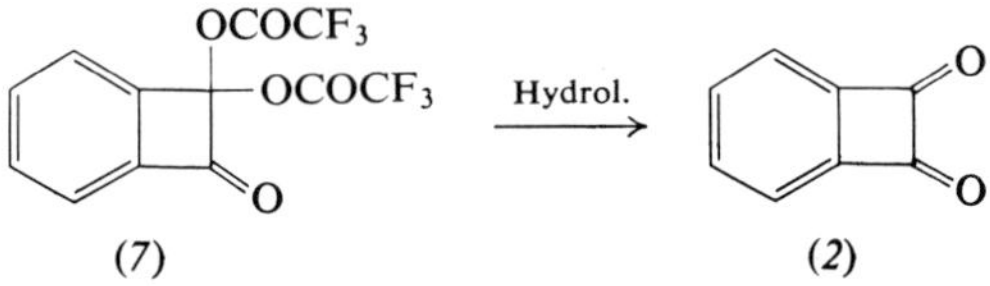

C. Unsuccessful Attempts to Prepare Benzocyclobutadienequinones

The formation of 1,2-disubstituted benzocyclobutenes by ring closure of intermediary *o*-quinodimethanes suggests the use of a similar method for the synthesis of benzocyclobutadienequinone. In the case of the quinone, a bis(ketene) intermediate is required. However, attempted generation of the bis(ketene) from both *cis*- (*1a*)[4] and *trans*-1,2-dihydrophthaloyl chloride (*1b*)[5a] by dehydrohalogenation with triethylamine gave no crystalline reaction products. The failure of the method is not surprising in view of similar unsuccessful attempts to synthesize simple cyclobutadienequinones via bis-(ketene) intermediates (see Chapter 4, Section C).

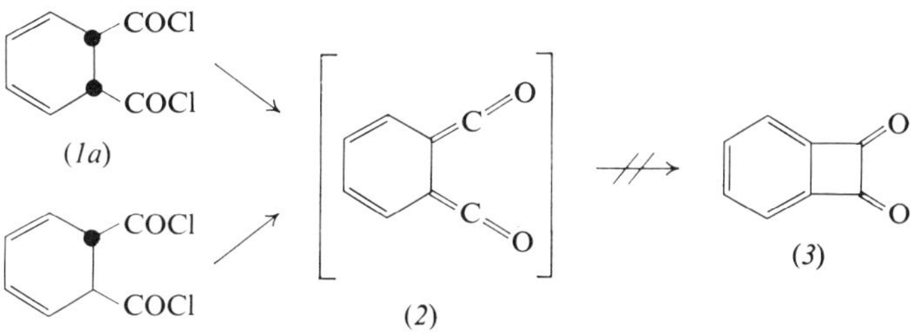

The formation of cyclobutadienequinones, e.g., 3,4-dimethylcyclobutadienequinone, by acid hydrolysis of the appropriate tetrafluoro and chlorotrifluoro compounds suggests the use of a similar method for benzocyclobutadienequinones. However, an attempt to apply the method to the preparation of a substituted benzocyclobutadienequinone (*4*) was unsuccessful.[10]

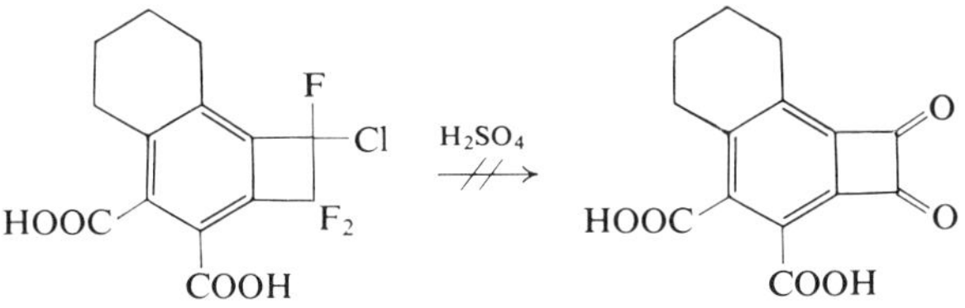

D. Physical Properties of Benzocyclobutadienequinone

Benzocyclobutadienequinone forms pale yellow prismatic crystals, m.p. 132°–135°, when it is recrystallized from methylene chloride–petroleum ether. The crude material can also be purified either by sublimation *in vacuo* (80°–85° at 0.5 mm) or by chromatography with benzene on grade II acidic alumina.

7. 1,2-BENZOCYCLOBUTADIENEQUINONE

The quinone has a sweet but pungent odor somewhat similar to that of p-benzoquinone.

The infrared spectrum of benzocyclobutadienequinone (Fig. 7.1) shows

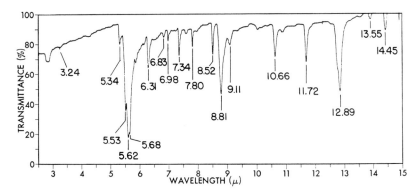

FIG. 7.1. Infrared spectrum of benzocyclobutadienequinone (potassium bromide disc) calibrated against polystyrene.

relatively few strong bands as a result of the symmetry of the molecule. The spectrum of crystalline benzocyclobutadienequinone (potassium bromide disc) is well resolved; the carbonyls absorb as a triplet at 5.53, 5.62, and 5.68 μ. The ultraviolet spectrum of the quinone in ethanol, shown in Fig. 7.2, exhibits four

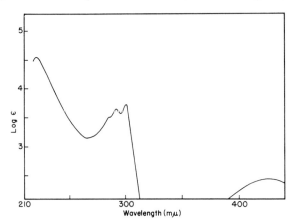

FIG. 7.2. Ultraviolet absorption spectrum of benzocyclobutadienequinone (in ethanol).[5a]

strong maxima at 220 mμ (log $\epsilon = 4.56$), 286 (3.51), 292 (3.69), and 301 (3.72) and a weak absorption band centered at 427 mμ (log $\epsilon = 2.44$) which is responsible for the yellow color of the compound.[5a]

E. Chemistry of Benzocyclobutadienequinone

The delocalization energy of benzocyclobutadienequinone was calculated by the molecular orbital method to be 3.13 β, or about 62 kcal/mole (assuming that $\beta = 20$ kcal/mole).[8] A comparison of this value with the experimentally determined resonance energy of naphthalene (62 kcal/mole) leads to the interesting conclusion that benzocyclobutadienequinone and naphthalene represent bicyclic fused aromatic structures of approximately equal stability, at least from the point of view of electronic factors alone. In considering the actual chemistry of the quinone, however, the considerable strain imposed on the molecule by the fused four-membered ring must be taken into account; although the quinone is quite stable both to heat and to dilute mineral acid, it undergoes a number of reactions, particularly under basic conditions, which lead to the relief of strain in the four-membered ring through rupture of the 1,2-bond.

1. THERMAL STABILITY

Thermolysis of benzocyclobutene at 150° leads to the formation of *o*-quinodimethane, which may be trapped with dienophiles such as maleic anhydride. In contrast, under similar reaction conditions (*N*-phenylmaleimide, 2 days' refluxing in xylene at 140°), benzocyclobutadienequinone (*1*) was found to be unchanged.[9] This result attests to the extraordinary thermal stability of benzocyclobutadienequinone.*

2. OXIDATION AND REDUCTION

Benzocyclobutadienequinone is oxidized rapidly by hydrogen peroxide in acetic acid to give phthalic acid (*2*) in almost quantitative yield (96%).[5]

Lithium aluminum hydride reduction of benzocyclobutadienequinone affords *cis*-1,2-benzocyclobutenediol (*3*) in 30% yield.[5]

* Note added in proof: A number of reactions involving the photochemical cleavage of benzocyclobutadienequinone have been reported recently. [See R. F. C. Brown and R. K. Solly, *Tetrahedron Letters* 169 (1966) and H. A. Staab and J. Ipatschi, *ibid.*, 583 (1966).]

3. KETAL FORMATION

Benzocyclobutadienequinone reacts with excess ethylene glycol in the presence of *p*-toluenesulfonic acid to give the bis(ethyleneketal) (*4*) in 77% yield.[11] When only one equivalent of ethylene glycol was used, the monoethyleneketal (*5*) was obtained (78% yield).[11] Hydrolysis of either of the two ketals with hot aqueous hydrochloric acid results in regeneration of the quinone.

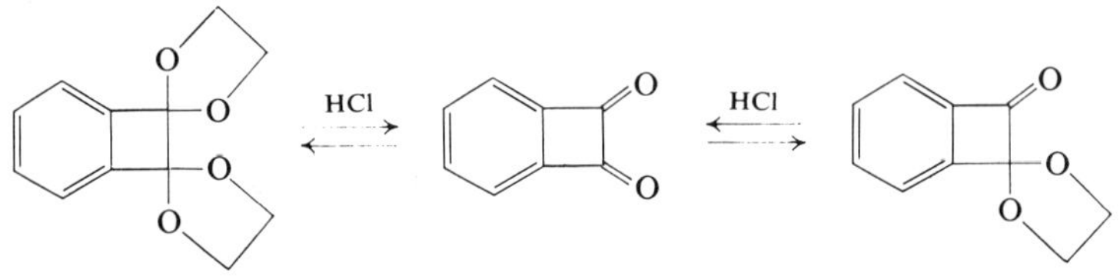

(*4*)   (*1*)   (*5*)

4. REACTION WITH HYDROXIDE ION

Benzocyclobutadienequinone is cleaved readily at room temperature by 5% aqueous methanolic sodium hydroxide to give the sodium salt of phthalaldehydic acid in 94% yield [(*1*) → (*6*)].[5] It is interesting to note that a benzilic acid rearrangement of the quinone does not take place; the product of such a rearrangement would be a derivative of the highly strained benzocyclopropene (*7*).

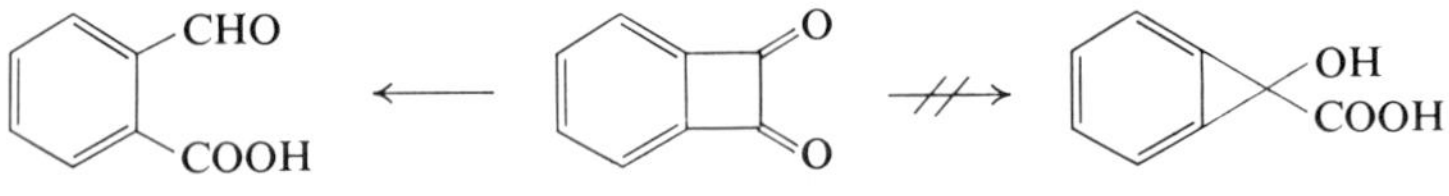

(*6*)   (*1*)   (*7*)

5. CARBANION-ADDITION REACTIONS

a. Grignard Reagents

Methylmagnesium bromide adds in a normal fashion to benzocyclobutadienequinone to give a mixture of *cis*- and *trans*-1,2-dimethylbenzocyclobutene-1,2-diol, (*8a*) and (*8b*), in 49% and 39% yield, respectively.[7a] 1,3-Diphenylisobenzofuran (*9*) was the only product isolated (60% yield) when the quinone was treated with phenylmagnesium bromide. It is probable that the diphenyldiol was formed initially and underwent acid-catalyzed rearrangement during the workup.[5a]

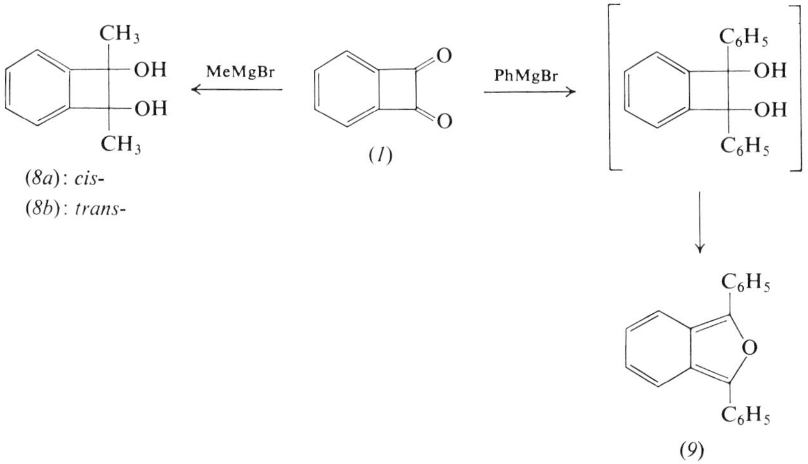

E. CHEMISTRY OF BENZOCYCLOBUTADIENEQUINONE

(8a): cis-
(8b): trans-

(1)

(9)

b. *Wittig Reagents* (*Phosphoranes*)

The reaction of benzocyclobutadienequinone with triphenylphosphine-methylene does not give dimethylenebenzocyclobutene (12) and the only identifiable product of the reaction is triphenylphosphine. The formation of the latter attests to an abnormal reaction course.[7a] Phosphoranes that are less

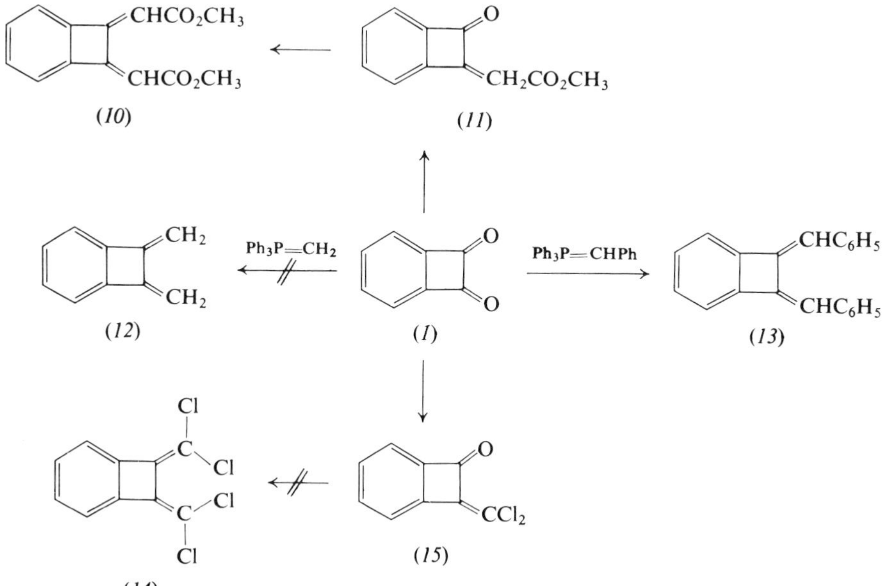

(10)

(11)

(12)

(1)

(13)

(14)

(15)

nucleophilic than triphenylphosphinemethylene react normally with benzocyclobutadienequinone. Thus, triphenylphosphinecarbomethoxymethylene gives either 1-keto-2-carbomethoxymethylenebenzocyclobutene [(*11*), 93% yield] or 1,2-bis(carbomethoxymethylene)benzocyclobutene [(*10*), 85% yield], depending upon whether one or two equivalents of the phosphorane is employed.[6]

The quinone reacts with excess triphenylphosphinedichloromethylene (generated from triphenylphosphine and carbon tetrachloride) to give 1-keto-2-dichloromethylenebenzocyclobutene (*15*) in 42% yield.[11] The failure of the reaction to give the bis condensation product (*14*) may be attributed to steric factors.

The moderately nucleophilic phosphorane, triphenylphosphinebenzylidine, reacts with the quinone to give the expected 1,2-dibenzylidinebenzocyclobutene (*13*), but only in poor yield (25%).[2]

6. Addition of Basic Nitrogen Functions

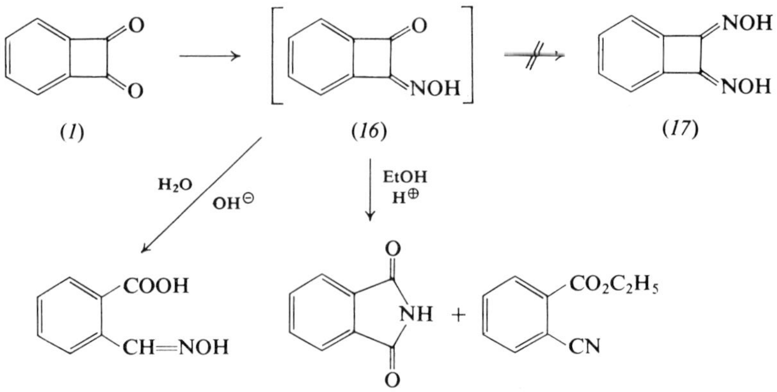

The condensation of benzocyclobutadienequinone with reagents containing a basic amino group proceeds only through attack of the reagent on one or on both carbonyls of the quinone. As a result, rearrangement products resulting from initial Michael addition are not observed as they are in the case of the cyclobutadienequinones (Chapter 4, Section D, 1, g). On the other hand, benzocyclobutadienequinone often gives unexpected condensation products in which the bond between the two original carbonyls has been cleaved. The reaction of the quinone with excess hydroxylamine affords neither the monooxime (*16*) nor the dioxime (*17*). Under basic conditions, phthalaldehydic acid oxime (*18*) is formed, while under acidic conditions the monooxime (*16*) is probably the initial product, but undergoes the Beckmann rearrangement by either the first-order or the second-order mechanism to give a mixture of phthalimide and ethyl o-cyanobenzoate.[11]

The quinone reacts with an excess of free hydrazine and also with hydrazine hydrochloride to give 1-phthalazone (23) rather than the normal hydrazone (21), which may be a short-lived intermediate in the reaction. Similarly, the quinone gives 2-tosyl-1-phthalazone (25) with one equivalent of tosylhydrazine but, surprisingly, it gives the normal bis(tosylhydrazone) (24) when treated with two equivalents of the reagent in the cold. In this instance it appears that the intermediary monotosylhydrazone (22) reacts with a second equivalent of tosylhydrazine at a rate competitive with that of the cleavage process.[11]

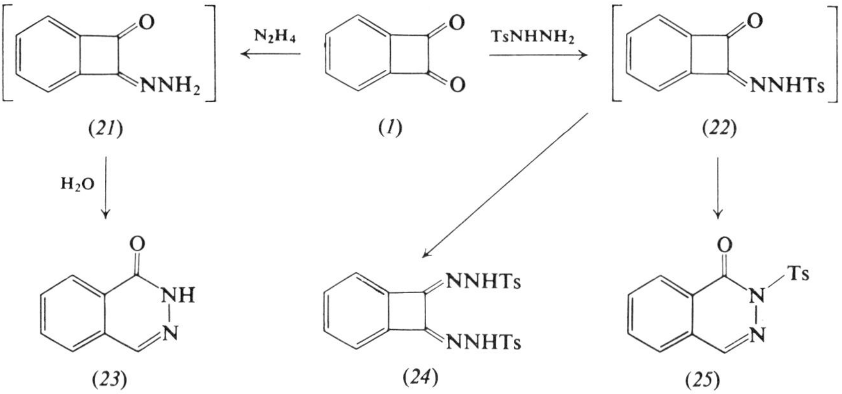

The reaction of 1,2-benzocyclobutadienequinone with excess aniline affords only a secondary cleavage product which is the anil of phthalaldehydic anilide (26).[11] In contrast, o-phenylenediamine gives the uncleaved product, 5,10-diazabenzo[b]biphenylene (27) in 63% yield.[5]

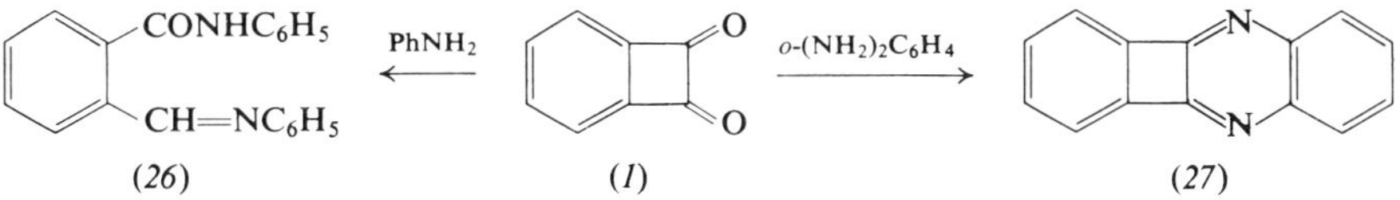

Semicarbazide,[11] phenylhydrazine,[11] and 2,4-dinitrophenylhydrazine[5] all react with the quinone to give uncleaved bis condensation products, i.e., (28), (29), and (30), respectively.

The factors which are conducive to cleavage of the four-membered ring of the quinone in some of the above reactions and not in others are not clearly discernible at this time. From the limited data available at present, it would

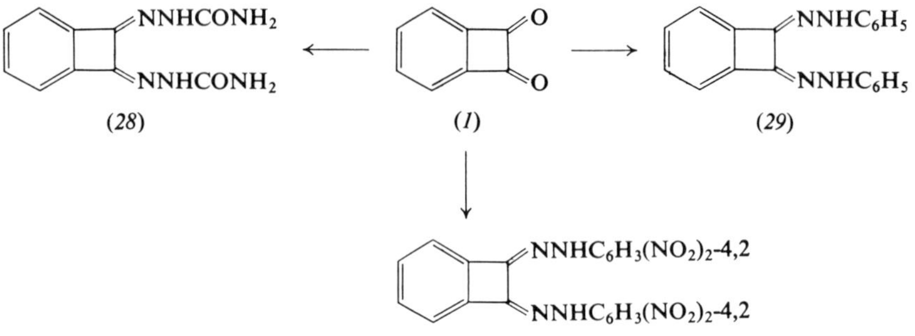

$$\text{(28)} \quad \begin{array}{c}\text{NNHCONH}_2\\ \text{NNHCONH}_2\end{array} \longleftarrow \text{(1)} \begin{array}{c}\text{O}\\ \text{O}\end{array} \longrightarrow \text{(29)} \begin{array}{c}\text{NNHC}_6\text{H}_5\\ \text{NNHC}_6\text{H}_5\end{array}$$

$$\downarrow$$

(30) NNHC$_6$H$_3$(NO$_2$)$_2$-4,2 / NNHC$_6$H$_3$(NO$_2$)$_2$-4,2

appear that the bis derivatives resist cleavage and rearrangement; on the other hand, the unknown mono derivatives appear to undergo cleavage or rearrangement very readily, probably by a process involving preliminary hydration of the remaining carbonyl group.

TABLE 7.1

NORMAL NITROGEN-CONTAINING CARBONYL DERIVATES OF BENZOCYCLOBUTADIENEQUINONE

Derivative	Physical properties[a]	Ref.
Quinoxaline	M.p. 238°–239° (w)	5a
Bis(2,4-dinitrophenylhydrazone)	M.p. 268°–270° (decomp.) (o)	5a
Bis(tosylhydrazone)	M.p. 215°–218° (decomp.) (w)	11
Bis(semicarbazone)	M.p. 300°–305° (decomp.) (w)	11
Bis(phenylhydrazone)	M.p. 193°–195° (y)	11

[a] Key: o (orange), w (white), y (yellow).

REFERENCES

1. Cava, M. P., *Bull. Soc. Chim. (France)* [5] 1744 (1959).
2. Cava, M. P., and Mangold, D., unpublished work.
3. Cava, M. P., and Mitchell, M. J., unpublished work.
4. Cava, M. P., and Muth, K., *J. Org. Chem.* **27**, 757 (1962).
5. (a) Cava, M. P., Napier, D. R., and Pohl, R. J., *J. Am. Chem. Soc.* **85**, 2076 (1963); (b) Cava, M. P., and Napier, D. R., *J. Am. Chem. Soc.* **79**, 3606 (1957).

6. Cava, M. P., and Pohl, R. J., *J. Am. Chem. Soc.* **82**, 5242 (1960).
7. (a) Cava, M. P., Pohl, R. J., and Mitchell, M. J., *J. Am. Chem. Soc.* **85**, 2080 (1963);
   (b) Cava, M. P., Mitchell, M. J., and Pohl, R. J., *Tetrahedron Letters* **No. 18,** 825 (1962);
   (c) Cava, M. P., and Pohl, R. J., *J. Am. Chem. Soc.* **82,** 5242 (1960).
8. Manatt, S. L., and Roberts, J. D., *J. Org. Chem.* **24**, 1336 (1959).
9. Pohl, R. J., Ph.D. diss., The Ohio State Univ. (1961).
10. Sharts, C. M., and Roberts, J. D., *J. Am. Chem. Soc.* **83**, 871 (1961).
11. Stein, R. P., Ph.D. diss., The Ohio State Univ. (1963).

CHAPTER 8

Methylene Analogs of 1,2-Benzocyclobutadienequinone

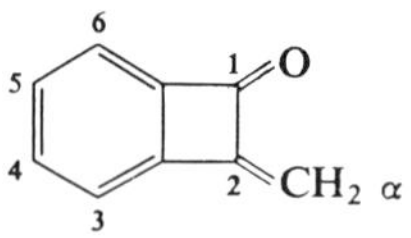

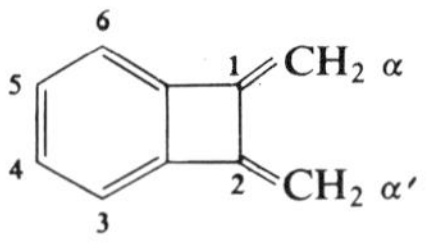

2-Methylenebenzocyclobutenone and 1,2-dimethylenebenzocyclobutene are, respectively, the parent o-quinomethide and the parent o-quinodimethane of the benzocyclobutadiene series. Like benzocyclobutadiene and 1,2-benzocyclobutadienequinone, both 2-methylenebenzocyclobutenone and 1,2-dimethylenebenzocyclobutene contain a four-membered ring consisting of sp^2-hybridized carbon atoms only.

A. History

There is no early history of attempts to prepare the methylene analogs of benzocyclobutadienequinone.

B. 2-Methylenebenzocyclobutenones

2-Methylenebenzocyclobutenone itself is not known, although molecular orbital calculations predict that it should have a delocalization energy of 3.17 β,[4] which is slightly larger than the value calculated for 1,2-benzocyclobutadienequinone (3.13 β).[4] The few substituted 2-methylenebenzocyclobutenones which have been prepared appear to be quite stable but their chemistry has not yet been studied in detail.

B. 2-METHYLENEBENZOCYCLOBUTENONES

1. SYNTHESIS OF 2-METHYLENEBENZOCYCLOBUTENONES

Several substituted 2-methylenebenzocyclobutenones have been prepared from benzocyclobutadienequinone by the Wittig reaction. Thus, 2-carbomethoxymethylenebenzocyclobutenone (*1*) is formed in 93% yield by the reaction of the quinone with one equivalent of triphenylphosphine carbomethoxymethylene at room temperature.[2a,c] Similarly, the reaction of 1,2-benzocyclobutadienequinone with triphenylphosphinedichloromethylene affords 2-dichloromethylenebenzocyclobutenone (*3*) in 42% yield.[6]

2-Carboxymethylenebenzocyclobutenone (*4*) was prepared in 72% yield by acid hydrolysis of the methyl ester (*1*).[2a,c]

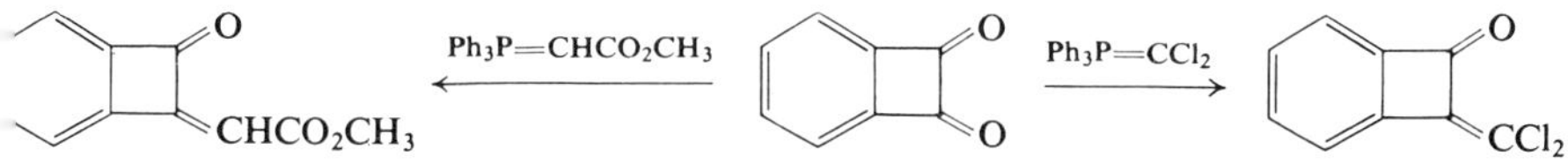

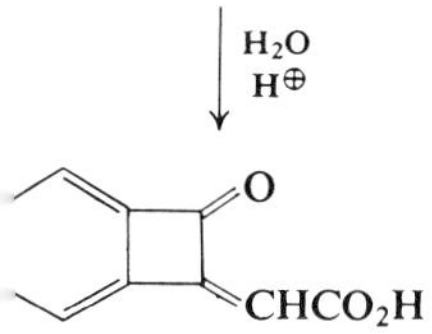

2. PHYSICAL PROPERTIES OF THE 2-METHYLENEBENZOCYCLOBUTENONES

The melting points and colors of the known 2-methylenebenzocyclobutenones are listed in Table 8.1.

The ketonic carbonyl group in each of the three known 2-methylenebenzocyclobutenones absorbs in the infrared at 5.62 μ (1780 cm^{-1}).[2a,c,6] The low-wavelength position of this absorption band is evidence that there is no appreciable conjugation of the ketonic carbonyl with the exocyclic double bond. 2-Dichloromethylenebenzocyclobutenone (*3*) in particular, would ordinarily be described in part by the ionic contributor (*3a*). However, structure (*3a*) is a high-energy benzocyclobutadienoid species, and consequently its contribution to the molecule is negligible.

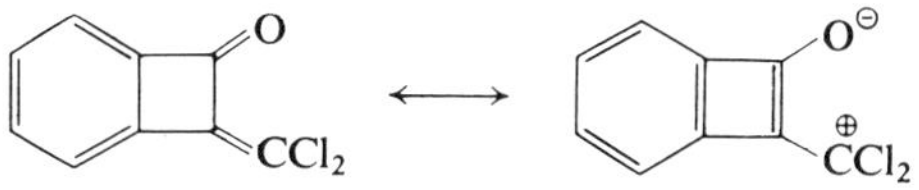

The ultraviolet spectra of both the keto ester $(1)^{2a,c}$ and the keto dichloride $(3)^6$ exhibit a strong absorption maximum at 246 mμ (log $\epsilon = 4.55$ and 4.59, respectively). [The spectrum of the keto dichloride (3) contains additional maxima at 268 mμ (log $\epsilon = 4.20$) and 279.4 mμ (log $\epsilon = 4.18$).] The 246-mμ band almost certainly corresponds to the strong maximum at 240 mμ observed in the spectrum of benzocyclobutenone.[1]

TABLE 8.1

PHYSICAL PROPERTIES OF SUBSTITUTED 2-METHYLENEBENZOCYCLOBUTENONES

2-Methylenebenzocyclobutenone	Physical properties[a]	Properties of derivatives	Ref.
α-Carbomethoxy-	M.p. 87°–88° (w)	DNP, m.p. 235° (decomp.) (y–o)	2a, c
α-Carboxy-	M.p. 216°–218° (w)	—	6
α,α-Dichloro-	M.p. 96°–98° (y)	—	6

[a] Key: w (white), y (yellow), y–o (yellow-orange).

3. CHEMISTRY OF THE 2-METHYLENEBENZOCYCLOBUTENONES

The ketonic function of 2-carbomethoxymethylenebenzocyclobutenone reacts with 2,4-dinitrophenylhydrazine to give a normal carbonyl derivative (5). It reacts also with triphenylphosphinecarbomethoxymethylene to give 1,2-bis(carbomethoxymethylene)benzocyclobutene (6) in 81% yield.[2a,c]

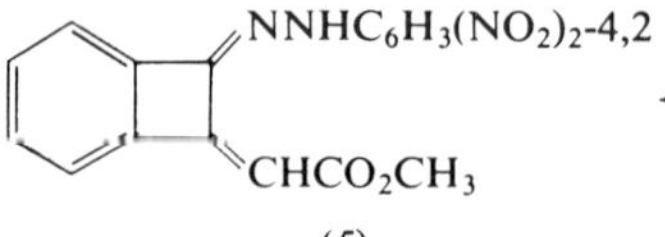

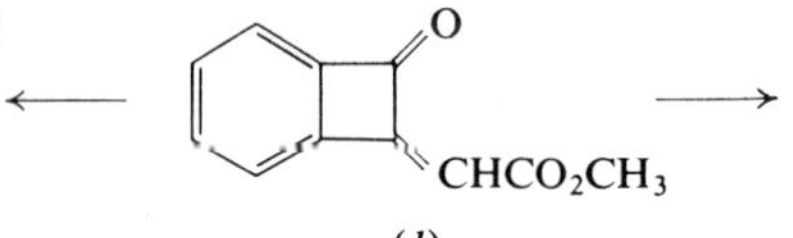

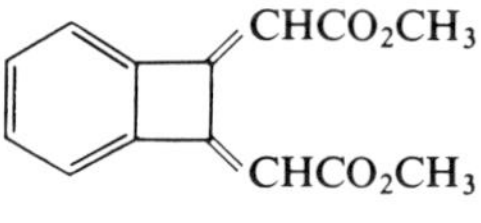

A substituted methylenebenzocyclobutenone (8) has been postulated as an intermediate in the chromic acid oxidation of 5,10-diphenylbenzo[b]biphenylene (7). The product of the reaction is 2-phenyl-2-(o-benzoylphenyl)indane-1,3-dione (9).[3]

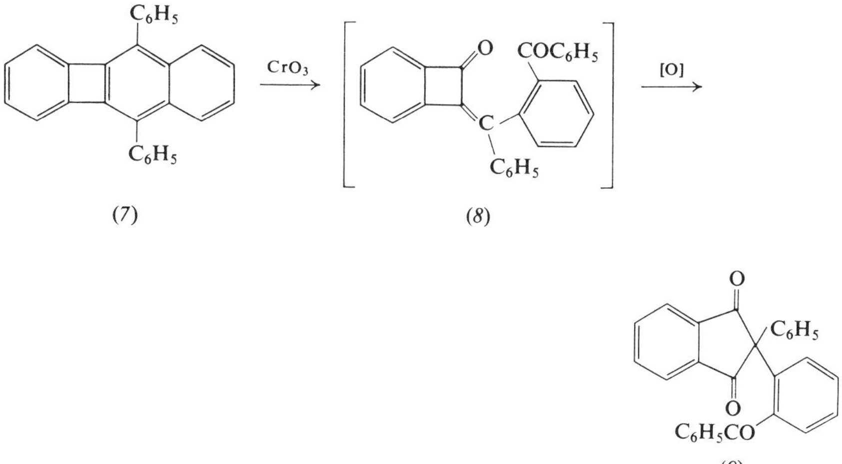

C. 1,2-Dimethylenebenzocyclobutenes

Molecular orbital calculations predict that 1,2-dimethylenebenzocyclobutene should have a delocalization energy (3.15 β) slightly larger than that of 1,2-benzocyclobutadienequinone (3.13 β).[4] The parent hydrocarbon, 1,2-dimethylenebenzocyclobutene, has been prepared from benzocyclobutadienequinone and is sufficiently stable to be isolated in a state of purity; however, it is prone to polymerization, and this suggests a high degree of free-valence character at the terminal methylene carbons.[2a,b] In contrast, the α,α'-dicarbomethoxy derivative of 1,2-dimethylenebenzocyclobutene is quite stable.[2a,c]

1. Synthesis of 1,2-Dimethylenebenzocyclobutenes

a. By the Wittig Reaction

The Wittig reaction has been employed in the synthesis of several α,α'-disubstituted 1,2-dimethylenebenzocyclobutenes; thus, 1,2-bis(carbomethoxymethylene)benzocyclobutene (1) was prepared in 85% yield by the reaction of benzocyclobutadienequinone (3) with two equivalents of triphenylphosphinecarbomethoxymethylene.[2a,c] The bis(carbomethoxymethylene) compound was also obtained in good yield (81%) when 2-(carbomethoxymethylene)benzocyclobutenone (2), prepared from benzocyclobutadienequinone and *one* equivalent of phosphorane, was treated with triphenylphosphinecarbomethoxymethylene. 1,2-Dibenzylidinebenzocyclobutene (4) was prepared in rather low yield (25%) by treating 1,2-benzocyclobutadienequinone with the more highly nucleophilic phosphorane, triphenylphosphinebenzylidene.[1]

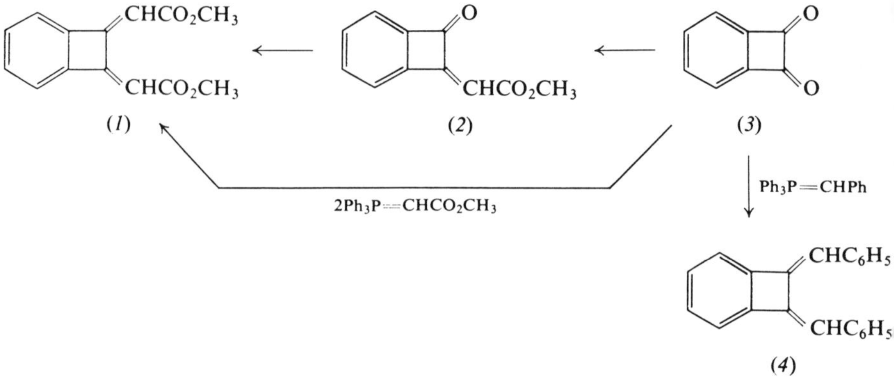

$$\text{(1)} \xleftarrow{} \text{(2)} \xleftarrow{} \text{(3)}$$

with (1) bearing =CHCO$_2$CH$_3$ groups, (2) bearing =O and =CHCO$_2$CH$_3$, and (3) bearing two =O groups.

Reagent from (3) to (2) and (1): 2Ph$_3$P=CHCO$_2$CH$_3$

From (3): Ph$_3$P=CHPh → (4) bearing =CHC$_6$H$_5$ groups.

b. By Pyrolysis of a 1,2-Dialkyl-1,2-diacyloxybenzocyclobutene

Vapors of the diacetate (5) of cis-1,2-dimethyl-1,2-benzocyclobutenediol (prepared from 1,2-benzocyclobutadienequinone and methylmagnesium bromide, followed by acetylation) were passed over Sterchamol firebrick at 300° to give the dimethylene compound (6) in 27% yield.[2a,b]

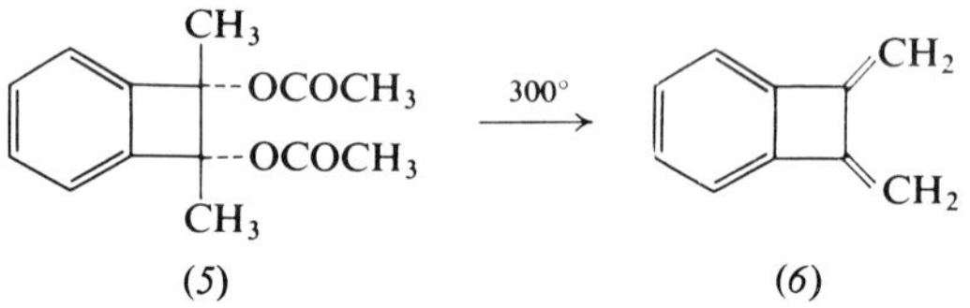

(5) [CH$_3$, OCOCH$_3$, OCOCH$_3$, CH$_3$ substituted benzocyclobutene] $\xrightarrow{300°}$ (6) [=CH$_2$, =CH$_2$ dimethylene benzocyclobutene]

c. By Cleavage of a Biphenylene Derivative

1,6,7,8-Tetramethoxybiphenylene-2,3-quinone-3-oxime (8) has been reported to give a highly substituted 1,2-dimethylenebenzocyclobutene. Thus, the monooxime (8) [prepared from the quinone (7)] underwent second-order Beckmann rearrangement when treated with tosyl chloride and base, and gave carboxycyanotetramethoxydimethylenebenzocyclobutene (9) in unspecified yield.[5] This reaction sequence constitutes the only reported example of the formation of a benzocyclobutene derivative by cleavage of a biphenylene.

2. Unsuccessful Attempts to Prepare 1,2-Dimethylenebenzocyclobutenes

1,2-Dimethylenebenzocyclobutene (6) was not obtained when 1,2-bis-(carboxymethylene)benzocyclobutene (10) was decarboxylated at elevated temperatures in the presence of soda lime, barium oxide, or quinoline–copper sulfate.[2a] Similarly, 1,2-dimethylenebenzocyclobutene was not produced,

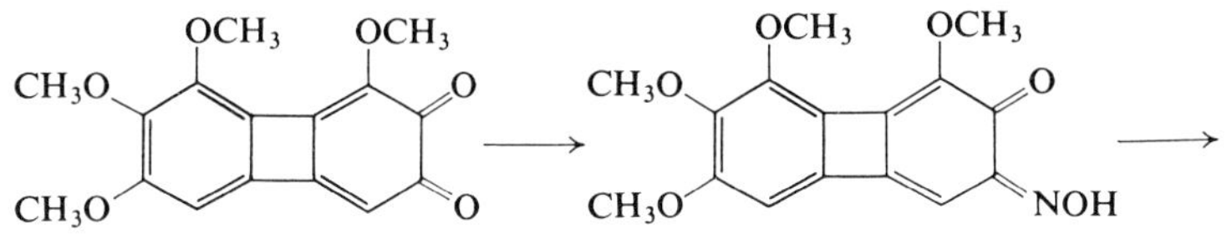

C. 1,2-DIMETHYLENEBENZOCYCLOBUTENES   237

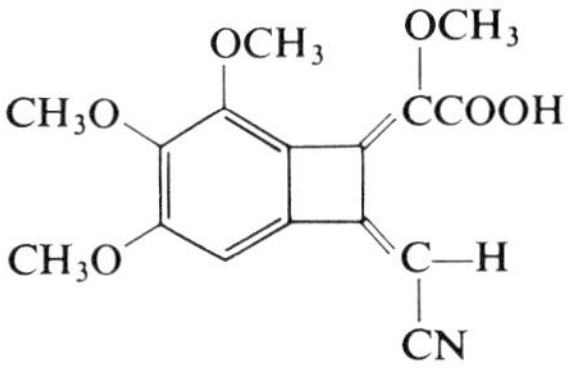

even in trace amounts, when 1,2-benzocyclobutadienequinone (3) was treated with the strongly nucleophilic phosphorane, triphenylphosphinemethylene. (The isolation of triphenylphosphine from the reaction mixture attests to an "abnormal" reaction course.[2a]) The failure of 1-keto-2-dichloromethylenebenzocyclobutene (11) to react with an excess of triphenylphosphine dichloromethylene to give 1,2-bis(dichloromethylene)benzocyclobutene (12) has been attributed to steric hindrance.[6] For two unsubstantiated dimethylenebenzocyclobutene structure assignments, see Chapter 6, Section C,7.

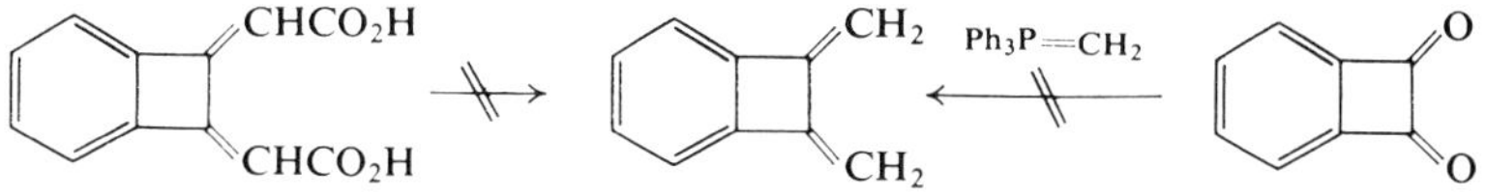

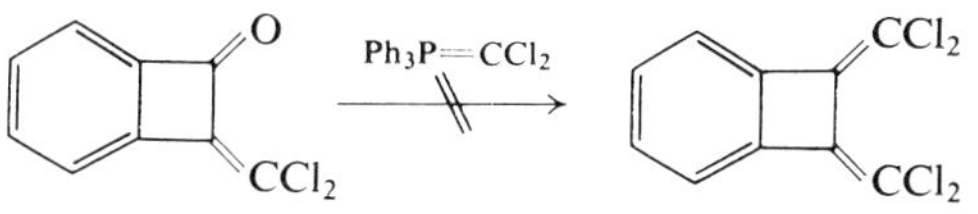

3. PHYSICAL PROPERTIES OF 1,2-DIMETHYLENEBENZOCYCLOBUTENES

1,2-Dimethylenebenzocyclobutene is a colorless liquid having a characteristic olefinlike odor. It can be frozen to give crystals melting at 15°–16° and, although fairly stable in the solid state, it polymerizes to a thick gum at room temperature within a few hours. It can be purified by gas chromatography at 150° on a special column of firebrick impregnated with silicone grease containing a small amount of *tert*-butylcatechol, but considerable loss through polymerization occurs during purification.[2a,b]

TABLE 8.2

Physical Properties of 1,2-Dimethylenebenzocyclobutenes

Benzocyclobutene	Method of preparation[a]	Physical properties[b]	Ref.
1,2-Dimethylene-	b	M.p. 15°–16° (cl)	2a, b
1,2-Di(carbomethoxymethylene)-	a	M.p. 125°–125° (w)	2a, c
1,2-Di(carboxymethylene)-	a	M.p. 313°–315° (decomp.) (w)	2a
1,2-Dibenzylidene-	a	M.p. 108°	1
1-Cyanomethylene-2-(α-methoxy-α-carboxy)methylene-3,4,5-trimethoxy-	c	—	5

[a] Method a is "By the Wittig Reaction," Section C, 1, a; Method b is "By Pyrolysis of a 1,2-Dialkyl-1,2-diacyloxybenzocyclobutene," Section C, 1, b; and Method c is "By Cleavage of a Biphenylene Derivative," Section C, 1, c.
[b] Key: cl (colorless), w (white).

The infrared spectrum of 1,2-dimethylenebenzocyclobutene, recorded in Fig. 8.1 shows strong methylene absorption bands at 11.42 μ (olefinic C—H

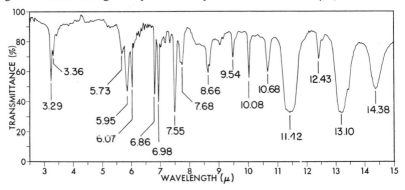

FIG. 8.1. Infrared spectrum of 1,2-dimethylenebenzocyclobutene.

bending vibration) and at 5.92 and 5.95 μ (olefinic C═C stretching vibrations).[2a,b] The ultraviolet spectrum of a solution of the hydrocarbon in ethanol, shown in Fig. 8.2, exhibits six well-defined bands between 230 and 329 mμ[2a,b]; the general appearance of the spectrum, and especially of the longer wavelength portion, matches that of biphenylene (see Fig. 10.2, Chapter 10, Section F).

In 1,2-bis(carbomethoxymethylene)benzocyclobutene the conjugated exocyclic double bond appears at 6.04 μ in the infrared. The ultraviolet spectrum

C. 1,2-DIMETHYLENEBENZOCYCLOBUTENES

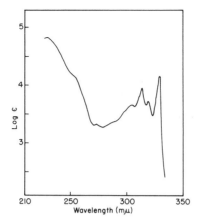

FIG. 8.2. Ultraviolet absorption spectrum of 1,2-dimethylenebenzocyclobutene (in ethanol).[2a]

of a solution of the diester in ethanol contains only two broad absorption bands, at 256 mμ (log $\epsilon = 4.64$) and at 290.5 mμ (log $\epsilon = 4.10$).[2a,c]

4. CHEMISTRY OF 1,2-DIMETHYLENEBENZOCYCLOBUTENES

a. Thermal Stability

The parent hydrocarbon, 1,2-dimethylenebenzocyclobutene, polymerizes readily at room temperature. At 700°, in the gas phase, it undergoes an unusual isomerization to give naphthalene in 18 % yield [(6) → (13)].[2a,b] α-Substituents which enter into conjugation with the methylene double bonds stabilize the 1,2-dimethylenebenzocyclobutene system. Thus, in contrast to the parent hydrocarbon, both α,α'-diphenyl-1,2-dimethylenebenzocyclobutene[1] and α,α'-dicarbomethoxy-1,2-dimethylenebenzocyclobutene[2a,c] are stable at room temperature.

b. Stability to Dienophiles, Acids, and Bases

When treated with tetracyanoethylene, a solution of 1,2-dimethylenebenzocyclobutene in benzene gives an orange color suggestive of complex formation, but no complex could be isolated. On standing, only products of a polymeric nature were formed.[2a,b] It is understandable that a normal Diels–

8. METHYLENE ANALOGS OF 1,2-BENZOCYCLOBUTADIENEQUINONE

Alder addition, which would have produced a high-energy benzocyclobutadiene species (*14*) did not take place, but it is surprising that a cycloaddition product (*15*) was not isolated.* When the more stable derivative, 1,2-bis(carbomethoxymethylene)benzocyclobutene (*1*) was heated with tetracyanoethylene in boiling toluene for 24 hours,[2a,c] it was recovered unchanged (93% recovery).

The exocyclic diene system of diester (*1*) is stable to both acids and bases. Alkaline hydrolysis of the ester affords 1,2-bis(carboxymethylene)benzocyclobutene (*10*) in 99% yield. Esterification of the diacid (*10*) with methanol–sulfuric acid proceeds normally to give 1,2-bis(carbomethoxymethylene)benzocyclobutene in 92% yield.[2a]

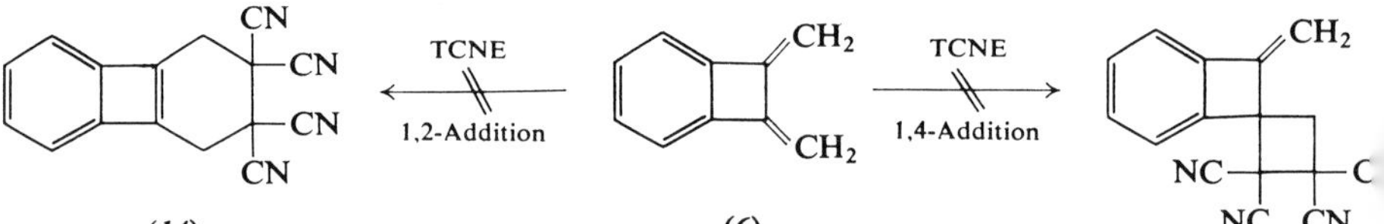

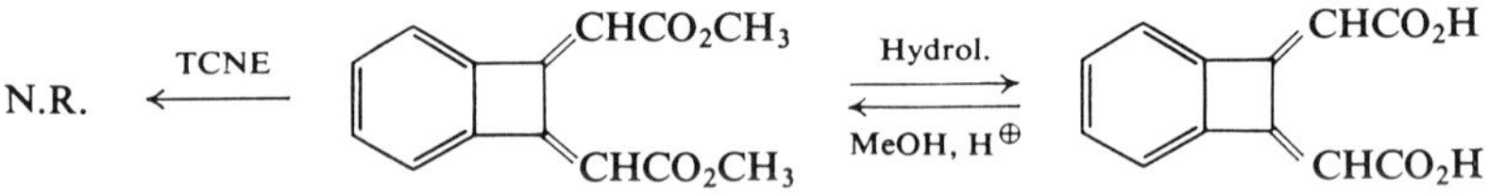

c. Other Reactions

1,2-Dimethylenebenzocyclobutene (*6*) is reduced readily by hydrogen in the presence of palladium-on-charcoal catalyst to give 1,2-dimethylbenzocyclobutene (*16*), presumably the cis isomer. In a similar manner, 1,2-bis(carbomethoxymethylene)benzocyclobutene (*1*) affords *cis*-1,2-bis(carbomethoxymethyl)benzocyclobutene (*17*) in 92% yield.[2a,c]

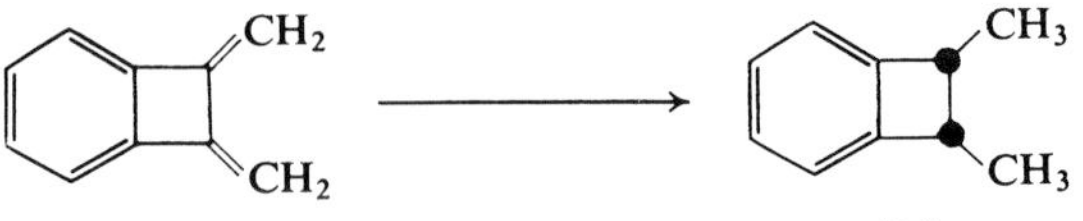

* The simple analogs, 3,4-dimethylenecyclobutene and 1,2-diphenyl-3,4-dimethylenecyclobutene, react with tetracyanoethylene to give an indeterminate product and a cycloaddition product, respectively. See Chapter 5, Section B,4,b.

References

1. Cava, M. P., and Mangold, D., unpublished work.
2. (a) Cava, M. P., Pohl, R. J., and Mitchell, M. J., *J. Am. Chem. Soc.* **85**, 2080 (1963); (b) Cava, M. P., Mitchell, M. J., and Pohl, R. J., *Tetrahedron Letters* **No. 18**, 825 (1962); (c) Cava, M. P., and Pohl, R. J., *J. Am. Chem. Soc.* **82**, 5242 (1960).
3. Cava, M. P., and Pohlke, R., *J. Org. Chem.* **27**, 1564 (1962).
4. Manatt, S. L., and Roberts, J. D., *J. Org. Chem.* **24**, 1336 (1959).
5. McOmie, J. F. W., *Rev. Chim. Acad. Rep. Populaire Roumaine* **7**, 1071 (1962).
6. Stein, R. P., Ph.D. diss., The Ohio State Univ. (1963).

CHAPTER 9

Higher Aromatic Analogs of Benzocyclobutadiene

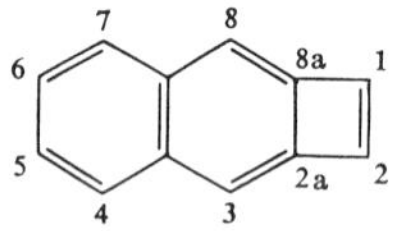

Naphtho[*b*]cyclobutadiene 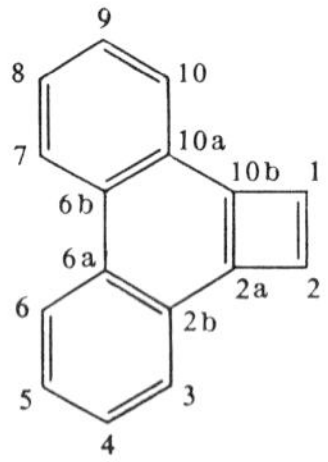   Phenanthro[*l*]cyclobutadiene

Benzocyclobutadiene is the simplest member of the series of completely unsaturated, polynuclear hydrocarbons derived by fusion of a benzenoid system with one side of a cyclobutadiene ring. The members of the series are isomeric with biphenylene and its benzologs (Chapters 10 and 11), but unlike the biphenylenes they always contain a terminally fused cyclobutadiene ring.

Up to the present time, the only experimental studies made of the higher aromatic analogs of benzocyclobutadiene have dealt with derivatives of naphtho[*b*]cyclobutadiene and phenanthro[*l*]cyclobutadiene. Both experiment and theory indicate that naphtho[*b*]cyclobutadiene is a more stable aromatic system than benzocyclobutadiene and that, in contrast, phenanthro[*l*]cyclobutadiene is less stable than benzocyclobutadiene.

Molecular orbital descriptions of naphtho[*a*]cyclobutadiene[10] and anthra-[*b*]cyclobutadiene[11] have been published. The results predict that naphtho[*a*]cyclobutadiene should be more reactive than benzocyclobutadiene and that anthra[*b*]cyclobutadiene should be less reactive than naphtho[*b*]cyclobutadiene.

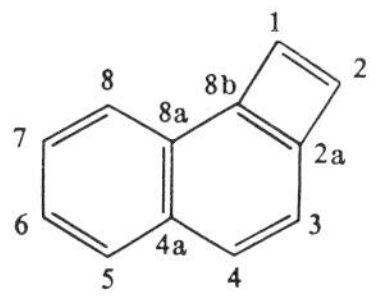

Naphtho[a]cyclobutadiene 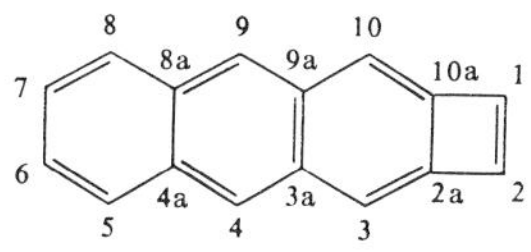   Anthra[b]cyclobutadiene

A. History

There is no early history of attempts to prepare higher aromatic analogs of benzocyclobutadiene.

B. Naphtho[b]cyclobutadiene

1. Generation (and Synthesis) of Naphtho[b]cyclobutadienes

Naphtho[b]cyclobutadienes have been generated by three methods: (a) the dehalogenation of 1,2-dihalonaphtho[b]cyclobutenes, (b) the dehydrohalogenation of 1-halonaphtho[b]cyclobutenes, and (c) the reverse Diels–Alder reaction.

a. Dehalogenation of 1,2-Dihalonaphtho[b]cyclobutenes

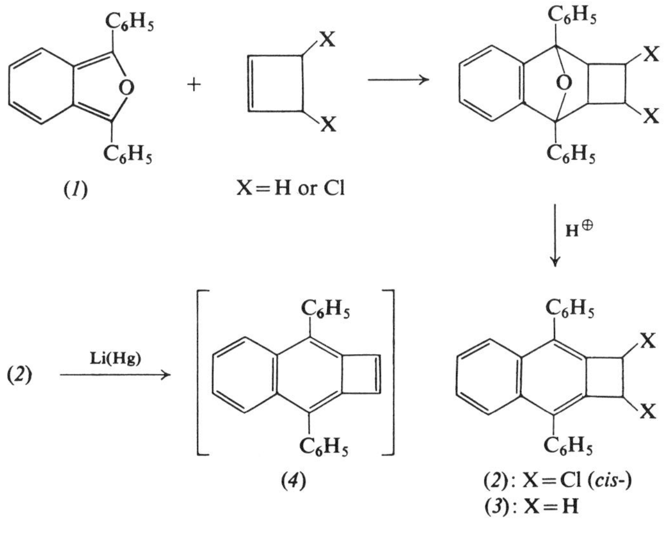

3,8-Diphenylnaphtho[b]cyclobutadiene (4) was generated by the action of lithium amalgam on *cis*-1,2-dichloro-3,8-diphenylnaphtho[b]cyclobutene (2) and 1,2-dibromo-3,8-diphenylnaphtho[b]cyclobutene (5).[2] 3,8-Diphenylnaphtho[b]cyclobutadiene is unstable but its formation as a transient intermediate in these reactions was demonstrated by trapping it with 1,3-diphenylisobenzofuran (see Section B, 3). The stable 1,2-dibromo analog (7) of 3,8-diphenylnaphtho[b]cyclobutadiene was synthesized by the action of activated zinc dust on 1,1,2,2-tetrabromo-3,8-diphenylnaphtho[b]cyclobutene (6).[5] The halides required as starting materials in these reactions were prepared from 1,3-diphenylisobenzofuran (1) by the reactions outlined below.

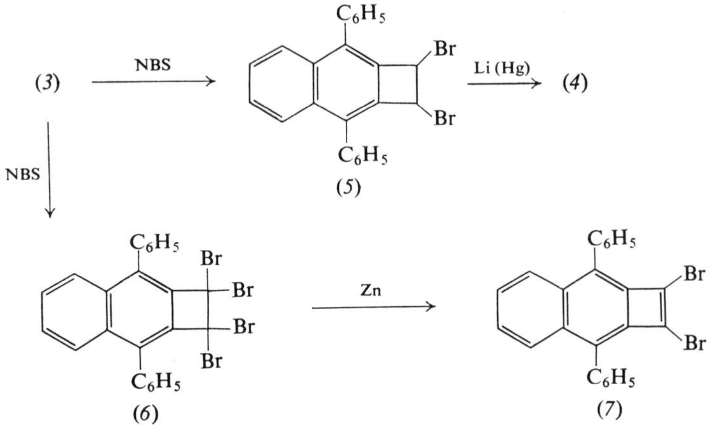

1,2-Diphenylnaphtho[b]cyclobutadiene (10) was synthesized by the action of zinc dust on 1,2-dichloro-1,2-diphenylnaphtho[b]cyclobutene (8) and 1,2-dibromo-1,2-diphenylnaphtho[b]cyclobutene (9).[6] [The dichloride (8) and the dibromide (9) were obtained from o-phthaldehyde and from α-chloro-2,3-dibenzylnaphthalene, respectively, by the two routes outlined below.[7]]

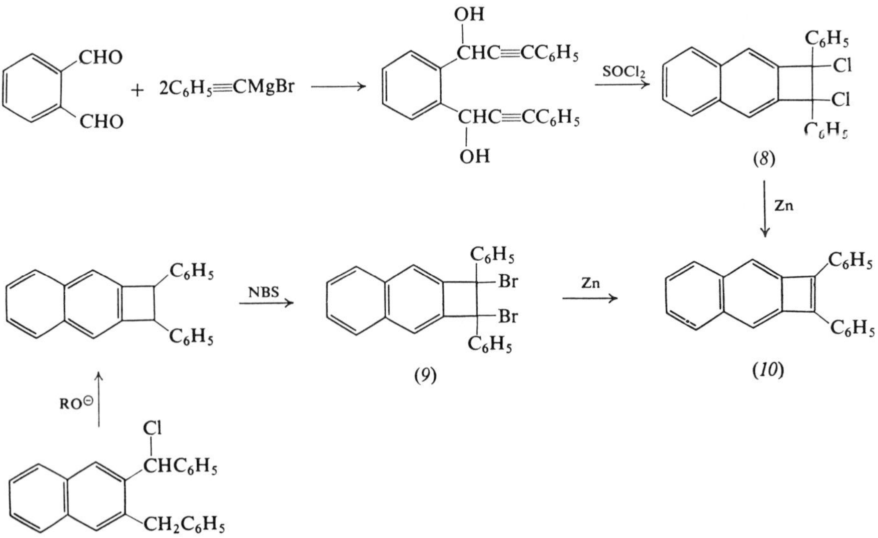

b. Dehydrohalogenation of 1-Halonaphtho[b]cyclobutenes

1-Bromo-3,8-diphenylnaphtho[b]cyclobutadiene (*12*) was generated by the reaction of potassium *tert*-butoxide with either 1,1-dibromo-3,8-diphenylnaphtho[b]cyclobutene (*11*) or 1,2-dibromo-3,8-diphenylnaphtho[b]cyclobutene (*5*).[5] [The 1,1-dibromide (*11*) was obtained as a side-product in the bromination of 3,8-diphenylnaphtho[b]cyclobutene.[5]] 1-Bromo-3,8-diphenylnaphtho[b]cyclobutadiene is unstable but its formation as a transient intermediate in this reaction was demonstrated by trapping it with 1,3-diphenylisobenzofuran.

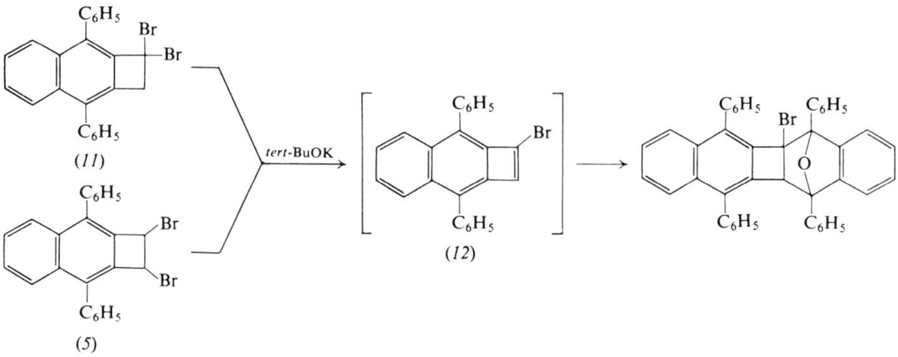

c. Reverse Diels–Alder Reaction

3,8-Diphenylnaphtho[b]cyclobutadiene (*4*) was generated by the thermolysis of a pentacyclic diester (*14*) at 270°.[12] [The diester (*14*) was prepared from the adduct (*13*) of dimethyl acetylenedicarboxylate and cyclooctatetraene by the sequence of reactions outlined below.] The formation of 3,8-diphenylnaphtho[b]cyclobutadiene was demonstrated by trapping it with 1,3-diphenylisobenzofuran (DPIBF).

2. Physical Properties of the Naphtho[b]cyclobutadienes

3,8-Diphenylnaphtho[b]cyclobutadiene and 1-bromo-3,8-diphenylnaphtho[b]cyclobutadiene are known only as transient reaction intermediates and have not been characterized. 1,2-Diphenylnaphtho[b]cyclobutadiene crystallizes as bright red needles, m.p. 136°–137°. It forms a sparingly soluble, black 2,4,7-trinitrofluorenone complex, m.p. 182°–183°, from which it can easily be regenerated by chromatography on alumina. The hydrocarbon is remarkably stable to air, heat, and light in the crystalline state. In the molten state, it retains its scarlet color up to approximately 260°.[6]

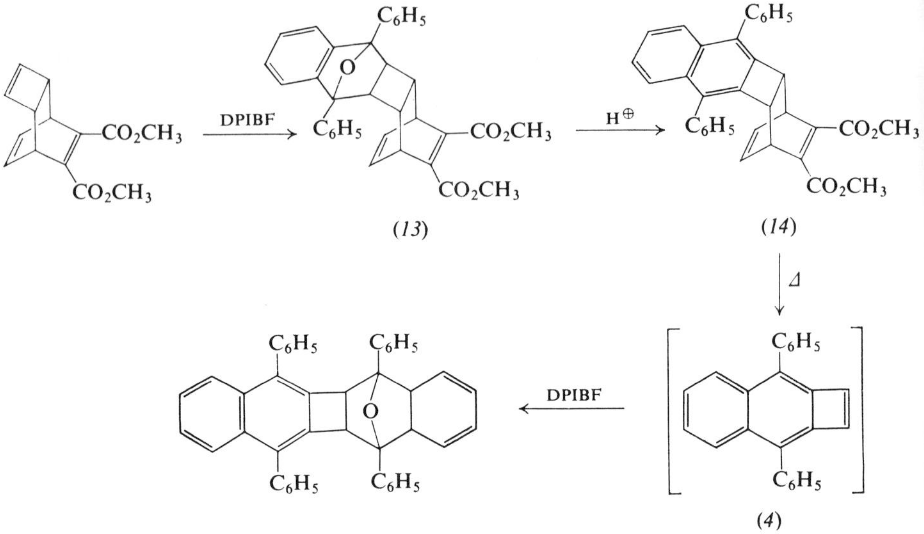

The ultraviolet-visible absorption spectrum of 1,2-diphenylnaphtho[b]-cyclobutadiene (Fig. 9.1) is complex and contains six bands of rather high intensity. Values for the absorption maxima and extinction coefficients (log ϵ) have been recorded: 209 (4.71), 257 (4.68), 289 (4.85), 300 (4.87), 436 (3.71), and 455 mμ (3.71) (in ethanol).[6]

The infrared spectrum of 1,2-diphenylnaphtho[b]cyclobutadiene (potassium bromide disc) is shown in Fig. 9.2. As a result of the symmetry of the molecule, the spectrum shows relatively few strong bands.

In addition to a large peak due to 14 protons in the usual aromatic region, the NMR spectrum of 1,2-diphenylnaphtho[b]cyclobutadiene shows a sharp peak in the olefinic region due to only two protons ($\delta-6.50$). The latter is attributed to the protons at C-3 and C-8, and corresponds very closely to the peak in the spectrum of cis-stilbene due to the olefinic protons ($\delta = 6.55$).[6] The high-field position of the C-3 and C-8 protons is indicative of an unusually large amount of double-bond character in the 2a,3- and 8,8a-bonds.

1,2-Dibromo-3,8-diphenylnaphtho[b]cyclobutadiene forms orange-yellow crystals which melt with decomposition at 162°–165°. The ultraviolet spectrum of 1,2-dibromo-3,8-diphenylnaphtho[b]cyclobutadiene shows the following absorption maxima (or shoulders) and extinction coefficients (log ϵ): 211 (4.52, shoulder), 247 (4.61), and 270 mμ (4.51, shoulder) (in ethanol).[4] The absence of longer-wavelength bands in the spectrum of 1,2-dibromo-3,8-diphenyl-naphtho[b]cyclobutadiene may be ascribed to the fact that its two phenyl substituents can be only slightly conjugated with the naphthocyclobutadiene

B. NAPHTHO[b]CYCLOBUTADIENE

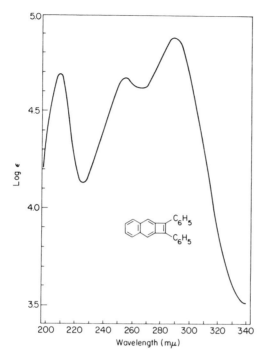

FIG. 9.1. Ultraviolet-visible absorption spectrum of 1,2-diphenylnaphtho[b]cyclobutadiene (in ethanol).[4]

system since, as in the case of 1-phenylnaphthalene, they are twisted out of the plane of the naphthalene ring.

3. Chemistry of the Naphtho[b]cyclobutadienes

Molecular orbital calculations predict that the unknown parent hydrocarbon,

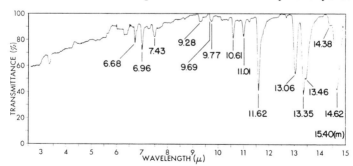

FIG. 9.2. Infrared spectrum of 1,2-diphenylnaphtho[b]cyclobutadiene (potassium bromide disc).[4]

naphtho[b]cyclobutadiene (15), should be stabilized by a remarkably large delocalization energy, i.e., 4.20 β.[9,11] The difference of about 1.1 β between this value and that of naphthalene is substantially greater than the corresponding difference (0·38 β) between the delocalization energies of benzocyclobutadiene and benzene.

Four canonical resonance forms may be written for naphtho[b]cyclobutadiene (15a)–(15d). Of these, structure (15a) may be expected to be the most important contributor to the hybrid because, in addition to being the most important Kekulé structure, it is a dimethylenecyclobutene structure as well. Conversely, structures (15c) and (15d) contain the electronically unfavorable cyclobutadiene nucleus and should therefore be minor contributors to the hybrid. Structure (15d) would appear to be particularly unfavorable because, in addition to being a cyclobutadiene structure, it is a 2,3-naphthoquinonoid structure, as well. It is interesting to note that the π-bond orders in naphtho-[b]cyclobutadiene as deduced by this simple qualitative treatment are in accord with those calculated from molecular orbital theory.[9,11]

(15a) ⟷ (15b) ⟷ (15c) ⟷ (15d)

Although the physical and chemical properties of the parent hydrocarbon naphtho[b]cyclobutadiene are as yet unknown, two substituted naphtho[b]cyclobutadienes, viz., 1,2-dibromo-3,8-diphenylnaphtho[b]cyclobutadiene (7)[5] and 1,2-diphenylnaphtho[b]cyclobutadiene (10)[6] have been isolated in crystalline form. The two substances are the only known isolable compounds which contain a cyclobutadiene nucleus having only one pair of neighboring carbons fused to an aromatic system and not stabilized by metal complexing. As far as stability is concerned, the naphtho[b]cyclobutadienes appear to occupy a borderline position between the unstable cyclobutadienes and benzocyclobutadienes on the one hand and the stable biphenylenes and benzobiphenylenes on the other.

The fact that 1,2-diphenylnaphtho[b]cyclobutadiene (10) is an isolable compound has been the impetus for a detailed theoretical study of the molecule. Theory predicts a high delocalization energy (9.12 β) for the molecule, as well as a singlet ground state and low free-valence indices at all positions (0.492 or less).[1] In accord with the presence of a sharp peak at δ=6.50 in the NMR corresponding to two olefinlike protons the calculated π-bond orders predict a considerable degree of π-bond fixation in the 2a,3- and the 8,8a-bonds.[1]

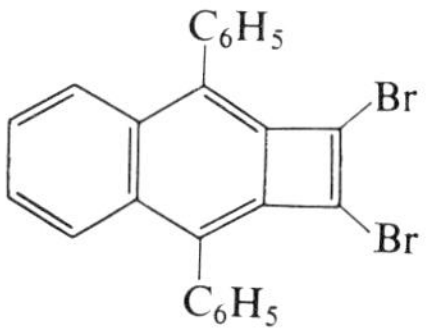

B. NAPHTHO[b]CYCLOBUTADIENE

(7) 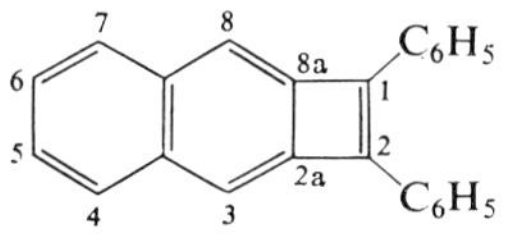   (10)

a. Polymer Formation

An amorphous polymer $(C_{24}H_{16})_x$, was the only product isolated when 3,8-diphenylnaphtho[b]cyclobutadiene (4) was generated by the thermolysis of a pentacyclic ester (14) at 270° or by lithium amalgam dechlorination of cis-1,2-dichloro-3,8-diphenylnaphtho[b]cyclobutene (2) at room temperature.[2] It is noteworthy that neither of these reactions produced a dimer of 3,8-diphenylnaphtho[b]cyclobutadiene or, by abstraction of hydrogen from the solvent, the known 3,8-diphenylnaphtho[b]cyclobutene (3).

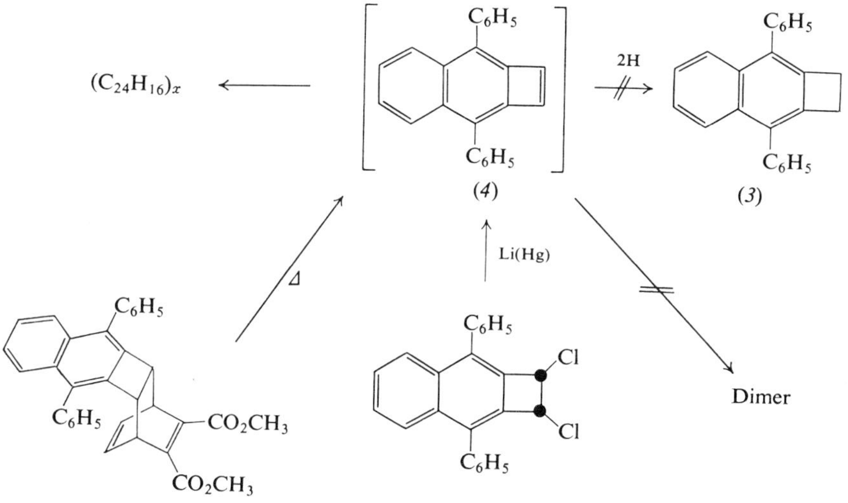

b. Diels–Alder Adduct Formation

The unstable 3,8-diphenyl analog (4) of naphtho[b]cyclobutadiene behaves as a reactive dienophile when generated in the presence of 1,3-diphenylisobenzofuran and gives the expected adduct (16). The adduct is formed when either of two methods is used to generate 3,8-diphenylnaphtho[b]cyclobutadiene, viz., thermolysis of a pentacyclic ester (14) at 270°[12] or dehalogenation of cis-1,2-dichloro-3,8-diphenylnaphtho[b]cyclobutene (2) and 1,2-dibromo-3,8-diphenylnaphtho[b]cyclobutene (5) with lithium amalgam.[2]

In a similar manner, the unstable monobromide, 1-bromo-3,8-diphenylnaphtho[b]cyclobutadiene (12), was trapped with 1,3-diphenylisobenzofuran

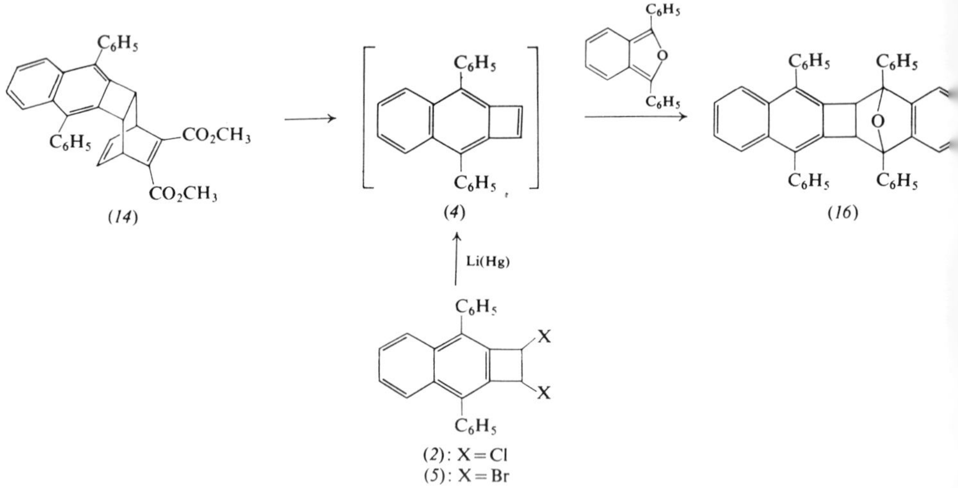

(DPIBF) to give the expected adduct (17). The monobromide (12) was generated by the action of potassium *tert*-butoxide on either 1,2-dibromo-3,8-diphenylnaphtho[b]cyclobutene (5) or 1,1-dibromo-3,8-diphenylnaphtho[b]cyclobutene (11).[5]

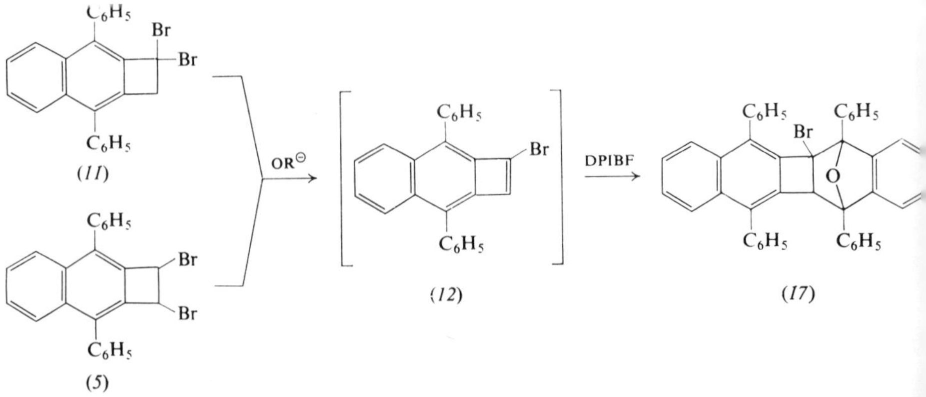

The isolable compounds 1,2-diphenylnaphtho[b]cyclobutadiene (10) and 1,2-dibromo-3,8-diphenylnaphtho[b]cyclobutadiene (7), were also found to be

reactive dienophiles. Both compounds reacted with 1,3-diphenylisobenzofuran at room temperature to give the corresponding adducts, (18)[6] and (19).[5]

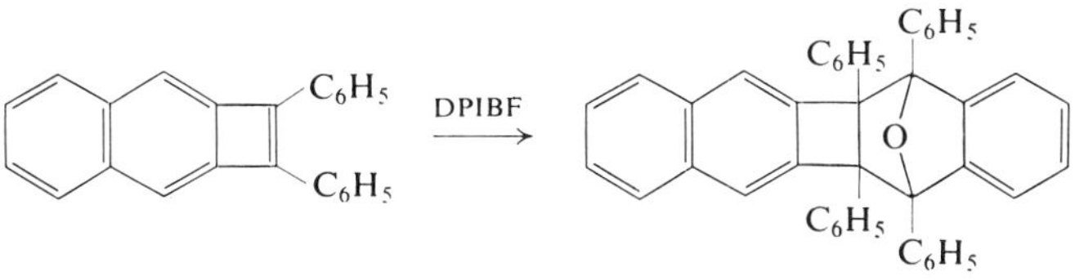

c. Other Reactions

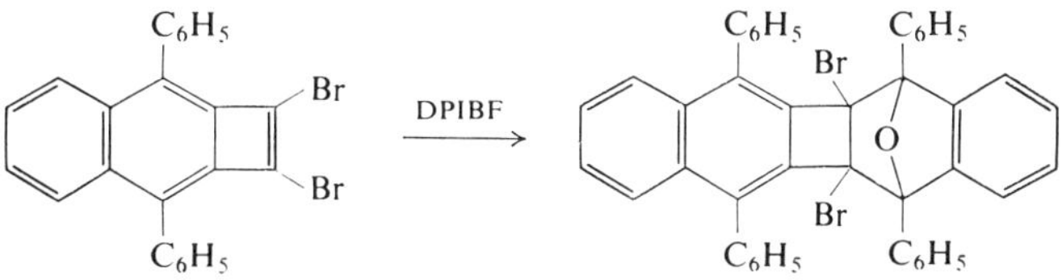

Catalytic reduction of 1,2-diphenylnaphtho[b]cyclobutadiene (10) gave cis-1,2-diphenylnaphtho[b]cyclobutene (20), while potassium permanganate oxidation yielded 2,3-dibenzoylnaphthalene (21).[6] Bromine was found to add to (10) to give 1,2-dibromo-1,2-diphenylnaphtho[b]cyclobutene (9).[4] These reactions all involve preferential attack of the reagent at the 1,2-unsaturated linkage of 1,2-diphenylnaphtho[b]cyclobutadiene (10).

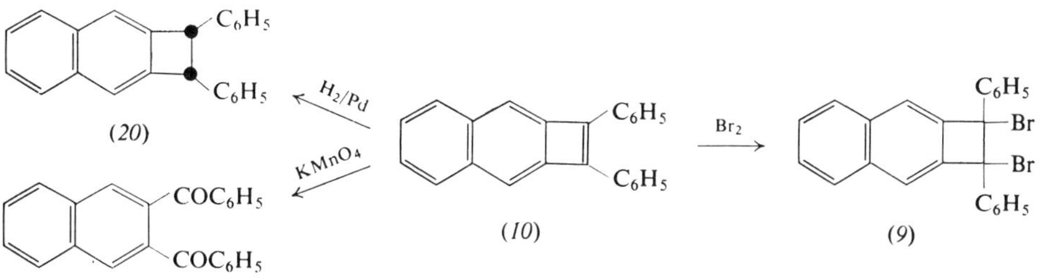

In a similar manner, catalytic reduction of 1,2-dibromo-3,8-diphenylnaphtho[b]cyclobutadiene (7) gave 3,8-diphenylnaphtho[b]cyclobutene (3), while addition of bromine afforded 1,1,2,2-tetrabromo-3,8-diphenylnaphtho[b]cyclobutene (6).[5]

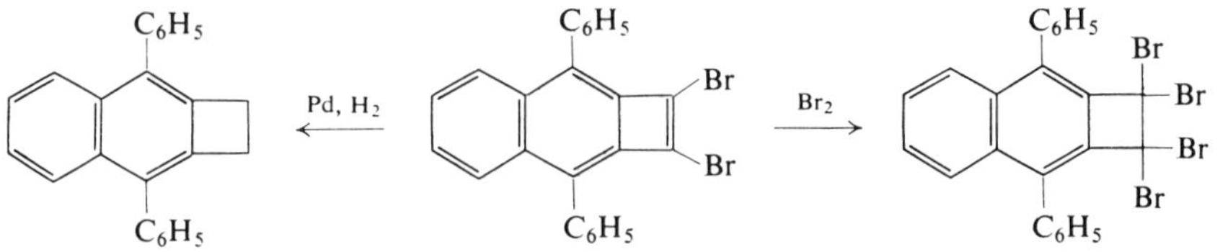

When 1,2-diphenylnaphtho[b]cyclobutadiene is dissolved in one of the common organic solvents and irradiated with ultraviolet light, a dimer of as yet undetermined structure is produced.[4]

4. Unsuccessful Approaches to Some 3,8-Diazanaphtho[b]cyclobutadienes

Unsuccessful attempts have been made to synthesize three substituted 3,8-diazanaphtho[b]cyclobutadienes, viz., 1,2-dimethyl- (22),[14] 1-phenyl- (23),[13] and 1,2-diphenyl-3,8-diazanaphtho[b]cyclobutadiene (24),[3] by reaction of the appropriate substituted cyclobutadienequinones with o-phenylenediamine. The failure of the reaction to give rise to the desired quinoxaline analogs (22), (23), and (24) must be attributed to the marked tendency of cyclobutadienequinones to undergo 1,4-addition and ring cleavage [e.g., (25) → (26)].* The question of the degree of stability of 3,8-diazanaphtho[b]cyclobutadienes is in no way answered by these unsuccessful reactions.

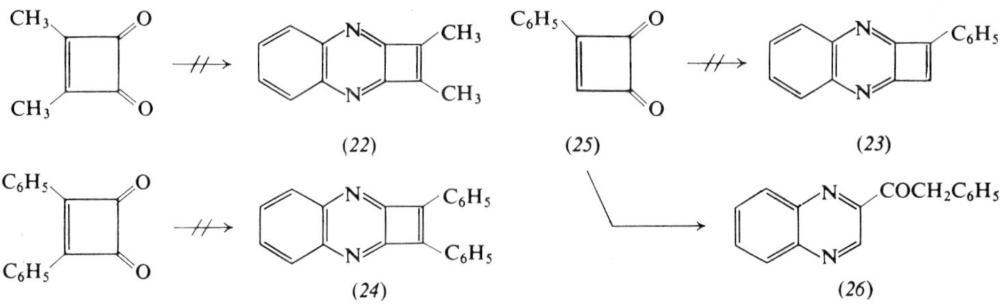

C. Phenanthro[l]cyclobutadiene

1. Generation of a Phenanthro[l]cyclobutadiene

1,2-Diphenylphenanthro[l]cyclobutadiene (1) was generated by the debromination of 1,2-dibromo-1,2-diphenylphenanthro[l]cyclobutene (2); both

* For a more detailed discussion of the action of o-phenylenediamine on cyclobutadienequinones, see Chapter 4, Section B, 1, g.

C. PHENANTHRO[*l*]CYCLOBUTADIENE   253

zinc and nickel carbonyl were used as dehalogenating agents.[8] The dibromide (2) required as the starting material was prepared from the readily available dihydrophencyclone (3)[10] by the two-step reaction sequence outlined below.[8]

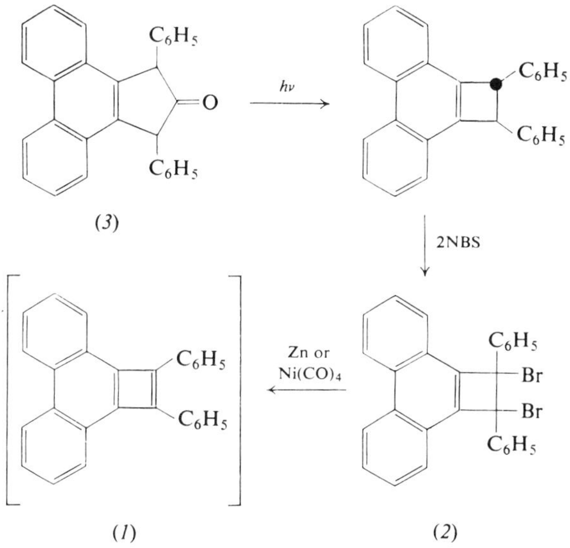

2. CHEMISTRY OF 1,2-DIPHENYLPHENANTHRO[*l*]CYCLOBUTADIENE

Molecular orbital calculations predict that the unknown parent hydrocarbon phenanthro[*l*]cyclobutadiene (4), should have a singlet ground state and an appreciable delocalization energy (5.69 β).[11] On the other hand, the π-bond orders at the 1,2-bond (0.933) and 2a,10b-bond (0.682) are rather large and suggest that the system will be unstable because of considerable cyclobutadiene character in the four-membered ring.[11]

1,2-Diphenylphenanthro[*l*]cyclobutadiene is also predicted to have a closed-shell configuration, but since the highest filled bonding molecular orbital of the hydrocarbon is separated from its lowest unfilled antibonding molecular orbital by only 9–10 kcal/mole, it is likely that the hydrocarbon exists as a thermally populated triplet species (see Chapter 12). 1,2-Diphenylphenanthro[*l*]cyclobutadiene (*1*) is quite unstable and all attempts to prepare it have resulted in the formation of an amorphous high-melting polymer. When the hydrocarbon (*1*) was generated in the presence of air, 9,10-dibenzoylphenanthrene (5) was formed in 85% yield, probably by way of a cyclic peroxide (6). 1,2-Diphenylphenanthro[*l*]cyclobutadiene does not yield a defiinte dimer nor does it react as a dienophile with diphenylisobenzofuran.[8]

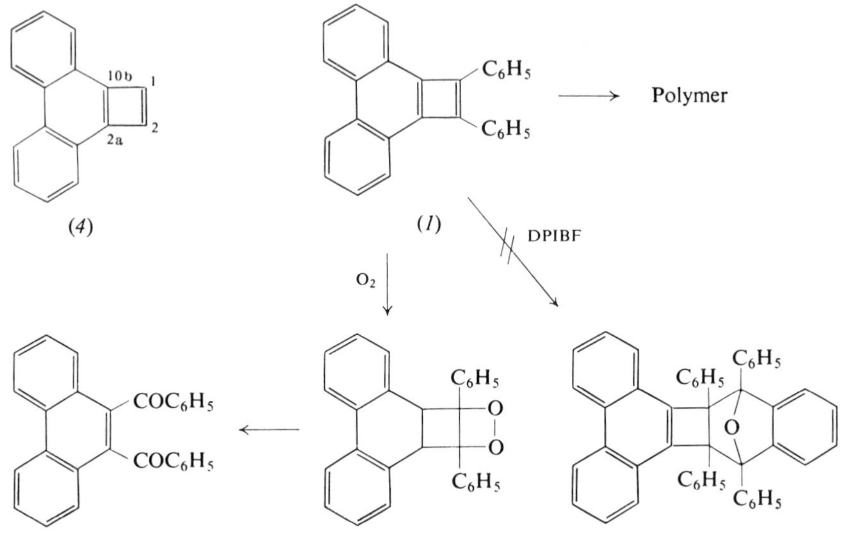

References

1. Anastassiou, A. G., private communication.
2. Avram, M., Dinulescu, I. G., Elian, M., Farcasiu, M., Marica, E., Mateescu, G., and Nenitzescu, C. D., *Ber.* **97**, 372 (1964).
3. Blomquist, A. T., and LaLancette, E. A., *J. Am. Chem. Soc.* **84**, 220 (1962).
4. Cava, M. P., and Hwang, B. Y., unpublished work.
5. Cava, M. P., and Hwang, B. Y., *Tetrahedron Letters* 2297 (1965).
6. Cava, M. P., Hwang, B. Y., and Van Meter, J. P., *J. Am. Chem. Soc.* **85**, 4032 (1963).
7. Cava, M. P., Hwang, B. Y., and Van Meter, J. P., *J. Am. Chem. Soc.* **85**, 4031 (1963).
8. Cava, M. P., and Mangold, D., *Tetrahedron Letters* 1751 (1964).
9. Coulson, C. A., and Poole, M. D., *Tetrahedron* **20**, 1859 (1964).
10. Dilthey, W., Horst, J. ter, and Schommer, W., *J. prakt. Chem.* **251**, 189 (1935).
11. Lee, H. S., *Chemistry (Taipei)* 22 (1963).
12. Nenitzescu, C. D., Avram, M., Dinulescu, I. G., and Mateescu, G., *Ann.* **653**, 79 (1962).
13. Smutny, E. J., Caserio, M. C., and Roberts, J. D., *J. Am. Chem. Soc.* **84**, 220 (1962).
14. Veirling, R. A., *Dissertation Abstr.* **23**, 79 (1962).

CHAPTER 10

Biphenylene*

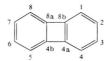

Biphenylene is formally the dibenzo derivative of cyclobutadiene. However, because of its unusual physical and chemical properties, it is more accurately viewed as the parent member of a unique series of polynuclear aromatic hydrocarbons. Biphenylene is also referred to in the literature as "diphenylene." Unfortunately, a similar name, "diphenyline," is applied to an unrelated compound, 2,4-diaminobiphenyl, and the similarity between the two names has been a source of considerable confusion.† The name "cyclobutadibenzene" was used by the editors of *Chemical Abstracts* from 1937 until 1946, but it is no longer sanctioned. "Biphenylene" has been approved as official usage by the IUPAC.[81]

The biphenylene literature has been reviewed in whole or in part by Baker and McOmie,[19, 20, 21] Baker,[8, 9] McOmie,[96] Cava,[43] Gelin,[66] and Vol'pin.[119]

A. History

The first attempt to synthesize biphenylene, presented unequivocably as such, was reported by Niementowski in 1901. A solution of the double diazonium salt of 2,2′-diaminobiphenyl in hydrochloric acid was treated with freshly precipitated copper, but the only product that could be isolated from the reaction mixture was carbazole. Niementowski also suggested the use of the Ullmann reaction for the preparation of biphenylene from 2,2′-dihalobiphenyls and implied that such work (never published) was then in progress.[100]

* For a discussion of the benzobiphenylenes, see Chapter 11. Partially reduced benzobiphenylenes which have an intact biphenylene ring system, e.g., 1,2,3,4-tetrahydrobenzo[*b*]biphenylene, are classified as biphenylenes and are discussed in the present chapter.

† See, for example, *Chem. Abstr.*, **55**, 282s (1961), wherein for "biphenylene, 511*a*," read "diphenyline, 511*a*."

In 1911, Dobbie and co-workers claimed that they had synthesized biphenylene by the action of sodium on 2,2'-dibromobiphenyl.[57] Their work was allegedly supported by the subsequent preparation of several transformation products[58] and by an independent synthesis of biphenylene from purpurotannin, claimed by Nierenstein.[101] However, because all subsequent efforts to duplicate their work have been unsuccessful[88, 93, 105, 109, 110] and because the chemical and physical properties of the product which they obtained are almost wholly different from those accepted for biphenylene, the claims of Dobbie and Nierenstein must be discounted.

Attempts to prepare biphenylene continued to be made during the next quarter of a century without abatement (see Table 10.6). Finally, in 1941, Lothrop published his now well-known paper in which convincing experimental evidence was presented for the first time in support of an author's claim of the synthesis of biphenylene.[88] [It is interesting to note that biphenylene was also prepared at about this time by Rapson and Shuttleworth as a minor by-product of the oxidation of biphenyl-2,2'-dimagnesium dibromide with cupric chloride (Krizewski–Turner reaction)[105b]; however, the English workers did not establish the identity of their product until several years later.[105a]] Contemporary authors[21, 88] cite the work of two early German chemists which might have been directed toward the synthesis of biphenylene, viz., (a) the reaction of dibenzofuran with sulfuric acid, reported by Hoffmeister,[77] and (b) the reaction of o-dibromobenzene with sodium in boiling ether, reported by Hosaeus.[78] It must be borne in mind, however, that neither Hoffmeister not Hosaeus mentioned biphenylene as the desired product. Indeed, the alleged conversion of diphenyl ether into biphenyl, cited by Hoffmeister [K. List and H. Limpricht, *Ann.* **90**, 190 (1854)], which might have served him as a model in his work with dibenzofuran, was in fact shown to be in error by Hoffmeister himself.[77]

It should be mentioned in passing that the synthesis of several dibenzobiphenylenes was claimed (erroneously) shortly before the publication of Lothrop's paper on biphenylene. Actually, the first synthesis of a compound of the benzobiphenylene type, 5-bromobenzo[a]biphenylene, was effected unknowingly by Finkelstein in 1910, but the identity of Finkelstein's product was not established until many years after Lothrop's report had been published (see Chapter 11).

B. Synthesis of Biphenylenes

1. From 2,2'-Dihalobiphenyls

Biphenylene was first synthesized by heating 2,2'-dihalobiphenyls with cuprous oxide at 350° (Lothrop,[88] 1941). The reaction appears to be closely related to the Ullmann reaction but differs significantly from the usual form of the latter in that it succeeds only when *cuprous oxide* rather than elemental copper is used as the dehalogenating agent.* Poor yields were obtained by Lothrop when the dihalo compound employed was 2,2'-dibromobiphenyl (*1*)—due partly to product isolation difficulties—but better yields were obtained with 2,2'-diiodobiphenyl (*2*) and with biphenyleneiodonium iodide (*3*).[89] [The two diiodides (*2*) and (*3*) are entirely equivalent in the Lothrop synthesis, since the latter rearranges to the former at temperatures well below the reaction temperature.] A detailed study of the reaction has recently led to a procedure which gives biphenylene in approximately 21% yield from a mixture of the diiodides.[15] However, high yields (e.g., 43%[20]) are rarely obtained by inexperienced workers and seem to be dependent on the quality of the cuprous oxide used ("aged samples" give better results)[25,46,121] and probably on a number of undetermined factors as well. Nevertheless, in its improved form, Lothrop's synthesis has remained the best method for the preparation of biphenylene up to the present time.†

In view of this fact, the most convenient routes for preparing diiodobiphenyl and biphenyleneiodonium iodide will be discussed briefly. (a) The classical reaction sequence[20,88] begins with the conversion of *o*-nitrochlorobenzene into 2,2'-dinitrobiphenyl by the Ullmann reaction [(*4*) → (*5*)], followed by reduction of the dinitro compound to 2,2'-diaminobiphenyl (*6*). Tetraazotization of the diamine, followed by reaction with potassium iodide, gives a mixture of 2,2'-diiodobiphenyl and biphenyleneiodonium iodide, (*2*) and (*3*), which need

* Note added in proof: Contrary to all previous reports, the reaction of several 2,2'-diiodobiphenyls with copper gives the corresponding biphenylenes in yields superior to those obtained with cuprous oxide. [See J. C. Salfeld and E. Baume, *Tetrahedron Letters* 3365 (1966).]

† But see also Abstract 10.15 in the Appendix.

not be separated. (b) By an alternate route, 2,2′-diaminobiphenyl is obtained in one step [(7) → (6)] by a Schmidt reaction on diphenic acid (available commercially); the reaction has been conducted satisfactorily on a molar scale.[45] (c) Finally, a rather different approach to the preparation of the diiodo compound consists in the conversion of 2-aminobiphenyl (available commercially) into 2-iodobiphenyl via the Sandmeyer reaction [(8) → (9)]. Oxidation of the iodobiphenyl with peracetic acid to give 2-iodosobiphenyl is followed by cyclization of the iodoso compound with sulfuric acid–acetic anhydride to give biphenyleneiodonium bisulfate [(9) → (10) → (11)], which is converted into biphenyleneiodonium iodide.[50] The last method has been used to prepare biphenylene in 30-gm lots in an overall yield of 26% (based on 2-aminobiphenyl).[2]

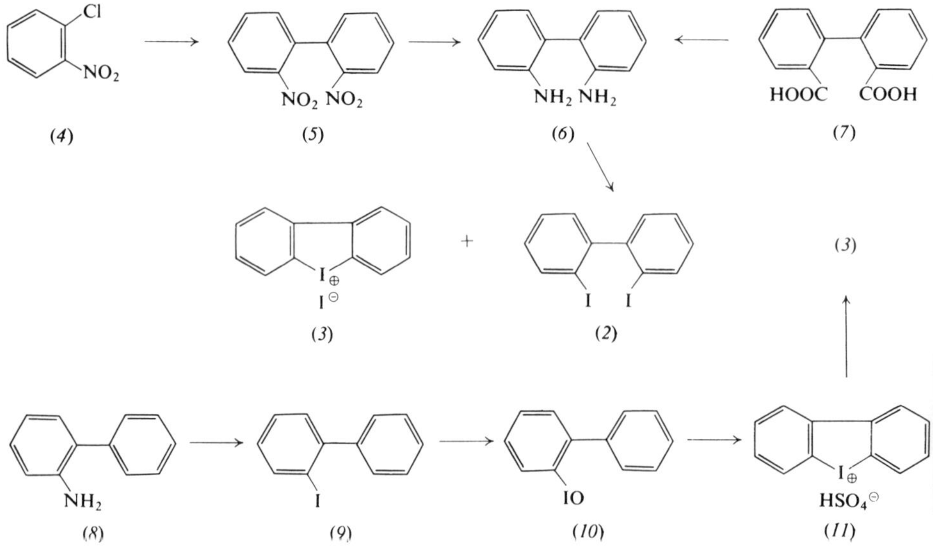

In the hands of Baker, McOmie, and their collaborators, Lothrop's method has become a remarkably effective tool for the synthesis of substituted biphenylenes. Of course, a number of these compounds were obtained only in minute yield, but it is interesting to note that complete failure of the method has yet to be reported.* Further, only three examples have been recorded in which the reaction took an unexpected course, i.e., (a) in the conversion of 3-bromobiphenyleneiodonium iodide, which gave biphenylene along with the expected product, 2-bromobiphenylene (12),[10] (b) in the conversion of an octahydrobinaphthaleneiodonium iodide (13), which gave dibenzo[b,h]biphenylene (14)

* For an unsuccessful attempt to prepare a *benzo*biphenylene by this method, see dibenzo[a,g]biphenylene, Chapter 11, Section C.

TABLE 10.1
Substituted Biphenylenes Prepared by Lothrop's Method

Biphenylene	Starting material	Yield (%)	Ref.
2-Bromo-	2-Bromodibenzoiodolium iodide	10	10
1,8-Dimethoxy-	1,9-Dimethoxydibenzoiodolium iodide and 2,2'-diiodo-6,6'-dimethoxybiphenyl (mixture)	—	10
2,3-Dimethoxy-	2,3-Dimethoxydibenzoiodolium iodide	6.1[a]	32
2,7-Dimethoxy-	3,7-Dimethoxydibenzoiodolium iodide	2	89
	2,2'-Diiodo-5,5'-dimethoxybiphenyl	42	13
2,7-Dimethoxy-1,3,6,8-tetramethyl-	2,2'-Diiodo-4,4'-dimethoxy-3,5,3',5'-tetramethylbiphenyl	28	13
1,8-Dimethyl-	1,9-Dimethyldibenzoiodolium iodide	"Poor"	89
2,7-Dimethyl-	2,8-Dimethyldibenzoiodolium iodide	6[a]	88
	3,7-Dimethyldibenzoiodolium iodide	4.5	88
1,8-Dimethyl-3,4,5,6-tetramethoxy-	2,2'-Diiodo-3,3'-dimethyl-5,6,5',6'-tetramethoxybiphenyl	27	16
2,6-Dinitro-	3,8-Dinitrodibenzoiodolium iodide	0.4[a]	10
	2,2'-Diiodo-4,5'-dinitrobiphenyl	0.4[a]	10
1,2,3,6,7,8-Hexamethoxy-	2,2'-Diiodo-3,4,5,3',4',5'-hexamethoxybiphenyl	22	16
1,2,4,5,7,8-Hexamethoxy-	2,2'-Diiodo-3,4,5,3',5',6'-hexamethoxybiphenyl	30	16
1,6,7,8-Hexamethoxy-2,3-quinone	2,2'-Diiodo-3,4,5,3',4',5'-hexamethoxybiphenyl	0.5	16
1-Methoxy-	1-Methoxydibenzoiodolium iodide and 2,2'-diiodo-6-methoxybiphenyl (mixture)	69[a]	10
2-Methoxy-	3-Methoxydibenzoiodolium iodide	8	10
1-Methyl-	1-Methyldibenzoiodolium iodide	44[a]	10
	2,2'-Diiodo-6-methylbiphenyl	41[a]	10
2-Methyl-	3-Methyldibenzoiodolium iodide	—	10
	2,2'-Diiodo-4-methylbiphenyl	—	10
2-Nitro-	3-Nitrodibenzoiodolium iodide	16[a]	10
1,2,3,4,7,8,9,10-Octahydrodibenzo[b,h]-(?)	1,2,3,4,8,9,10,11-Octahydrobinaphtho[2,3-b:2',3'-d]iodolium iodide	1.0[a]	121
1,2,3,4,5,6,7,8-Octamethoxy-	2,2'-Dibromo-3,4,5,6,3',4',5',6'-octamethoxybiphenyl	4	16
2,3,6,7-Tetramethoxy-	2,2'-Diiodo-4,5,4',5'-tetramethoxybiphenyl	31	13
	2,2'-Dibromo-4,5,4',5'-tetramethoxybiphenyl	4	13

[a] Calculated from weight data recorded in the literature.

instead of 1,2,3,4,7,8,9,10-octahydrodibenzo[b,h]biphenylene,[56] and (c) in the conversion of 2,2'-diiodo-3,4,5,3',4',5'-hexamethoxybiphenyl which gave 3,4,5,3',4',5'-hexamethoxybiphenyl and 1,6,7,8-hexamethoxybiphenylene-2,3-quinone (15) in addition to the expected product 1,2,3,6,7,8-hexamethoxybiphenylene.[16] (See Table 10.1.*)

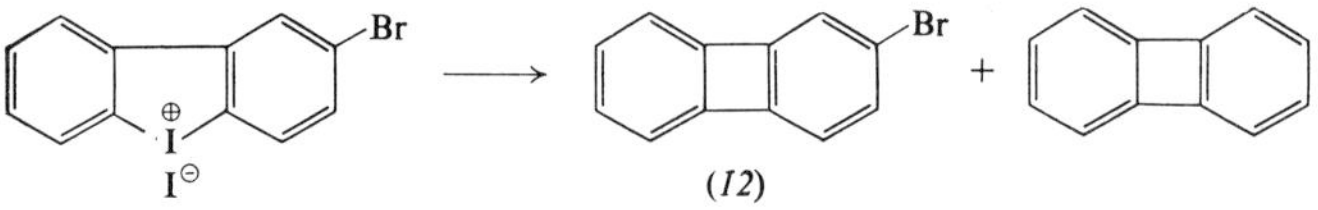

+ 1,2,3,6,7,8-Hexamethoxybiphenylene

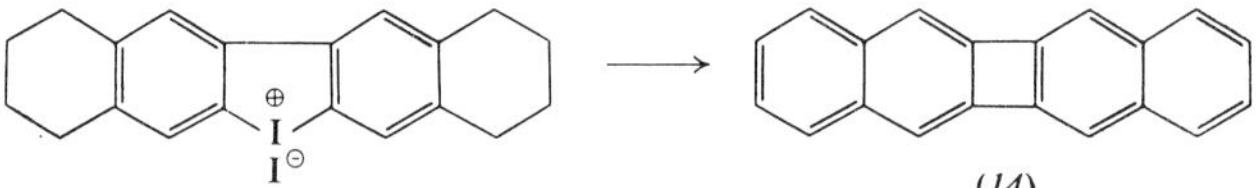

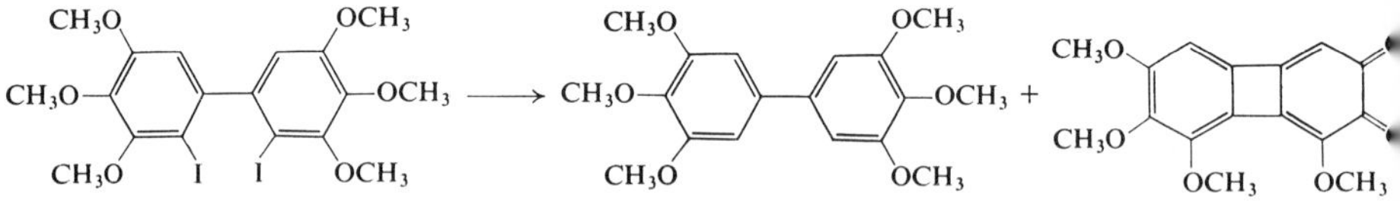

A "double" Lothrop reaction between o-diiodobenzene and cuprous oxide failed to give biphenylene (see Section C, "Unsuccessful Approaches to the Biphenylenes").[15]

2. Via Benzynes

Of the biphenylenes prepared from readily accessible starting materials, the number prepared by dimerization of benzynes is second only to the number prepared by Lothrop's method. Biphenylene itself has been obtained in

* Annellated biphenylene iodonium salts are difficult to name according to the "biphenyleneiodonium" nomenclature. For the sake of uniformity, the *Chemical Abstracts* "iodolium" nomenclature has been used in compiling the tables and the index. However, in order to avoid burdening the casual reader with a nomenclature with which he may not be familiar, only "biphenyleneiodonium" names are used in the text. The numbering systems for the biphenyleneiodonium and the dibenzoiodolium nomenclatures are compared below.

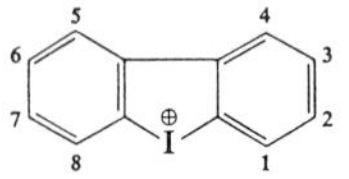

Biphenyleneiodonium ion 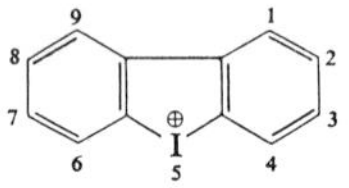   Dibenzoiodolium ion 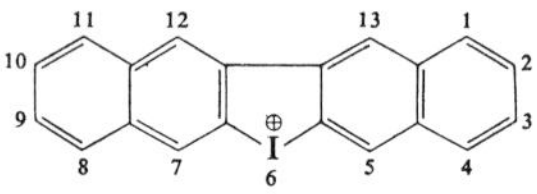   Dinaphtho[2,3-b:2',3'-d] iodolium ion

B. SYNTHESIS OF BIPHENYLENES

this way as the product or by-product of a number of reactions in which benzyne is believed to be generated as a transient intermediate.* Thus, (a) the dehalogenation of o-fluorobromobenzene (16) with lithium amalgam in ether gives biphenylene in 24% yield.[128,136] Similarly, (b) the dehalogenation of o-dibromobenzene with sodium amalgam in tetrahydrofuran[132] or with lithium amalgam in ether[136] gives biphenylene in 4.5% and 9% yield, respectively, while (c) the dehalogenation of o-bromoiodobenzene (17) with magnesium gives the hydrocarbon in 3.5% yield.[73] (d) Flash photolysis of a

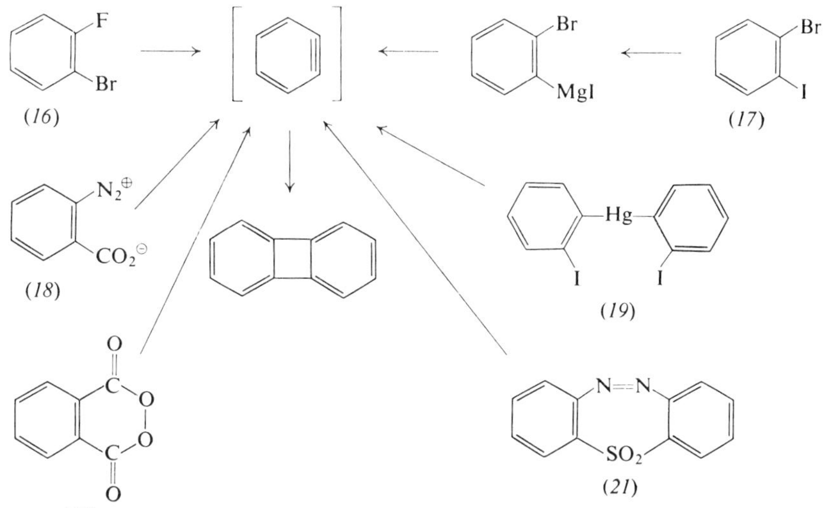

thin film of solid o-benzenediazonium carboxylate (18) generates benzyne (detected by its ultraviolet absorption spectrum) which rapidly dimerizes to biphenylene.[27] (e) The low-pressure, gas-phase, flash pyrolysis of bis(o-iodophenyl)mercury (19) and phthaloyl peroxide (20) at 600° gives biphenylene in yields of 54% and 27%, respectively.[130] (f) In contrast, thermolysis of 2,3,6,7-dibenzo-1-thia-4,5-diazacycloheptatriene-1,1-dioxide (21) gives the hydrocarbon in trace amounts only,[130] while (g) the flash pyrolysis of biphenyl-iodonium-2-carboxylate at 325° gives biphenylene[85] in 5% yield.[86] (h) Pyrolysis of 1,2,3-benzothiadiazole-1,1-dioxide (22) in the presence of nitrous oxide affords biphenylene in 52% yield, together with a small amount of 2-nitrobiphenylene.[134] (i) Dehalogenation of o-diiodobenzene in the gas phase gives biphenylene in 21% yield when nitrogen is the carrier gas and in 7% yield when hydrogen is the carrier gas.[69] (j) Biphenylene is formed in the positive column

* Reviewed in refs. 37, 72, 90, 96, 127, and 128.

of a special cold-discharge tube (Schüler tube) containing the vapors of benzene,[106b] bromobenzene,[106b] chlorobenzene,[106] fluorobenzene,[106a] iodobenzene,[106b] or phenylacetylene.[106a] The biphenylene is believed to be formed by the generation and subsequent dimerization of benzene.*,[106b,107] (k) Biphenylene is also formed in the thermolysis of 1,2-diiodobenzene,[62] and by treating o-dilithiobenzene with cupric chloride.[126] (1) Treatment of o-fluorotoluene with phenyllithium gave 1,5(or 8)-dimethylbiphenylene in 35% yield.†,[74]

Several substituted biphenylenes have been prepared by treatment of the appropriately substituted o-dihalobenzenes with dehalogenating agents. Treatment of 2-fluoro-3-bromo-1,4,5-trimethylbenzene with lithium amalgam afforded 1,2(or 3),4,5,6,8-hexamethylbiphenylene in 16% yield[131]; and treatment of 3-bromo-4-iodotoluene (23) with magnesium gave 2,6-dimethylbiphenylene (24), as did treatment of the dihalide (23) with n-butyllithium followed by chlorodiethylphosphine.[71]

The formation of the 2,6-isomer in this reaction may have theoretical implications, especially if the 2,7-isomer is not also formed. The most obvious implication is that the reaction does in fact proceed by dimerization of a benzyne intermediate and not by a less direct pathway.‡ (See also Section B, 3, "By the Ullmann Reaction.") A second implication, which stems from the first, is that the formal bond of 4-methylbenzyne is polarized to a significant degree and is responsible for the complementary arrangement of the two halves of the resulting 2,6-dimethyl isomer. Finally, a third implication is that substituted biphenylenes such as 2,7-dimethylbiphenylene, which have the noncomplementary arrangement, cannot be prepared by the benzyne dimerization method.§

Whether or not these implications have any basis in fact remains to be

---

* Compare the formation of biphenylene by the cold-discharge method from biphenyl and 2-fluorobiphenyl, Section B, 9.
† Calculated from weight data recorded in the literature.
‡ For which see refs. 65 and 79.
§ Compare Abstract 10.24 in the Appendix.

B. SYNTHESIS OF BIPHENYLENES

demonstrated. Indeed, it must be recognized that the argument in support of the above implications depends on the fact that the 2,7-dimethyl isomer is not also formed in the reaction. [Because the yield of biphenylenes produced in benzyne dimerization reactions is usually small (the 2,6-dimethyl isomer was formed in only 0.4% yield), it is not impossible that the 2,7-isomer was also formed in the reaction and that it was overlooked in the complex mixture of products.]

The method is not without its limitations. Because benzyne reacts readily with both nucleophilic and electrophilic reagents, it must be generated in the absence of such reagents and in aprotic solvents if it is to undergo dimerization. [The dehydrohalogenation of fluorobenzene with phenyllithium, for example, gives biphenyl (25) and not biphenylene.[136]] However, in one case a good yield of a substituted biphenylene was obtained under seemingly unfavorable conditions. Thus, 2-fluorotoluene was dehydrohalogenated with phenyllithium in the presence of N,N-dimethylbenzylamine in an attempt to study the amination of toluyne with the latter; the product actually formed was found to be 1,5(or 8)-dimethylbiphenylene (26), obtained in 35% yield.*,[74]

In the case of biphenylene itself, the two most serious disadvantages of the method are (a) the low yield and (b) the difficulty of separating biphenylene from triphenylene, which is also formed in the reaction. However, separation of the two hydrocarbons can be effected through the trinitrobenzene complexes. (The trinitrobenzene complex of biphenylene is less stable than that of triphenylene.[130])

Finally, it must be mentioned that benzyne and substituted benzynes generated by decomposition of the appropriate o-phenylenediazonium carboxylates fail to give biphenylenes.†,[65] Other unsuccessful attempts to pre-

---

* The example of 2,6-dimethylbiphenylene (see above) suggests a 1,5-disubstitution pattern for this product. Compare the physical properties of the 1,5(or 8)-compound with those of known 1,8-dimethylbiphenylene in Table 10.14.

† See, however, Abstract 10.15 in the Appendix.

pare biphenylenes by the benzyne dimerization method are included in Table 10.6.

TABLE 10.2

Substituted Biphenylenes Prepared via Benzynes

Biphenylene	Reactants	Yield (%)	Ref.
1,5(or 8)-Dimethyl-	o-Fluorotoluene and phenyllithium	35[a]	74
2,6-Dimethyl-	3-Bromo-4-iodotoluene and n-butyllithium, followed by chlorodiethylphosphine	8.8[a]	71
	3-Bromo-4-iodotoluene and magnesium	0.4	75
1,2(or 3),4,5,6,8-Hexamethyl-	2-Fluoro-3-bromo-1,4,5-trimethylbenzene and lithium amalgam	16	131

[a] Calculated from weight data recorded in the literature.

3. By the Ullmann Reaction

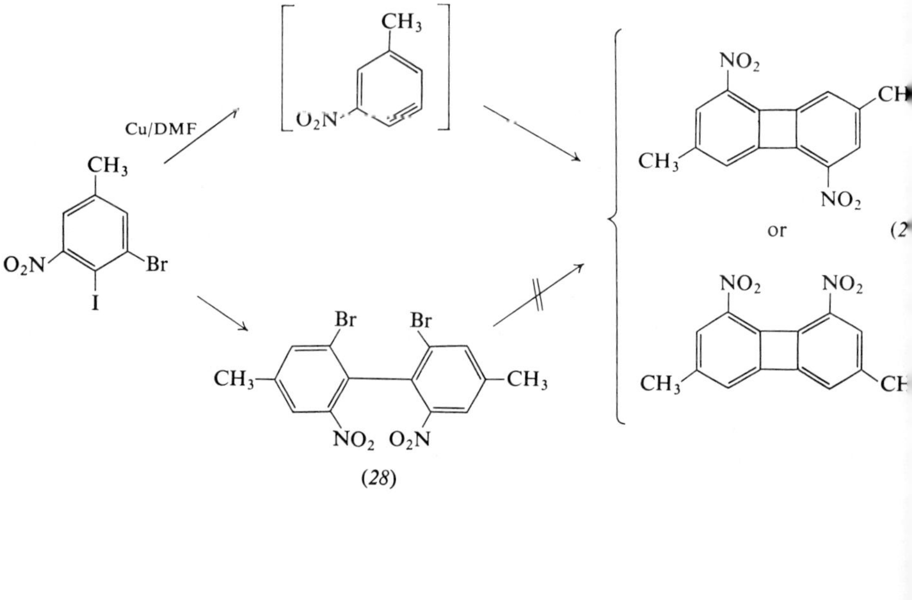

As a rule, biphenylenes cannot be obtained through the Ullmann reaction, but several instances are known in which substituted biphenylenes were prepared in this way (i.e., by heating the appropriate o-dihalobenzenes with copper powder). The method of Corbett and Holt,[52] which employs dimethylformamide as the solvent, holds great promise and no doubt merits broader application.

It has been suggested that the reaction does not proceed by two successive couplings as might have been supposed, but by benzyne formation and dimerization[52] (see Section B, 2, above). In support of this suggestion, it has been pointed out that 2,2'-dibromo-4,4'-dimethyl-6,6'-dinitrobiphenyl (*28*), which is produced as a side-product in the preparation of 3,7(or 6)-dimethyl-1,5(or 8)-dinitrobiphenylene (*29*) from 3-bromo-4-iodo-5-nitrotoluene, "is not an intermediate in the formation of the biphenylene derivative."[52]

It is noteworthy that a trace of 2,3,6,7-tetramethoxybiphenylene was produced as a side-product in the synthesis of 2,2'-dibromo-4,5,4',5'-tetramethoxybiphenyl from 4-bromo-5-iodoveratrole by the Ullmann reaction (see also Section B, 2). It is possible that the tetramethoxybiphenylene was formed via a benzyne intermediate, but it is interesting to note that when an effort was made to improve the generation of the benzyne by substituting magnesium for the copper used in the Ullmann reaction, the tetramethoxybiphenylene was not obtained.[13]

TABLE 10.3

BIPHENYLENES PREPARED BY THE ULLMANN REACTION

Biphenylene	Reactants	Yield (%)	Ref.
3,7(or 6)-Dimethoxy-1,5(or 8)-dinitro-	3,4-Dibromo-5-nitroanisole and copper	5	97
3,7(or 6)-Dimethyl-1,5(or 8)-dinitro-	3-Bromo-4-chloro-5-nitrotoluene and copper bronze	6	52
	3-Bromo-4-iodo-5-nitrotoluene and copper bronze	2.5	52
	3,4-Dibromo-5-nitrotoluene and copper bronze	40^a	52a
2,3,6,7-Tetramethoxy-	4-Bromo-5-iodoveratrole and copper	0.3^b	13

^{a} Yields are variable (e.g., 7–10%) and are probably dependent on the quality of the copper bronze (ref. 52b).
^{b} Calculated from weight data recorded in the literature.

4. FROM ORGANOMETALLIC BIPHENYLS

The decomposition of organometallic biphenyls is reported to give biphenylene in good yield, and it is almost certain that the usefulness of the reaction has not been fully realized. (a) When biphenylenemercury [in reality, cyclic tetramer (*30*)] is heated with silver powder at 300°, biphenylene is produced in 54% yield.[133] The starting material is readily available from 2,2′-diiodobiphenyl via 2,2′-dilithiobiphenyl. (b) In a related reaction, o-phenylenemercury [in reality, cyclic hexamer (31)][68] gives biphenylene in 55% yield when heated with silver powder at 260°; biphenylenemercury (30) is said to be an intermediate in the reaction, but proof is lacking.[130] (o-Phenylenemercury is available in 41% yield by treatment of o-dibromobenzene with sodium amalgam.)

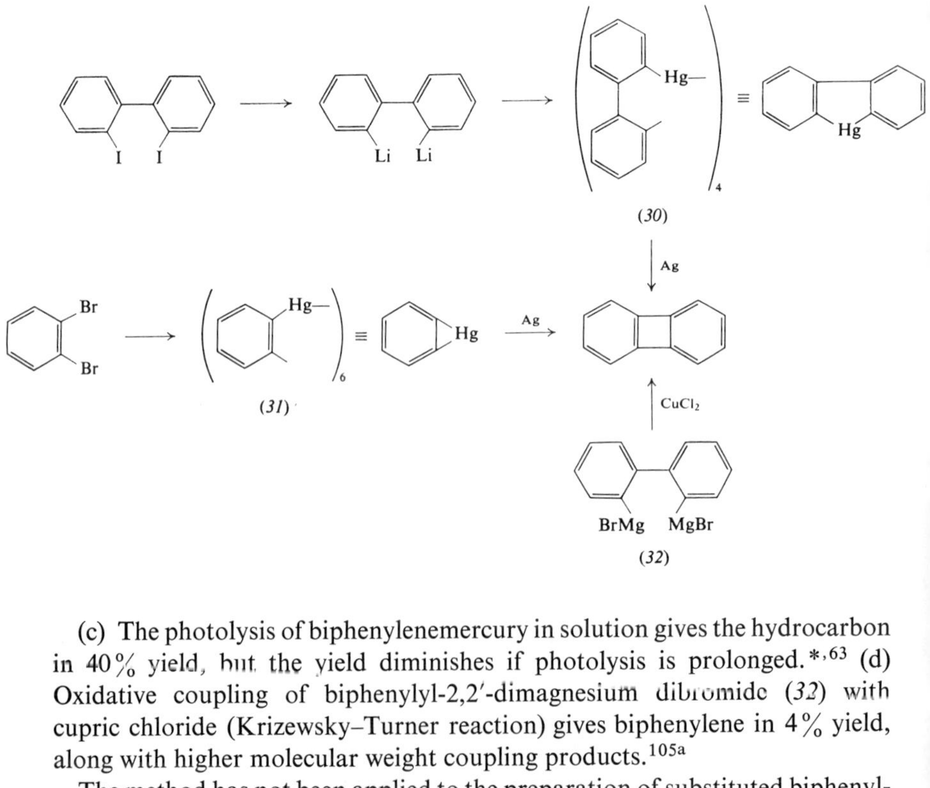

(c) The photolysis of biphenylenemercury in solution gives the hydrocarbon in 40% yield, but the yield diminishes if photolysis is prolonged.*,[63] (d) Oxidative coupling of biphenylyl-2,2′-dimagnesium dibromide (32) with cupric chloride (Krizewsky–Turner reaction) gives biphenylene in 4% yield, along with higher molecular weight coupling products.[105a]

The method has not been applied to the preparation of substituted biphenylenes. Unsuccessful attempts to prepare biphenylene via organometallic biphenyls are recorded in Table 10.6.

5. From Reduced Biphenylenes

The preparation of biphenylenes from reduced biphenylenes is without practical value at the present time, because the latter are either difficult to

---

* Biphenylene itself is not affected by irradiation with ultraviolet light (refs. 1 and 83).

obtain or are resistant to aromatization. However, (a) a tetrabromotetrahydrobiphenylene, possibly 2,3,4,4a-tetrabromo-2,3,4,4a-tetrahydrobiphenylene* [(33), prepared in 4% yield by treating biphenylene with excess bromine],[72] was converted into biphenylene by dehalogenation with zinc,[20] and (b) an "octahydrobiphenylene," $C_{12}H_{16}$ [prepared from 4-bromocyclohexene (34) by treatment with silver fluoroborate or with silver fluoroantimonate in sulfur dioxide, followed by hydrolysis of the resulting salt], was aromatized by selenium to give biphenylene.[102]

In this connection, it is interesting to note that 1,4-oxido-1,4,4a,8b-tetrahydrophenylene (35) failed to give biphenylene when treated with hydrochloric acid,[44] and that 2,3,6,7-tetramethyl-4a,4b,8a,8b-tetrahydrobiphenylene-1,4,5,8-bis(quinone) (36) could not be aromatized beyond the dihydro stage.[36, 51] (But compare the behavior of the reduced benzo- and dibenzobiphenylenes, some of which are readily aromatized, p. 325ff.)

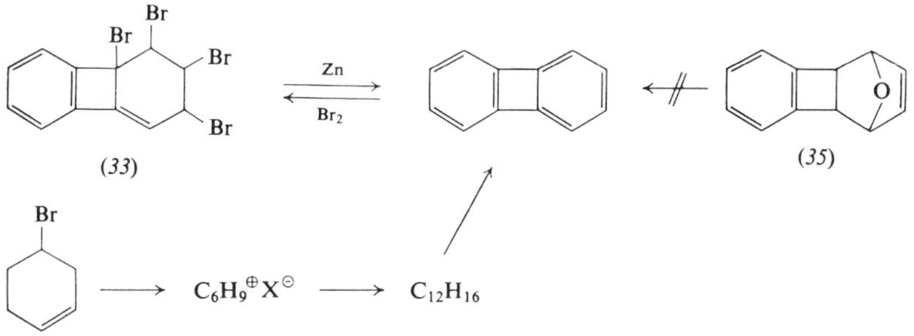

(c) A mixture of reaction products obtained by the alkali metal reduction of biphenylene in liquid ammonia underwent disproportionation during distillation to regenerate a small amount of biphenylene.[4]

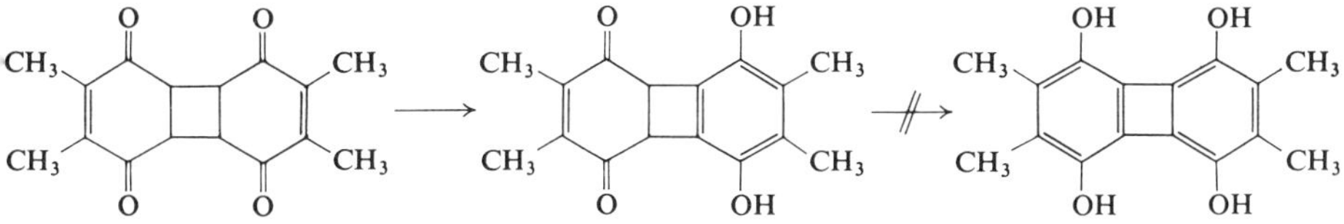

6. FROM A NONAROMATIC PRECURSOR

1,3,7,9-Cyclododecatetrayne (37), an unstable substance prepared by the oxidative coupling of 1,5-hexadiyne, was isomerized to biphenylene in 25%

* See Abstract 10.3 in the Appendix.

yield at room temperature by potassium *tert*-butoxide. A by-product of the isomerization, cyclododecatetraenediyne (*38*), is not an intermediate in the reaction because further treatment of the tetraenediyne with base does not convert it into biphenylene.[137]

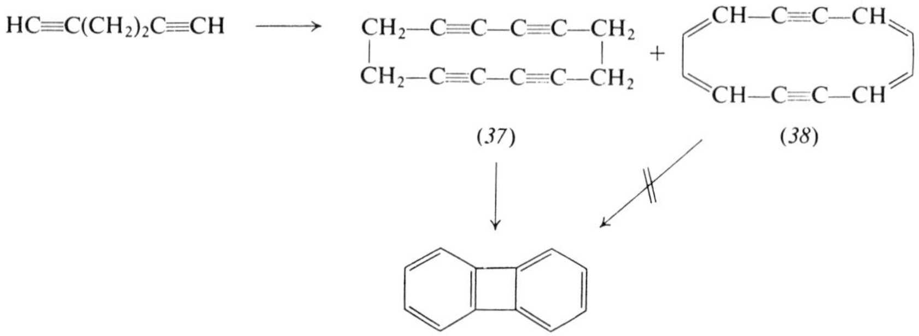

$$HC\equiv C(CH_2)_2C\equiv CH \longrightarrow \begin{array}{c} CH_2-C\equiv C-C\equiv C-CH_2 \\ | \quad\quad\quad\quad\quad\quad\quad | \\ CH_2-C\equiv C-C\equiv C-CH_2 \end{array} + \begin{array}{c} CH-C\equiv C-CH \\ \diagup \quad\quad\quad\quad \diagdown \\ \diagdown \quad\quad\quad\quad \diagup \\ CH-C\equiv C-CH \end{array}$$

(*37*)   (*38*)

7. By Direct Substitution

Biphenylene and a number of substituted biphenylenes have been subjected to direct substitution reactions to give the derivatives listed in Table 10.4. The chemistry of the reactions is discussed in Section G.

TABLE 10.4

Biphenylene Derivatives Prepared by Direct Substitution

Biphenylene	Reactants and reaction conditions	Yield (%)	Ref.
2-Acetamido-3-bromo-	2-Acetamidobiphenylene and bromine in acetic acid	65	23a
2-Acetoxy-[a]	Biphenylene and lead tetraacetate	—	15
2-Acetoxymercuri-	Biphenylene and mercuric acetate	89[b,c]	11
2-Acetyl-	Biphenylene (Friedel–Crafts)	58	15
		67[b]	11
2-Acetyl-3-hydroxy	2-Acetoxybiphenylene (Fries)	—	30
3-Acetyl-2-hydroxy-7-methoxy-	2,7-Dimethoxybiphenylene (Friedel–Crafts)	—	99
2-Amino-3-phenylazo-	2-Aminobiphenylene hydrochloride (coupling)	—	33
2-Benzoyl-	Biphenylene (Friedel–Crafts)	59	31
2-(2′-Biphenylenyl)-	Biphenylene and nitric acid	3.3	11
2-Bromo-	Biphenylene and bromine in carbon tetrachloride	49[d]	11
2-(β-Carbomethoxypropionyl)-	Biphenylene and β-carbomethoxypropionyl chloride (Friedel–Crafts)	59	14

B. SYNTHESIS OF BIPHENYLENES

TABLE 10.4 (continued)

Biphenylene	Reactants and reaction conditions	Yield (%)	Ref.
2-Chloro-	Biphenylene and iodine monochloride	58	11
	Biphenylene and chlorine	—	22
2-(p-Chlorophenylazo)-3-hydroxy-	2-Hydroxybiphenylene (coupling)	—	31
2,6-Diacetamido-3,7-dibromo-	2,6-Diacetamidobiphenylene and bromine–acetic acid with sodium acetate	—	23a
2,6-Diacetyl-	Biphenylene (Friedel–Crafts)	14	15
		65	11
1,4-Dibromo--2,3-quinone	Biphenylene-2,3-quinone and bromine in acetic acid	—	31
Dichloro- (?)	Biphenylene and chlorine	—	22
2,6(or 7)-Di(iodomethyl)-	Biphenylene and methyl chloromethyl ether; hydriodic acid	29a	11
2,6-Dinitro-	Biphenylene and nitric acid–sulfuric acid	4	11
—2,6(?)-Disulfonic acid[e]	Biphenylene and sulfuric acid	—	15
2-Hydroxy-3-phenylazo-	2-Hydroxybiphenylene (coupling)	—	31, 96
2-Hydroxy-3-(p-sulfophenyl)azo-[a]	2-Hydroxybiphenylene (coupling)	—	31
2-Iodo-	Biphenylene and iodine–hydriodic acid	10[b,c]	11
	Biphenylene and iodine–sodium persulfate	24[b,c]	11
1-Keto-1,2,3,4-tetrahydrobenzo[b]-	γ-(2-Biphenylenyl)butyric acid and polyphosphoric acid	63	14
2-Nitro-	Biphenylene and nitric–sulfuric acid	2.3[b]	11
2,3,6,7-Tetramethyl-(?)	2,6(or 7)-Di(iodomethyl)biphenylene and zinc–acetic acid (reduction–dismutation?)	21[b]	11

[a] Not isolated.
[b] Calculated from weight data recorded in the literature.
[c] Based on unrecovered starting material.
[d] Isolated as the TNF complex.
[e] Isolated as the S-benzylthiuronium salt.

8. By Transformation of Functional Groups

The transformation of substituents on the biphenylene nucleus is straightforward. Such complications as arise are usually traceable to the high strain energy of the ring system, which leads to cleavage of the four-membered ring in some instances and to saturation of one of the six-membered rings in others (see Section E, 4). Biphenylene itself has been prepared by the hydrogenolysis of 2-iodobiphenylene.[11] Unsuccessful attempts to prepare substituted biphenylenes by the transformation of functional groups are included in Table 10.6. Some 32 substituted biphenylenes prepared by the transformation of functional groups are listed in Table 10.5.

TABLE 10.5

SUBSTITUTED BIPHENYLENES PREPARED BY TRANSFORMATION OF FUNCTIONAL GROUPS

Biphenylene	Reactants and reaction conditions	Yield (%)	Ref.
2-Acetamido-	2-Acetylbiphenylene oxime (Beckmann) and		
	(a) polyphosphoric acid	60	15
		65	23a
	(b) phosphorus pentachloride	4[a]	23a
	(c) boron trifluoride	53	23a
2-Acetamido-3-phenylazo-	2-Amino-3-phenylazobiphenylene and acetyl chloride–pyridine	61[a]	33
2-Amino-	2-Nitrobiphenylene and stannous chloride	73[a]	11
	2-Acetylbiphenylene (Schmidt)	43[a]	11
	2-Acetamidobiphenylene (hydrolysis)	70[a]	15
2-Amino-3-bromo-	2-Acetamido-3-bromobiphenylene[b]	—	23a
2-Amino-3-hydroxy- (sulfate dihydrate)	2-Hydroxy-3-(p-sulfoxyphenyl)azo-biphenylene and sodium dithionite	—	31
2-Benzamido-	2-Benzoylbiphenylene oxime (Schmidt)	70[a]	31
2-Benzoyloxy-	2-Benzoylbiphenylene (Bayer–Villiger)	47–90	31
2-(2′-Biphenylenyl)-	2-Iodobiphenylene and hydrazine hydrate (palladium)	8.6	11
	2-Iodobiphenylene and magnesium	—	11
2-Bromo-	2-Acetoxymercuribiphenylene and bromine	78[a]	11
	2-Amino-3-bromobiphenylene (deamination)	10	23a
—2-Carboxylic acid	2-Acetylbiphenylene and sodium hypochlorite	89[a]	15
	2-Acetylbiphenylene and potassium permanganate	15	98
	2-(β-Carboxypropionyl)biphenylene and alkaline potassium permanganate	—	14
	2-Cyanobiphenylene (base hydrolysis)	88[a]	11
	2-Iodobiphenylene (Grignard)	—	11
2-(β-Carboxypropionyl)-	2-(β-Carbomethoxypropionyl)biphenylene	75	14
2-Chloro-	2-Acetoxymercuribiphenylene and hydrochloric acid–potassium chlorate	68[a]	11
2-Cinnamoyl-	2-Acetylbiphenylene and benzaldehyde	85[a]	98
2-Cyano-	2-Iodobiphenylene and cuprous cyanide–pyridine	77[a]	11
2,6-Diacetamido-	2,6-Diacetylbiphenylene oxime (Beckmann)	62	23a
		70	23a
2,6-Diamino-	2,6-Diacetamidobiphenylene	73[a]	23a
1,2-Dihydrobenzo[b]-	1-Hydroxy-1,2,3,4-tetrahydrobenzo[b]biphenylene[b] and phosphoryl chloride	38	14
2,3-Dihydroxy-	Biphenylene-2,3-quinone and sodium dithionite	—	31

B. SYNTHESIS OF BIPHENYLENES

TABLE 10.5 (continued)

Biphenylene	Reactants and reaction conditions	Yield (%)	Ref.
2,3-Dimethoxy-	2,3-Dihydroxybiphenylene and methyl sulfate	20	31
6,7-Dimethoxy--2,3-quinone	2,3,6,7-Tetramethoxybiphenylene and nitric acid	"Good yield"	99
2-Formyl-	Biphenylene-2-glyoxylic acid and dimethylaniline (decarboxylation)	21	14
2-Glyoxylic acid	2-Acetylbiphenylene and potassium permanganate	49	98
2-Hydroxy-	2-Acetoxybiphenylene[b] (from biphenylene and lead tetraacetate)	1.5^a	15
	2-Methoxybiphenylene and pyridine hydrochloride	31^a	10
	2-Benzoyloxybiphenylene (alkaline hydrolysis)	50	15, 31
1-Hydroxy-1,2,3,4-tetra-hydrobenzo[b]-[b]	1-Keto-1,2,3,4-tetrahydrobenzo[b]biphenylene and sodium borohydride	—	14
2-Iodo-	2-Acetoxymercuribiphenylene and iodine	16^a	11
—2,3-Quinone	2-Hydroxybiphenylene and potassium nitrosodisulfonate	75	31, 96
	2-Amino-3-hydroxybiphenylene and chromium trioxide	34	31
1,2,3,4-Tetrahydro-benzo[b]-	1-Keto-1,2,3,4-tetrahydrobenzo[b]biphenylene and amalgamated zinc–hydrochloric acid	43^a	14
1,6,7,8-Tetramethoxy--2,3-quinone	1,2,3,6,7,8-Hexamethoxybiphenylene and nitric acid	—	96
2,3,6,7-Tetramethyl- (?)	2,6(or 7)-Di(iodomethyl)biphenylene and zinc–acetic acid (reduction and dismutation)	21^a	11
Other Substituted Biphenylenes			
β-(2-Biphenylenyl)acrylic acid	Biphenylene-2-aldehyde	—	98
β-(Biphenylenyl)acrylophenone	Biphenylene-2-aldehyde	—	98
γ-(2-Biphenylenyl)butyric acid	2-(β-Carboxypropionyl)biphenylene	58	14

[a] Calculated from weight data recorded in the literature.
[b] Not isolated.

9. By Electron Bombardment, Ultraviolet Irradiation, and Thermolysis of Biphenyls and Terphenyls

Biphenylene is formed in the positive column of a special cold-discharge tube (Schüler tube) containing the vapors of biphenyl [106b] or fluorobiphenyl. [106a]

γ-Radiolysis of biphenyl with a Co^{60} source leads to the formation of trace amounts of biphenylene and a large number of other hydrocarbons (identified by mass spectrometry).[91] The formation of biphenylene by ultraviolet irradiation of 2-iodobiphenyl[84] has been reinvestigated and has been found to be in error.[82, 83]

Thermolysis of *m*-terphenyl in the presence of an aluminum oxide–cuprous oxide catalyst at 800°F under 250 psi hydrogen pressure gives 1.1% biphenylene, together with a number of other hydrocarbons, as shown by mass spectrometric analysis of the crude reaction product.[118]

Finally, the mass spectrum of anthraquinone (obtained with an electron-bombardment ion source producing 50-eV electrons), contains metastable peaks at masses 155.8 and 126.3, corresponding to the transitions

$$208^+ \rightarrow 180^+ + 28,$$
$$180^+ \rightarrow 152^+ + 28,$$

which are due to biphenylene ion (152^+), fluorenone ion (180^+), and carbon monoxide (M.W. 28).[29, 108]

10. From Biphenylene–Metal Complexes

Biphenylene is regenerated in unspecified yield when its mono- and bis-(tricarbonylmolybdenum) complexes (*39*) and (*40*), respectively, are heated *in vacuo* at 60°–140° with 1,2-bis(diphenylphosphine)ethane.[48]

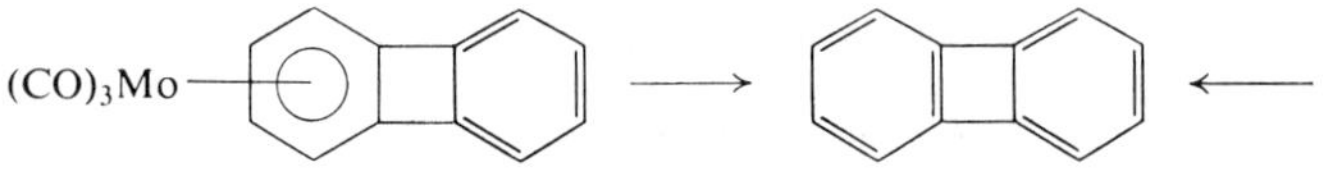

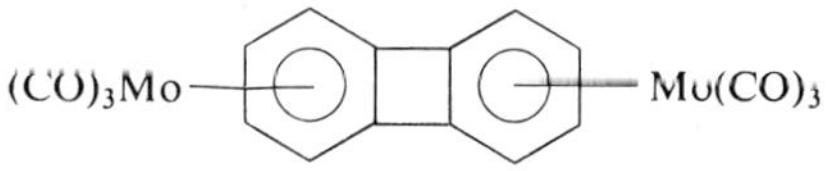

C. Unsuccessful Approaches to the Biphenylenes

The literature contains reports of some 42 unsuccessful attempts to prepare biphenylene and the substituted biphenylenes. The majority of these are attempted syntheses of biphenylene from simple aromatic starting materials. Only 12 of these attempts gave identifiable products and of these the majority gave biphenyl. (See Table 10.6.)

In addition, the literature contains a report on the formation of 1,5(or 8)-diisopropyl-4,8(or 5)-dimethylbiphenylene in 40% yield by the reaction of

C. UNSUCCESSFUL APPROACHES TO THE BIPHENYLENES 273

TABLE 10.6

ATTEMPTED SYNTHESES OF BIPHENYLENE, BIPHENYLENE–METAL COMPLEXES, AND SUBSTITUTED BIPHENYLENES

Expected product	Reactants and reaction conditions	Products obtained	Ref.
Biphenylene	Biphenylyl-2,2′-bis(diazonium chloride) and copper	—	100
	Biphenylyl-2,2′-dilithium and cuprous or manganous chloride	Biphenyl and quaterphenyl	135
	2-Bromobiphenyl (γ-radiolysis)	Biphenyl and other products	103
	Bromoiodobenzene and		
	(1) copper	—	93
	(2) copper in dimethylformamide	Biphenyl	104
	o-Dibromobenzene and hydrazine–palladium catalyst	Biphenyl and benzene	38
	2,2′-Dibromodiphenyl and		
	(1) aluminum oxide	S.M.[a]	88
	(2) calcium oxide	S.M.[a]	88
	(3) cuprous oxide	Biphenyl	88
	(4) copper	—	88, 93
	(5) cupric oxide	S.M.[a]	88
	(6) hydrogen	Biphenyl	88
	(7) lithium	Biphenyl	88
	(8) magnesium	"Dense oil"	88
	(9) phenyllithium	Biphenyl and triphenylene	24
	(10) potassium	Biphenyl	88
	(11) sodium	Biphenyl	88, 93, 105, 109, 110
		"Biphenylene"	57, 58
	(12) zinc oxide	S.M.[a]	88
	2,2′-Dichlorobiphenyl and sodium	—	93
	o-Diiodobenzene and		
	(1) copper	Tars	58
	(2) cuprous oxide	—	15
	2,2′-Diiodobiphenyl and sodium	—	93

Table 10.6 (*continued*)

Expected product	Reactants and reaction conditions	Products obtained	Ref.
	Diphenic acid and		
	(1) copper chromite	—	15
	(2) cuprous oxide	—	15
	Diphenic acid silver salt (pyrolysis)	Biphenyl and 2-hydroxybiphenyl-2-carboxylic acid lactone	58
	Diphenic acid sodium salt (electrolysis)	—	15
	Fluorobenzene and phenyllithium	Biphenyl	136
	2-Iodobiphenyl (γ-radiolysis)	Biphenyl and other products	103
	N-Nitroso-2-acetamidobiphenyl (pyrolysis)	—	15
	1,4-Cxido-1,4,4a,8b-tetrahydrobiphenylene	$C_{12}H_{11}Cl(?)$	44
	o-Phenylphenol and sulfuric acid (dehydration)	Dibenzofuran	55
	o-Phenylenedilithium and phosgene	Biphenyl	129
	Purpurotannin and zinc (dry distillation)	"Biphenylene"	101
Biphenylene–Metal Complexes			
Chromium tricarbonyl	Biphenylene and chromium hexacarbonyl at 225° for 40 hr	Bis(biphenylene) and fluorenone	5
Iron tricarbonyl	Biphenylene and iron pentacarbonyl (UV irradiation)	S.M.^{a}	47
Nickel carbonyl	Biphenylene and nickel tetracarbonyl at		
	(1) 40° for 3 hr	S.M.^{a}	47
	(2) 140° for 12 hr	S.M.^{a}	47
	Biphenylene and nickel tricarbonyl triphenylphosphine	S.M.^{a}	47
	Biphenylene and nickel dicarbonyl bis(triphenylphosphine) at 100° for		
	(1) 1 hr	S.M.^{a}	47
	(2) 7 hr	Tetraphenylene	47
Nickel cyclopentadiene	Biphenylene and cyclopentadienylnickel	S.M.^{a} and a black solid	47
Platinum bis(triphenylphosphine)	Bip enylene, platinum dichloride bis(triphenylphosphine), and hydrazine	S.M.^{a}	47
	2,2′-Dilithiobiphenyl and platinum dichloride bis(triphenylphosphine)	S.M.^{a}	47
	Biphenyl-2,2′-bis(magnesium iodide) and platinum dichloride bis(triphenylphosphine)	S.M.^{a}	47

C. UNSUCCESSFUL APPROACHES TO THE BIPHENYLENES

Substituted Biphenylenes			
2-Acetoxy-	2-Acetylbiphenylene (Bayer–Villiger)	S.M.[a]	15
Bromo-	Biphenylene and bromine[b]	"A mixture"	15
	Biphenylene and N-bromosuccinimide (benzoyl peroxide)	—	21
2-Chloro-	Biphenylene and chlorine[b]	"A mixture"	11, 15
—2-Diazonium salt	2-Aminobiphenylene and nitrous acid	"Insoluble solid"	11
Dibromo-	"Biphenylene" and bromine	"Dibromobiphenylene"	58
2,7-Dihydroxy-	2,7-Dimethoxybiphenylene (hydrolysis)	—	89
2,6-Dimethyl-	3-Bromo-4-iodotoluene and lithium	4,4′-Dimethylbiphenyl	75
Dinitro-	"Biphenylene" and nitric acid	"Dinitrobiphenylene"	58
2-Formyl-	Biphenylene and dimethylformamide–phosphorus oxychloride	—	98
2-Hydroxy-	2-Methoxybiphenylene and hydrobromic–acetic acid	—	17
Nitro-	Biphenylene and nitric acid–sulfuric acid[b]	—	15
1,2,3,4,7,8,9,10-Octahydro-dibenzo[b,h]-	6-Bromo-7-fluoro-1,2,3,4-tetrahydronaphthalene and lithium amalgam	S.M.[a]	122
	1,2,3,4,8,9,10,11-Octahydrobinaphtho[2,3-b:2′,3′-d]-iodolium iodide and cuprous oxide	Dibenzo[b,h]biphenylene	56, 121
	5,6,7,8,5′,6′,7′,8′-Octahydro-2,2′-binaphthyl-3,3′-tetra-zonium chloride and cuprous oxide	Dinaphthofuran	56a, 121
		Dibenzo[b,h]biphenylene	56
	5,6,7,8,5′,6′,7′,8′-Octahydro-2,2′-binaphthyl-3,3′-tetra-zonium sulfate and cuprous oxide	Octahydrodibenzocarbazole	56a, 121
1,4,5,8-Tetraacetoxy-2,3,6,7-tetramethyl-	5,8-Dihydroxy-2,3,6,7-tetramethyl-4a,8b-dihydrobi-phenylene-1,4-quinone (acetylation)	"A diacetate"	36
2,3,6,7-Tetramethoxy-	4-Bromo-5-iodoveratrole and magnesium	—	13
Tetranitro-	"Biphenylene" and nitric acid–sulfuric acid	"Tetranitrobiphenylene"	58
1,2,3-Triacetoxy-	Biphenylene-2,3-quinone and acetic anhydride–sulfuric acid	2,3,4a,8b-Tetraacetoxy-4a,8b-dihydrobiphenylene	31

[a] Starting material.
[b] The method was later found to be successful. Compare Table 10.4.

2-cymylmagnesium bromide with excess acetophenone.[117] However, in the absence of confirmatory evidence, this report must be viewed with skepticism on mechanistic grounds.

D. Proof of Structure of Biphenylene

1. BY CHEMICAL METHODS

In 1941, Lothrop published his now well-known paper in which convincing experimental evidence was presented for the first time to support an author's claim that he had prepared the elusive hydrocarbon, biphenylene.[88] Lothrop demonstrated through elemental analysis and a molecular weight determination that the compound which he had obtained had the molecular formula of biphenylene, $C_{12}H_8$; that on oxidation with chromic oxide it gave phthalic acid (*1*); and that on reduction with hydrogen over a "red hot copper" catalyst it gave biphenyl (*2*).

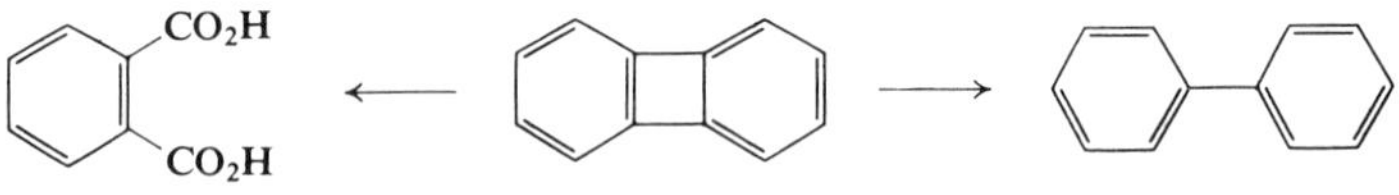

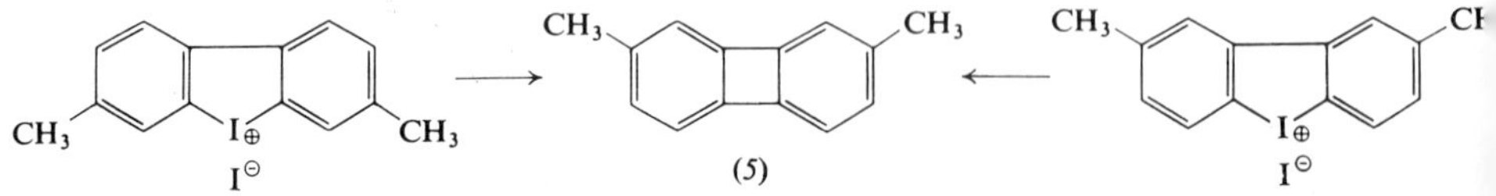

The proposed structure is in accord with these results, since (a) the molecular formula $C_{12}H_8$ eliminates tetraphenylene from consideration; (b) the formation of phthalic acid indicates that the starting material, 2,2'-dihalobiphenyl underwent carbon-to-carbon bond formation (ring closure) at an ortho position; and (c) the formation of biphenyl by reductive cleavage makes it unlikely that the product resulted from a profound skeletal alteration of the starting material.

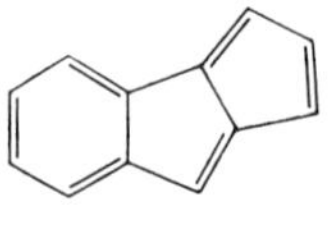

Since the two aromatic rings of the biphenylene system are symmetrically disposed about the central four-membered ring, it should be possible to establish the symmetry of the biphenylene system by preparing the same

D. PROOF OF STRUCTURE OF BIPHENYLENE

substituted biphenylene from either of two appropriately substituted 2,2'-dihalobiphenyls. This Lothrop was able to do with 2,7-dimethylbiphenyleneiodonium iodide (3) and 3,6-dimethylbiphenyleneiodonium iodide (4), both of which gave the same product, 2,7-dimethylbiphenylene (5).[88] Although Lothrop's interpretation of the evidence appears in retrospect to be clearly correct, the suggestion was made in 1942 that the hydrocarbon was not biphenylene but rather benzopentalene (6), largely because it was found to absorb 3 moles of hydrogen at atmospheric pressure in the presence of palladium.[7] This result was held to be more in keeping with the expected behavior of benzopentalene than of biphenylene because a precedent for the palladium-catalyzed reduction of a benzene ring was not known at the time. More recently, it was found that benzocyclobutene readily absorbs 3 moles of hydrogen to give bicyclo[4.2.0]octane in the presence of a palladium catalyst which does not reduce o-xylene.[44] The latter result suggests that the product of the biphenylene hydrogenation is a reduced biphenylene (7) and not "a mixture of phenylcyclohexane and phenylcyclohexenes."[15]

The erroneous benzopentalene structure assignment found temporary support (a) in a theoretical paper (1942) in which the calculated strain energy of biphenylene was reported to be some 100 kcal/mole greater than that calculated for benzopentalene[53] and (b) in a later publication (1950) in which the experimentally determined ultraviolet absorption spectrum was said to be in closer agreement with that calculated for benzopentalene.[80] In the light of current knowledge, of course, the benzopentalene structure is untenable and is only of historical interest.

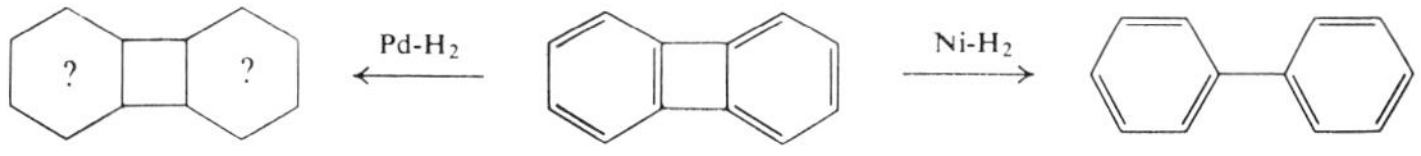

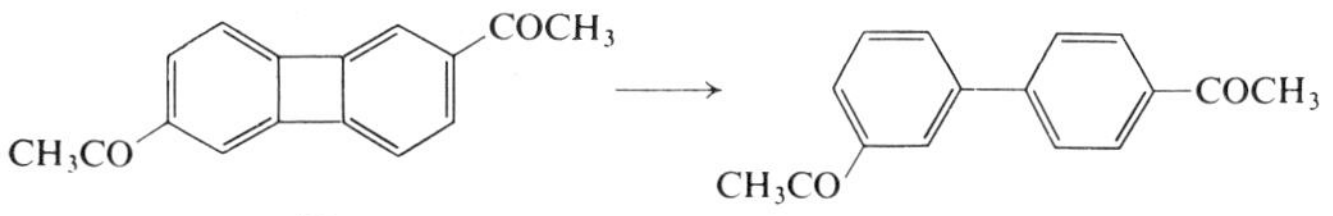

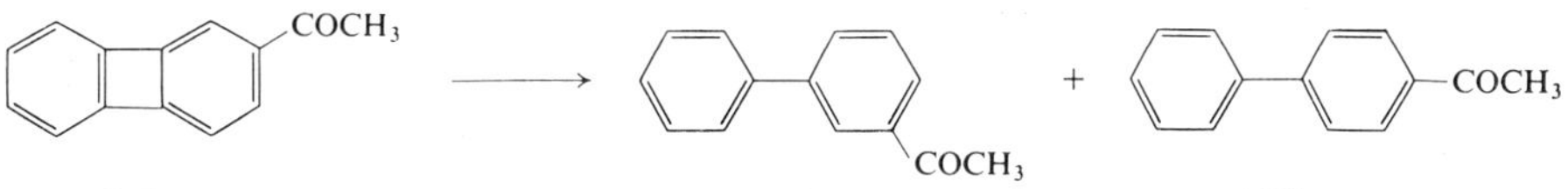

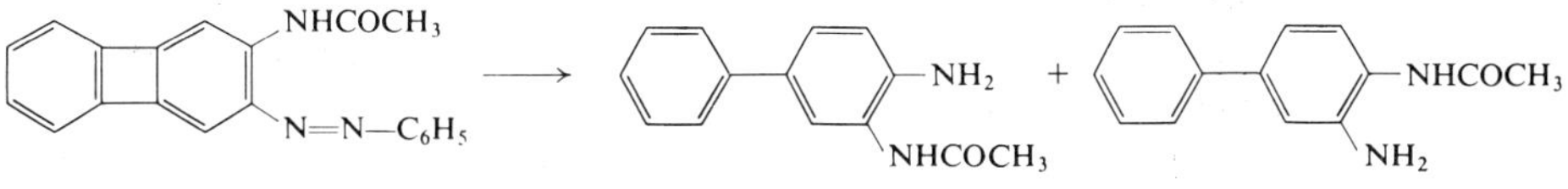

Unlike palladium, activated Raney nickel cleaves the four-membered ring of the biphenylene system to give biphenyl (8).[8,15] Similarly, a "symmetrically" substituted biphenylene, such as 2,6-diacetylbiphenylene (9), gives the expected biphenyl, 3,4'-diacetylbiphenyl (10), but an "unsymmetrically" substituted biphenylene gives a mixture of two biphenyl isomers. This is because the bonds that unite the two six-membered rings of the biphenylene system are equally susceptible to cleavage; thus, 2-acetylbiphenylene (11) gives a mixture of 3- and 4-acetylbiphenyl [(12a) and (12b)].[15,20]

Since the preparation of substituted biphenyls for comparison purposes is not difficult, reductive cleavage is an invaluable tool for the determination of substitution patterns in biphenylenes. In addition to (a) 2-acetylbiphenylene, (b) biphenylene, and (c) 2,6-diacetylbiphenylene, reductive cleavage with Raney nickel has been carried out on (d) 2-acetamido-3-phenylazobiphenylene [(13), which gives 3-acetamido-4-aminobiphenyl (14a) and 4-acetamido-3-aminobiphenyl (14b) by hydrogenolysis of both the four-membered ring and the azo group[33]] and on (e) biphenylene-2-carboxylic acid (which gives only unidentifiable products).[15] In contrast, 2-acetamidobiphenylene is unaffected by Raney nickel.[15]

2. BY PHYSICAL METHODS

The structure of biphenylene was confirmed by electron diffraction studies of the vapor phase[124] and by X-ray analysis of the crystalline material.[123] These studies, conducted in 1943 and 1944, established conclusively that biphenylene has the structure of a dibenzocyclobutadiene. They also showed that the central bonds joining the two six-membered rings are significantly longer than the bonds that make up the six-membered rings. Recently (1962), a more refined X-ray analysis has established the complete planarity of the biphenylene molecule and the exact bond distances and angles between all of the carbon atoms.[97] The data are summarized in Section F, 1 and are discussed in Chapter 12.

E. Chemistry of Biphenylene

The biphenylene molecule has been the subject of a number of theoretical calculations, the results of which lead to the prediction that all substitution reactions, whether nucleophilic, electrophilic, or free radical in nature,[35,60,61] will take place exclusively at the 2-position. The prediction has been amply confirmed in the case of electrophilic substitution and to some extent in the case of free-radical substitution.

It is interesting to note that simple resonance theory leads to the erroneous conclusion that the 1,2-bond has more double-bond character than the

2,3-bond. Of course, the bond orders were calculated on the assumption that all five canonical resonance forms of biphenylene contribute equally to the resonance hybrid; however, if the contrary assumption is made, i.e., that the cyclobutadienoid resonance forms (D and E, Chart 10.1) are high-energy forms which contribute very little to the hybrid, the calculated bond orders and bond lengths are in accord with the experimentally determined values, as shown in Table 10.7. (See also Chapter 12.)

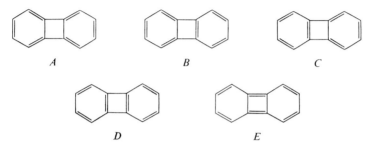

CHART 10.1. Canonical resonance forms of biphenylene.

TABLE 10.7

BOND LENGTHS IN BIPHENYLENE[a]

Bond	Calculated bond length, (Å)			Measured bond length, (Å)[c]
	According to resonance theory		According to molecular orbital theory	
	Complete canon	Selected canon[b]		
A	1.41	1.37	1.38	1.35
B	1.38	1.43	1.40	1.42
C	1.41	1.37	1.38	1.38
D	1.41	1.43	1.41	1.38
E	1.45	1.54	1.47	1.52

[a] After Mak and Trotter, ref. 92.
[b] Cyclobutadienoid forms excluded.
[c] X-ray diffraction method, ref. 92.

Consideration of the free-valence indices of biphenylene leads to another erroneous prediction. In many polycyclic aromatic hydrocarbons (naphthalene, anthracene, phenanthrene) the position of greatest free valence is also the position of highest free-radical reactivity. In biphenylene the 1-position has a slightly larger free-valence index (0.43) than the 2-position (0.41), and this leads to the erroneous conclusion that the 1-position is more reactive to free-radical attack than the 2-position. The reliability of free valence as a criterion of reactivity has been questioned,[54] but in the case at hand the free-valence values are so similar that further discussion is pointless.

The problem of relative rates of electrophilic substitution at the 1- and 2-positions of biphenylene has been attacked experimentally by Streitwieser and Schwager, who found that tritiation proceeds some 64 times faster at the 2-position than at the 1-position ($k_1 = 4.32 \times 10^{-5}$ sec^{-1} and $k_2 = 2.75 \times 10^{-3}$ sec^{-1}; $k_2/k_1 = 64$; in 96.9:3.1 v/v trifluoroacetic acid–70% perchloric acid).[115] The ratio of the rate constants agrees qualitatively—but not quantitatively—with the results of most molecular orbital calculations. (For example, localization energy calculations predict $k_2/k_1 \cong 4$; see also ref. 113 and Chapter 12.)

1. Electrophilic Substitution

(a) *Biphenylene* undergoes electrophilic substitution somewhat more readily than benzene. The acetoxymercuri, acetyl, benzoyl, bromo, chloro, iodo, and nitro groups have been introduced by direct substitution, usually in good yield and always at the 2-position (see Table 10.4). The selectivity of the substitution reaction can be understood by comparing the group of canonical-resonance-stabilized carbonium ions that would be formed through α attack on the one hand with the group formed through β attack on the other.

Carbonium ions formed through α attack:

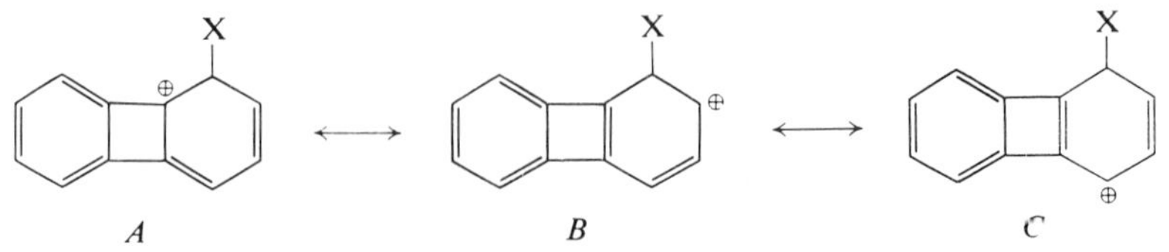

A   B   C

Carbonium ions formed through β attack:

D   E   F

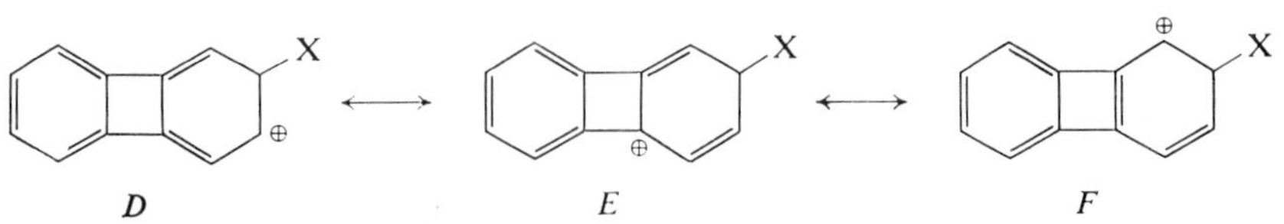

Chart 10.2. Resonance-stabilized carbonium ions of biphenylene.

If all cyclobutadienoid forms (*B*, *C*, and *F*, Chart 10.2) are assumed to be of minor importance, the remaining forms (*A*, *D*, and *E*) are seen to be distributed in two-to-one fashion between the β and α hybrids. In other words,

E. CHEMISTRY OF BIPHENYLENE 281

β attack has the merit of generating two important resonance structures (D and E) and therefore is presumably more favorable (has lower energy requirements) than α attack, which generates only one such favorable structure (A). The argument takes on added cogency when it is realized that contributor D is an analog of dimethylenebenzocyclobutene—a system of considerable delocalization energy (see Chapter 8, Section C). Indeed, for all practical purposes, resonance structure D may be viewed as the *only* intermediate structure generated in the electrophilic substitution reactions of biphenylene.

(b) *β-Substituted biphenylenes* have also been subjected to the action of electrophilic reagents. The results lead to the generalization that *an entering electrophilic group is directed to the 3-position by an electron-donating β substituent and to the 6-position by an electron-withdrawing β substituent* (see Table 10.4). A fuller understanding of the directive influences can be gained from the arguments used to explain the substitution behavior of biphenylene itself. For example, if the β substituent is electron-donating, the major resonance contributor (dimethylenebenzocyclobutene analog A, Chart 10.3)

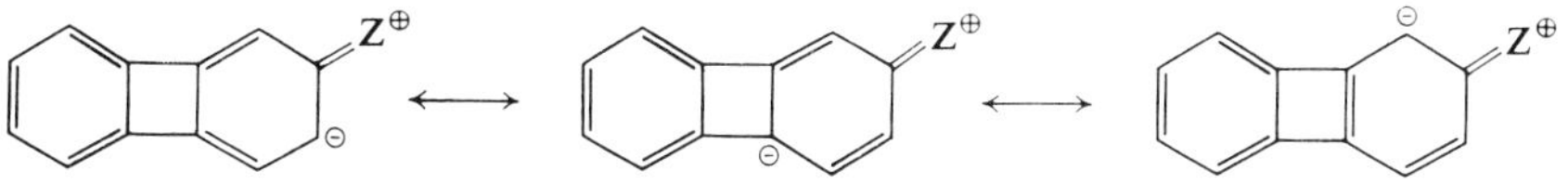

CHART 10.3. Resonance stabilization of biphenylene with an electron-donating β substituent.

must accept an entering electrophilic species exclusively in the 3-position. The bromination of 2-acetamidobiphenylene[23] and the coupling of 2-hydroxybiphenylene,[33, 96] with diazonium salts are examples of this kind of substitution. (The coupling reaction was first suggested by Longuet-Higgins as a test of the directive influences in β-substituted biphenylenes.[87])

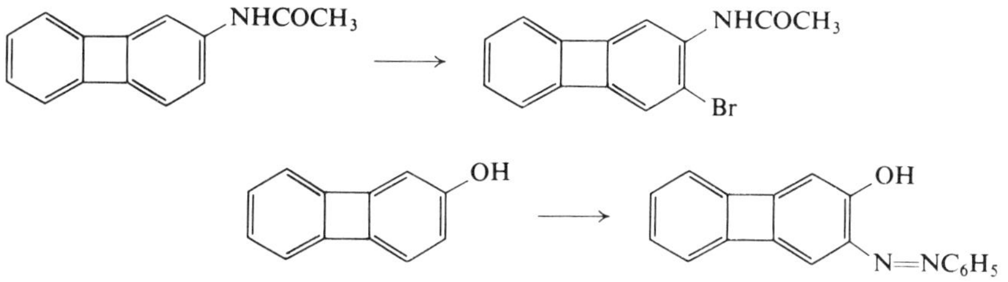

On the other hand, if the β substituent is electron-withdrawing, the 3- and 4a-positions will be deactivated in the classical ortho–para fashion, leaving the β positions (6 and 7) of the *other* six-membered ring open to attack. A comparison of the two canonical groups of resonance-stabilized carbonium ions that would be formed through attack at the 6-position on the one hand and at the 7-position on the other shows that the latter group involves three high-energy structures in which two positive charges are either in conjugation with each other (*A* and *B*, Chart 10.4) or are adjacent to each other (*C*) and that, in contrast, attack at the 6-position leads to a group of resonance-stabilized carbonium ions of lower energy requirements in which the two positive charges are cross-conjugated (*D*, *E*, and *F*) and are never adjacent. Thus, in the dinitration of biphenylene, the second nitro group enters at the 6-position, as does the second acetyl group in the diacetylation reaction.

Carbonium ions formed through attack at the 7-position:

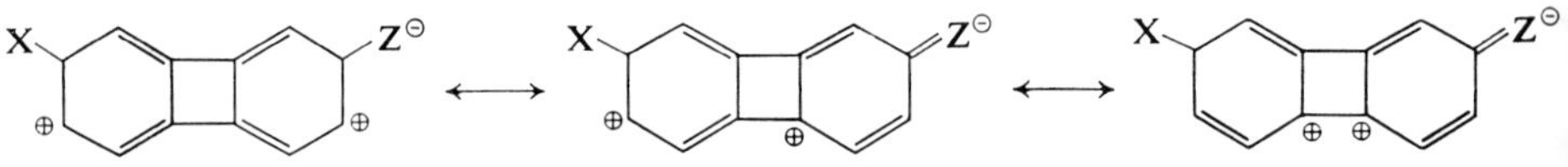

$$A \quad\quad B \quad\quad C$$

Carbonium ions formed through attack at the 6-position:

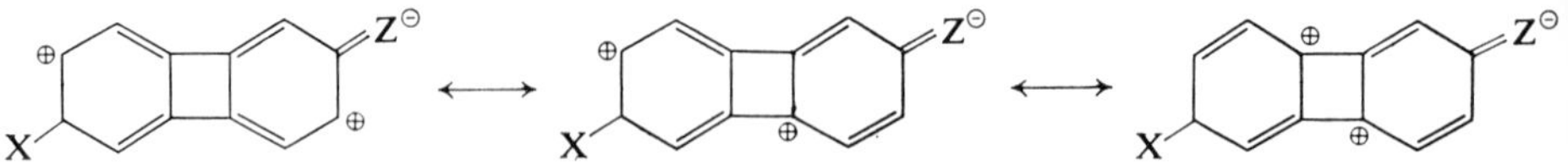

$$D \quad\quad E \quad\quad F$$

CHART 10.4. Resonance stabilization of biphenylene with an electron-withdrawing β substituent.

2. NUCLEOPHILIC SUBSTITUTION

No example of nucleophilic substitution has been reported for biphenylene.

3. FREE-RADICAL SUBSTITUTION

The prediction that biphenylene would be almost as resistant to free-radical attack as benzene[34] has been confirmed experimentally: thus, (a) biphenylene is not attacked by *N*-bromosuccinimide under conditions which result in the bromination of naphthalene (benzoyl peroxide initiator)[21] and (b) reaction of biphenylene with lead tetraacetate proceeds with difficulty to give 2-acetoxybiphenylene in poor yield (about 1.5%).[15] The suggestion has been made[21] that the formation of bis(biphenylenyl) as a side-product in the nitration of biphenylene[11] is an example of free-radical substitution; it is also possible that the reaction may be an example of free-radical *coupling*.

4. ADDITION

a. Hydrogen

The addition of 1 mole of hydrogen to biphenylene is well known; the reaction consists in the reductive cleavage of the four-membered ring to give biphenyl. Lothrop, who discovered the reaction, used hydrogen over a "red hot copper" catalyst,[88] but it was subsequently found that the reaction proceeds smoothly at room temperature in the presence of activated Raney nickel with ethanol as solvent.[8, 15] The value of the reaction lies in its utility in determining the orientation of substituents in substituted biphenylenes (see Section D, 1). Biphenylene is not reduced by amalgamated zinc–hydrochloric acid,[15] by cyclohexadiene and palladium,[22] by hydrazine–palladium,[22] by diborane,[3] or by diimide.[4] Sodium in liquid ammonia has been reported to cleave the four-membered ring of biphenylene to give biphenyl (*1*),[15] but simple molecular orbital theory predicts that the reduction will occur at the 2- and 4a-positions to give 2,4a-dihydrobiphenylene (*2*).[116] More recently, it has been found that alkali metal reduction of biphenylene affords a mixture of reduction products which undergo disproportionation on distillation to regenerate a small amount of biphenylene.[4]

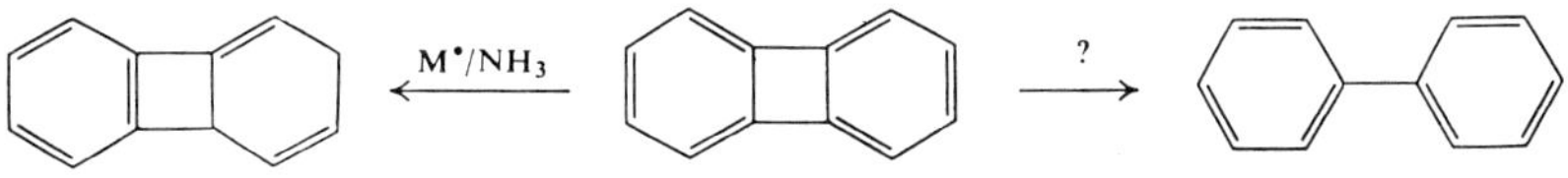

b. Bromine

Biphenylene usually reacts with halogens to give 2-halobiphenyls, but two cases are known in which addition rather than substitution occurs. In the first of these, biphenylene was treated with excess bromine to give a tetrabromo derivative in 4% yield.[20,22] The latter was assigned the tentative structure of

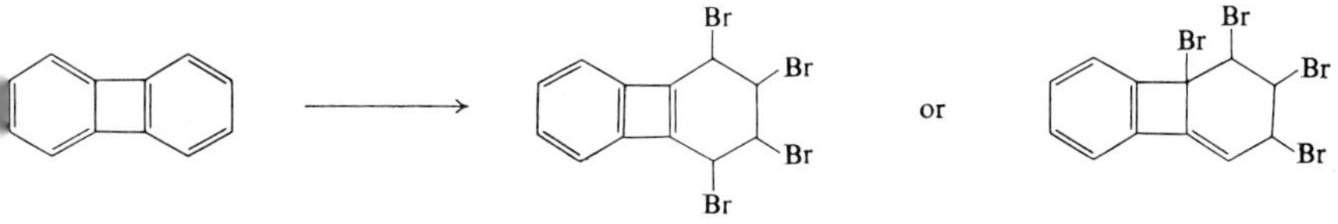

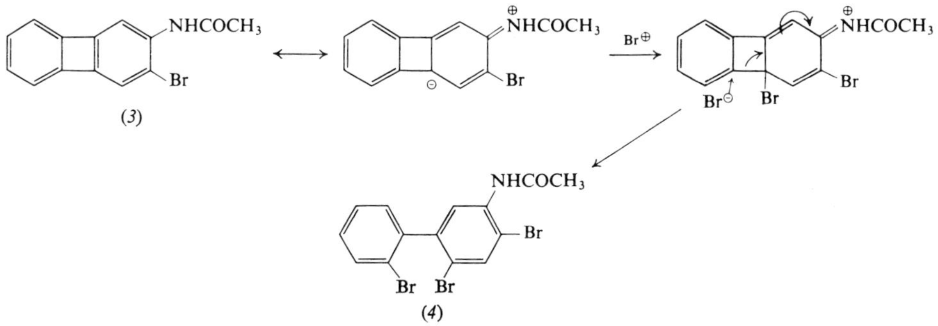

1,2,3,4-tetrabromo-1,2,3,4-tetrahydrobiphenylene (*2a*), which is a benzocyclobutadiene derivative. [In view of the known instability of the benzocyclobutadienes, the isomeric methylenebenzocyclobutene structure (*2b*) would appear to be more likely.*]

In the second example of bromine addition, a substituted biphenylene, 2-acetamido-3-bromobiphenylene (*3*), underwent ring cleavage on treatment with bromine to give 5-acetamido-2,4,2'-tribromobiphenyl (*4*).[23] Apparently the 2-acetamido group activates the 4a-position sufficiently to bring about the observed transformation.†

c. Miscellaneous Reagents

Biphenylene reacts with ozone,[15] osmium tetraoxide,[22] and ethyl diazoacetate,‡,[22] but the products have not been fully characterized. Oxidation of biphenylene with chromic oxide gives phthalic acid.[88]

d. Complexing Agents

Benzene solutions of biphenylene give a yellow color indicative of complex formation when they are treated with maleic anhydride, but an adduct cannot be formed even under forcing conditions (140°, no solvent).[12]

Biphenylene reacts with some of the transition metal carbonyls to give various products. Thus, when equimolar amounts of biphenylene and molybdenumtricarbonyldiglyme are heated together at 100° for 6½ hours, biphenylenetricarbonylmolybdenum (*5*), orange crystals (from petroleum ether), m.p. 163°–185° (decomp.), was formed in 21.5% yield.[48] Biphenylene reacts with an excess of molybdenumtricarbonyldiglyme to give μ-biphenylenebis(tricarbonylmolybdenum) (*6*), a scarlet solid, m.p. 195°–204°. (The trans arrangement of the metal atoms with respect to the organic ligand was inferred from X-ray crystallographic data.) The bis(tricarbonylmolybdenum) complex is stable in air; it dissolves sparingly in hot organic solvents to give solutions which decompose rapidly on standing.[48]

In contrast, biphenylene reacts with nickel dicarbonyl–bis(triphenylphosphine) to give tetraphenylene (*7*)[48] and with chromium hexacarbonyl to give bis(biphenylene)ethylene (*8*) and fluorenone.[5] Iron pentacarbonyl-, nickel tetracarbonyl-, and platinum dichloride–bis(triphenylphosphine) and hydrazine have no effect on biphenylene.[48]

* But see Abstract 10.3 in the Appendix.
† But see also the contrasting behavior of the 2-acetamido compound towards Raney nickel, Section D, 1.
‡ See also Abstract 10.19 in the Appendix.

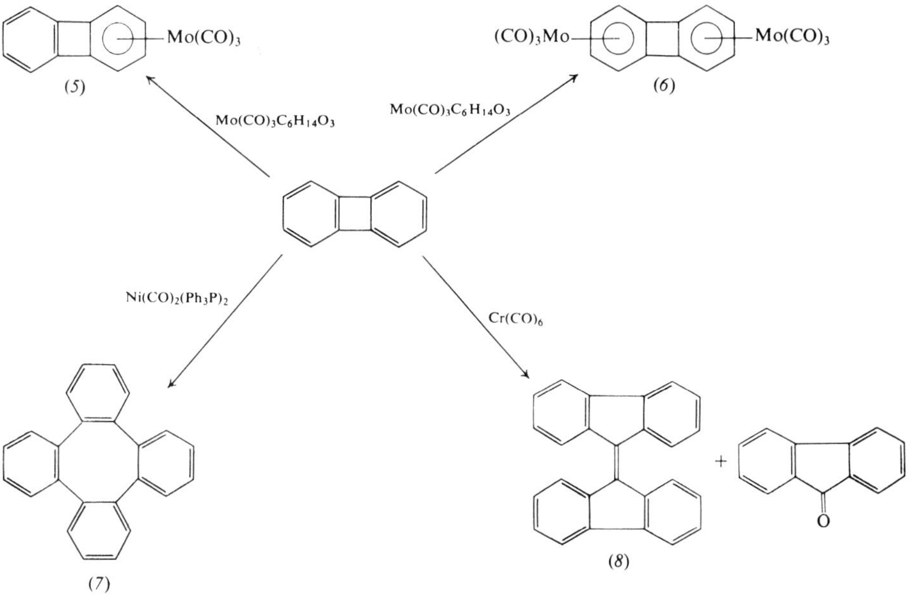

F. PHYSICAL PROPERTIES OF BIPHENYLENE   285

F. Physical Properties of Biphenylene

Biphenylene, $C_{12}H_8$, $M=152.18$, forms pale yellow ("straw-colored") prisms, m.p. 110°–111°, having a faint odor similar to that of biphenyl. It is volatile with steam, sublimes readily, and forms a crystalline complex with picric acid (scarlet crystals, m.p. 122°, from ethanol),[88, 105a, 131] with trinitrofluorenone [scarlet needles, m.p. 154° (decomp.), from ethanol],[15] and with trinitrobenzene (orange needles, m.p. 127°–128°, from ethanol).[130] Solutions of the tetracyanoethylene complex have been studied spectroscopically.[59] Biphenylene is stable (but volatile) in air, and it is not easily decomposed by heat (it does not char below 750°).[125] The melting points of biphenylene and other even-numbered cyclic homologs of o-phenylene have been compared and found to increase regularly with increasing complexity[135]: biphenylene m.p. 111°; tetraphenylene, 235°; hexaphenylene, 335°; octaphenylene, 425°.

1. Crystallographic Data

An X-ray analysis of crystalline biphenylene was first made by Waser and Lu[123] (see also Section D, 2). Recently, a more refined treatment has given the following data[92]: Monoclinic prisms (obtained by recrystallization from 1-propanol), elongated along the c axis; $a=19.66\pm0.06$, $b=10.57\pm0.04$, $c=5.85\pm0.01$ Å, $\beta=91.0\pm0.5°$. Unit cell volume, 1215.5 Å^3. Density: calcd., 1.240 gm/cm^3 (for six molecules per unit cell); found, 1.24 gm/cm^3 (by flotation). Absorption coefficient for X-rays, $\lambda = 1.542$ Å, $\mu = 6.46$ cm^{-1}. Electrons per unit cell $= F(000) = 480$. Space group, $C_{2h}^5 P2_1/a$. Absent spectra: $h0l$ when h is odd, $0k0$ when k is odd. Molecular symmetry: point group mmm in crystal $\bar{1}$ and 1. The intermolecular contacts in crystalline biphenylene correspond to the normal van der Waals interactions, and the molecule is completely planar. The calculated mean bond distances and valence angles are summarized in Fig. 10.1. The results indicate that the D bonds are under compression strain[67] and that biphenylene is closer structurally to a tetramethylenecyclobutane than to a tetramethylcyclobutadiene. The use of optical transforms in crystal-structure analysis has been illustrated with biphenylene.[70]

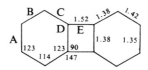

FIG. 10.1. Calculated mean bond distances (Ångstroms) and valence angles (degrees) of biphenylene.

When the experimentally determined bond lengths of biphenylene are compared with the hybrid bond of benzene (1.39 ± 0.02 Å) and with the saturated C—C bond (1.54 Å), it is apparent that bond A has slightly more, and bond B has slightly less, double-bond character than the aromatic linkage in benzene. Bond E is remarkably elongated, being almost as long as the normal single bond. (See also Table 10.7.)

2. SPECTROSCOPIC DATA

The ultraviolet spectrum of biphenylene[59] contains two sets of sharply resolved bands and differs markedly from that of biphenyl, which has only one absorption maximum (Fig. 10.2; see also Table 10.9). The more intense of the two sets of bands occurs at 235–260 mμ, while the less intense is found at 330–370 mμ. The two sets of bands have been ascribed to the second and first electronic transitions in biphenylene, respectively.[106a] The longer-wavelength bands are considered to be made up of two overlapping band systems, one of slightly higher and the other of slightly lower frequency,[40] and are considered to indicate that there is some degree of resonance interaction (π-orbital overlap) between the two benzenoid rings despite the unusual length of the E bonds which unite them (Fig. 10.1).[21, 40]

An intensive investigation has been made of the ultraviolet absorption spectrum of biphenylene in the vapor phase at 100° and in a rigid glass of 5:5:1 ether–isopentane–alcohol at $-183°$.[76] The lowest-energy absorption

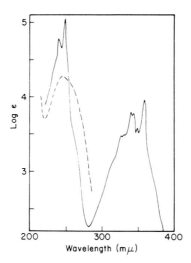

FIG. 10.2. Ultraviolet spectrum of biphenylene (———) and biphenyl (----) in ethanol (after Baker and McOmie[58]).

band in the spectrum of the vapor occurs at 25,041 cm^{-1} (399.3 mμ) and has been ascribed to a symmetry-forbidden electronic transition.*

The long-wavelength absorption of biphenylene has been found to be intermediate with those of naphthalene and anthracene, in keeping with the number of formal double bonds (Table 10.8).[135]

TABLE 10.8

ULTRAVIOLET ABSORPTION CHARACTERISTICS OF REPRESENTATIVE AROMATIC SYSTEMS[a,b]

Compound	Number of double bonds	λ_{max} (mμ)	log ϵ
Naphthalene	5	314	2.5
Biphenylene	6	358	4.0
Anthracene	7	380	3.9

[a] Absorption maxima are for the band of maximum absorption in the long-wavelength region.
[b] After Baker and McOmie, ref. 21.

The infrared spectrum of crystalline biphenylene has been recorded by Wittig[135] and infrared spectral data have been recorded by Curtis.[56] The

* For a disputed relation between ultraviolet absorption and polarographic half-wave potential, see Section F, 4. For the theory of biphenylene spectroscopy, see Chapter 12.

single intense band of *o*-disubstituted benzenes (e.g., *o*-xylene and tetralin) which usually occurs at 13.5 μ is split in the spectrum of biphenylene. The resulting couplet occurs at 13.34 and 13.63 μ.[56] (See Fig. 10.3.)

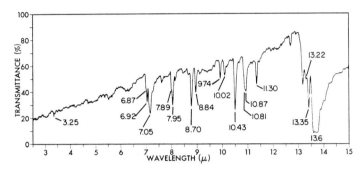

FIG. 10.3. Infrared spectrum of crystalline biphenylene (potassium bromide disc) calibrated against polystyrene.

3. Thermochemical Data

The heat of combustion of crystalline biphenylene was first determined by Cass, Springall, and Quincey ($-\Delta H_c^\circ = 1481 \pm 5$ kcal/mole),[42] but a recent, more refined treatment has afforded a slightly higher value (1486.3 ± 0.7 kcal/mole).[25] The latent heat of sublimation ($L_g^{st} = 30.8$ kcal/mole) was calculated from the vapor pressure which, in the range 98°–100°, is expressed by the equation,

$$\log p_{mm} = -6755.53/T + 18.26.$$

From the observed $-\Delta H_{cq}^\circ$ value for biphenylene and the corresponding thermochemical data, the standard-state heat of formation ($-\Delta H_f^\circ = -84.4$ kcal/mole) was calculated; further calculation led to the atomic heat of formation of gaseous biphenylene ($-\Delta H_{fa} = 2352.3$ kcal/mole). Calculation of the resonance energy, E_r, of biphenylene from $-\Delta H_c^\circ$ and L_g^{st} yields a value of 17.1 kcal/mole. Subtracting this value from the resonance energy calculated for biphenyl (81.4 kcal/mole) affords a value of 64.3 kcal/mole for the strain energy of biphenylene.[25] In other words, the strain energy of biphenylene is estimated to be almost four times greater than the resonance energy.

4. Electrochemical and Electromagnetic Data

Bergman has examined the polarographic behavior of more than a hundred hydrocarbons and has found that they may be separated into five or six families of structurally related compounds such that within any given family the polarographic half-wave potential bears a simple relationship to the wavenumber ($\tilde{\nu}$) of the "first p band" in the ultraviolet absorption spectrum.[26] For biphenylene, the relation is expressed by the equation,

$$E_{1/2} = -1.41 \times 10^{-4}\tilde{\nu} + 2.22,$$

which is applicable to benzo[a]biphenylene and to dibenzo[a,i]biphenylene as well (see Fig. 10.4). [For biphenylene, the pertinent data are $E_{1/2} = 1.73$ V (first

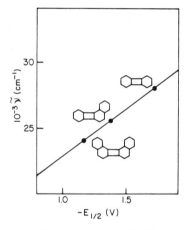

FIG. 10.4. p-Band absorption (?) as a function of half-wave potential for biphenylene, benzo[a]biphenylene, and dibenzo[a,i]biphenylene (after Bergman[26a]).

stage) and 2.02 V (second stage), and $\tilde{\nu} = 28{,}000$ cm^{-1} (358 mμ)]. Wallenberger[120] has disputed Bergman's use of Clar's p-band terminology[49] and has suggested that in the absence of any theoretical reason, the relation between half-wave potential and absorption maximum can be viewed only as an empirical one. Indeed, Hochstrasser's subsequent investigation of the spectrum of biphenylene has shown that the 28,000-cm^{-1} band is polarized along the long axis[76] and hence, by definition, that the 28,000-cm^{-1} band cannot be a p band.

In another polarographic study, biphenylene and benzo[b]biphenylene were found to deviate from the well-established relation between half-wave potential and m_{m+1} (the energy of the lowest vacant molecular orbital in the HMO approximation).[114] The deviation of biphenylene can be rationalized by invoking the "long-bonds" concept, but that of benzo[b]biphenylene cannot.

The nuclear magnetic resonance spectrum of biphenylene is made up of a multiplet of lines centered at 3.52 τ ($J_{1,2} = 0.13$ ppm).[64]

The average magnetic susceptibility of biphenylene was determined using random-packed crystalline powders and an examination of anisotropy was made with bundles of oriented needles (larger crystals could not be grown). The observed molar susceptibility[6] ($K = -41.2 \times 10^{-6}$) is less than half the value calculated with the Pascal relations,[112] but the observed anisotropy ($\Delta K = 0.0 \pm 0.1$) is in agreement with the value calculated from MO theory ($\Delta K \cong -0.15$).[28]

The electron spin resonance spectra of the radical anion and the radical cation of biphenylene have been recorded.[95] The spectra were found to consist of five lines in each case (25 lines theoretically possible) with hyperfine interactions due to four equivalent protons. The overall extent of the radical cation and anion spectra are 18 gauss and 2.75 gauss, respectively.[95] Subsequently, a completely resolved ESR spectrum of the radical anion was recorded with a Varian ESR spectrometer (12-inch magnet) with 100-kc/second modulation (see Fig. 10.5).[94]

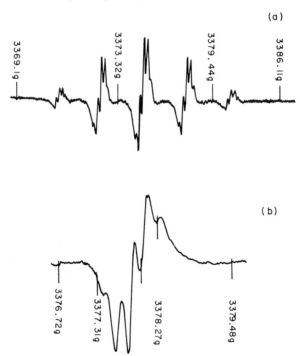

FIG. 10.5. ESR spectrum of the biphenylene radical anion in glyme (after McDowell and Paulus[94]).

The anion and cation ESR spectra were also recorded by Carrington and Santos, who used them to evaluate the Hückel calculations for biphenylene.[41]

G. Chemical and Physical Properties of Substituted Biphenylenes

1. ALKYL AND ARYL DERIVATIVES

1-Methyl-,[10] 2-methyl-,[10] 1,8-dimethyl-,[89] and 2,7-dimethylbiphenylene[88] were prepared by Lothrop's method (see Section A, 1, and Table 10.14). With the exception of 1-methylbiphenylene, which is reported to be an oil, the compounds are all pale yellow crystalline solids at room temperature. 2,6-Dimethylbiphenylene[71] and 1,2(or 3),4,5,6,8-hexamethylbiphenylene (bright yellow crystals)[131] were prepared through the corresponding benzyne intermediates, as was 1,5-dimethylbiphenylene[74]; the latter is undoubtedly a constituent, possibly along with the solid 1,8-isomer, in the liquid "dimethylbiphenylene" formed when o-fluorotoluene is treated with phenyllithium.[74] 2,3,6,7-Tetramethylbiphenylene (2) has been suggested for the structure of a crystalline, straw-colored hydrocarbon, m.p. 222.5°–223.5°, obtained from biphenylene by an unusual two-step reaction sequence[11]: thus, (a) attempted iodomethylation of biphenylene with methyl chloromethyl ether–hydriodic acid afforded an unstable yellow solid which could not be purified but which, on the basis of an elemental analysis of the bis(thiuronium) dipicrate, appears to be the expected 2,6(or 7)-bis(iodomethyl)biphenylene (1). (b) Reduction of

the crude bis(iodomethyl) compound (1) with zinc and hydrogen chloride in acetic acid gave the presumed tetramethylbiphenylene (2), "which must have been formed as the result of a type of dismutation involving migration of iodomethyl groups."[11] An alternative explanation is that sufficient reduction by hydriodic acid occurred during the first step to convert part of the bis(iodomethyl) compound [2,6-di(iodomethyl)biphenylene?] into the dimethylbiphenylene (3); further iodomethylation of the dimethylbiphenylene—also during the first step of the sequence—would give a bis(iodomethyl)dimethylbiphenylene (4?). Reduction of the latter in the unpurified reaction mixture with zinc would then give the product 2,3,6,7-tetramethylbiphenylene (2).

TABLE 10.9

ULTRAVIOLET ABSORPTION CHARACTERISTICS OF BIPHENYLENE AND THE ALKYL-, ALKENYL-, AND ARYLBIPHENYLENES[a]

Compound	Absorption maxima, mμ, and extinction coefficients (log ϵ)	Ref.
Biphenylene	241 (4.60); 250 (4.90); 340 (3.63); 345 (3.59); 360 (3.80)	59
Alkylbiphenylenes		
2-(γ-Carboxypropyl)-	252 (4.86); 334 (3.39); 346 (3.62); 362 (3.82)	14
1,8-Dimethyl-	245 (4.88); 254 (5.15); 348 (3.95); 367 (4.08)	59
2,6-Dimethyl-[b]	243 (4.7); 253 (4.49); 349 (3.9); 368 (4.1)	71
2,7-Dimethyl-	248 (4.78); 257 (5.10); 340 (3.58); 345 (3.58); 360 (3.68)	59
1,2(or 3),4,5,6,8-Hexamethyl-[b,c]	257 (4.7); 268 (5.0); 350 (3.5); 367 (3.6)	131
1-Methyl-[d]	242 (4.66); 251 (4.91); 339 (3.58); 343 (3.57); 358 (3.75)	11
2-Methyl-[e]	242 (4.74); 250 (4.93); 343 (3.80); 361 (3.88)	11
1,2,3,4,7,8,9,10-Octahydrodibenzo[b,h]- (?)	249 (5.05); 258 (5.4); 266, 288, 295, 300, 319, 339, 356 (5.1); 375 (5.3); 393 (5.0)	121
1,2,3,4-Tetrahydrobenzo[b]-	246 (4.58); 255 (4.87); 300 (2.92); 348 (3.78); 368 (3.95)	14
2,3,6,7-Tetramethyl-	246 (4.80); 256 (5.02); 372 (4.09)	11
Alkenylbiphenylene		
1,2-Dihydrobenzo[b]-	254 (4.83); 266 (4.63); 274 (4.60); 349 (3.92); 367 (4.12); 381 (3.98)	14
Arylbiphenylene		
2-(2'-Biphenylenyl)-	247 (4.25); 280 (4.45); 289 (4.34); 358 (3.79); 372 (3.90)	11

[a] Spectra determined in ethanol unless otherwise noted.
[b] Values interpolated from a spectral curve in the literature.
[c] Spectrum determined in cyclohexane.
[d] Two trivial bands are omitted from the table: λ_{max}^{EtOH} 326 mμ (3.29) and 331 (3.29).
[e] A trivial band, λ_{max}^{EtOH} 298 mμ (3.13), is omitted from the table.

Like biphenylene, the alkylbiphenylenes form crystalline complexes with picric acid and with trinitrofluorenone. The tetracyanoethylene complexes of 1,8- and 2,7-dimethylbiphenylene have been studied spectrophotometrically.[59]

An examination of the ultraviolet absorption data reveals that the introduction of methyl groups onto the biphenylene nucleus causes a small but detectable bathochromic shift (Table 10.9 and Fig. 10.6) the magnitude of which is affected by orientation.

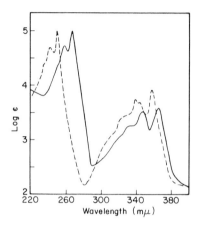

FIG. 10.6. Ultraviolet absorption spectrum of hexamethylbiphenylene (——) and of biphenylene (----) in cyclohexane (after Wittig and Härle[131a]).

2. CARBONYL-CONTAINING DERIVATIVES OF BIPHENYLENE

Several acyl derivatives of biphenylene have been prepared by the Friedel–Crafts method. Thus, (a) 2-acetylbiphenylene (5) was prepared in good yield (67%) when biphenylene was treated with one equivalent of acetyl chloride and a large excess of aluminum chloride (4.4 equivalents).[11] (b) 2,6-Diacetylbiphenylene (6) was prepared in good yield (65%) by using a large excess of both acetyl chloride (six equivalents) and aluminum chloride (three equivalents);[11] it was also formed in appreciable yield (14%) when only 1.1 equivalents of acetyl chloride was employed.[15] Like many other derivatives of biphenylene, the mono- and diacetyl compounds are yellow, crystalline substances which are readily cleaved by Raney nickel to the corresponding biphenyls (see Section D, 1). The Friedel–Crafts method has also been used to prepare (c) 2-benzoyl- (7)[31] and (d) 2-(β-carbomethoxypropionyl)biphenylene (8).[14] The latter was hydrolyzed to the free acid which was reduced to γ-(2-biphenylenyl)butyric acid; ring-closure gave another acyl derivative, 1-keto-1,2,3,4-tetrahydrobenzo[b]biphenylene (9).[14]

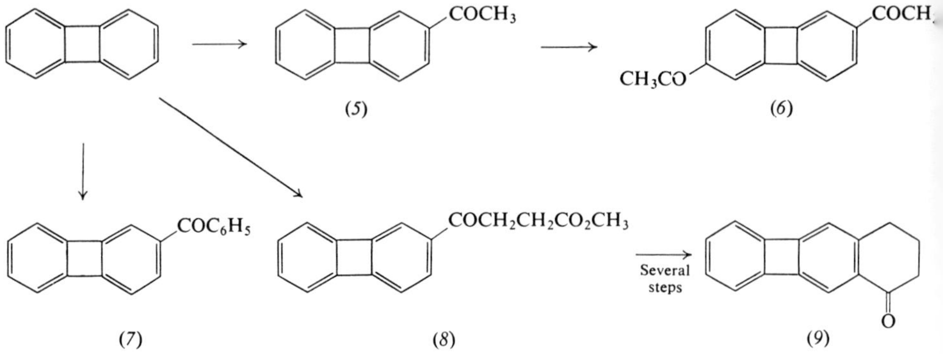

The value of acyl and aroyl derivatives as synthetic intermediates is well-known, but it is diminished in the biphenylene series by the fact that (a) 2-aminobiphenylene (*12*) [prepared from the oxime of 2-acetylbiphenylene by the sequence: oxime (*10*) → 2-acetamidobiphenylene (*11*) → 2-aminobiphenylene (*12*)][15] is not amenable to transformation by the Sandmeyer reaction[11] and (b) 2-acetylbiphenylene does not give the 2-acetoxy derivative (*13*) in the Bayer–Villiger reaction.[15] On the other hand, 2-benzoylbiphenylene (*7*) readily undergoes the Bayer–Villiger reaction, and the resulting 2-benzoyloxy compound is easily hydrolyzed to 2-hydroxybiphenylene [(*14*) → (*15*)].[31]

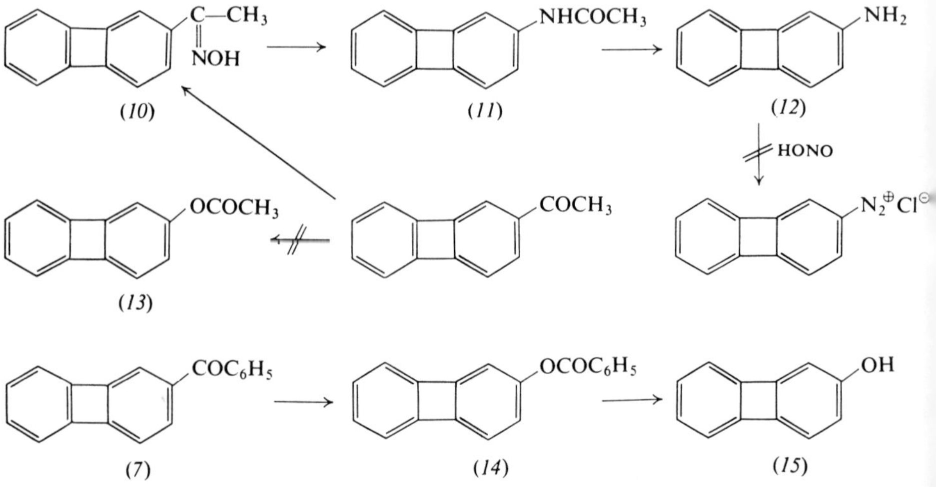

Biphenylene-2-carboxylic acid has been obtained in good yield by the oxidation of 2-acetylbiphenylene with sodium hypochlorite [(*5*) → (*16*)][15] and by several other methods (see Table 10.5). 2-Formylbiphenylene (*18*) was synthesized in 21% yield by the controlled permanganate oxidation of

TABLE 10.10

ULTRAVIOLET ABSORPTION CHARACTERISTICS OF CARBONYL-CONTAINING BIPHENYLENES AND DERIVATIVES[a]

Biphenylene	Absorption maxima, mμ, and extinction coefficients (log ϵ)	Ref.
2-Acetyl-	236 (4.61); 348 (3.63); 363 (3.80)	15
Dinitrophenylhydrazone[b]	295 (4.20); 375 (4.31); 415 (4.35)	15
Oxime	237.5 (4.40); 263 (4.67); 355 (4.86); 370 (3.90)	15
2-Benzoyl-	260 (4.35); 354.5 (2.82); 368.5 (2.98)	31
2-(β-Carboxypropionyl)-	263 (4.64); 349 (3.62); 365 (3.77)	14
—2-Carboxylic acid	256.5 (4.87); 347 (3.75); 362 (3.91)	15
2-Formyl-	240 (4.47); 249 (4.65); 270 (4.41); 345 (3.56)	98
1-Keto-1,2,3,4-tetrahydrobenzo[b]-	233 (3.40); 270 (4.75); 310 (3.25); 327 (3.45); 345 (3.70); 363 (3.89)	14
2-Cinnamoyl-	237 (4.48); 243 (4.50); 265 (4.34); 281 (4.34); 313 (4.43); 369 (3.95); 394 (3.94)	98
2,6-Diacetyl-	219 (3.87); 261 (4.24); 286 (4.34); 359 (3.62); 376 (3.79)	11
β-(2-Biphenylenyl)acrylic acid	239 (4.43); 248 (4.48); 283 (4.59); 383 (4.02)	98

[a] All spectra determined in ethanol unless otherwise noted.
[b] Spectrum determined in benzene.

2-acetylbiphenylene to biphenylene-2-glyoxylic acid [(5) → (17), 49% yield, not purified] which was decarboxylated at 110°–120° in dimethylaniline.[98] The aldehyde was condensed with malonic acid and with acetophenone to give β-(2-biphenylenyl)acrylic acid (19) and 1-(2′-biphenylenyl)-2-benzoylethylene (20). For spectral comparison with the latter, the isomeric ketone, 2-cinnamoylbiphenylene was synthesized from benzaldehyde and 2-acetylbiphenylene.[98] (See Table 10.10.)

An attempt to prepare 2-formylbiphenylene directly from biphenylene failed because the hydrocarbon does not react with dimethylformamide and phosphorus oxychloride.[98] The result is in agreement with the suggestion that nonpericondensed aromatic polycyclic hydrocarbons do not undergo formylation if the free-valence index is less than 0.510.[39] (The free-valence indices of biphenylene at the 1- and 2-position are 0.43 and 0.41, respectively.)

The ultraviolet absorption spectrum of 2-acetylbiphenylene is similar to that of 2-aminobiphenylene and 2-hydroxybiphenylene[15] (see Fig. 10.7).

3. Halogen Derivatives of Biphenylene

Direct substitution of halogen on biphenylene is complicated by the tendency of the hydrocarbon to undergo addition as well as substitution (see Section E, 4). Thus, chlorination of biphenylene with elemental chlorine gave unresolvable mixtures,[11] but the desired product, 2-chlorobiphenylene (21), was obtained unexpectedly and in good yield in an attempt to prepare the 2-iodo compound from biphenylene and iodine monochloride.[23] [Subsequently, 2-chlorobiphenylene (21) and a dichlorobiphenylene were also prepared, albeit in low yield, by careful chlorination of biphenylene.[22]] The orientation of the chloro substituent in 2-chlorobiphenylene was confirmed by an independent synthesis from 2-acetoxymercuribiphenylene by treatment with sodium chlorate–hydrochloric acid [(23) → (21)].[11] 2-Bromobiphenylene (22) was prepared in acceptable yield (49%) by careful bromination of biphenylene[11] and by two other methods[11,23a] (see Section G, 4). In keeping with the

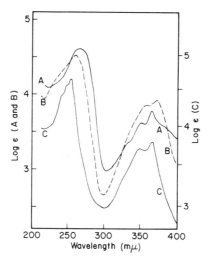

FIG. 10.7. Ultraviolet spectra of 2-acetyl- (A), 2-amino- (B), and 2-hydroxybiphenylene (C) (after Baker et al. [15]).*

prediction that biphenylene is resistant to free-radical attack, it has been found that N-bromosuccinimide has no effect on the hydrocarbon, even when benzoyl peroxide is used as initiator [21] (Section E, 3).

Biphenylene was found to undergo direct iodination with iodine–iodic acid to give 2-iodobiphenylene in 5% yield; the yield was raised to 24% when sodium persulfate was substituted for the iodic acid.[11] The halogen derivatives of biphenylene are all yellow, crystalline solids with similar melting points (64°–68°).

* The right-hand ordinate of Fig. 10.7 was inadvertently omitted from the literature (ref. 15). We are indebted to Prof. Wilson Baker and to Dr. J. F. W. McOmie for a corrected copy of their paper.

TABLE 10.11

ULTRAVIOLET ABSORPTION CHARACTERISTICS OF HALOBIPHENYLENES[a]

Biphenylene	Absorption maxima, mμ, and extinction coefficients (log ε)				Ref.
2-Bromo-	245 (4.59)	253 (4.79)	344 (3.61)	363 (3.76)	11
2-Chloro-	244 (4.49)	252 (4.72)	344 (3.54)	363 (3.66)	11
2-Iodo-		257 (4.82)	345 (3.79)	364 (3.95)	11

[a] All spectra determined in ethanol.

4. NITROGEN-CONTAINING DERIVATIVES OF BIPHENYLENE

2-Acetylbiphenylene oxime and 2,6-diacetylbiphenylene dioxime undergo the Beckmann rearrangement to give 2-acetamido- (10)[15, 23a] and 2,6-diacetamidobiphenylene,[23a] respectively. Similarly, 2-benzoylbiphenylene oxime gives 2-benzamidobiphenylene.[31] The amides have been converted into 2-amino-[11, 15] and 2,6-diaminobiphenylene[23a] by acid hydrolysis. 2-Aminobiphenylene is more conveniently prepared by a modified Schmidt reaction (sodium azide in trichloroacetic acid–polyphosphoric acid) on 2-acetylbiphenylene.[11]

Diazotization of 2-aminobiphenylene was unsuccessful and only an uncharacterized, insoluble precipitate was obtained.[11] On the other hand, 2-amino-3-bromobiphenylene [prepared by the bromination of 2-acetamidobiphenylene (10), followed by hydrolysis of the resulting bromoamide (24)] was found to be readily diazotized; deamination of the resulting diazonium compound gave 2-bromobiphenylene.[23a] 2-Aminobiphenylene (12) is evidently highly electrophilic, for it couples readily with benzenediazonium chloride to give 2-amino-3-phenylazobiphenylene (25).[33] In forming a C-azo derivative rather than a diazoamino compound, it resembles β-naphthylamine rather than aniline.

2,6-Diaminobiphenylene darkens in air, but it may be converted into a stable maroon-colored dihydrochloride; the diamine was prepared by hydrolysis of the 2,6-diacetamido compound [(27) → (26)].[23a] The latter undergoes bromination to give 2,6-diacetamido-3,7-dibromobiphenylene (28).[23a] The 2-amino derivative was also prepared by the reduction of 2-nitrobiphenylene, which was obtained in 24% yield by direct nitration of biphenylene with nitric acid in acetic acid at 0°.[11] 2,6-Dinitrobiphenylene was isolated in low yield (4%), along with 2,2'-bis(biphenylenyl), as a by-product of the mononitration of biphenylene in sulfuric acid.[11]

G. CHEMICAL AND PHYSICAL PROPERTIES OF SUBSTITUTED BIPHENYLENES 299

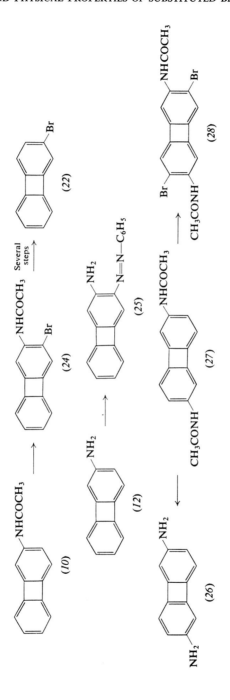

TABLE 10.12

ULTRAVIOLET ABSORPTION CHARACTERISTICS OF NITROGEN-CONTAINING BIPHENYLENES[a]

Biphenylene	Absorption maxima, mμ, and extinction coefficients (log ε)	Ref.
2-Acetamido-	237 (4.36); 259 (4.81); 349.5 (3.97); 368 (4.07)	15
2-Amino-	258.5 (4.53); 357 (3.85); 371 (3.93)	15
Hydrochloride[b]	239.5 (4.49); 248 (4.73); 330.5 (3.30); 339 (3.53); 357 (3.65)	15
2-Cyano-	252 (5.00); 259 (5.11); 334 (3.56); 347 (3.84); 362 (4.01); 352 (3.93)	11
3,7(or 6)-Dimethyl-1,5(or 8)-dinitro-	224 (4.6); 286 (4.0); 421 (3.93)	52
2,6-Dinitro-	258 (3.88); 304 (3.67); 365 (3.34); 382 (3.45)	11
2-Nitro-	232 (4.19); 239 (4.18); 264 (4.14); 328 (3.17); 348 (3.32); 364 (3.51); 400 (3.44)	11

[a] All spectra determined in ethanol except as noted.
[b] Spectrum determined in 1N hydrochloric acid. (See also Fig. 10.8.)

The ultraviolet absorption spectrum of 2-aminobiphenylene resembles that of 2-acetylbiphenylene (Fig. 8.6); not surprisingly, the spectrum of the amine hydrochloride is virtually superimposable on that of biphenylene itself (Fig. 10.8). (See also Table 10.12.)

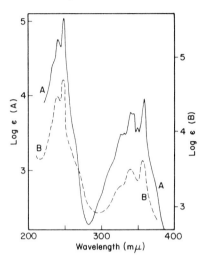

FIG. 10.8. Ultraviolet absorption spectrum of biphenylene in ethanol (———) and 2-aminobiphenylene in 1N hydrochloric acid (----).

5. ACETOXYMERCURI AND SULFONIC ACID (SULFO) DERIVATIVES

Biphenylene reacts with mercuric acetate in acetic acid to give 2-acetoxymercuribiphenylene.[11] The latter is a valuable intermediate, especially in view of the fact that 2-aminobiphenylene is not amenable to transformation by the Sandmeyer reaction.[11] Thus, 2-acetoxymercuribiphenylene has been used in the preparation of 2-bromo-, 2-chloro-, and 2-iodobiphenylene.[11] Biphenylenedisulfonic acid, isolated as the bis(S-benzylthiuronium) salt, was formed by the sulfonation of biphenylene at room temperature with concentrated sulfuric acid.[15] The orientation of the sulfonic acid groups has not been established, but it is quite probable that they occupy the 2- and 6-positions (see Section E, 1).

6. PHENOLS, PHENOL ETHERS, AND ESTERS

2-Hydroxybiphenylene (15) was obtained (a) by the alkaline hydrolysis of 2-acetoxybiphenylene (13), prepared by the direct acetoxylation of biphenylene with lead tetraacetate,[15] (b) by the alkaline hydrolysis of 2-benzoyloxybiphenylene (prepared from 2-benzoylbiphenylene by the Bayer–Villiger reaction),[15, 31] and (c) by the demethylation of 2-methoxybiphenylene (29)

302   10. BIPHENYLENE

TABLE 10.13

ULTRAVIOLET ABSORPTION CHARACTERISTICS OF THE BIPHENYLENE PHENOLS, ETHERS, AND ESTERS[a]

Biphenylene	Absorption maxima, mμ, and extinction coefficients (log ϵ)	Ref.
2-Benzoyloxy-	244 (4.93); 251 (5.14); 344 (3.96); 362 (4.06)	31
1,8-Dimethoxy-	256 (4.73); 265 (4.97); 303 (3.02); 318 (3.09); 335 (3.05); 350 (2.82)	11
2,3-Dimethoxy-	255 (4.96); 362 (4.09); 368 (4.16)	31
2,7-Dimethoxy-	245 (4.37); 254 (4.92); 360 (4.96)	13
	245 (5.03); 254 (5.21); 342 (4.16); 360 (4.25);	11
	246 (4.91); 256 (5.10); 342 (4.08); 362 (4.15)	59
2,7-Dimethoxy-1,3,6,8-tetramethyl-[b]	253 (4.67); 263 (3.46); 337 sh, 351 (3.79); 372 (3.93)	13
1,8-Dimethyl-3,4,5,6-tetramethoxy-	251 (4.41); 257 (4.27); 280 (3.21); 316 (3.07); 340 (3.11)	16
1,2,3,6,7,8-Hexamethoxy-	270 (4.50); 346 (4.19); 365 (4.22)	16
1,2,4,5,7,8-Hexamethoxy-	272 (4.74); 309 (3.35); 336 (3.19)	16
2-Hydroxy-	253 (4.64); 347 (3.72); 364 (3.82)	11
	253 (4.70); 346.5 (3.78); 364 (3.87)	15
1-Methoxy-	253 (4.01); 345 (3.02); 360 (3.07); 363 (3.07)	11
2-Methoxy-	253 (4.61); 275 (4.22); 306 (3.27); 324 (3.08); 338 (3.18); 356 (3.17)	11
1,2,3,4,5,6,7,8-Octamethoxy-	268 (4.68); 310 (4.12); 351 (3.98); 376 (4.07)	16

[a] All spectra determined in ethanol.
[b] Key: sh (shoulder).

with pyridine hydrochloride at 230°.[13] Although the conversion of 2-benzoylbiphenylene into 2-benzoyloxybiphenylene by the Bayer–Villiger reaction was found to proceed without difficulty, the conversion of 2-acetyl- into 2-acetoxybiphenylene could not be effected.[15]

Like 2-aminobiphenylene, 2-hydroxybiphenylene conforms to theory (Section E, 1) and couples readily with diazonium salts at the 3-position to give 3-azo-2-hydroxy derivatives. In this manner, 3-(p-chlorophenylazo)-2-hydroxy- (30),[31] 2-hydroxy-3-phenylazo- (31),[31,96] and 2-hydroxy-3-(p-sulfophenylazo)biphenylene (32) have been prepared.[31] The position of coupling was demonstrated by converting the p-sulfophenylazo compound into 2,3-biphenylenequinone.[31]

Although 2,3-dihydroxybiphenylene is certain to be an intermediate in the conversion of 2,3-biphenylenequinone into 2,3-dimethoxybiphenylene (by reduction of the quinone with sodium dithionite, followed by methylation of the resulting phenol with methyl sulfate),[31] neither the diol nor any other polyhydroxy derivative of biphenylene has been characterized directly.* 2,3-Dimethoxybiphenylene was also synthesized by Lothrop's method from 4,5-dimethoxybiphenylene-2,2'-iodonium iodide, which was itself obtained by three different routes.[32] Similarly, 1,8-dimethoxybiphenylene was synthesized from 4,5-dimethoxybiphenyleneiodonium iodide and 2,2'-diiodo-6,6'-dimethoxybiphenyl,[10] and 2,7-dimethoxybiphenylene was synthesized from 2,7-dimethoxybiphenyleneiodonium iodide.[89] It is noteworthy that a trace of 2,3,6,7-tetramethoxybiphenylene (34) was produced as a side-product in the synthesis of 2,2'-dibromo-4,5,4',5'-tetramethoxybiphenyl (33) from 4-bromo-5-iodoveratrole by the Ullmann reaction. It is possible that the tetramethoxybiphenylene was formed via a benzyne intermediate, but when an effort was made to improve the generation of benzyne by substituting magnesium for the copper used in the Ullmann reaction, the tetramethoxybiphenylene was not obtained.[13] Lothrop's method was also used in the synthesis of 2,7-dimethoxy-1,3,6,8-tetramethyl-[177] and 2,3,6,7-tetramethoxy biphenylene.[13]

Attempts to hydrolyze the 2,7-dimethoxybiphenylene to 2,7-dihydroxybiphenylene (35) were either without effect or resulted in extensive decomposition, depending on the severity of the reaction conditions.[89]

The ultraviolet spectrum of 2-hydroxybiphenylene is similar to that of 2-amino- and 2-acetylbiphenylene (see Fig. 10.7 and Table 10.13).

7. BIPHENYLENEQUINONES

Structures can be drawn for six biphenylenequinones, of which three (1,2-,

* A number of the polyhydroxybiphenylenes are known in the form of their methyl ethers (see Table 10.13). See also tetrahydroxydibenzo[b,h]biphenylene, p. 358.

[Scheme showing structures (33), (34), (35)]

1,4-, and 2,3-biphenylenequinone) contain one intact benzenoid ring of the Kekulé type and three do not (1,6-, 1,8-, and 2,7-biphenylenequinone). (See Chart 10.5.) The latter are probably unstable high-energy structures.* Two of the Kekulé structures, viz., 1,2- and 1,4-biphenylenequinone, are in a formal sense derivatives of benzocyclobutadiene, which is known to be a very reactive species. In contrast, the third Kekulé structure, 2,3-biphenylenequinone, is formally analogous to 1,2-bis(carbomethoxymethylene)benzocyclobutene, which is known to be a stable compound. Indeed, the chemical properties of 2,3-biphenylenequinone indicate that it is a remarkably stable quinone (see below). Experimental data are not available as yet for the other biphenylenequinones.

It has been pointed out that it is likely that 2,3,6,7-biphenylenebis(quinone) will resemble 2,3-biphenylenequinone in being a stable *o*-quinone.[96]

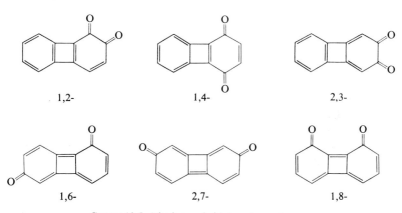

CHART 10.5. The isomeric biphenylenequinones.

* A report appeared after the manuscript of this chapter had been prepared, in which the synthesis of 1,3,6,8-tetramethoxybiphenylene-2,7-quinone was described (see ref. 17, this chapter, and Abstract 10.2 in the Appendix). MO calculations indicate that the 2,7-quinone is the least stable of the six possible unsubstituted biphenylenequinones (see ref. 111).

The delocalization energies of all six biphenylenequinones have been calculated.[111] The relative stabilities are as follows: 2,3- > 1,4- > 1,2- > 1,8- > 1,6- > 2,7-biphenylenequinone. The calculations indicate (a) that in the case of the known 2,3-biphenylenequinone, nucleophilic substitution should take place at the 4a,8b-positions, while free-radical and electrophilic substitution should occur at the 1,4-positions and (b) that most of the quinones should have a marked tendency to undergo reactions at one of the atoms of the four-membered ring.[111] The predictions are confirmed in part by the results of the Thiele acetylation and by the bromination of 2,3-biphenylenequinone (see p. 307).

2,3-Biphenylenequinone was synthesized from 2-hydroxybiphenylene by two routes: (a) the phenol (*10*) was coupled with diazotized sulfanilic acid, the resulting azo dye (*32*) was reduced directly to 2-amino-3-hydroxybiphenylene (*36*) (isolated as the bisulfate dihydrate), which was oxidized with chromic acid to give the desired product, 2,3-biphenylenequinone (*37*);[31] (b) direct oxidation of 2-hydroxybiphenylene with potassium nitrosodisulfonate gave the quinone in 75% yield [(*10*) → (*37*)].[31, 96]

The structure of the quinone was proven by reduction with sodium dithionite to give 2,3-dihydroxybiphenylene (not isolated), which was methylated *in situ* to give the known derivative, 2,3-dimethoxybiphenylene [(*37*) → (*38*)].[31] Unlike most *o*-quinones, 2,3-biphenylenequinone is quite stable, witness its mode of preparation, chromic acid oxidation of 2-amino-3-hydroxybiphenylene. Furthermore, the quinone shows no tendency to dimerize or polymerize, and it does not react with maleic anhydride [which might have produced a benzocyclobutadiene derivative (*39*)][31]; in this respect it behaves like 1,2-dimethylenebenzocyclobutene (see Chapter 8, Section C, 4, b). Indeed, the quinone is a cyclic conjugated analog of dimethylenebenzocyclobutene, and its 12 carbon atoms have a total of 10 π electrons, which satisfies the Hückel requirement. (According to Vol'pin, the Hückel rule is applicable to fused ring

G. CHEMICAL AND PHYSICAL PROPERTIES OF SUBSTITUTED BIPHENYLENES

systems as well as to monocyclic systems, but only if no atoms common to more than two rings are involved.[119]

2,3-Biphenylenequinone reacts readily with *o*-phenylenediamine to give a

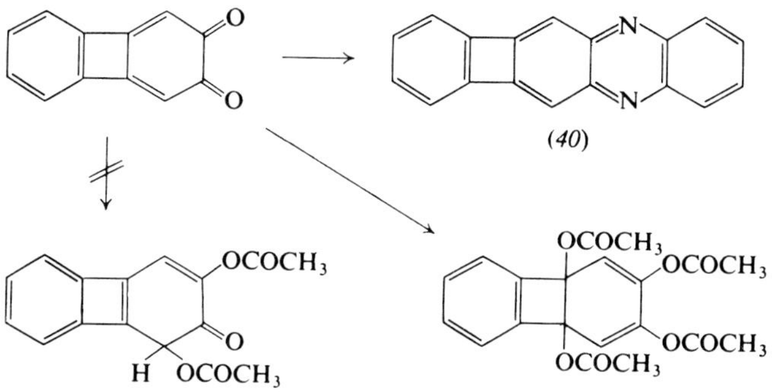

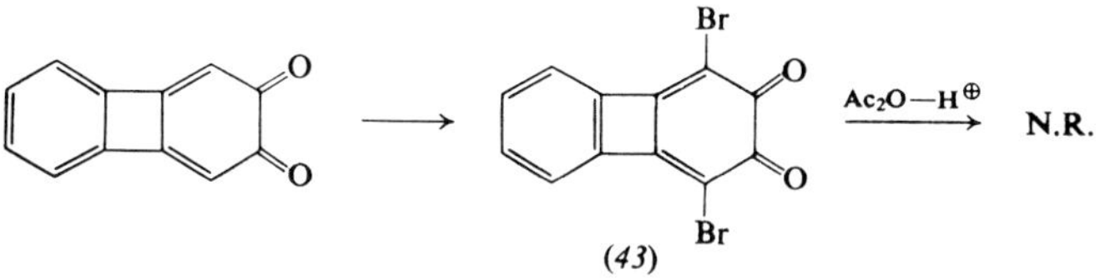

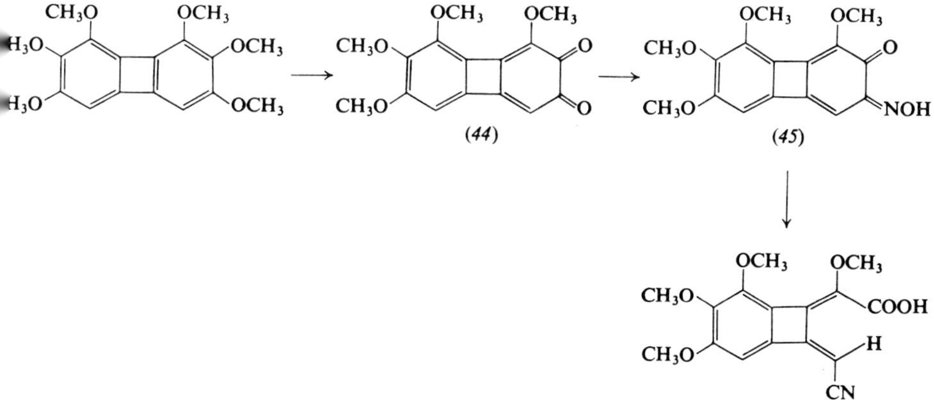

quinoxaline derivative, benzo[3,4]cyclobuta[1,2-*b*]phenazine (*40*).[31] The reaction of the quinone with acetic anhydride in the presence of an acid catalyst (Thiele acetylation) does not give a diacetoxybiphenylene (*41*) as expected; rather it gives a tetraacetoxy compound, $C_{20}H_{18}O_8$, formulated as a reduced biphenylene, as shown above (*42*).[31] Formation of the "normal" Thiele acetylation product (*41*) involves an unfavorable benzocyclobutadienoid intermediate (see also Chapter 6, Section C, 1).

Direct bromination of 2,3-biphenylenequinone gives a substitution product formulated as 1,4-dibromo-2,3-biphenylenequinone (*43*).[31] Unlike the parent quinone, the dibromo derivative remains unchanged when it is subjected to Thiele acetylation.[31]

6,7-Dimethoxy-2,3-biphenylenequinone[99] and 1,6,7,8-tetramethoxy-2,3-biphenylenequinone (*44*)[96] have been prepared by oxidative demethylation of 2,3,6,7-tetramethoxybiphenylene and of 1,2,3,6,7,8-hexamethoxybiphenylene, respectively. The methoxybiphenylenequinones do not react with acetic anhydride under Thiele acetylation conditions, but they react normally with *o*-phenylenediamine to give the corresponding quinoxalines.[18] The tetramethoxyquinone has been converted into a monooxime (probably the anti isomer), which was converted in turn into a cyanoacid by treatment with tosyl chloride and base [(*44*) → (*45*) → (*46*)].[96] The formation of the acid by a second-order Beckmann rearrangement constitutes the only known case of the formation of a benzocyclobutene derivative through cleavage of a biphenylene.

Ultraviolet absorption data have been recorded for biphenylene-2,3-quinone [λ_{max}^{EtOH} 268 mμ (log ϵ = 4.62), 372 (4.21)] and 1,4-dibromobiphenylene-2,3-quinone [222 (4.21), 286 (4.52), 389 (4.17)].[31] (See also ref. 17 and Abstract 10.2 in the Appendix.)

G. CHEMICAL AND PHYSICAL PROPERTIES OF SUBSTITUTED BIPHENYLENES

TABLE 10.14

PHYSICAL PROPERTIES OF SUBSTITUTED BIPHENYLENES

Biphenylene	Method of preparation[a]	Physical properties[b]	Properties of derivatives	Ref.
2-Acetamido-	8	M.p. 146°–147° (p-y)	—	15, 23a
2-Acetamido-3-bromo-	7	M.p. 211°–212° (c-c)	TNF, m.p. 179°–181° (b)	23a
2-Acetamido-3-phenylazo-	8	M.p. 227°–228°	—	33
2-Acetoxy-[c]	7	M.p. 139°–140°	—	15
2-Acetoxymercuri-	7	M.p. 176°–177° (y)	—	11
2-Acetyl-	7	M.p. 134°–135° (y)	DNP, m.p. 257°–258°; oxime, m.p. 176° (p-y)	11, 15, 23a
2-Acetyl-3-hydroxy-	7	—	—	30
3-Acetyl-2-hydroxy-7-methoxy-	7	—	—	99
2-Amino-	8	M.p. 123°–124° (y), subl.	—	11, 12
2-Amino-3-bromo-[c]	8	—	—	23a
2-Amino-3-hydroxy-[c]	8	—	Sulfate dihydrate (w)	31
2-Amino-3-phenylazo-	7	M.p. 175°–178°	—	33
2-Benzamido-	8	M.p. 209°–210° (p-y)	—	31
2-Benzoyl-	7	M.p. 116°–117.5° (y)	Oxime, m.p. 160° (y)	31
2-Benzoyloxy-	8	M.p. 150°–151° (s-c)	—	31
2-(2'-Biphenylenyl)-	7, 8	M.p. 242.5°–243.5° (y)	—	11, 30
2-Bromo-	1, 7, 8	M.p. 64°–65° (c-c)	TNF, m.p. 135°–136°	10, 11, 23a
—2-Carboxylic acid	8	M.p. 223°–224° (y)	Me ester, m.p. 114°–115° (p-y), subl.	11, 14, 15, 98
2-(β-Carboxypropionyl)-	8	M.p. 215° (decomp.) (y)	Me ester, m.p. 130°–132° (y)	14
2-Chloro	7, 8	M.p. 67.5°–68.5° (p-y)	TNF, m.p. 133°–134° (s)	11, 22
2-(p-Chlorophenylazo)-3-hydroxy-	7	M.p. 221°–223° (d-r)	—	31

Table 10.14 (continued)

Biphenylene	Method of preparation[a]	Physical properties[b]	Properties of derivatives	Ref.
2-Cinnamoyl-	8	M.p. 142°–144° (y)	—	98
2-Cyano-	8	M.p. 99°–100° (y); subl.	—	11
2,6-Diacetamido-	8	M.p. 328°–330° (decomp.) (o)	—	23a
2,6-Diacetamido-3,7-dibromo-	7	M.p. 260° (decomp.) (y)	—	23a
2,6-Diacetyl-	7	M.p. 247°–248° (y)	Dioxime, m.p. 290°–291° (decomp.) (y)	11, 15, 23a
2,6-Diamino-	8	M.p. 218°–220° (decomp. at 205°) (y)	Dihydrochloride, subl. 215° (m)	23a
1,4-Dibromo--2,3-quinone	7	M.p. 278°–279° (decomp.) (y)	—	31
1,2-Dihydrobenzo[b]-	8	M.p. 214°–215° (y)	—	14
2,3-Dihydroxy-[c]	8	—	—	31
2,6(or 7)-Di(iodomethyl)-	7	M.p. 86° (decomp.) (y)	Dithiuronium picrate, m.p. 204°–206° (o)	11
1,8-Dimethoxy-	1	M.p. 107°–108° (y); subl. 130° (12 mm)	—	10
2,3-Dimethoxy-	1, 8	M.p. 87°–88° (y)	Picrate, m.p. 125° (d-r)	31, 32
2,7-Dimethoxy-	1	M.p. 107.5°–109°	TCNE complex (UV spect.)	13, 89
				59
3,7(or 6)-Dimethoxy-1,5(or 8)-dinitro-6,7-Dimethoxy--2,3-quinone	3	—	—	97
	8	—	—	99
2,7-Dimethoxy-1,3,6,8-tetramethyl-	1	M.p. 116°–116.5° (p-y)	—	13
1,5(or 8)-Dimethyl-	2	B.p. 123°–140° (12 mm); 108°–115° (0.8 mm)	—	74

G. CHEMICAL AND PHYSICAL PROPERTIES OF SUBSTITUTED BIPHENYLENES

Compound			
1,8-Dimethyl-	1	M.p. 79°–80° (w)	Picrate, m.p. 126° (r)	89
2,6-Dimethyl-	2	M.p. 138°–140°	TCNE complex (UV spect.)	59
		B.p. 118°–124° (0.8 mm), m.p. 139°–141°	Picrate (decomp.) (r)	75, 71
2,7-Dimethyl-	1	M.p. 110° (s-c); subl.	Picrate, m.p. 122° (s)	88
			TCNE complex (UV spect.)	59
3,7(or 6)-Dimethyl-1,5(or 8)-dinitro-	3	M.p. 302° (y)	—	52
1,8-Dimethyl-3,4,5,6-tetramethoxy-	1	M.p. 132°–133°, subl. (110°–120°/0.1 mm)	—	16
2,6-Dinitro-	1, 7	Subl. > 260°	—	10, 11
—2,6(?)-Disulfonic acid	7		Bis-(S-benzylthiuronium) salt, m.p. 248°–249° (decomp.)	15
2-Formyl-	8	M.p. 78°–79° (y), subl.	DNP, m.p. 312°–314° (d-r)	98
—2-Glyoxylic acid	8	"An orange solid"	Semicarbazone, m.p. 222°–224° (decomp.)	98
1,2,3,6,7,8-Hexamethoxy-	1	M.p. 91°–91.5° (y)	Picrate, m.p. 128°–129° (b)	16, 97
1,2,4,5,7,8-Hexamethoxy-	1	M.p. 119°–119.5° (y)	Picrate, m.p. 145°–146° (b)	16
1,2(or 3),4,5,6,8-Hexamethyl-	2	M.p. 177.5°–178°	Picrate, m.p. 183°–184° (decomp.)[d]	131
2-Hydroxy-	8	M.p. 140°–141° (y),[e] 139°–140°[f]	—	10, 11, 15, 31
2-Hydroxy-3-phenylazo-	7	M.p. 208°–210° (r)	—	31
2-Hydroxy-3-(p-sulfophenylazo)-[c]	7		—	31
1-Hydroxy-1,2,3,4-tetrahydrobenzo[b]-[c]	8	—	—	14
2-Iodo-	7, 8	M.p. 64.5°–65.5° (p-y)	TNF, m.p. 132.5°–134° (d-r)	11
1-Keto-1,2,3,4-tetrahydrobenzo[b]-	7	M.p. 134°–135° (y)	—	14
1-Methoxy-	1	M.p. 41°–42° (p-y)	TNF, m.p. 157.5°–159° (m)	10
2-Methoxy-	1	M.p. 69°–70° (y), vol. st., subl.	TNF, m.p. 140°–141° (p)	10
1-Methyl-	1	"A yellow oil"	Picrate, m.p. 91°–92° (s); TNF, m.p. 162°–163.5° (d-r)	10
2-Methyl-	1	M.p. 45°–46° (s-c); subl.	Picrate, m.p. 86°–87° (s); TNF, m.p. 150°–151° (p)	10

Table 10.14 (continued)

Biphenylene	Method of preparation[a]	Physical properties[b]	Properties of derivatives	Ref.
2-Nitro-	1, 2, 7	M.p. 105°–106.5° (d-y); vol. st.	—	11, 134
1,2,3,4,7,8,9,10-Octahydrodibenzo[b,h]- (?)	1	M.p. 224°–226° (y)	—	121
1,2,3,4,5,6,7,8-Octamethoxy- -2,3-Quinone	1 8	M.p. 138°–139° (y) M.p. 216°–217° (o)	Picrate, m.p. 168°–169°, (b) See the dibromoquinone	16 31
1,2,3,4-Tetrahydrobenzo[b]-	8	M.p. 109.5°–111° (y)	—	14
2,3,6,7-Tetramethoxy-	1, 3	M.p. 211°–212° (y), subl.	—	13
1,6,7,8-Tetramethoxy- -2,3-quinone	1, 8	M.p. 214°–215°	Oxime, m.p. 220°–221° (y)	16, 17, 96
2,3,6,7-Tetramethyl- (?)	7, 8	M.p. 222.5°–223.5° (s-c)	TNF, m.p. 171°–172° (br)	11
Other Substituted Biphenylenes				
β-(2-Biphenylenyl)acrylic acid	8	M.p. 228°–229° (y)	—	98
β-(2-Biphenylenyl)acrylophenone	8	M.p. 128°–129° (y)	—	98
γ-(2-Biphenylenyl)butyric acid	8	M.p. 118.5°–119.5° (p-y)	Me ester, m.p. 130°–132° (y)	14

[a] Method 1 is "From 2,2′-Dihalobiphenyls," Section B,1; Method 2 is "Via Benzynes," Section B,2; Method 3 is "By the Ullmann Reaction," Section B, 3; Method 7 is "By Direct Substitution," Section B, 7; and Method 8 is "By Transformation of Functional Groups," Section B, 8.
[b] Key to abbreviations: b (black), br (brown), c-c (cream colored), d-r (deep red), d-y (deep yellow), m (maroon), o (orange), p (purple), p-y (pale yellow), s (scarlet), s-c (straw colored), subl. (sublimes), vol. st. (volatile with steam), w (white), y (yellow).
[c] Not isolated.
[d] Reference 131a.
[e] Reference 10.
[f] Reference 31.

References

1. Atkinson, E. R., private communication.
2. Atkinson, E. R., Bruni, R. J., Whaley, W. M., and Kerridge, K. A., private communication.
3. Atkinson, E. R., and Gatti, A. R., private communication.
4. Atkinson, E. R., and Granchelli, F. E., private communication.
5. Atkinson, E. R., Levine, P. L., and Dickelman, T. E., private communication.
6. Atkinson, E. R., and Simon, I., private communication.
7. Baker, W., *Nature* **150**, 210 (1942).
8. Baker, W., *J. Chem. Soc.* 258 (1945).
9. Baker, W., in Todd, A., (ed.), "Perspectives in Organic Chemistry," Interscience, New York (1956), p. 28.
10. Baker, W., Barton, J. W., and McOmie, J. F. W., *J. Chem. Soc.* 2658 (1958).
11. Baker, W., Barton, J. W., and McOmie, J. F. W., *J. Chem. Soc.* 2666 (1958).
12. Baker, W., Barton, J. W., and McOmie, J. F. W., unpublished work reported in ref. 21.
13. Baker, W., Barton, J. W., McOmie, J. F. W., Penneck, R. J., and Watts, M. L., *J. Chem. Soc.* 3986 (1961).
14. Baker, W., Barton, J. W., McOmie, J. F. W., and Searle, R. J. G., *J. Chem. Soc.* 2633 (1962).
15. Baker, W., Boarland, M. P. V., McOmie, J. F. W., *J. Chem. Soc.* 1476 (1954).
16. Baker, W., McLean, N. J., and McOmie, J. F. W., *J. Chem. Soc.* 922 (1963).
17. Baker, W., McLean, N. J., and McOmie, J. F. W., *J. Chem. Soc.* 1067 (1964).
18. Baker, W., McLean, N. J., and McOmie, J. F. W., unpublished work cited in ref. 96.
19. Baker, W., and McOmie, J. F. W., in Cook, J. W., (ed.), *Prog. Org. Chem.*, Academic Press, New York (1955), Vol. 3, p. 44.
20. Baker, W., and McOmie, J. F. W., *Chem. Soc. Special Publication No. 12*, 49 (1958).
21. Baker, W., and McOmie, J. F. W., in Ginsburg, D., "Non-Benzenoid Aromatic Compounds," Interscience, New York (1959), p. 43.
22. Baker, W., and McOmie, J. F. W., unpublished work cited in ref. 21.
23. (a) Baker, W., McOmie, J. F. W., Preston, D. R., and Rogers, V., *J. Chem. Soc.* 414 (1960); (b) Baker, W., McOmie, J. F. W., and Rogers, V., *Chem. Ind. (London)* 1236 (1958).
24. Barton, J. W., and McOmie, J. F. W., *J. Chem. Soc.* 796 (1956).
25. Bedford, A. F., Carey, J. G., Millar, I. T., Mortimer, C. T., and Springall, H. D., *J. Chem. Soc.* 3895 (1962).
26. (a) Bergman, I., *Trans. Faraday Soc.* **52**, 690 (1956); (b) *ibid.*, **50**, 829 (1954); (c) *ibid.*, *Experientia* **18**, 46 (1962).
27. Berry, R. S., Spokes, G. N., and Stiles, R. M., *J. Am. Chem. Soc.* **82**, 5240 (1960).
28. Berthier, G., Mayot, M., and Pullman, B., *J. Phys. Radium* **12**, 717 (1951).
29. Beynon, J. H., Lester, G. R., and Williams, A. E., *J. Phys. Chem.* **63**, 1861 (1959).
30. Blatchly, J. M., and McOmie, J. F. W., unpublished work cited in ref. 96.
31. Blatchly, J. M., McOmie, J. F. W., and Thatte, S. D., *J. Chem. Soc.* 5090 (1962).
32. Blatchly, J. M., McOmie, J. F. W., and Watts, M. L., *J. Chem. Soc.* 5085 (1962).
33. Bosshard, H. H., and Zollinger, H., *Helv. Chim. Acta* **44**, 1985 (1961).
34. Brown, R. D., *Trans. Faraday Soc.* **45**, 300 (1949).
35. Brown, R. D., *Trans. Faraday Soc.* **46**, 146 (1950).
36. Bruce, J. M., *J. Chem. Soc.* 2782 (1962).
37. Bunnett, J. F., *J. Chem. Ed.* **38**, 278 (1961).
38. Busch, M., and Weber, W., *J. prakt. Chem.* **146**, 1 (1936).

39. Buu-Hoï, N. P., Lavit, D., and Chalvet, O., *Tetrahedron* **8**, 7 (1960).
40. Carr, E. P., Pickett, L. W., and Voris, D., *J. Am. Chem. Soc.* **63**, 3231 (1941).
41. Carrington, A., and Santos Veiga, J. dos, *Mol. Phys.* **5**, 285 (1962).
42. Cass, R. C., Springall, H. D., and Quincey, P. G., *J. Chem. Soc.* 1188 (1955).
43. Cava, M. P., *Am. Chem. Soc., Div. Petrol. Chem., Preprints* **1**, No. 4, 11 (1956).
44. Cava, M. P., and Mitchell, M. J., unpublished work. See Mitchell, M. J., Ph.D. diss., The Ohio State Univ. (1960).
45. Cava, M. P., Stein, G., and Arora, S. S., unpublished work.
46. Cava, M. P., and Stucker, J. F., *J. Am. Chem. Soc.* **77**, 6022 (1955).
47. Chatt, J., private communication.
48. Chatt, J., Guy, R. G., and Watson, H. R., *J. Chem. Soc.* 2332 (1961). See also ref. 20.
49. Clar, E., "Aromatische Kohlenwasserstoffe," 2nd ed., Springer, Berlin (1952), p. 25 *ff.*
50. Collette, J., McGreer, D., Crawford, R., Chubb, F., and Sandin, R. B., *J. Am. Chem. Soc.* **78**, 3819 (1956).
51. Cookson, R. C., Cox, D. A., Hudec, J., *J. Chem. Soc.* 4499 (1961).
52. (a) Corbett, J. F., and Holt, P. F., *J. Chem. Soc.* 4261 (1961); (b) *ibid.*, private communication.
53. Coulson, C. A., *Nature* **150**, 577 (1942).
54. Crawford, V. A., and Coulson, C. A., *J. Chem. Soc.* 1990 (1948).
55. Cullinane, N. M., Morgan, N. M. E., and Plummer, C. A. J., *Rec. trav. Chim.* **56**, 627 (1937).
56. (a) Curtis, R. F., and Viswanath, G., *J. Chem. Soc.* 1670 (1959); (b) *ibid.*, *Chem. Ind. (London)* 1174 (1954). *Cf.* ref. 121.
57. Dobbie, J. J., Fox, J. J., and Guage, A. J. H., *J. Chem. Soc.* 683 (1911).
58. Dobbie, J. J., Fox, J. J., and Guage, A. J. H., *J. Chem. Soc.* 36 (1913).
59. Farnum, D. G., Atkinson, E. R., and Lothrop, W. C., *J. Org. Chem.* **26**, 3024 (1961).
60. Fernández Alonso, J. I., and Domingo, R., *Anales Real Soc. Espan. Fis. Quim. (Madrid)* **51B**, 447 (1958).
61. Fernández Alonso, J. I., and Peradejordi, F., *Anales Real Soc. Espan. Fis. Quim. (Madrid)* **50B**, 253 (1954).
62. Fisher, I. P., and Lossing, F. P., *J. Am. Chem. Soc.* **85**, 1018 (1963).
63. Fonken, G. J., *Chem. Ind. (London)* 716 (1961).
64. Fraenkel, G., Asahi, Y., Mitchell, M. J., and Cava, M. P., *Tetrahedron* **20**, 1179 (1964).
65. Friedman, L., *Chem. Eng. News* **41**, (24), 46 (1963). See also Friedman, L., and Logullo, F. M., *J. Am. Chem. Soc.* **85**, 1549 (1963).
66. Gelin, S., *Chim. Mod.* **8**, 241 (1963).
67. Goldish, E., *J. Chem. Ed.* **36**, 408 (1959).
68. Grdenić, D., *Ber.* **92**, 231 (1959).
69. Günther, H., *Ber.* **96**, 1801 (1963).
70. Hargreaves, A., *Advan. X-Ray Anal.* **4**, 1 (1961).
71. Hart, F. A., and Mann, F. G., *J. Chem. Soc.* 3939 (1957).
72. Heany, H., *Chem. Rev.* **62**, 81 (1962).
73. Heany, H., Mann, F. G., and Millar, I. T., *J. Chem. Soc.* 3930 (1957).
74. Hellman, H., and Unseld, W., *Ann.* **631**, 82 (1960).
75. Hinton, R. C., Mann, F. G., and Millar, I. T., *J. Chem. Soc.* 4704 (1958).
76. (a) Hochstrasser, R. M., *Can. J. Chem.* **39**, 765 (1961); (b) *ibid.*, *J. Chem. Phys.* **33**, 950 (1960).
77. Hoffmeister, W., *Ann.* **159**, 191 (1871).
78. Hosaeus, W., *Monatsh.* **14**, 323 (1893).

79. Huisgen, R., in Zeiss, H., "Organometallic Chemistry," Reinhold, New York (1960), p. 63.
80. Ilse, F. E., Z. Naturforsch. **5A**, 469 (1950).
81. IUPAC, "1957 Report of the Commission on the Nomenclature of Organic Chemistry," J. Am. Chem. Soc. **82**, 5545 (1960), rule A-21.2.
82. Kampmeier, J. A., and Hoffmeister, E., J. Am. Chem. Soc. **84**, 3787 (1962).
83. Kharasch, N., private communication.
84. Kharasch, N., Wolf, W., Erpelding, T. J., Naylor, P. G., and Tokes, L., Chem. Ind. (London) 1720 (1962).
85. Le Goff, E., J. Am. Chem. Soc. **84**, 3786 (1962).
86. Le Goff, E., private communication.
87. Longuet-Higgins, H. C., Proc. Chem. Soc. 161 (1957).
88. Lothrop, W. C., J. Am. Chem. Soc. **63**, 1187 (1941).
89. Lothrop, W. C., J. Am. Chem. Soc. **64**, 1698 (1942).
90. Lüttringhaus, A., and Schubert, K., Naturwissenschaften **42**, 17 (1955).
91. MacGregor, J. R., private communication. [Calif. Research Corp., progress letter for May–June, 1961, to the U.S. Atomic Energy Commission, Project AT(04-3)-248.]
92. (a) Mak, T. C. W., and Trotter, J., J. Chem. Soc. 1 (1962); (b) ibid., Proc. Chem. Soc. 163 (1961).
93. Mascarelli, L., and Gatti, D., Gazz. Chim. Ital. **63**, 661 (1933).
94. McDowell, C. A., and Paulus, K. F., Can. J. Chem. **40**, 828 (1962).
95. McDowell, C. A., and Rowlands, J. R., Can. J. Chem. **38**, 503 (1960).
96. McOmie, J. F. W., Rev. Chim. Acad. Rep. Populaire Roumaine **7**, 1071 (1962).
97. McOmie, J. F. W., and Smith, J. P., unpublished work cited in ref. 96.
98. McOmie, J. F. W., and Thatte, S. D., J. Chem. Soc. 5298 (1962).
99. McOmie, J. F. W., and Watts, M. L., unpublished work cited in ref. 96.
100. Niementowski, S. von, Ber. **34**, 3325 (1901).
101. Nierenstein, M., Ann. **386**, 318 (1911).
102. Olah, G. A., and Tolgyesi, W. S., J. Am. Chem. Soc. **83**, 5031 (1961).
103. Parrack, J. D., Swan, G. A., and Wright, D., J. Chem. Soc. 911 (1962).
104. Pearson, B. D., Chem. Ind. (London) 899 (1960).
105. (a) Rapson, W. S., Shuttleworth, R. G., and Niekerk, J. N. van, J. Chem. Soc. 326 (1943); (b) Rapson, W. S., and Shuttleworth, R. G., J. Chem. Soc. 487 (1941).
106. (a) Schüler, H., Spectrochim. Acta **15**, 981 (1959); (b) Schüler, H., and Lutz, E., Z. Naturforsch. **16A**, 57 (1961).
107. Schüler, H., Preprints of Papers, 5th International Symposium on Free Radicals, Uppsala 62–1 (1961). [Chem. Abstr. **59**, 5094b (1963).]
108. Schumaker, E., Helv. Chim. Acta **46**, 1295 (1963).
109. Schwechten, H.-W., cited by Mascarelli, L., and Gatti, D., Gazz. Chim. Ital. **63**, 661 (1933).
110. Schwechten, H.-W., Ber. **65**, 1605 (1932).
111. Schweizer, H. R., Helv. Chim. Acta **45**, 1934 (1962).
112. Selwood, P. W., "Magnetochemistry," 2nd ed., Interscience, New York (1956), p. 91 ff.
113. Streitwieser, A., "Molecular Orbital Theory for Organic Chemists," Wiley, New York p. 345.
114. Streitwieser, A., and Schwager, I., J. Phys. Chem. **66**, 2316 (1962) and the references cited therein.
115. Streitwieser, A., and Schwager, I., J. Am. Chem. Soc. **85**, 2855 (1963).
116. Streitwieser, A., and Suzuki, S., Tetrahedron **16**, 153 (1961).
117. Strübell, W., and Baumgärtel, H., J. prakt. Chem. **11**, 20 (1960).

118. U.S. Atomic Energy Commission, *Brit. Patent* 901,684, July 25, 1962.
119. Vol'pin, M. E., *Usp. Khim.* **29**, 147 (1960). [*Russ. Chem. Rev.* **29**, 142 (1960).]
120. (a) Wallenberger, T. F., *Experientia* **16**, 83 (1960); (b) *ibid.* **18**, 46 (1962).
121. Ward, E. R., and Pearson, B. D., *J. Chem. Soc.* 1676 (1959).
122. Ward, E. R., and Pearson, B. D., *J. Chem. Soc.* 515 (1961).
123. Waser, J., and Lu, C.-S., *J. Am. Chem. Soc.* **66**, 2035 (1944), discussed in Kitaigorodskii, A. I., "Organic Chemical Crystallography," Consultants Bureau, New York (1955), p. 417.
124. Waser, J., and Schomaker, V., *J. Am. Chem. Soc.* **65**, 1451 (1943).
125. West, W. W., "The Radiolysis of Prospective Organic Reactor Coolants," Report No. 13, Calif. Research Corp., AECU-4295 (1959). [*Chem. Abstr.* **54**, 4184g (1960).]
126. Winkler, H. J. S., and Wittig, G., *J. Org. Chem.* **28**, 1733 (1963).
127. Wittig, G., *Suomen Kemistilehti* **29A**, 283 (1956).
128. Wittig, G., *Angew. Chem.* **69**, 245 (1957).
129. Wittig, G., and Bickelhaupt, F., *Ber.* **91**, 883 (1958).
130. (a) Wittig, G., and Ebel, H. F., *Ann.* **650**, 20 (1961); (b) *ibid.*, *Angew. Chem.* **72**, 564 (1960).
131. (a) Wittig, G., and Härle, H., *Ann.* **623**, 17 (1959); (b) Härle, H., Ph.D. diss., Tübingen (1957).
132. Wittig, G., Hahn, E., and Tochtermann, W., *Ber.* **95**, 431 (1962).
133. Wittig, G., and Herwig, W., *Ber.* **87**, 1511 (1954).
134. Wittig, G., and Hoffmann, R. W., *Ber.* **95**, 2718 (1962).
135. Wittig, G., and Lehman, G., *Ber.* **90**, 875 (1957).
136. Wittig, G., and Pohmer, L., *Ber.* **89**, 1334 (1956).
137. Wolovsky, R., and Sondheimer, F., *J. Am. Chem. Soc.* **84**, 2844 (1962).

CHAPTER 11

The Benzobiphenylenes*

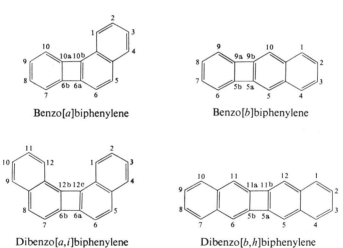

Benzo[a]biphenylene

Benzo[b]biphenylene

Dibenzo[a,i]biphenylene

Dibenzo[b,h]biphenylene

Like biphenylene, the benzobiphenylenes are all formal derivatives of cyclobutadiene. Two monobenzobiphenylenes are possible, benzo[a]- and benzo[b]-biphenylene, both of which are known (see above). Five dibenzobiphenylenes are possible, dibenzo[a,c]-, dibenzo[a,g]-, dibenzo[a,h]-, dibenzo[a,i]-, and dibenzo[b,h]biphenylene, of which only two are known, viz., dibenzo[a,i]- and

* Partially reduced benzobiphenylenes that have an intact biphenylene ring system, e.g., 1,2,3,4-tetrahydrobenzo[b]biphenylene, are classified as biphenylenes and are discussed in Chapter 10.

dibenzo[b,h]biphenylene.* Similarly, two tribenzobiphenylenes and one tetrabenzobiphenylene may be written, but none of these has been prepared (see Chart 11.1).

The benzobiphenylenes have been reviewed by Baker and McOmie[10] and, in part, by Gelin.[37]

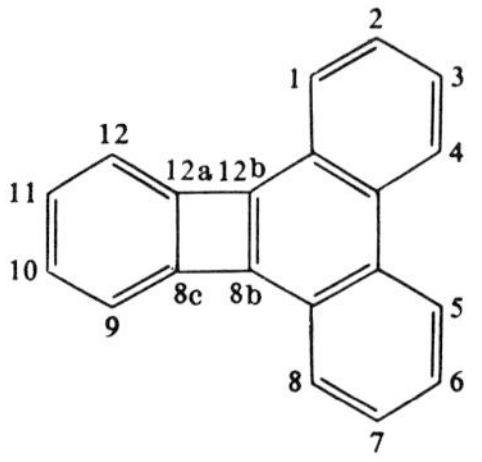

Dibenzo[a,c]biphenylene

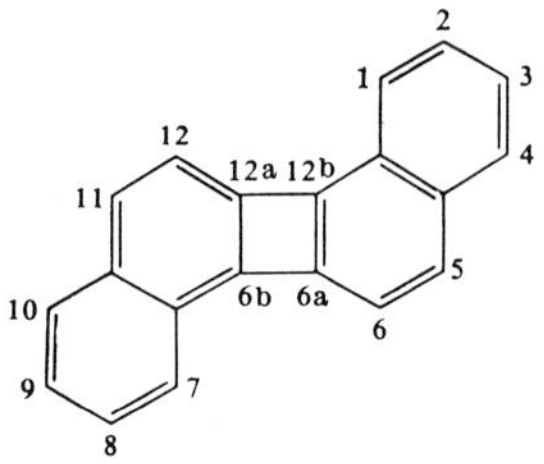

Dibenzo[a,g]biphenylene

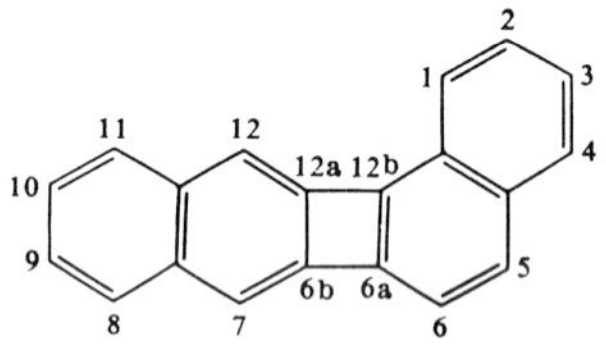

Dibenzo[a,h]biphenylene

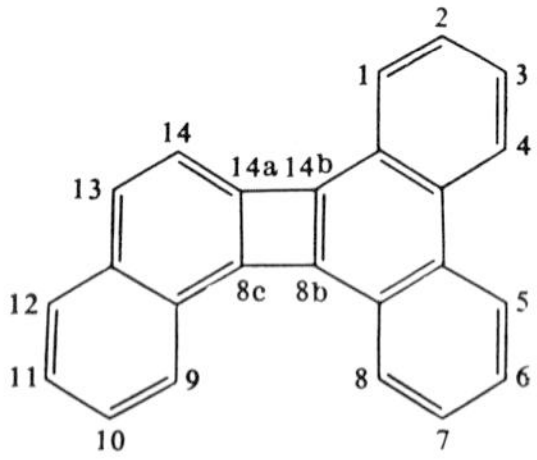

Tribenzo[a,c,g]biphenylene

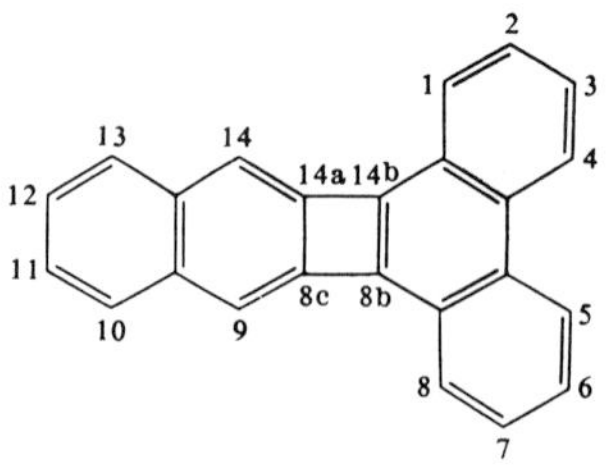

Tribenzo[a,c,h]biphenylene

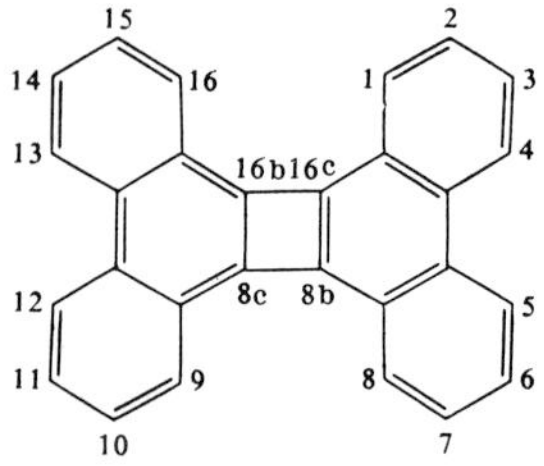

Tetrabenzo[a,c,g,i]biphenylene

CHART 11.1. Unknown benzobiphenylenes.

A. History

Braun and Kirschbaum were the first to claim the synthesis of a benzobiphenylene, viz., dibenzo[a,g(or i)]biphenylene [(3a) or (3b)].[13] According to

* See Abstracts 11.2 and 11.3 in the Appendix for the synthesis of dibenzo[a,g]- and dibenzo[a,c]biphenylene, respectively.

these authors, 1,2-dihydronaphthalene underwent dimerization in the presence of sulfuric acid to give a hydrocarbon which was said to be 6a,6b,12a,12b-tetrahydrodibenzo[a,g(or i)]biphenylene [(2a) or (2b)]; aromatization of the tetrahydro compound with lead oxide gave the alleged dibenzobiphenylene [(3a) or (3b)]. The same compound, $C_{20}H_{12}$, m.p. 165°, was obtained by Dansi and Ferri by treating tetrahydronaphthalene with aluminum chloride.[35] However, Orchin and co-workers were subsequently able to show that the product is not a dibenzobiphenylene but benzo[j]fluoranthene (1).[42]

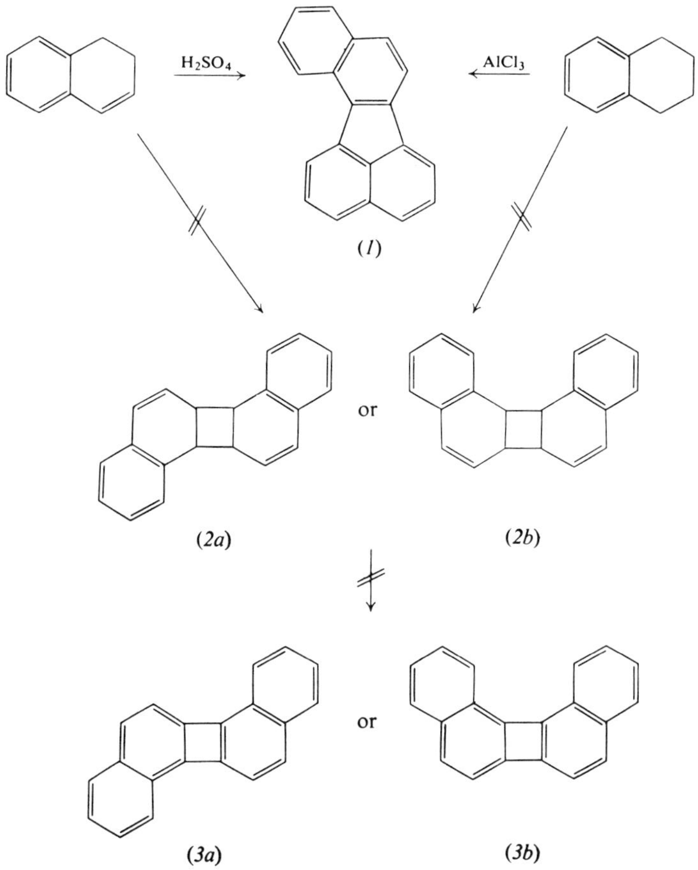

In 1937, Rosenhauer and co-workers claimed the synthesis of dibenzo[b,h]-biphenylene-5,6,11,12-bis(quinone) (4) and the parent hydrocarbon (5), but their claim was soon withdrawn when it was discovered that the compounds are actually triphthaloylbenzene (6) and 2,3-trinaphthylene (7), respectively.[44, 45] Bell and Hunter tried unsuccessfully to prepare dibenzo[b,h]biphenylene by using the Ullmann reaction, the Krizewski–Turner reaction, and several other methods (see Table 11.4).[11]

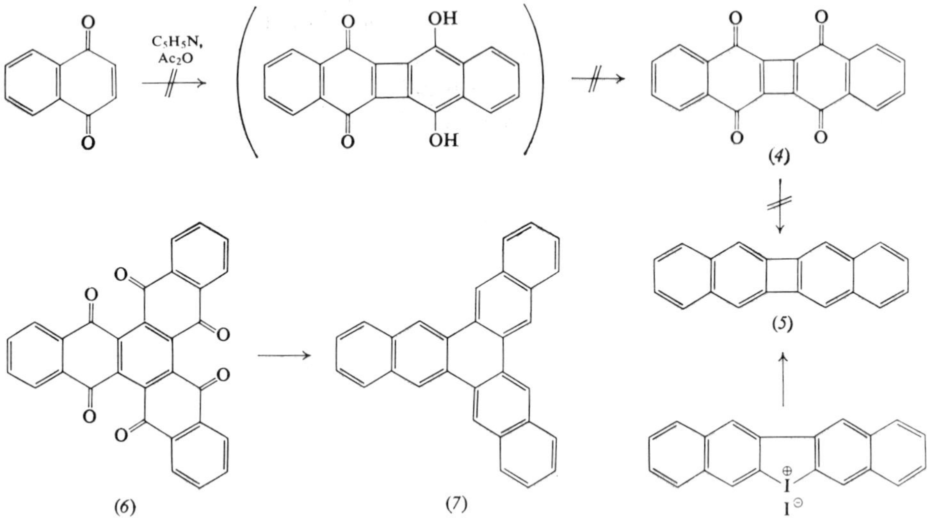

Credit for the first report—though not for the first synthesis—of an authentic benzobiphenylene must go to Curtis and Viswanath, who in 1954 prepared dibenzo[b,h]biphenylene (5) by Lothrop's method from the corresponding iodonium compound.[33] The first synthesis of a benzobiphenylene or, for that matter, of a biphenylene of any type, was carried out unknowingly by Thiele and Finkelstein almost half a century earlier during the course of their studies on the chemistry of 1,2-dibromobenzocyclobutene (8)[36]; however, the identity of their product, 5-bromobenzo[a]biphenylene (9), was not recognized until 1957.[21] (See also Chapter 6.)

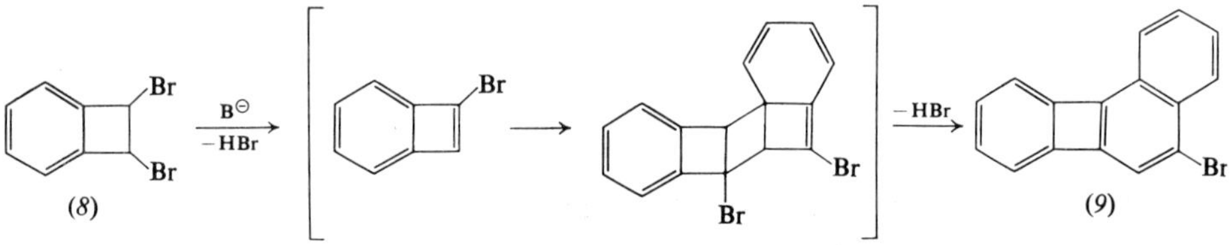

In recent years benzobiphenylenes have been prepared by several other methods, which are described in detail in the following section.

B. Synthesis of Benzo- and Dibenzobiphenylenes

To a significant degree, the synthetic methods used to prepare the benzo- and dibenzobiphenylenes parallel those used to prepare biphenylene and its derivatives (Chapter 10, Sections B, 1–10). However, a number of substituted benzo[a]- and benzo[b]biphenylenes have been prepared in good yield by special methods that find no analogy among the simple biphenylenes (see, for example, Section B, 5).

One case has been reported in which a dibenzobiphenylene was formed unexpectedly, i.e., in the pyrolysis of 5,6,7,8,5',6',7',8'-octahydro-2,2'-binaphthyl-3,3'-tetrazonium chloride (1a), which gave dibenzo[b,h]biphenylene

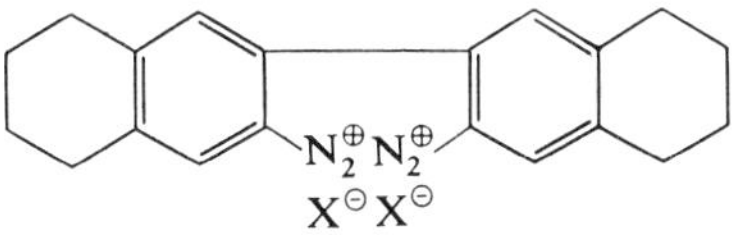

(1a): X=Cl
(1b): X=HSO₄

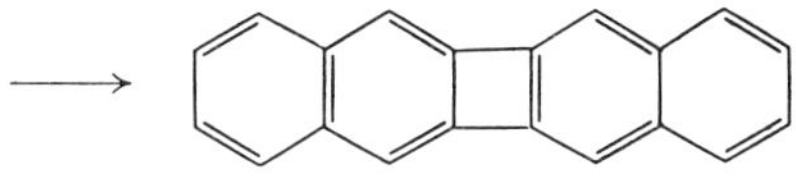

(2) instead of 1,2,3,4,7,8,9,10-octahydrodibenzo[b,h]biphenylene.[33b] It is interesting to note that the pyrolysis of the tetrazonium *bisulfate* (1b) under similar reaction conditions gave only octahydrodibenzocarbazole (3).[49]

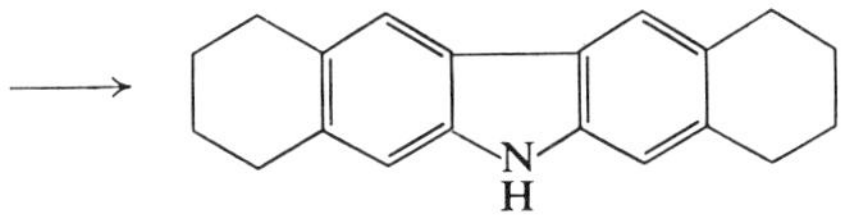

1. FROM 2,2'-DIHALOBIARYLS

Lothrop's cuprous oxide method has been applied to the preparation of relatively few benzo- and dibenzobiphenylenes because the dihalobiaryls which are required as starting materials are difficult to obtain. Although the reaction conditions are drastic (pyrolysis at about 300°), the product usually sublimes from the reaction mixture as it is formed, and so the method may be used even in the preparation of benzobiphenylenes of lesser stability such as dibenzo[a,i]biphenylene (4).[28] (See Table 11.1.)

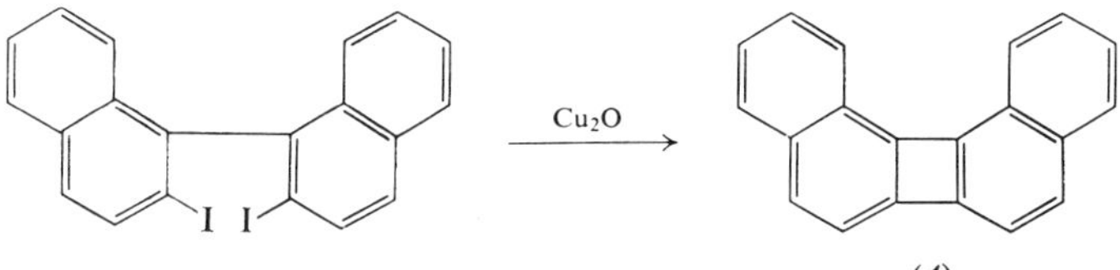

TABLE 11.1

Benzo- and Dibenzobiphenylenes Prepared by Lothrop's Method*

Biphenylene	Starting material	Yield (%)	Ref.
Benzo[a]-	1-(2'-Iodophenyl)-2-iodonaphthalene	63	28
	Benzo[b]naphth[1,2-d]iodolium iodide	6.4	28
5-Bromobenzo[a]-	4-Bromo-1-(2'-iodophenyl)-2-iodonaphthalene	22[a]	21, 29
Dibenzo[a,i]-	2,2'-Diiodo-1,1'-binaphthyl	5.65[a]	28
Dibenzo[b,h]-	2,2'-Binaphthyl-3,3'-tetrazonium chloride	"Low yield"	33
	Dinaphth[2,3-b:2',3'-d]iodolium iodide	3	33a
	5,6,7,8,5',6',7',8'-Octahydro-2,2'-binaphthyl-3,3'-tetrazonium chloride[b]	—	33a
	1,2,3,4,8,9,10,11-Octahydrobinaphth-[2,3-b:2',3'-d]iodolium iodide[b]	29[c]	33a
		4.5	49
1,2,3,4-Tetrahydro-dibenzo[b,h]- (?)	1,2,3,4,8,9,10,11-Octahydrobinaphth-[2,3-b:2',3'-d]iodolium iodide[b]	2.6	49

[a] Isolated in the form of the trinitrofluorenone complex.
[b] Cyclization accompanied by aromatization.
[c] Calculated from weight data recorded in the literature.

2. Via 1,2- and 2,3-Naphthalynes

Because the 2,2'-dihalobiaryls which are required as starting materials for Lothrop's method are often difficult to obtain, numerous attempts have been made to prepare the benzobiphenylenes by the more direct naphthalyne dimerization method. (See also Section B, 3.) Unfortunately, only two benzobiphenylenes could be prepared in this way, viz., dibenzo[b,h]biphenylene and

* Annellated biphenylene iodonium salts are difficult to name according to the "iodonium" nomenclature and are named in Table 11.1 and the index as iodolium compounds (see also ref. 43 and Table 10.1). The naming and numbering of three representative annellated iodolium ions is illustrated below.

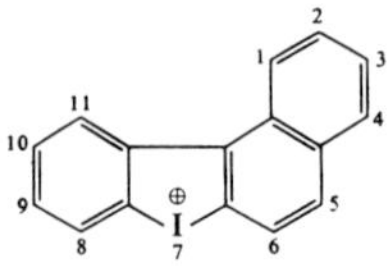

Benzo[b]naphth[1,2-d]iodolium ion

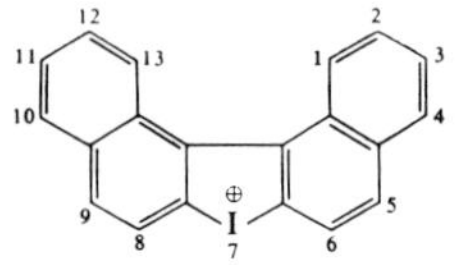

Dinaphth[2,1-b:1',2'-d]iodolium ion

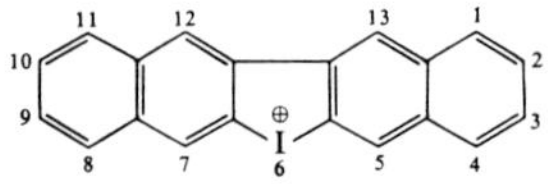

Dinaphth[2,3-b:2',3'-d]iodolium ion

B. SYNTHESIS OF BENZO- AND DIBENZOBIPHENYLENES

dibenzo[a,g]biphenylene. (The identity of the latter has not been established unambiguously, but its ultraviolet spectrum is consistent with the benzobiphenylene structure.[49]) Thus, (a) dibenzo[b,h]biphenylene (2) was prepared in 0.2% yield* by treating 2,3-dibromonaphthalene (5) with lithium amalgam in ether.[51] (b) Similarly, treating 1-bromo-2-iodonaphthalene (6) with magnesium in ether gave a hydrocarbon, orange plates, m.p. 271°, in low yield (about 0.2%), which may be the hitherto unreported dibenzo[a,g]biphenylene (7).[49] (But see also Section D, 3.)

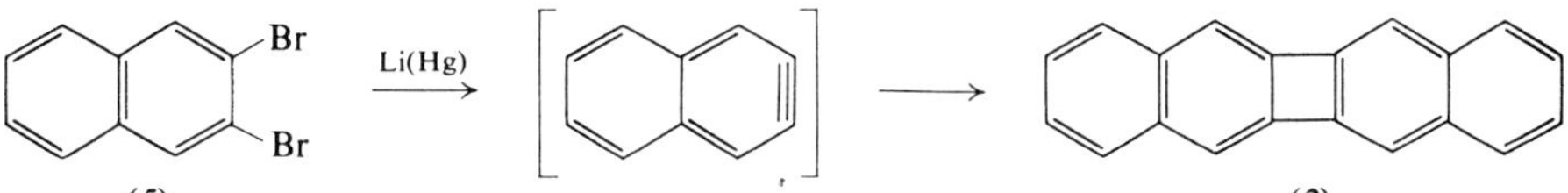

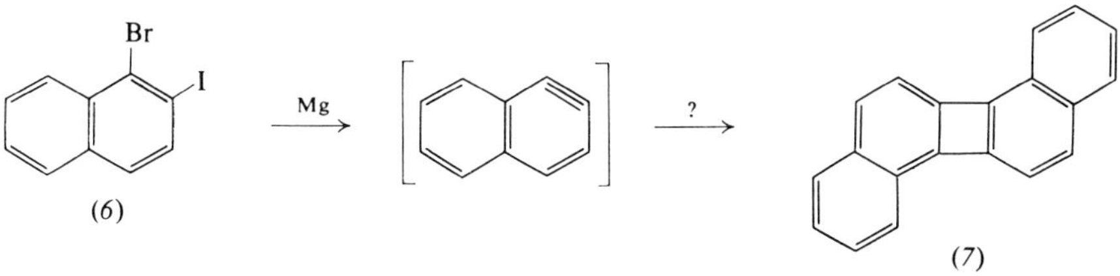

3. BY THE ULLMANN REACTION

Biphenylenes and benzobiphenylenes are usually not formed when o-dihaloaryls are subjected to the conditions of the Ullmann reaction (compare Chapter 10, Section B, 3). However, two instances are known in which benzobiphenylenes were produced in this way, viz., (a) that of dibenzo[b,h]biphenylene (2) in which the dibenzobiphenylene was produced in about 1% yield by treating either 2,3-diiodonaphthalene[47] or 2-bromo-3-iodonaphthalene (8)[51] with copper bronze in dimethylformamide, and (b) that of 5,11(or 6)-dinitrodibenzo[b,h]biphenylene [(11a) or (11b)] in which the dibenzobiphenylene was produced in low yield (1-3%) by treating either 2,3-dibromo-1-nitronaphthalene (9) or 3-bromo-2-iodo-1-nitronaphthalene (10) with copper bronze in dimethylformamide.[47] It is interesting to note that 2,3-dibromo-5-nitronaphthalene (12), an isomer of dibromide (9), failed to give a benzobiphenylene under similar reaction conditions.[48]

It has been suggested that the formation of biphenylenes and benzobiphenylenes from o-dihaloaryls proceeds through benzyne or naphthalyne intermediates, but adequate proof is lacking (see Section B, 2 and Chapter 10, Sections B, 2-3). An attempt to trap a naphthalyne intermediate with furan in

---

* Calculated from weight data recorded in ref. 51.

the conversion of 2,3-diiodonaphthalene into dibenzo[b,h]biphenylene (see above) failed.[47]

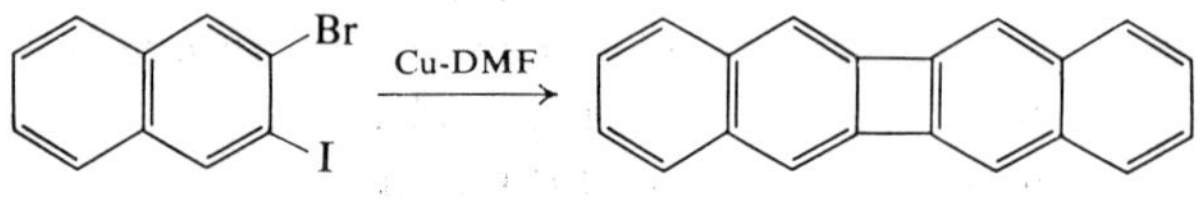

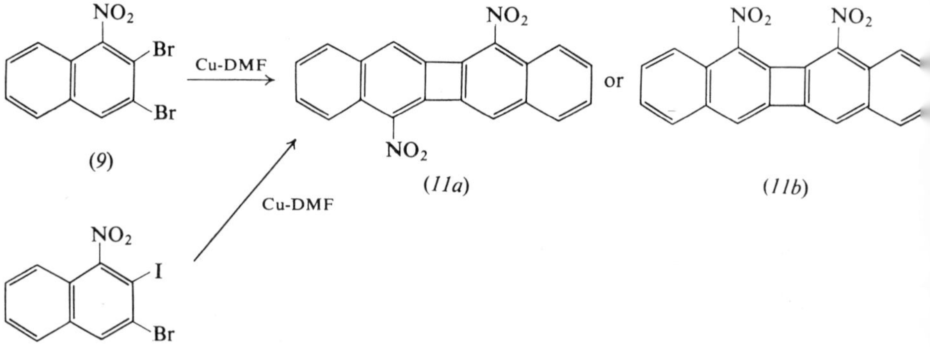

4. From Organometallic Biaryls

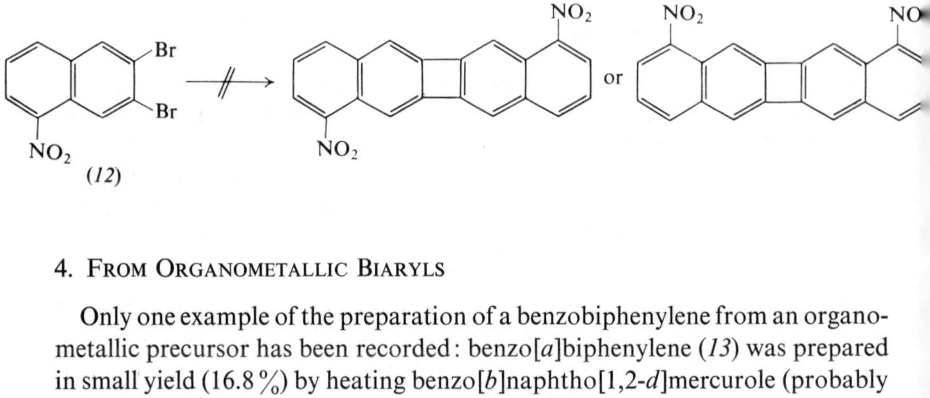

Only one example of the preparation of a benzobiphenylene from an organometallic precursor has been recorded: benzo[a]biphenylene (*13*) was prepared in small yield (16.8%) by heating benzo[b]naphtho[1,2-d]mercurole (probably in the form of a tetramer) with silver powder.[28a]

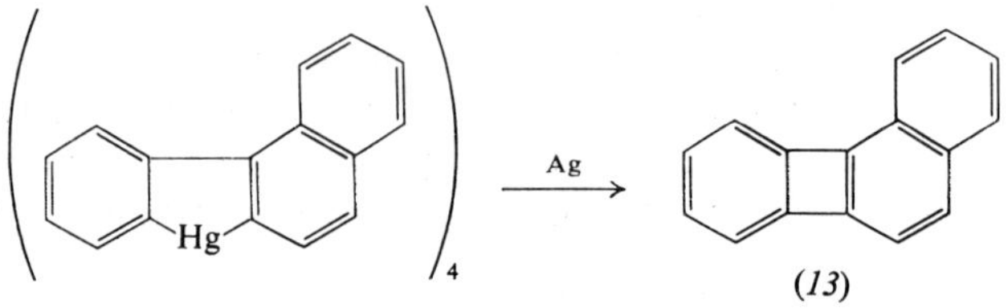

5. VIA REDUCED BENZOBIPHENYLENES

It was pointed out earlier (Chapter 10, Section B, 5) that the preparation of biphenylenes from reduced biphenylenes is without practical value at the present time because the latter are largely inaccessible. In the case of many benzo- and dibenzobiphenylenes, however, the reduced compounds required as starting materials are readily available in the form of adducts and dimers of benzocyclobutadiene, naphthocyclobutadiene, naphthoquinone, etc. Thus, (a) 6a,10b-dihydrobenzo[*a*]biphenylene (*15*), prepared from 1,2-dihalobenzocyclobutenes (*14a*) or (*14b*) via benzocyclobutadiene was aromatized with *N*-bromosuccinimide to give benzo[*a*]biphenylene (*13*).[21] Similarly, (b) 5-10-diphenyl-5,10-oxido-5,5,a9b,10-tetrahydrobenzo[*b*]biphenylene (*16*), prepared in good yield by generating benzocyclobutadiene in the presence of diphenylisobenzofuran,[23] was readily dehydrated with acetic anhydride and a trace of mineral acid to give a benzo[*b*]biphenylene derivative, 5,10-diphenylbenzo[*b*]biphenylene (*17*).[6] (c) 5,6a(or 10b)-Dibromo-6a,10b-dihydrobenzo[*a*]biphenylene (*18a*) and 5,6a(or 10b)-diiodo-6a,10b-dihydrobenzo[*a*]biphenylene (*18b*) undoubtedly are intermediates in the dehydrohalogenation of 1,2-dibromo- and 1,2-diiodobenzocyclobutene; the dihalodihydrobenzobiphenylenes are not isolable since they are aromatized by excess base *in situ* to give 5-bromo- (*19a*)[21,19,36] and 5-iodobenzo[*a*]biphenylene (*19b*).[27]

5,6-Dibromobenzo[*a*]biphenylene (*21*) was prepared by treating the reduced benzobiphenylene 5,5,6,6a,10b-pentabromobenzo[*a*]biphenylene (*20*) with potassium *tert*-butoxide.[8] The latter was itself prepared from 5-bromobenzo[*a*]biphenylene (*19a*) by taking advantage of the fact that the four-membered ring of the benzo[*a*]biphenylene system has considerable cyclobutadienoid character and tends to add bromine at the 6a,10b-positions.

5,6-Dibromobenzo[*a*]biphenylene (*21*) and 5,10-dibromobenzo[*b*]biphenylene (*27*) were both formed when α,α,α',α'-tetrabromo-*o*-xylene was treated with excess base.[9,39] Formation of the benzo[*a*] isomer undoubtedly proceeds by way of a reduced benzobiphenylene intermediate [(*22*) → (*26*) → (*21*)][20,39]; the origin of the benzo[*b*] isomer is obscure [(*29*) → (*30*) → (*28*) → (*27*)?],[20] but it does not arise through either 1,2-dibromobenzocyclobutadiene (*22*) or through tetrabromide (*23*), which is also formed in the dehydrohalogenation of tetrabromo-*o*-xylene. Indeed, the former (*22*) is known to give exclusively 5,6-dibromobenzo[*a*]biphenylene (*21*) when it is generated by the dehydrohalogenation of tribromobenzocyclobutene (*25*),[20] while the latter gives only 5,10-dibromoindeno[2,1-*a*]indene (*24*).[25] The conversion of tetrabromide (*23*) (formerly believed to be tetrabromodibenzotricyclooctadiene) into 5,10-dibromobenzo[*b*]biphenylene (*27*), reported by Jensen and Coleman[39] and confirmed by Baker *et al.*,[9] has been shown to be in error.[25]

(d) 5,6,11,12-Tetraphenyldibenzo[*b,h*]biphenylene (*34a*) was prepared in

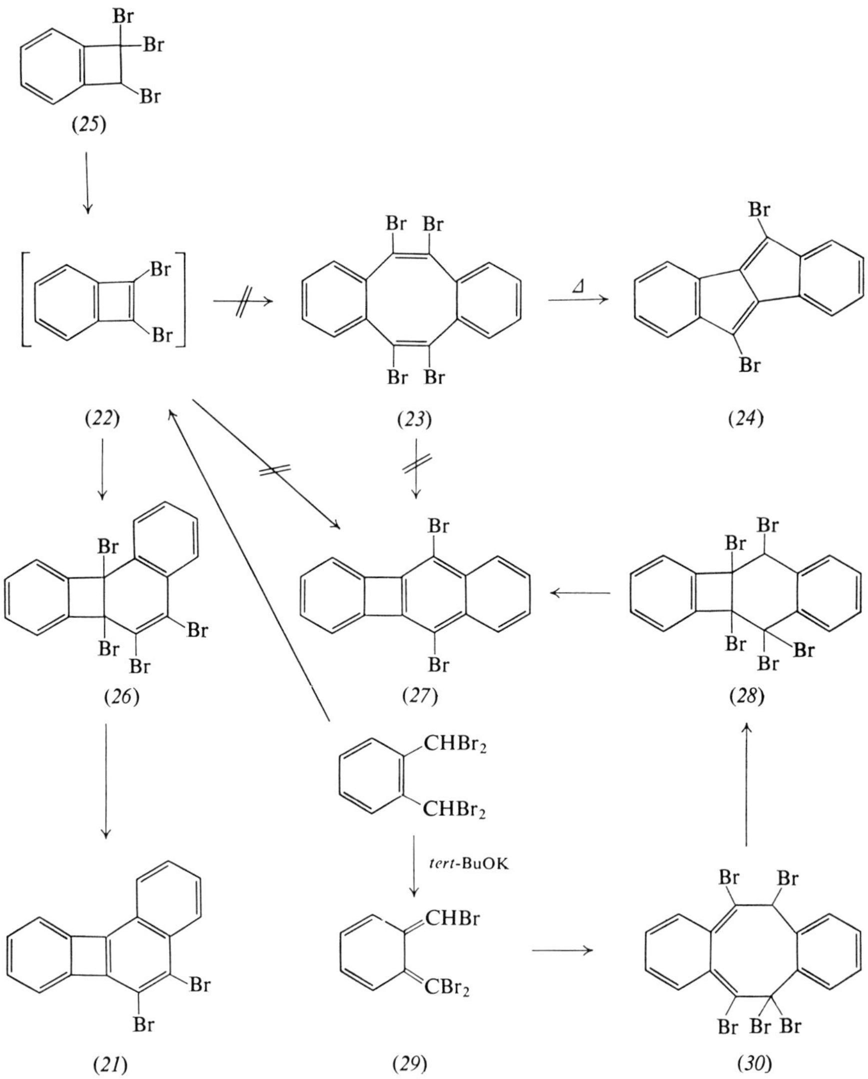

B. SYNTHESIS OF BENZO- AND DIBENZOBIPHENYLENES   327

88% yield by aromatization of the reduced compound 5,6,11,12-tetraphenyl-5,12-oxido-5,5a,11b,12-tetrahydrodibenzo[b,h]biphenylene (32), which was itself prepared by generating diphenylnaphthocyclobutadiene (31) in the presence of diphenylisobenzofuran.*,41 (e) Similarly, 5,6,11,12-tetrahydroxydi-

* For the generation of 3,8-diphenylnaphtho[b]cyclobutadiene by another method, see Chapter 9, Section B, 1, a.

TABLE 11.2

Benzo- and Dibenzobiphenylenes Prepared from (via) Reduced Benzo- and Dibenzobiphenylenes

Biphenylene	Reactants and reaction conditions[a]	Yield (%)	Ref.
Benzo[a]-	6a,10b-Dihydrobenzo[a]biphenylene and NBS	48	21
5-Bromobenzo[a]-	5-Bromo-6a,10b-dihydrobenzo[a]biphenylene and NBS	22[b,c]	21
	1,1-Dibromobenzocyclobutene and base (via 1-bromobenzocyclobutadiene)	75	19
	1,2-Dibromobenzocyclobutene (via 1-bromobenzocyclobutadiene)	86.1	29, 36
	5,6a,10b-Tribromo-6a,10b-dihydrobenzo[a]biphenylene and sodium iodide	95	20
5,6-Dibromobenzo[a]-	5,6-Dibromo-6a,10b-dihydrobenzo[a]biphenylene and NBS	52	20
	5,5,6a,10b-Pentabromo-5,6,6a,10b-tetrahydrobenzo[a]biphenylene (20) and base	—	9
	1,1,2,2-Tetrabromobenzocyclobutene and sodium iodide (via 1,2-dibromobenzocyclobutadiene)	41	20
	5,6,6a,10b-Tetrabromo-6a,10b-dihydrobenzo[a]biphenylene (26) and potassium tert-butoxide	61	20
	α,α,α',α'-Tetrabromo-o-xylene and potassium tert-butoxide (via 1,2-dibromobenzocyclobutadiene)	4	9
	1,1,2-Tribromobenzocyclobutene and base (via 1,2-dibromobenzocyclobutadiene)	85	20
5-Iodobenzo[a]-	1,2-Diiodobenzocyclobutene and potassium tert-butoxide (via 1-iodobenzocyclobutadiene)	71	27
Benzo[b]-	1,2-Dihydrobenzo[b]biphenylene and		
	(1) selenium	38	9
	(2) chloranil	15	9
	1,2,3,4-Tetrahydrobenzo[b]biphenylene and selenium	20	9
5,10-Dibromobenzo[b]-	Tetrabromo-o-xylene and excess base	—	9, 25, 39

Compound	Conditions		
5,10-Diphenylbenzo[b]-			
5,10-Diphenyl-5a,9b-dibromo-5,10-oxido-5,5a,9b,10-tetrahydrobenzo[b]biphenylene and			
	(1) zinc in ethanol[d]	63	23
	(2) lithium amalgam in ether[d]	31	23
5,10-Diphenyl-5a-iodo-5,10-oxido-5,5a,9b,10-tetrahydrobenzo[b]biphenylene and			
	(1) hydrochloric acid–acetic acid	97	23
	(2) potassium tert-butoxide in dimethylsulfoxide	74	23
5,10-Diphenyl-5,10-oxido-5,5a,9b,10-tetrahydrobenzo[b]biphenylene and hydrochloric acid in			
	(1) acetic anhydride	—	6
	(2) ethanol	97[b]	23
5,10-Diphenyl-5,10-oxido-5,5a,9b,10-tetrahydrobenzo[b]biphenylene and phosphoric acid		—	30
5,6,11,12-Tetracetoxydibenzo[b,h]-	5a,5b,11a,11b-Tetrahydrodibenzo[b,h]biphenylene-5,6,11,12-bis(quinone) and sodium acetate–acetic anhydride	81	14
5,6,11,12-Tetrahydroxydibenzo[b,h]-	5a,5b,11a,11b-Tetrahydrodibenzo[b,h]biphenylene-5,6,11,12-bis(quinone) and base	48	14
5,6,11,12-Tetramethoxydibenzo[b,h]-	5a,5b,11a,11b-Tetrahydrodibenzo[b,h]biphenylene-5,6,11,12-bis(quinone) and base followed by dimethyl sulfate	47	14
5,6,11,12-Tetraphenyldibenzo[b,h]-	5,6,11,12-Tetraphenyl-5,12-oxido-5,5a,11b,12-tetrahydrodibenzo[b,h]biphenylene (32) and hydrochloric acid–acetic acid	88	7, 41

[a] Italicized numbers in parentheses refer to the structural formulas in Section B, 4.
[b] Calculated from weight data recorded in the literature.
[c] Isolated as the TNF complex.
[d] Dehalogenation and "reduction" (removal of oxygen).

benzo[b,h]biphenylene (*34b*) was prepared from 5a,5b,11a,11b-tetrahydro-dibenzo[b,h]biphenylene-5,6,11,12-bis(quinone) (*33*),[14] which is the photodimer of 1,4-naphthoquinone.[46] (See also Table 11.2.)

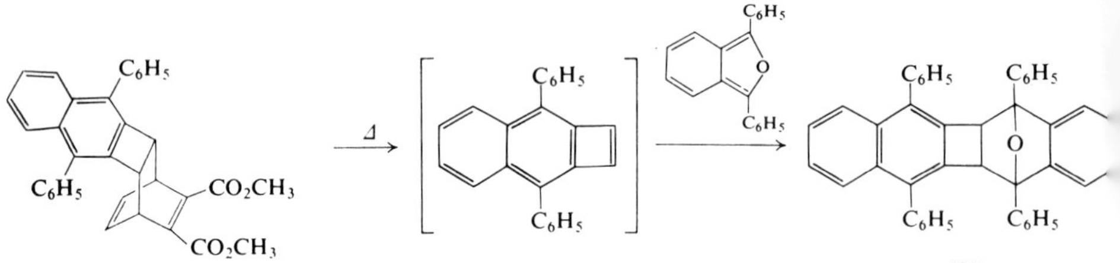

(*34a*): R = C$_6$H$_5$
(*34b*): R = OH

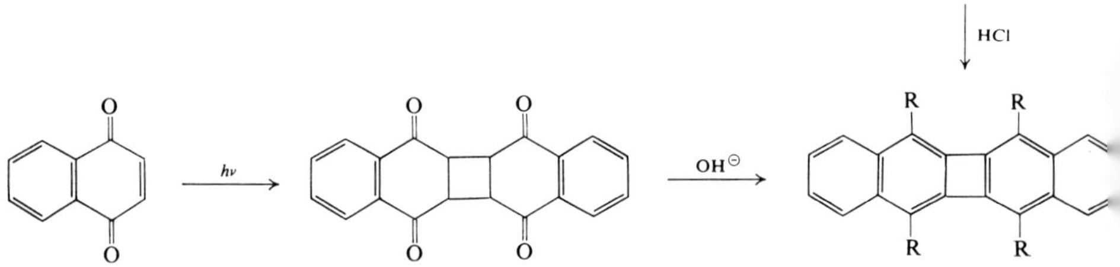

6. From Nonaromatic Precursors

No example of the preparation of a benzo- or a dibenzobiphenylene directly from a nonaromatic precursor has been reported. (Compare Chapter 10, Section B, 6.)

7. By Direct Substitution

No example of the preparation of a benzo- or a dibenzobiphenylene by direct substitution has been recorded. (But see 5-acetylbenzo[a]biphenylene and dinitrodibenzo[b,h]biphenylene in Table 11.4.)

8. By Transformation of Functional Groups

(a) A number of 5-substituted benzo[a]biphenylenes were prepared starting from the 5-bromo and 5-iodo derivatives, which are readily available by the dehydrohalogenation of 1,2-dihalobenzocyclobutenes (see also Section D,1,c).[27] (b) Similarly, both benzo[b]biphenylene and benzo[b]biphenylene-5,10-dicarboxylic acid were prepared from 5,10-dibromobenzo[b]biphenylene via the 5,10-dilithio derivative[39]; the dibromide is readily available by the dehydrohalogenation of α,α,α',α'-tetrabromo-*o*-xylene.[39] (c) Other than a

B. SYNTHESIS OF BENZO- AND DIBENZOBIPHENYLENES

TABLE II.5

SUBSTITUTED BENZO- AND DIBENZOBIPHENYLENES PREPARED BY TRANSFORMATION OF FUNCTIONAL GROUPS

Biphenylene	Reactants and reaction conditions	Yield (%)	Ref.
Benzo[a]-	5-Bromobenzo[a]biphenylene and n-butyllithium, followed by methanol	19.1[a]	29
5-Acetylbenzo[a]-	Benzo[a]biphenylene-5-carbonyl chloride and dimethylcadmium	93	27
Benzo[a]-	5-Cyanobenzo[a]biphenylene and methylmagnesium bromide	11[a]	27
-5-carbonyl chloride[b]	Benzo[a]biphenylene-5-carboxylic acid and thionyl chloride–pyridine	—	27
Benzo[a]- -5-carboxamide	Benzo[a]biphenylene-5-carbonyl chloride and ammonia	76	27
Benzo[a]- -5-carboxylic acid	5-Benzo[a]biphenylenylmagnesium bromide and carbon dioxide	39	27
	5-Bromobenzo[a]biphenylene and n-butyllithium, followed by carbon dioxide	37[c]	27
	5-Iodobenzo[a]biphenylene and n-butyllithium, followed by carbon dioxide	56.5[c]	27
5-Carbomethoxybenzo[a]-	Benzo[a]biphenylene-5-carbonyl chloride and methanol	94	27
5-Cyanobenzo[a]-	Benzo[a]biphenylene-5-carboxamide and thionyl chloride	70	27
	5-Iodobenzo[a]biphenylene and cuprous cyanide	84	27
5-Grignard (iodide)[b] of benzo[a]-	5-Iodobenzo[a]biphenylene	—	27
5-Lithiobenzo[a]-[b]	5-Bromobenzo[a]biphenylene and n-butyllithium	—	27, 29
	5-Iodobenzo[a]biphenylene	—	27
Benzo[b]-	5,10-Dibromobenzo[b]biphenylene and n-butyllithium, followed by methanol	79	31, 39
Benzo[b]- -5,10-dicarboxylic acid	5,10-Dibromobenzo[b]biphenylene and n-butyllithium, followed by carbon dioxide	57.4	31, 39
5-Bromobenzo[b]-[b]	5,10-Dibromobenzo[b]biphenylene and n-butyllithium, followed by methanol	67	31
5-Bromo-10-lithiobenzo[b]-[b]	5,10-Dibromobenzo[b]biphenylene and n-butyllithium	—	31
5,10-Dilithiobenzo[b]-	5,10-Dibromobenzo[b]biphenylene and n-butyllithium	—	31, 39
5,6,11,12-Tetraacetoxydibenzo[b,h]-	Tetrahydroxydibenzo[b,h]biphenylene	95[d]	14

[a] Obtained through the TNF complex.
[b] Not isolated.
[c] Based on 5-halobenzo[a]biphenylene.
[d] Based on the precursor of the tetrahydroxy compound, 5a,5b,11a,11b-tetrahydrodibenzo[b,h]biphenylene-5,6,11,12-bis(quinone).

TABLE 11.4

ATTEMPTED SYNTHESES OF BENZO- AND DIBENZOBIPHENYLENES

Expected biphenylene	Reactants and reaction conditions	Products obtained	Ref.
5-Acetylbenzo[a]-	Benzo[a]biphenylene (Friedel–Crafts)	Oil	27
Benzo[b]-	"5,1(=Diacetoxy-5,5a,9b,10-tetrahydrobenzo[b]biphenylene"	See refs. 5, 24	4
5,10-Dibromobenzo[b]-	"Tet'abromopentacyclohexadecahexaene" and		
	(1) sodium iodide	$C_{16}H_8Br_2$[a]	39
	(2) heat	$C_{16}H_8Br_2$[a]	9
Dibenzo[a,g]-	Dinaphth[1,2-b:1',2'-d]iodolium iodide and cuprous oxide	—	11, 50
Dibenzo[a,g(or i)]-	1-Bromo-2-fluoronaphthalene and lithium amalgam		
	(1) in ether	$C_{20}H_{24}Hg_2$[b] and "a polymer"	51
	(2) in tetrahydrofuran	2-Fluoronaphthalene	51
	1,2-Dibromonaphthalene and		
	(1) copper (double Ullmann)	"Tar"	11
	(2) lithium amalgam in ether	$C_{20}H_{24}Hg_2$[b]	51
	(3) lithium amalgam in tetrahydrofuran	2,2'-Binaphthyl	51
	(4) magnesium followed by cuprous chloride	—	11
	(5) naphthalene (Friedel–Crafts)	—	11
	(6) potassium	—	11
	(7) sodium	—	11
Dibenzo[a,g(or i)]-	"6a,6b,12a,12b-Tetrahydrodibenzo[a,g(or i)]biphenylene"	Benzo[j]fluoranthene	13, 42
	Tetrahydronaphthalene	Benzo[j]fluoranthene	35, 42
Dibenzo[a,h(or i)]-	2,2'-Binaphthyl-1-diazonium salt (Pschorr)	Dibenzocarbazole	11
	1-Bromo-2,2'-binaphthyl (Friedel–Crafts) "in various solvents"	—	11
Dibenzo[a(or b),h]-	2,3-Dibromonaphthalene and naphthalene (Friedel–Crafts)	—	11

B. SYNTHESIS OF BENZO- AND DIBENZOBIPHENYLENES   333

Dibenzo[a,i]-	1,1-Dibromo-2,2′-binaphthyl and		
	(1) copper (Ullmann)	—	11
	(2) sodium	—	11
Dibenzo[b,h]-	2-Bromo-3-iodonaphthalene and magnesium	S.M.[c]	49
	"Dibenzo[b,h]biphenylene-5,6,11,12-bis(quinone)" (triphthaloylbenzene)	2,3-Trinaphthylene	44, 45
	2,3-Dibromonaphthalene and		
	(1) copper (Ullmann)	—	11
	(2) magnesium followed by cuprous chloride	—	11
	(3) potassium	—	11
	(4) sodium	—	11
Dibenzo[b,h]--5,6,11,12-bis(quinone)	1,4-Naphthoquinone and pyridine–acetic anhydride	Triphthaloylbenzene	45
1,7(or 10)-Dinitrodibenzo[b,h]-	2,3-Dibromo-5-nitronaphthalene and copper bronze in dimethylformamide	—	48
Dinitrodibenzo[b,h]-	Dibenzo[b,h]biphenylene and nitric acid	—	48
Tetrabenzo-	9-Bromo-10-fluorophenanthrene and lithium amalgam	Polymer	1, 52
	9-Bromo-10-iodophenanthrene and		
	(1) magnesium in ether	9,9′-Biphenanthryl	1
	(2) copper bronze in dimethylformamide	—	1
	9-Bromophenanthrene and potassium amide in liquid ammonia	Tetrabenzophenazine and 9-phenanthrylamine	1
	9,10-Dibromophenanthrene and magnesium in		
	(1) ether	S.M.[c]	1
	(2) tetrahydrofuran	S.M.[c]	1
	9,10-Dichlorophenanthrene and magnesium in tetrahydrofuran	Hexabenzotriphenylene	1, 16

[a] 5,10-Dibromoindeno[2,1-a]indene.
[b] Dibenzo[a,h]phenomercurin.
[c] Starting material.

tetraacetate ester and a tetramethyl ether of 5,6,11,12-tetrahydroxydibenzo-[b,h]biphenylene,[14] no substituted dibenzobiphenylene has been synthesized by functional group transformation.

The chemistry of the above transformations is described in greater detail in Section D. (See also Table 11.3.)

C. Unsuccessful Approaches to the Benzobiphenylenes

The literature contains reports of some 28 unsuccessful attempts to prepare the benzo-, dibenzo-, and tetrabenzobiphenylenes. The largest number of these are attempted syntheses of dibenzo[a,g(or i)]biphenylene [(1a) or (1b)] by the benzyne (naphthalyne) dimerization method. Similarly, 9,10-phenanthryne, a potential precursor of tetrabenzobiphenylene, has been generated under a variety of conditions, but the tetrabenzo compound (2) was not formed.[1, 15, 52] Any prospective scheme for synthesizing a benzobiphenylene must take into account not only the suitability of the method to be employed, but the stability (or instability) of the product, as well. Thus, the failure of the attempts to prepare tetrabenzobiphenylene via 9,10-phenanthryne may be due either to the low reactivity ("coolness") of the phenanthryne[38] or to the instability of the tetrabenzo compound (see Table 11.4 and Chapter 12).[1]

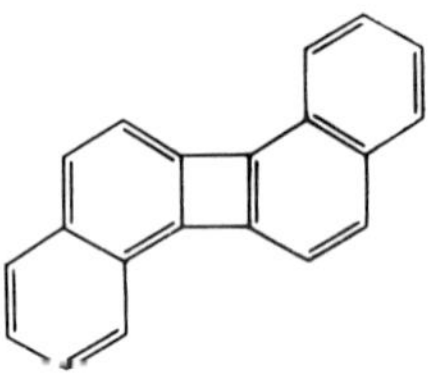

(1a) 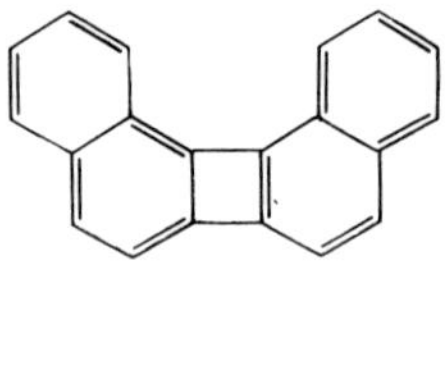 (1b) 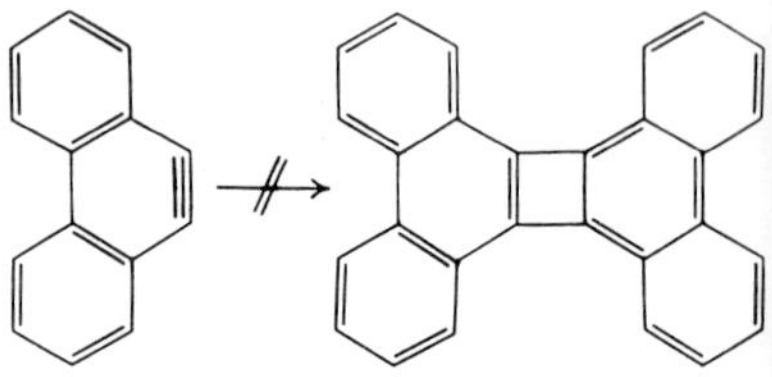 (2)

The reported transformation of a compound believed to be 1,2,5,6-tetrabromodibenzotricyclo[4.2.0.0^{2,5}]octadiene (3)* into 5,10-dibromobenzo[a]biphenylene (4)[9, 39] has been shown to be incorrect. The correct structure of the tetrabromide is that of 1,2,5,6-tetrabromodibenzo[a,e]cyclooctatetraene (5)† and that of the product is 5,10-dibromoindeno[2,1-a]indene (6).[25]

* Chemical Abstracts designation, 2,3,10,11-tetrabromopentacyclo[10.4.0.0^{2,11}.0^{3,10}.0^{4,9}]-hexadeca-4,6,8,12,14,16-hexaene.
† For details of the method of preparation and the proof of structure of (5), see Chapter 6, Section D, 2.

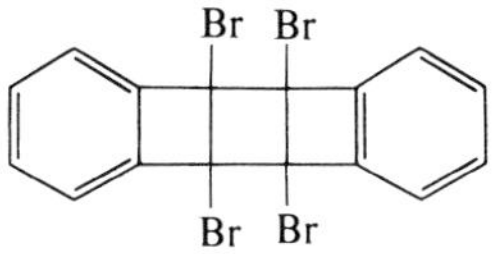

D. PROPERTIES OF THE BENZO- AND DIBENZOBIPHENYLENES 335

(3)

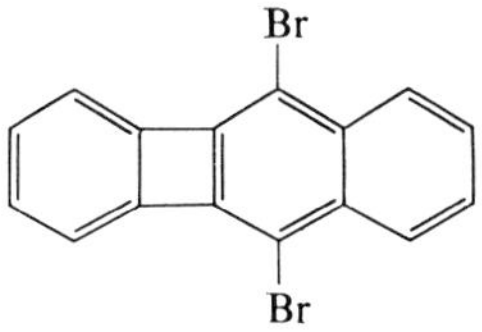

(4)

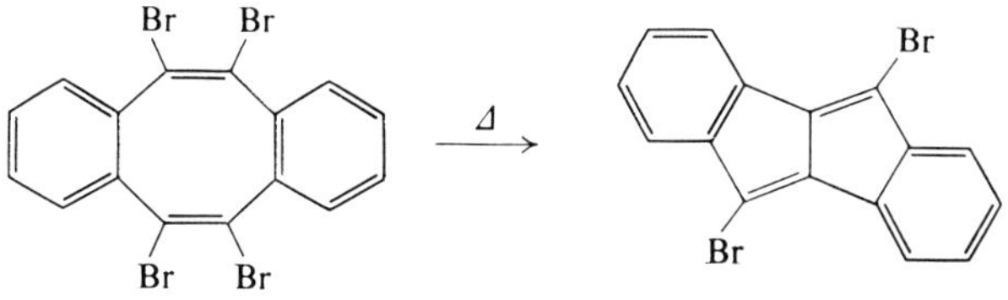

(5)

(6)

D. Chemical and Physical Properties of the Benzo- and Dibenzobiphenylenes

The striking disparity between the properties of dibenzo[a,i]- and dibenzo-[b,h]biphenylene first noted by Cava and Stucker[28] and later commented on by other workers[33a,34,49,51b] has been the impetus for four theoretical studies,[1,2,3,40] the results of which indicate that *linearly annellated benzo-biphenylenes, e.g., dibenzo[b,h]biphenylene, should be more stable than biphenylene itself and that angularly annellated benzobiphenylenes, e.g., dibenzo-[a,g(or i)]biphenylene, should be less stable than biphenylene*. These conclusions are in accord with what is known about the stability, reactivity, and ease of preparation of various benzobiphenylenes. For example, (a) dibenzo[b,h]-biphenylene (1) is an exceedingly stable substance (it sublimes unchanged at 350°)[33] whereas dibenzo[a,i]biphenylene (2) decomposes rapidly at 160° and is readily bleached by ultraviolet light.[28] (b) Biphenylene itself undergoes normal substitution at the 2-position when treated with bromine under carefully controlled conditions;[8] in contrast, two substituted benzo[a]biphenylenes, 5-bromo- and 5,6-dibromobenzo[a]biphenylene, (3a) and (5), were found to undergo *addition* of bromine, rather than substitution, to give 5,6a,10b-tribromo-6a,10b-dihydro-[20a] and 5,6,6a,10b-tetrabromo-6a,10b-dihydro-benzo[a]biphenylene,[20] (4) and (6), respectively.

(c) Biphenylene, substituted biphenylenes, and the linearly annellated benzobiphenylene, dibenzo[b,h]biphenylene, have all been prepared by the benzyne (or 2,3-naphthalyne) dimerization method; in contrast, of some eight attempts which have been made to prepare the two possible angularly annellated dibenzobiphenylenes (dibenzo[a,g(and i)]biphenylene) via 1,2-naphthalyne, none appears to have been successful (see Section D, 3). (d)

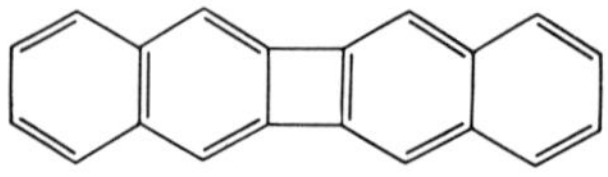

The linearly annellated biphenylene, 5,6,11,12-tetrahydroxydibenzo[b,h]biphenylene (7) is readily formed by aromatization of a reduced biphenylenebis(quinone);[14] in contrast, the aromatization of 2,3,6,7-tetramethyl-4a,4b,-8a,8b-tetrahydrobiphenylene-1,4,5,8-bis(quinone), does not proceed beyond the dihydro stage [(8) → (9)].[14]

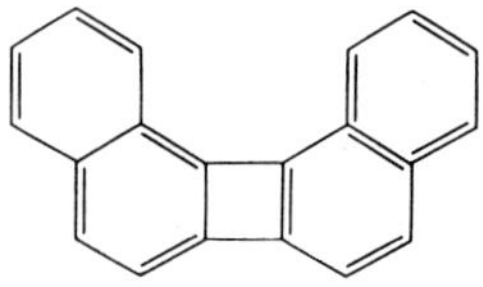

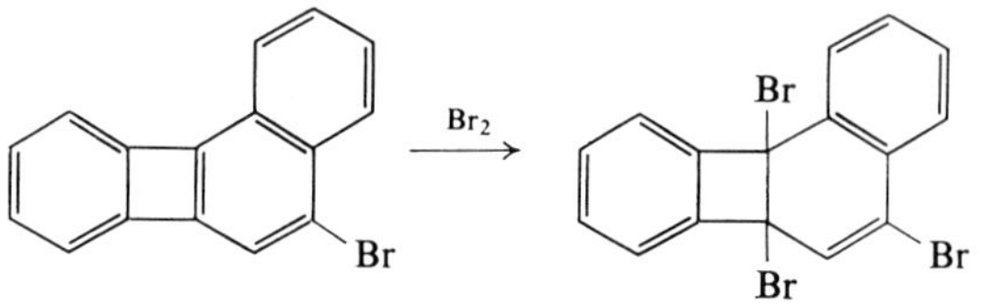

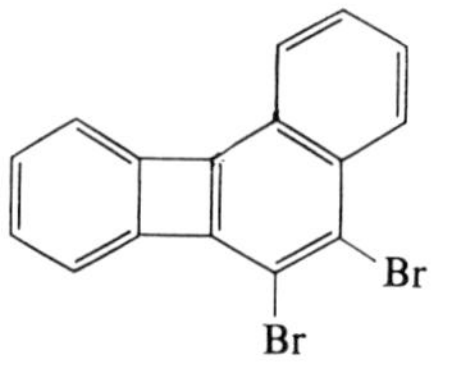

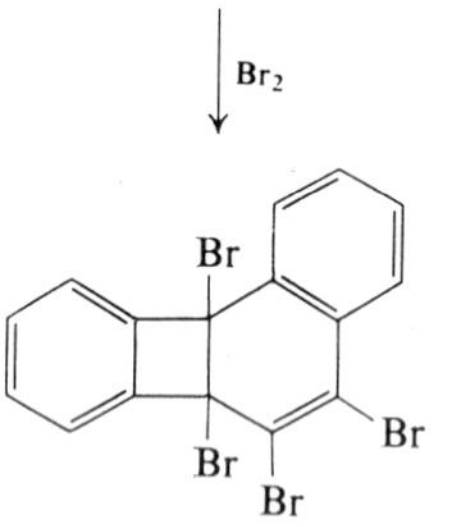

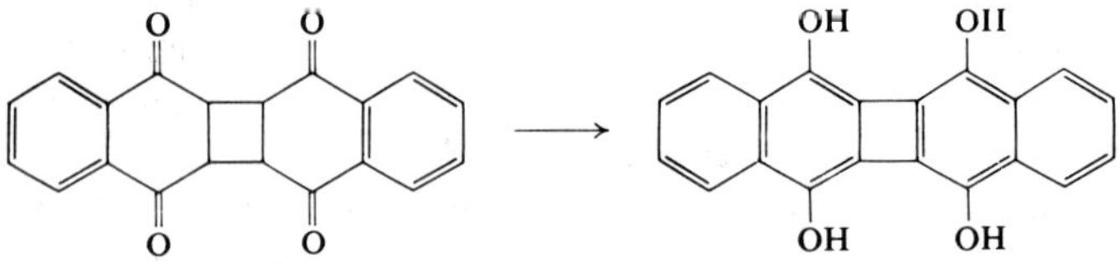

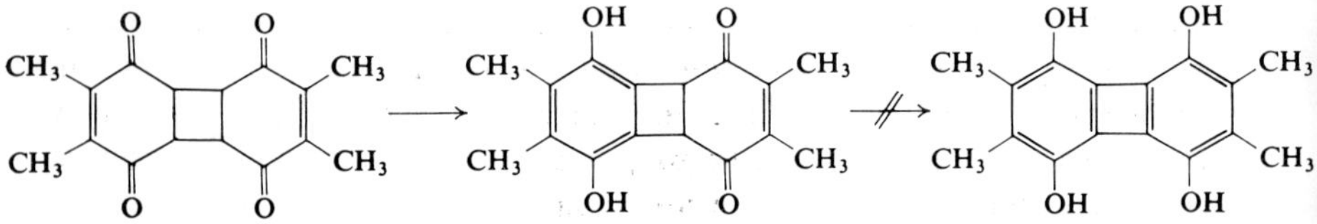

Similarly, (e) treatment of 5,6,11,12-tetraphenyl-5,12-oxido-5,5a,11b,12-tetrahydrodibenzo[b,h]biphenylene with hydrochloric acid gives the fully aromatic dibenzobiphenylene (10),[41] while the related (but not strictly analogous) biphenylene derivative, 1,4-oxido-1,4,4a,8b-tetrahydrobiphenylene (11), gives only a halogen-containing product of unknown structure under similar reaction conditions.[18]

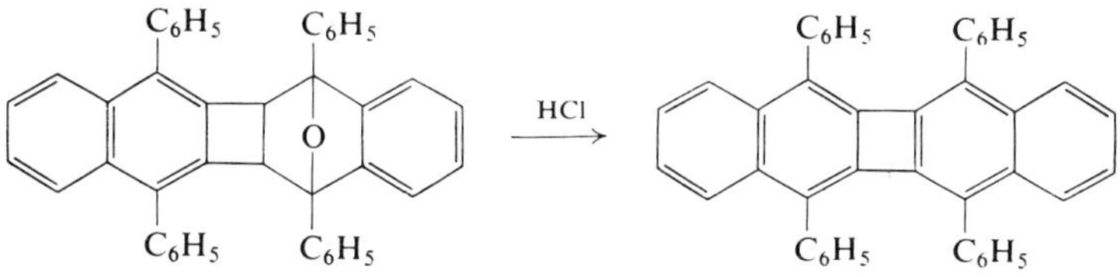

1. BENZO[a]BIPHENYLENE

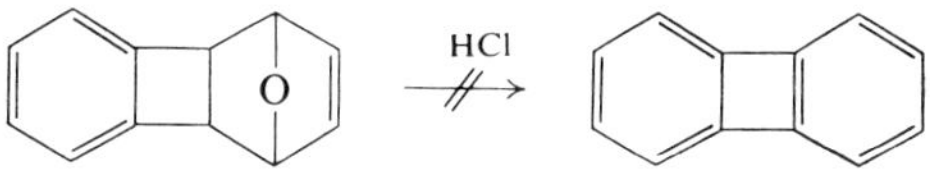

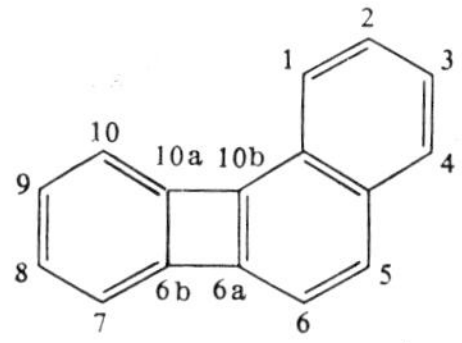

a. General Properties*

Benzo[a]biphenylene forms bright yellow needles, m.p. 72.0°–72.8°, having a characteristic roselike odor.[28] It is best isolated and purified as the sparingly soluble, black, 2,4,7-trinitrofluorenone derivative, m.p. 201.5°–202.5°. The hydrocarbon is moderately stable and can be kept indefinitely at room temperature in the absence of light; however, like dibenzo[a,i]biphenylene, it is slowly bleached by ultraviolet light,[28] possibly through a photochemical oxidative process.

The ultraviolet spectrum of benzo[a]biphenylene[28,33] (Fig. 11.1) resembles that of biphenylene, but the bands are shifted to longer wavelengths. The two spectra are alike in that each is made up of two distinct groups of absorption

---
* For a discussion of the polarographic behavior of benzo[a]biphenylene, see Chapter 10, Section F, 4.

11. THE BENZOBIPHENYLENES

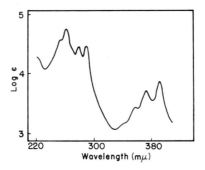

FIG. 11.1. Ultraviolet absorption spectrum of benzo[a]biphenylene (in ethanol).[33a]

bands; they differ in that the shorter-wavelength bands are more clearly resolved in the spectrum of benzo[a]biphenylene, while the longer-wavelength bands are more clearly resolved in the spectrum of biphenylene. Values for the ultraviolet absorption maxima and extinction coefficients (log ϵ) have been recorded: λ_{max}^{EtOH} 254 (4.58), 262 (4.77), 279 (4.47), 291 (4.48), 359 (3.41), 375 (3.71) 393 mμ (3.87).[33a]

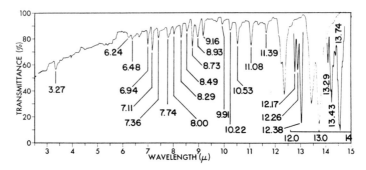

FIG. 11.2. Infrared spectrum (potassium bromide disc) of benzo[a]biphenylene, calibrated against polystyrene.

Infrared spectral values have been recorded for a carbon disulfide solution of benzo[a]biphenylene[33a]: $\lambda_{max}^{CS_2}$ 3.24, 3.43, 3.64, 7.12, 7.37, 7.76, 8.02, 8.33, 8.55, 8.76, 8.98, 9.18, 9.96, 10.28, 10.57, 11.07, 11.44, and 11.74 mμ (all of medium or weak intensity); 12.38, 13.41, 13.64 μ (all of high intensity). The infrared spectrum (potassium bromide disc) is shown in Fig. 11.2.

All of the known benzo[a]biphenylenes are more or less highly colored. The color ranges from lemon yellow in the parent hydrocarbon to bright red in benzo[a]biphenylene-5-carboxylic acid.

b. Methods of Synthesis

Benzo[a]biphenylene (13) was prepared in good yield (63%) from 1-(2'-iodophenyl)-2-iodonaphthalene (12) by Lothrop's cuprous oxide method.[33] [The iodonaphthalene derivative was obtained from 1-(2'-aminophenyl)-2-naphthylamine by the Sandmeyer reaction.] When the corresponding iodonium iodide (14) was used in place of the diiodo compound, the yield of benzo[a]biphenylene decreased to 16%.[33] The hydrocarbon was also synthesized by heating benzo[b]naphtho[1,2-d]mercurole (15) with silver powder.[33] [The mercurole (probably a tetramer) was prepared through the dilithio derivative.]

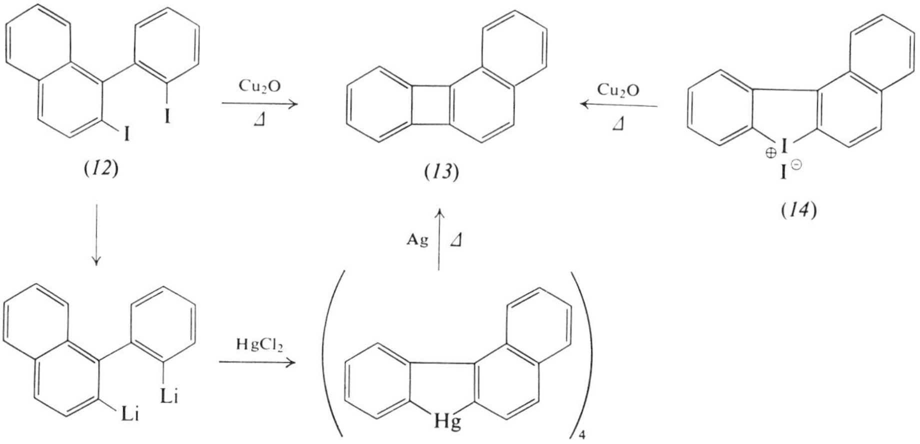

In addition, benzo[a]biphenylene was prepared by the protonation of two organometallic compounds (16) and (17), derived from 5-bromo- and 5-iodobenzo[a]biphenylene (3a) and (3b), which are readily available from 1,2-dibromo- and 1,2-diiodobenzocyclobutene (18a) and (18b).[27, 29] Aromatization of a reduced benzo[a]biphenylene (19), obtained in good yield by the dehalogenation of 1,2-dibromo- and 1,2-diiodobenzocyclobutene, also gave benzo[a]biphenylene.[21]

c. Substituted Benzo[a]biphenylenes

The direct introduction of a substituent into the benzo[a]biphenylene nucleus has not been reported, and no prediction of the site of maximum reactivity has been made. However, it is to be expected that direct substitution may be complicated by a tendency, already discernible in substituted benzo[a]biphenylenes, to *add* reagents at the 6a,10b-bond (see Section D, 1, d). The few derivatives of benzo[a]biphenylene that are known are all substituted at the 5- or the 5,6-positions as a consequence of their mode of preparation, i.e., by dimerization of 1-substituted or 1,2-disubstituted benzocyclobutadienes or by the simple transformation of the dimers. Thus, 5-bromo- and 5-iodobenzo[a]biphenylene were prepared from 1,2-dibromo-[29,36] and 1,2-diiodobenzocyclobutene[27] via the corresponding benzocyclobutadienes (see above). Similarly, 5,6-dibromo-benzo[a]biphenylene (5) was obtained as the major product, via 1,2-dibromo-benzocyclobutadiene, when 1,1,2-tribromobenzocyclobutene was dehydrobrominated with potassium tert-butoxide[20] and when 1,1,2,2-tetrabromobenzocyclobutene was dehalogenated with sodium iodide.[20] In still another synthesis, the dibromide (5) was obtained through the same intermediate, 1,2-

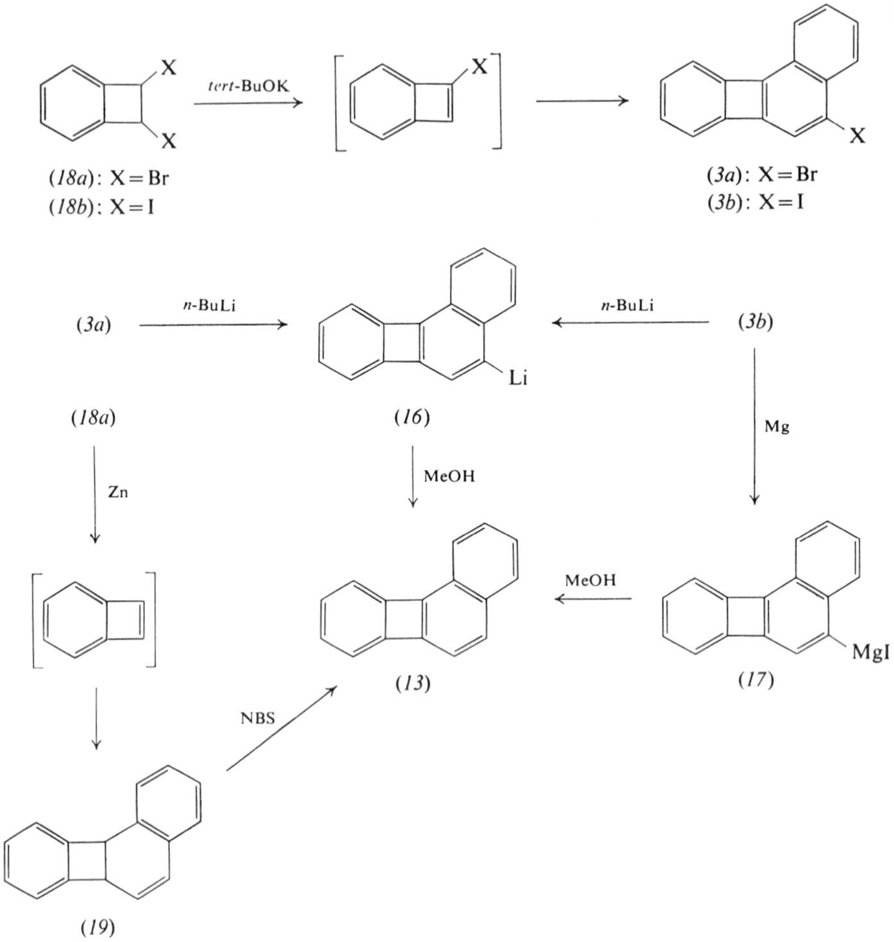

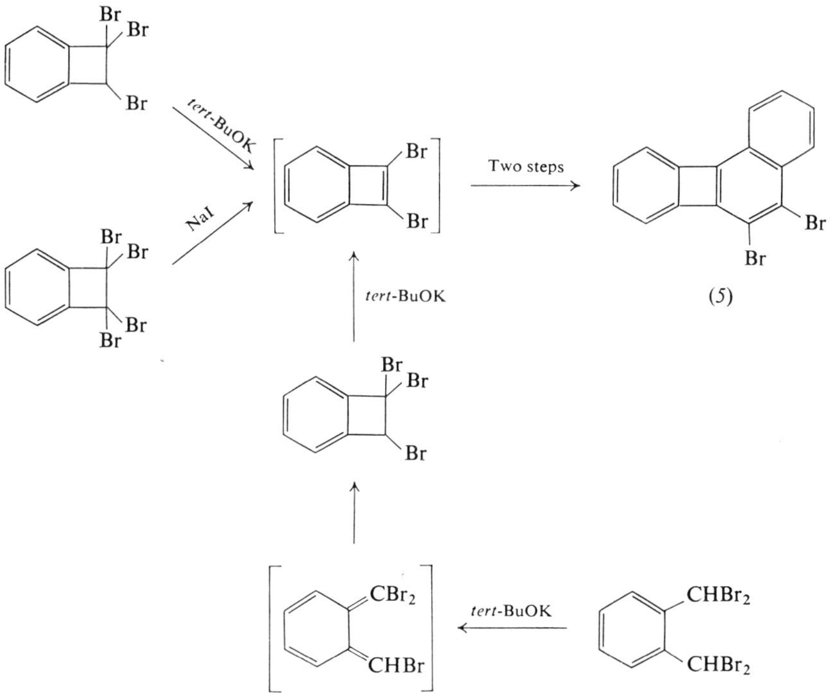

dibromobenzocyclobutadiene, but in lower yield (4%), when α,α,α′,α′-tetrabromo-o-xylene was treated with potassium tert-butoxide.[9]

The structure of the dibromide was confirmed by an unambiguous synthesis starting from 6a,10b-dihydrobenzo[a]biphenylene [(19) → (20) → (5)].[20] One of the intermediates in this sequence, 5-bromo-6a,10b-dihydrobenzo[a]-biphenylene (20) was aromatized with N-bromosuccinimide to give 5-bromobenzo[a]biphenylene (3a),[20] thereby confirming the structure of the product resulting from the dehydrohalogenation of 1,2-dibromobenzocyclobutene.[21]

The products formed by addition of bromine to the 6a,10b-bond of benzo[a]-biphenylene are easily debrominated to the fully aromatic precursors. Thus, tribromide (4), which is formed by the addition of bromine to 5-bromobenzo-[a]biphenylene (3a),[20a] is reconverted into (3a) in good yield (95%) by treatment with sodium iodide.[20] Tetrabromide (6) is *dehalogenated* rapidly by potassium tert-butoxide to give 5,6-dibromobenzo[a]biphenylene (5)[20]; it has been shown that elimination takes place by initial attack of alkoxide ion on one of the bromine atoms of the 6a,10b-dibromide system and that tetrabromide (6) does not exist in equilibrium with (5) in solution.[20] The direct conversion of pentabromide (21) into the aromatic dibromide (5) with potassium tert-butoxide undoubtedly proceeds via intermediate (6).[9]

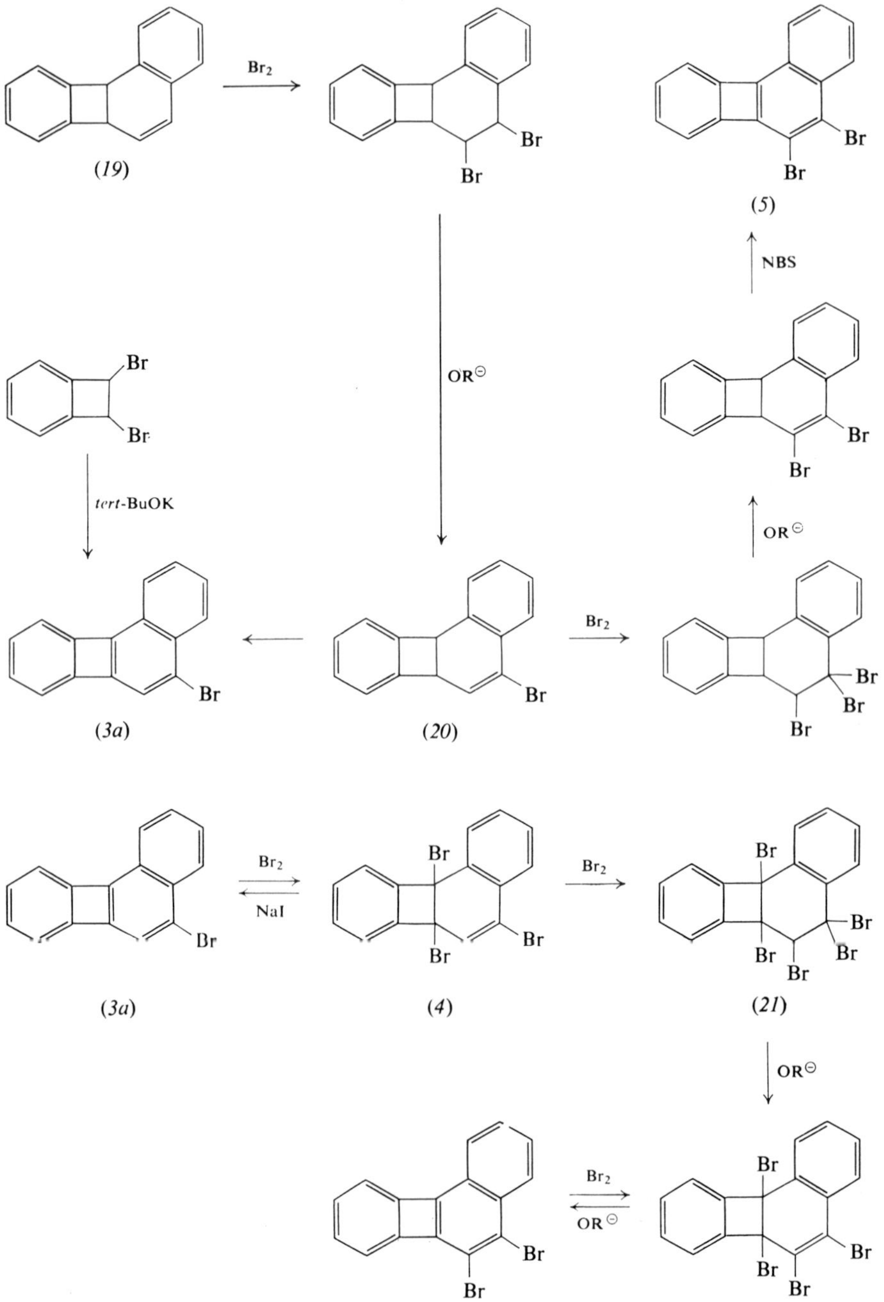

11. THE BENZOBIPHENYLENES

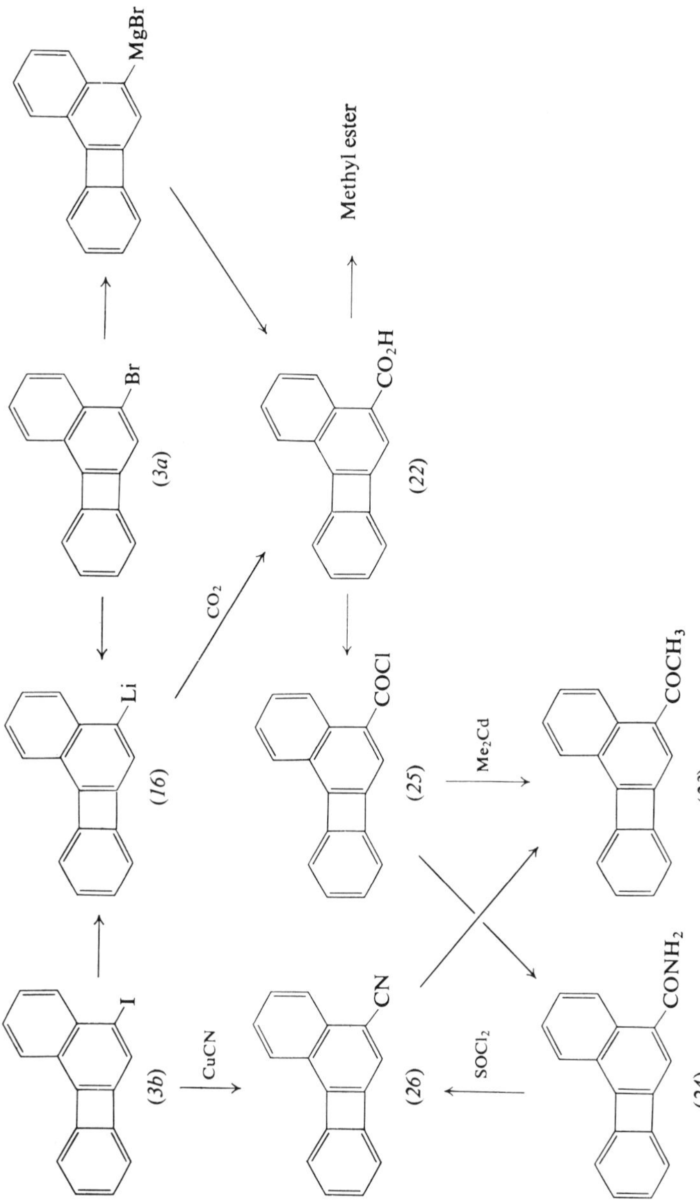
MgBr
Br
(3a)
Li
(16)
I
(3b)
CO2
CO2H
(22)
Methyl ester
COCl
(25)
CN
(26)
CuCN
Me2Cd
COCH3
(23)
CONH2
(24)
SOCl2

The ready accessibility of 5-bromo- and 5-iodobenzo[a]biphenylene has permitted the preparation of a number of transformation products. Thus, 5-lithiobenzo[a]biphenylene (16), prepared by treating 5-bromo- or 5-iodobenzo[a]biphenylene [(3a) or (3b)] with n-butyllithium, was carbonated to give benzo[a]biphenylene-5-carboxylic acid (22) in good yield (57%, based on the bromide).[27] The acid was also obtained through the Grignard reagent (prepared from the bromide). The acid was converted into the methyl ester and also into the amide (24) via the acid chloride (25).[27] Dehydration of the amide with thionyl chloride afforded the nitrile (26); the latter was also prepared by treating 5-iodobenzo[a]biphenylene with cuprous cyanide in dimethylformamide.[27] 5-Acetylbenzo[a]biphenylene (23) was prepared by standard methods from both the nitrile (26) and the acid chloride (25).[27]

The infrared spectrum of 5-bromobenzo[a]biphenylene contains strong absorption bands at 13.42 and 13.63 μ[25] which are almost identical with the bands at 13.41 and 13.64 μ in the spectrum of the parent hydrocarbon.[28a] 5-Bromobenzo[a]biphenylene: $\lambda_{max}^{CS_2}$ 3.30, 7.01, 7.09, 7.37, 8.02, 8.37, 9.09, 9.82, 10.25, 10.44, 11.00, 11.48, 11.55, and 12.08 μ (all of low or medium intensity).[29]

Ultraviolet absorption data for six substituted benzo[a]biphenylenes are recorded in Table 11.5.

d. Reactions of the Four-Membered Ring

Benzo[a]biphenylene behaves like biphenylene in undergoing reductive cleavage with Raney nickel. Both the 6a,6b- and the 10a,10b-bonds are susceptible to cleavage, with the result that a mixture of 1-phenyl- and 2-phenylnaphthalene is formed.[28]

Vigorous chromic acid oxidation of either benzo[a]biphenylene (13)[28a] or 5-bromobenzo[a]biphenylene (3a)[29] gives the dilactone of benzophenone-2,2'-dicarboxylic acid (28). A careful study of the oxidation of both 5-iodo- and 5-bromobenzo[a]biphenylene with chromic acid in one case and permanganate in another has shown that the initial attack of the oxidizing agent occurs at the central, cyclobutadienoid 6a,10b-bond; oxidative cleavage under controlled conditions yields an eight-membered diketone (27) which is further oxidized to the dilactone (28).[26] It is interesting to note that the oxidation of

TABLE 11.5
Ultraviolet Absorption Characteristics of Substituted Benzo[a]biphenylenes[a]

Benzo[a]biphenylene	Absorption maxima, mμ, and extinction coefficients (log ε)	Ref.
5-Acetyl-	246 (4.18); 267 (4.61); 297 (4.35)	27
5-Bromo-	225 (4.36); 264 (4.72); 283 (4.47); 294 (4.49); 381 (4.75); 400 (4.91);	29
	243 (4.65); 255 (4.61); 264 (4.73); 285 (4.44); 295 (4.45); 330 (2.84); 347 (3.08); 380 (3.75); 399 (3.89)	9
—5-Carboxylic acid	264 (4.69); 287.5 (4.30); 297 (4.42)	27
5-Cyano-	239 (4.43); 270 (4.73); 290 (4.42); 299.5 (4.42)	27
5,6-Dibromo-	229 (4.38); 266 (4.67); 292 (4.38); 304 (4.39); 384 (3.56); 402 (3.66)	9
5-Iodo-	226 (4.23); 266 (4.31)	27

[a] All spectra determined in ethanol.

5-bromobenzo[a]biphenylene to dilactone (28) was first recorded by Finkelstein in 1910,[36] and that the structures of the bromide (3a) and the neutral oxidation intermediate (27) were not elucidated until many years later.[26]

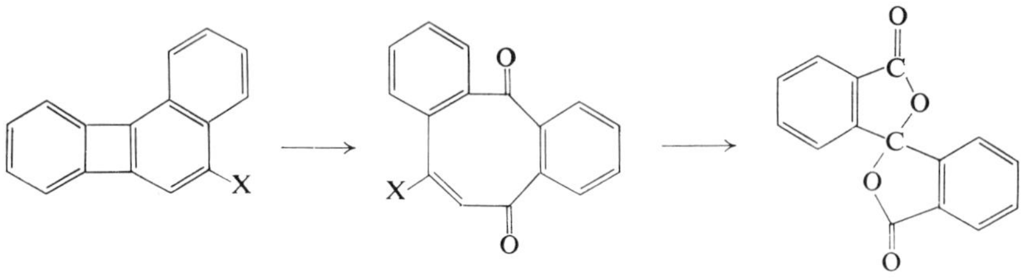

(3a): X=Br (27) (28)
(13): X=H

The cyclobutadienoid 6a,10b-bond of the benzo[a]biphenylene system adds bromine very easily and without cleavage. The reaction is reminiscent of the addition of bromine to the 9,10-positions of anthracene and phenanthrene and implies that the 6a,10b-bond may be the preferred site of attack by electrophilic reagents in general. A report to the contrary notwithstanding,[29] 5-bromobenzo[a]biphenylene (3a) has been shown to add bromine at 0° to give the pale yellow tribromide, 5,6a,10b-tribromo-6a,10b-dihydrobenzo[a]biphenylene (4) in 65% yield;[20] excess bromine gives 5,5,6,6a,10b-pentabromo-5,6,6a,10b-tetrahydrobenzo[a]biphenylene (21).[9] Similarly, 5,6-dibromobenzo[a]biphenylene (5) adds one equivalent of bromine rapidly in boiling chloroform to give 5,6,6a,10b-tetrabromo-6a,10b-dihydrobenzo[a]biphenylene (4) in 68% yield.[20]

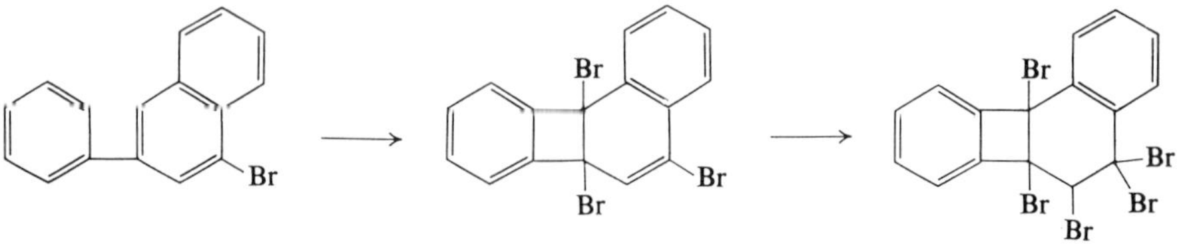

(3a) (4) (21)

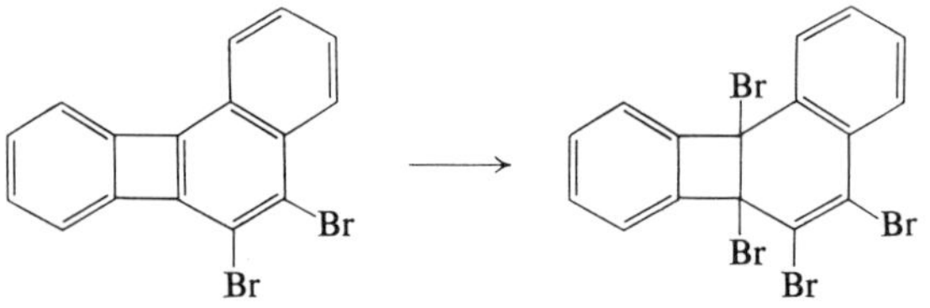

(5) (4)

2. BENZO[b]BIPHENYLENE

[structure diagram of benzo[b]biphenylene with numbered positions 1-10, 5a, 5b, 9a, 9b]

a. General Properties

Benzo[b]biphenylene, a pale yellow crystalline compound, m.p. 242.6°–243.2°,[39] forms a red 2,4,7-trinitrofluorenone derivative, m.p. 214°–216°.[9] The ultraviolet spectrum of the hydrocarbon (Fig. 11.3) is similar in some aspects to that of the benzo[a] isomer (see Fig. 11.1). Absorption maxima and extinction coefficients (log ϵ) for benzo[b]biphenylene have been recorded by Coleman: λ_{max}^{MeOH} 262 (4.92), 283 (4.38), 294 (4.55), 330 (3.45), 347 (3.57), 364 (3.76), 385 mμ (3.83)[31]; and by Baker et al.: λ_{max}^{EtOH} 255 (4.87), 264 (4.89), 285 (4.40), 296 (4.55), 331 (3.68), 348 (3.71), 367 (3.83), 386 mμ (3.78).[9]

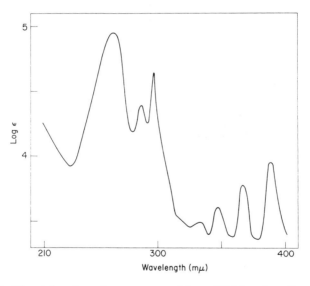

FIG. 11.3. Ultraviolet absorption spectrum of benzo[b]biphenylene (in ethanol).[17]

The infrared spectrum shows bands at 742 and 880 cm^{-1} (13.48 and 11.37 μ) characteristic of a 1,2-disubstituted and a 1,2,4,5-tetrasubstituted benzene ring[9] (Fig. 11.4).

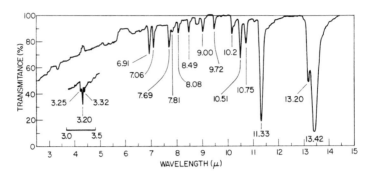

FIG. 11.4. Infrared spectrum of benzo[b]biphenylene (potassium bromide disc), calibrated against polystyrene.

b. *Methods of Synthesis*

Benzo[b]biphenylene (*31*) was prepared in good yield (79%) from 5,10-dibromobenzo[b]biphenylene (*30*); the latter is readily available through an unusual transformation of α,α,α',α'-tetrabromo-*o*-xylene.[39] Thus, treating α,α,α',α'-tetrabromo-*o*-xylene with potassium *tert*-butoxide in *tert*-butanol gave dibromide (*30*) along with tetrabromodibenzocyclooctatetraene (*29*) and 5,6-dibromobenzo[a]biphenylene (*5*).[39] Dibromide (*30*) was then treated with *n*-butyllithium to give the dilithio derivative which was protonated with methanol to give the desired hydrocarbon (*31*).[39]

Benzo[b]biphenylene (*31*) was also prepared by a multi-step reaction sequence starting from biphenylene.[9] Thus, biphenylene (*32*) was converted via 2-(β-carbomethoxypropionyl)biphenylene, 2-(γ-carboxypropyl)biphenylene, and 1-keto-1,2,3,4-tetrahydrobenzo[b]biphenylene into a mixture of 1,2-dihydrobenzo[b]biphenylene (*33*) and 1,2,3,4-tetrahydrobenzo[b]biphenylene (*34*). Both compounds (*33*) and (*34*) gave benzo[b]biphenylene (*31*) when they were dehydrogenated with selenium.[9]

c. *Substituted Benzo[b]biphenylenes*

Direct substitution in the benzo[b]biphenylene nucleus has not been reported and no prediction of the sites of maximum reactivity has been made. The few derivatives of benzo[b]biphenylene which are known are all substituted in the 5,10-positions as a consequence of their mode of preparation (see Section B, 5). Thus, 5,10-dibromobenzo[b]biphenylene was prepared in good yield by an unusual transformation of α,α,α',α'-tetrabromo-*o*-xylene*,[39] (see the preceding section), and was transformed into two other substituted benzo[b]biphenylenes:

* The transformation of tetrabromodibenzocyclooctatetraene (*29*) into 5,10-dibromobenzo[b]biphenylene reported in refs. 9 and 39 is erroneous. (See Section C and ref. 25.)

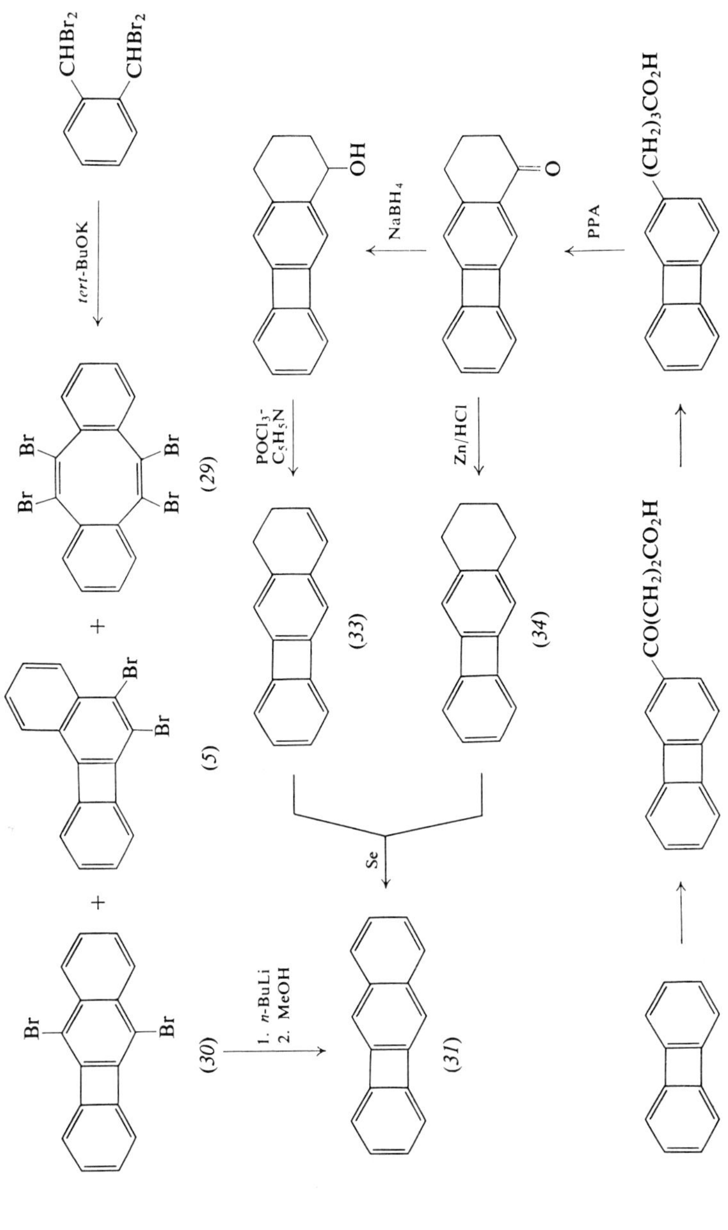

D. PROPERTIES OF THE BENZO- AND DIBENZOBIPHENYLENES

(a) metal exchange with one equivalent of n-butyllithium gave 5-bromo-10-lithiobenzo[b]biphenylene which was protonated with methanol to give 5-bromobenzo[b]biphenylene (35)[44]; and (b) metal exchange with two equivalents of n-butyllithium gave the dilithio compound which was carbonated with carbon dioxide to give benzo[b]biphenylene-5,10-dicarboxylic acid (36) in 57% yield.[39]

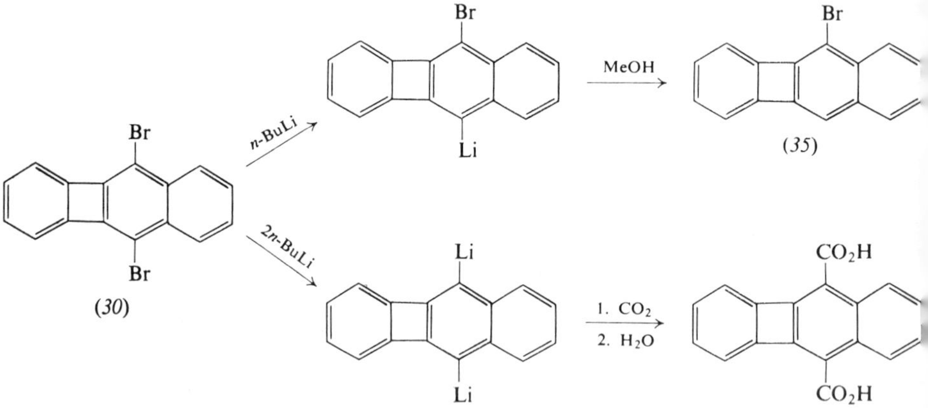

Another substituted benzo[b]biphenylene, 5,10-diphenylbenzo[b]biphenylene (38), was synthesized in good yield by the dehydration of 5,10-diphenyl-5,10-oxido-5,5a,9b,10-tetrahydrobenzo[b]biphenylene (37a) with hydrochloric acid in ethanol (93% yield)[23] or in acetic anhydride.[6] The epoxide (37a) was obtained by generating benzocyclobutadiene in the presence of 1,3-diphenylisobenzofuran. Dehydration of adduct (37a) with polyphosphoric acid gave a mixture of biphenylene (38) and a rearrangement product, diphenyldibenzopentalene (39); the biphenylene itself (38) was not rearranged by polyphosphoric acid.[30]

5,10-Diphenylbenzo[b]biphenylene (38) was also prepared from two halogen-substituted analogs of (37), viz., 5,10-diphenyl-5a-iodo- (37b) and 5,10-diphenyl-5a,9b-dibromo-5,10-oxido-5,5a,9b,10-tetrahydrobenzo[b]biphenylene (37c).[23] (See Table 11.2.)

Coleman has recorded the ultraviolet absorption maxima and extinction coefficients (log ϵ) of 5-bromobenzo[b]biphenylene: λ_{max}^{MeOH} 270 (4.92), 286 (4.32), 296 (4.66), 332 (3.18), 350 (3.35), 368 (3.58), 388 mμ (3.70);[31] and of 5,10-dibromobenzo[b]biphenylene: λ_{max}^{MeOH} 272 (4.89), 285 (4.53), 296 (4.69), 332 (3.52), 351 (3.54), 370 (3.63), 391 mμ (3.70).[31]

D. PROPERTIES OF THE BENZO- AND DIBENZOBIPHENYLENES

(37a): X = Y = H
(37b): X = I, Y = H
(37c): X = Y = Br

(38)

(39)

(37a) —PPA→ (38) + (39)

The ultraviolet spectrum of 5,10-diphenylbenzo[b]biphenylene (Fig. 11.5) does not show the separation into two systems of bands usually found in unsubstituted benzobiphenylenes.[6]

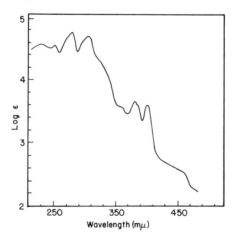

FIG. 11.5. Ultraviolet absorption spectrum of 5,10-diphenylbenzo[b]biphenylene (in ethanol).[6]

d. Reactions of the Four-Membered Ring

Benzo[b]biphenylene is cleaved by Raney nickel to give 2-phenylnaphthalene in 95% yield.[39] Similarly, 5,10-dibromobenzo[b]biphenylene (30) undergoes hydrogenolysis of both the four-membered ring and the halogen substituents in the presence of palladium-on-charcoal and triethylamine.[39] In contrast, 5,10-diphenylbenzo[b]biphenylene (*38*) is cleaved with difficulty, and the product is a mixture of 1,2,4-triphenylnaphthalene and partly reduced 1,2,4-triphenylnaphthalenes.[23] The two phenyl substituents are probably

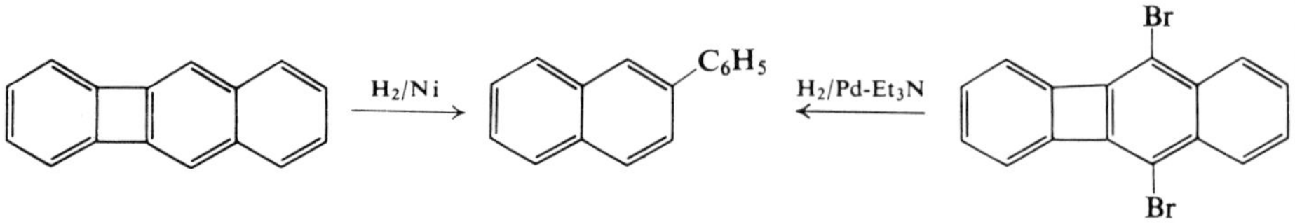

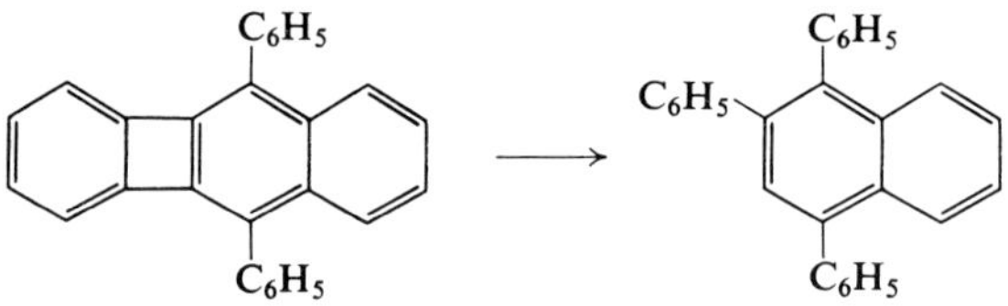

responsible for this behavior, since they are not coplanar with the rest of the molecule and probably hinder saturation of the 5a,5b-bond on the surface of the catalyst.

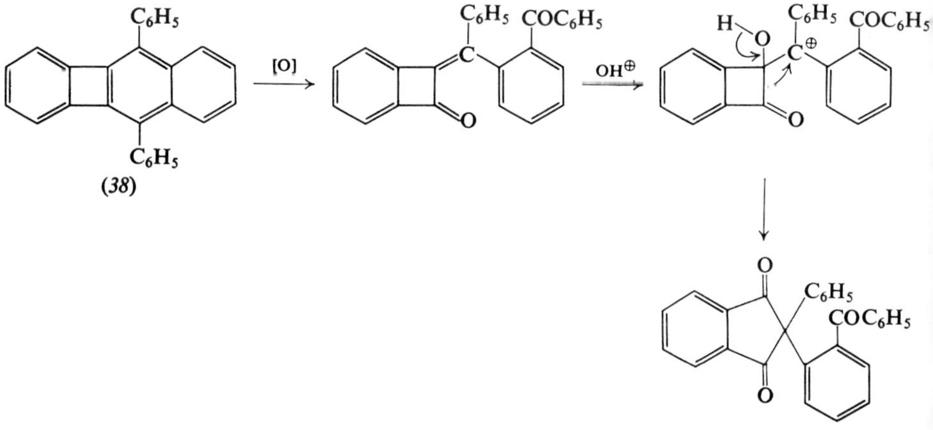

Oxidation of benzo[b]biphenylene has not been reported; oxidation of 5,10-diphenylbenzo[b]biphenylene (*38*) with sodium dichromate in sulfuric acid gave a substituted 1,3-indanedione (*40*).[23] Formation of the latter points to an initial attack by the oxidant on the 5,5a-bond; further attack by the oxidant (formally represented by "OH$^⊕$" in the chart below), followed by an unexceptional Wagner-Meerwein rearrangement, leads to the indanedione (40).

3. DIBENZO[a,g]BIPHENYLENE

A number of unsuccessful attempts have been made to synthesize either dibenzo[a,g]- or dibenzo[a,i]biphenylene by the dimerization of 1,2-naphthalyne (generated from 1,2-dihalonaphthalenes). (See Table 11.4.)

One of these attempts (the reaction of 1-bromo-2-iodonaphthalene with magnesium in ether) gave a hydrocarbon, m.p. 271°, in very low yield (about 0.2%), which crystallizes as orange plates and affords a red 2,4,7-trinitrofluorenone derivative, m.p. 267°. It was believed that the hydrocarbon might be dibenzo[a,g]biphenylene (41) and, indeed, its ultraviolet absorbtion spectrum was not inconsistent with such a formulation (Fig. 11.6). However, further proof of the structure of the hydrocarbon was not obtained because sufficient material was not available.*

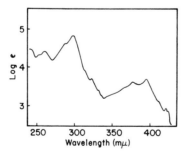

FIG. 11.6. Ultraviolet absorption spectrum of Ward and Pearson's hydrocarbon, m.p. 271° (in ethanol).[49]

* Note added in proof: Authentic dibenzo[a,g]biphenylene has now been synthesized (see Abstract 11.2 in the Appendix) and is not identical with the hydrocarbon described above.

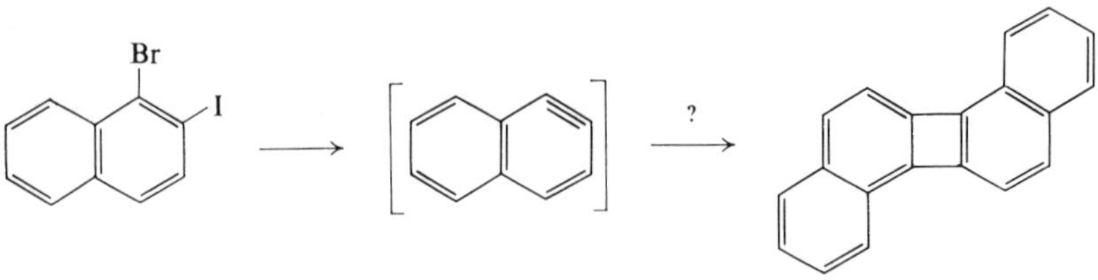

4. Dibenzo[a,i]biphenylene

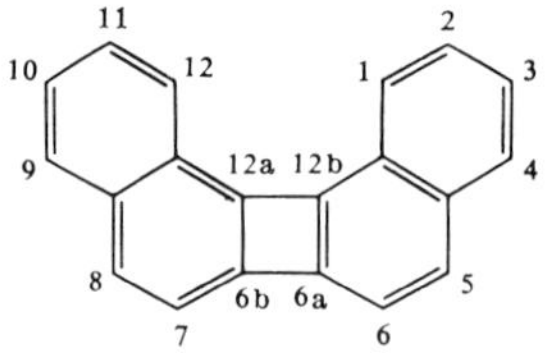

Dibenzo[a,i]biphenylene is commonly called "1,2-binaphthylene." However, since this name is equally applicable to the a,g isomer, it is probably better to affix *syn* or *anti* to the common name, as the case might be, in order to avoid confusion.

a. General Properties*

Dibenzo[a,i]biphenylene, deep red needles, m.p. 136.8°–138.9°, is best isolated and stored as the stable 2,4,7-trinitrofluorenone complex, dark purple needles, m.p. 205°–206°.[28] The hydrocarbon is thermally unstable (it decomposes to give a dark brown gum when heated at 160° *in vacuo*), but dilute solutions in the usual organic solvents are quite stable at room temperature when kept in the dark. Dibenzo[a,i]biphenylene is bleached by ultraviolet light far more easily than is the analogous monobenzo compound, benzo[a]biphenylene; the identity of the irradiation products is not known.

The ultraviolet spectrum of dibenzo[a,i]biphenylene (Fig. 11.7)[28a,33a] is made up of three sets of bands and is similar to the spectrum of benzo[a]-biphenylene; however, most of the bands in the former are better resolved and are shifted to longer wavelengths than those in the latter. The infrared spectrum of dibenzo[a,i]biphenylene is reproduced in Fig. 11.8. Infrared spectral values

---

* For a discussion of the polarographic behavior of dibenzo[a,i]biphenylene, see Chapter 10, Section F, 4.

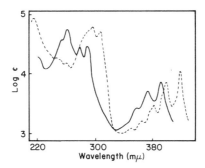

FIG. 11.7. Ultraviolet absorption spectrum of dibenzo[a,i]biphenylene (----) and benzo[a]biphenylene (———) (in ethanol).[28]

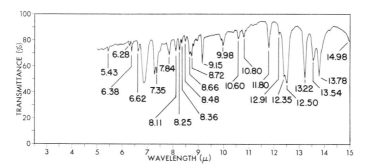

FIG. 11.8. Infrared spectrum of dibenzo[a,i]biphenylene (Nujol mull), calibrated against polystyrene.

have been recorded by Cava: λ_{max}^{Nujol} 5.43, 6.28, 6.38, 7.35, 7.84, 8.11, 8.25, 8.36, 8.48, 8.66, 8.72, 9.15, 9.98, 10.60, 10.80, 11.80, 12.19, 12.35, and 14.98 μ (all of medium or weak intensity); and 12.50, 13.22, 13.78 μ (all of high intensity).[28a]

b. *Methods of Synthesis*

Dibenzo[a,i]biphenylene (2) has been synthesized by Lothrop's method only.[28] The average yield in a series of 70 preparations was 5.65%, based on 2,2′-diiodo-1,1′-binaphthyl.

c. *Substituted Dibenzo[a,i]biphenylenes*

No substituted dibenzo[a,i]biphenylene is known. The site of maximum reactivity has been predicted to be the 5-position.[3]

d. *Reactions of the Four-Membered Ring*

Dibenzo[a,i]biphenylene is cleaved reductively by Raney nickel.[28] Unlike benzo[a]biphenylene, which gives both 1-phenyl- and 2-phenylnaphthalene on reductive cleavage, the dibenzo compound gives only 2,2'-binaphthyl, albeit in low yield.[28] Because it was shown that 1,1'-binaphthyl is not destroyed under the conditions used to cleave the dibenzobiphenylene, its absence from the cleavage product is of more than passing interest and may have theoretical implications.

No other reaction of the four-membered ring has been reported.

5. DIBENZO[b,h]BIPHENYLENE

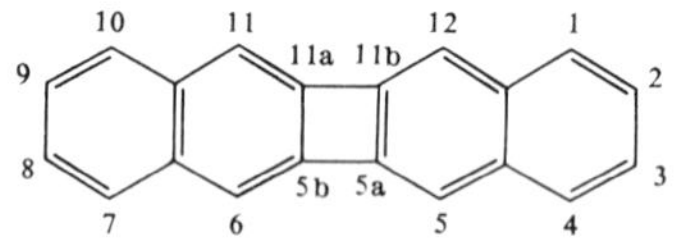

Dibenzo[b,h]biphenylene is commonly called 2,3-binaphthylene or 2,3:6,7-dibenzobiphenylene.

a. *General Properties*

The hydrocarbon crystallizes in the form of pale yellow plates, m.p. 376° ± 2° (sealed tube); it is thermally stable (it sublimes unchanged at 340°–345° on the Kofler block), and dilute solutions of the hydrocarbon in benzene exhibit a bright bluish fluorescence under ultraviolet light.[33a] The hydrocarbon is only sparingly soluble in the usual organic solvents. Its expected molecular weight (252 for $C_{20}H_{12}$) has been confirmed by mass spectrometry.[34]

The ultraviolet spectrum of dibenzo[b,h]biphenylene is made up of 14 absorption maxima which are separable into two distinct groups[33a, 41, 49]: λ_{max}^{EtOH} 217 (log ϵ = 4.56), 250 (4.23), 257 (4.41), 278 (4.93), 288 (5.23), 301 (4.47), 322 (4.31), 341 (3.61), 349 (3.56), 362 (3.63), 370 (3.57), 380 (4.15), 394 (3.63), 406 mμ (4.43).[33a] Its spectrum is similar to those of benzo[a]biphenylene and dibenzo[a,i]biphenylene (Fig. 11.9).[33a] The absorption maxima predicted by Crawford[32] are in fair agreement with the experimental values.

D. PROPERTIES OF THE BENZO- AND DIBENZOBIPHENYLENES

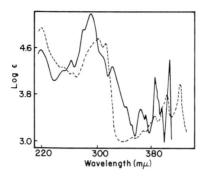

FIG. 11.9. Ultraviolet absorption spectra of dibenzo[b,h]biphenylene (———) and dibenzo-[a,i]biphenylene (----) (in ethanol).[33a]

Infrared spectral values (Fig. 11.10) for crystalline dibenzo[b,h]biphenylene (potassium bromide disc) have been recorded by Curtis: 6.05 (m), 6.20 (w), 6.66 (w), 6.89 (m), 7.91 (w), 8.07 (m), 8.51 (w), 8.78 (w), 8.84 (w), 9.76 (w), 10.50 (m), 11.39 (vs), 13.2 (m), 13.28 (m), 13.45 μ (s).[33a]

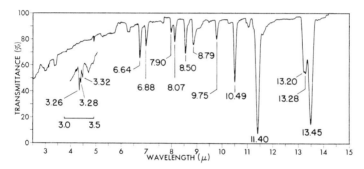

FIG. 11.10. Infrared spectrum of dibenzo[b,h]biphenylene (potassium bromide disc) calibrated against polystyrene.

b. Methods of Synthesis

Dibenzo[b,h]biphenylene was prepared by Lothrop's method. Thus, diazotization of 3,3'-diamino-2,2'-binaphthyl (42) and treatment of the resulting bis(diazonium) salt with sodium iodide gave the iodonium iodide. Pyrolysis of the crude iodonium iodide with cuprous oxide gave dibenzo[b,h]biphenylene (2). The hydrocarbon was also formed unexpectedly in an attempt to prepare its octahydro derivative from 1,2,3,4,8,9,10,11-octahydrodinaphth[2,3-b:2',3'-d]-iodolium iodide (43).[33,49] Evidently, in this case ring-closure is accompanied by dehydrogenation.

Dibenzo[b,h]biphenylene has also been obtained in low yield (less than 1%)

by the dehalogenation of 2,3-dibromonaphthalene (44); the reaction probably proceeds via 2,3-naphthalyne.[47,51] (See Section B, 2.)

c. *Substituted Dibenzo[b,h]biphenylenes*

The direct introduction of a substituent into the dibenzo[b,h]biphenylene nucleus has not been recorded,* but the site of maximum reactivity has been predicted to be the 5-position.[3] Of the few derivatives of dibenzo[b,h]biphenylene that are known, the majority are substituted in the 5,6,11,12-positions as a consequence of their mode of formation (from 1,4-disubstituted naphthalene derivatives). Thus, 5,6,11,12-tetrahydroxydibenzo[b,h]biphenylene (7) was obtained in 48% yield by the base-catalyzed isomerization of 1,4-naphthoquinone photodimer,[14] a report to the contrary notwithstanding.[46] Solutions of (7) in aqueous sodium hydroxide are colored orange-red; such solutions are attacked by oxygen, becoming progressively blue, green, and finally brown. Neutral solutions of (7) in hydroxylic solvents decompose slowly with the formation of a brown resin.[14] Two derivatives of (7), viz., 5,6,11,12-tetramethoxy- (45) and 5,6,11,12-tetraacetoxydibenzo[b,h]biphenylene (46) were obtained in 47% and 81% yield, respectively, from the photodimer.[14] The tetraacetate was also obtained in 95% yield from (7) by treatment with refluxing acetic anhydride in the presence of sodium acetate. A solution of the tetraacetate in acetic anhydride exhibits a blue fluorescence under ultraviolet light.[14] (See also Tables 11.6 and 11.7.)

5,6,11,12-Tetraphenyldibenzo[b,h]biphenylene (10) was obtained as the final product of an unusual reaction sequence starting from the adduct (47) of cyclooctatetraene and dimethyl acetylenedicarboxylate.[47] Thus, adduct (47) was treated with 1,3-diphenylisobenzofuran to give a new adduct (48). The latter was not isolated but was aromatized *in situ* with hydrochloric acid to

---

* For an attempted preparation of 5,11(or 6)-dinitrodibenzo[b,h]biphenylene by direct nitration, see Table 11.4.

D. PROPERTIES OF THE BENZO- AND DIBENZOBIPHENYLENES

TABLE 11.6

ULTRAVIOLET ABSORPTION CHARACTERISTICS OF TETRAHYDROXYDIBENZO[b,h]BIPHENYLENE AND DERIVATIVES

Dibenzo[b,h]biphenylene	Absorption maxima, mμ, and extinction coefficients (log ε)
5,6,11,12-Tetrahydroxy-	218 (4.36); 257 (4.33); 293 (4.85); 306 (5.14); 346 (4.15); 376.5 (3.41); 399 (3.72); 424 (3.90)
5,6,11,12-Tetraacetoxy-	220.5 (4.49); 284.5 (4.93); 296.5 (5.23); 333 (4.37); 361 (3.27); 380 (3.52); 404.5 (3.66)
5,6,11,12-Tetramethoxy-	219 (4.37); 259.5 (4.16); 291.5 (4.89); 303.5 (5.16); 342.5 (4.28); 389.5 (3.12); 412 (3.29)

TABLE 11.7

INFRARED ABSORPTION CHARACTERISTICS OF TETRAHYDROXYDIBENZO[b,h]BIPHENYLENE AND DERIVATIVES

Dibenzo[b,h]biphenylene	Absorption maxima (cm^{-1}) and intensities[a]
5,6,11,12-Tetrahydroxy-	3436b; 3246b; 3163b; 1661m; 1596m; 1520w; 1430m; 1366; 1298; 1266m; 1244; 1199; 1187; 1144m; 1093m; 1060w; 1029w; 976w; 947w; 919; 791m; 756; 720w
5,6,11,12-Tetraacetoxy-	1770; 1725m; 1662m; 1585w; 1565w; 1514m; 1464m; 1379; 1344; 1307w; 1287; 1212; 1195; 1182; 1153m; 1093; 1061m; 1039w; 1015; 995w; 945m; 899; 865w; 810w; 761; 740w; 733w; 720w; 700w
5,6,11,12-Tetramethoxy-	1622; 1606; 1581m; 1512; 1456; 1422w; 1402w; 1346; 1298w; 1286; 1215m; 1202m; 1181m; 1146w; 1100; 1082m; 1033m; 1017; 982w; 966; 920w; 872w; 791w; 771; 749m; 710m; 663m

[a] Key: b (broad), m (medium), w (weak).

give a naphthocyclobutene derivative (49), which is formally an adduct of 3,8-diphenylnaphthocyclobutadiene and dimethylphthalate. Thermolysis of (49) at 270° in the presence of 1,3-diphenylisobenzofuran gave 5,6,11,12-tetraphenyl-5,12-oxido-5,5a,11b,12-tetrahydrodibenzo[b,h]biphenylene (50) which was aromatized with hydrochloric acid to give 5,6,11,12-tetraphenyldibenzo[b,h]biphenylene (10).[41]

The ultraviolet spectrum of the tetraphenyl derivative has been compared with that of the parent hydrocarbon (Fig. 11.11).[41] Absorption maxima and extinction coefficients (log ϵ) for the tetraphenyl derivative are: λ_{max} 266 (4.41), 308 (5.06), 344 (4.30), 392 (3.81), 417 (3.92), 430 mμ (3.89) (ethylene dichloride).[41]

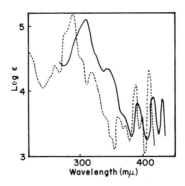

FIG. 11.11. Ultraviolet absorption spectrum of 5,6,11,12-tetraphenyldibenzo[b,h]biphenylene (——) and dibenzo[b,h]biphenylene (----) (in ethylene dichloride).[41]

Infrared values for 5,6,11,12-tetraphenyldibenzo[b,h]biphenylene have been recorded by Nenitzescu: ν_{max}^{Nujol} 700 (s), 733 (w), 755 (s), 770 (s), 840 (w), 870 (w), 890 (w), 906 (w), 974 (w), 1000 (w), 1021 (m), 1070 (m), 1095 (m), 1120 (w),

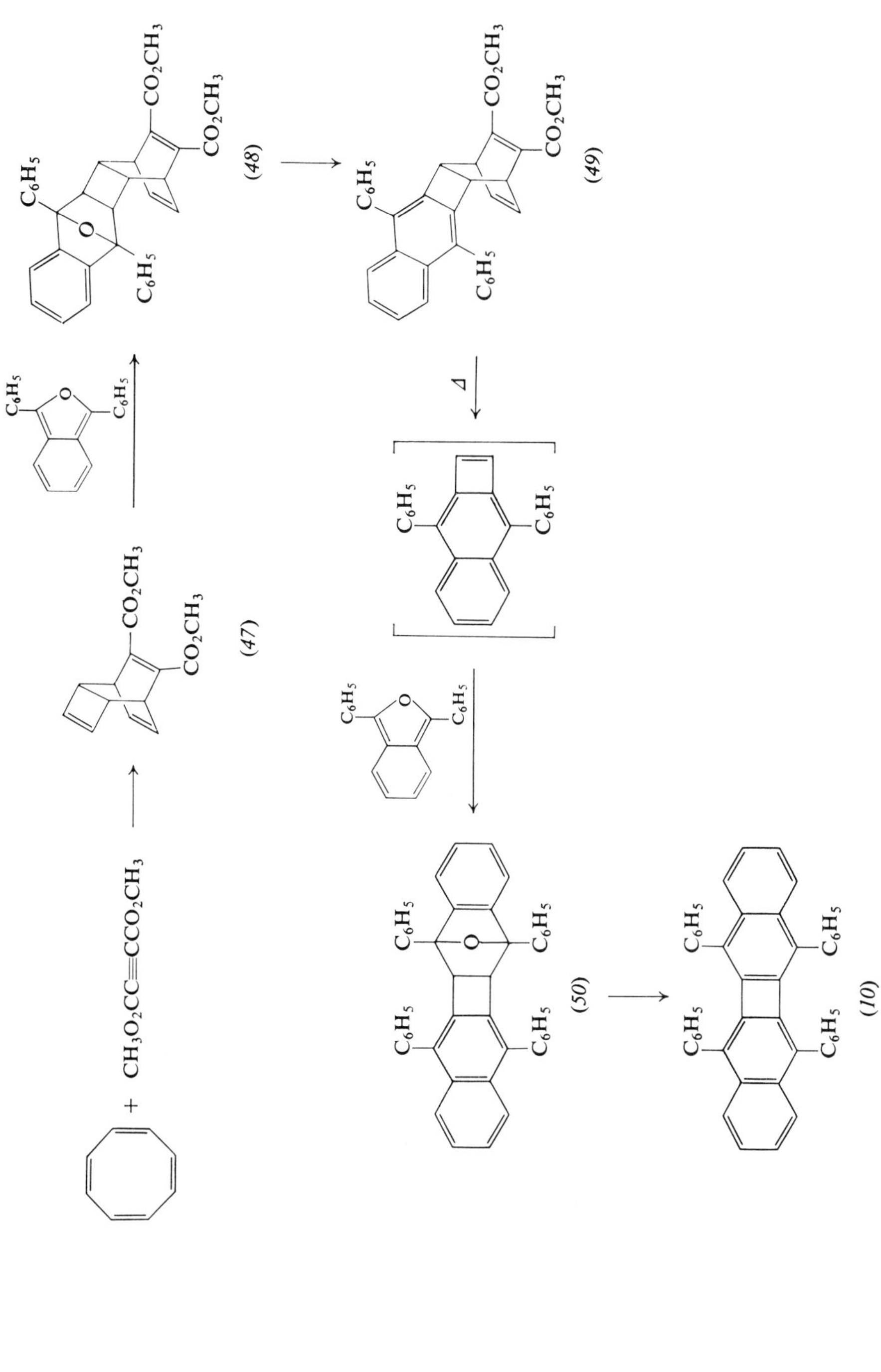

D. PROPERTIES OF THE BENZO- AND DIBENZOBIPHENYLENES

TABLE 11.8

Physical Constants of the Benzo- and Dibenzobiphenylenes

Biphenylene	Method of preparation[a]	Physical properties[b]	Properties of derivatives	Ref.
Benzo[a]-	1, 4, 5, 8	M.p. 72.0°–72.8° (b-y)	TNF, m.p. 201.5°–202.5° (b)	21, 28, 29
5-Acetylbenzo[a]-	8	M.p. 85.5°–86.5° (r)	TNF, (d-p)	27
5-Bromobenzo[a]-	1, 5	M.p. 124°–125° (o)	TNF, m.p. 202°–204° (br)	9, 19, 20, 21, 26, 29, 36
Benzo[a]- -5-carbonyl chloride[c]	8	"Crystalline"	—	27
Benzo[a]- -5-carboxamide	8	M.p. 273° (y)	—	27
Benzo[a]- -5-carboxylic acid	8	M.p. 257° (r)	Me ester, m.p. 126°–127° (o)	27
5-Cyanobenzo[a]-	8	M.p. 154°–155° (o)	—	27
5,6-Dibromobenzo[a]-	5	M.p. 149°–150° (o); subl.	—	9, 20
5-Grignard (iodide)[c] of benzo[a]-	8	(y)	—	27
5-Iodobenzo[a]-	5	M.p. 131°–132° (o)	—	27
5-Lithiobenzo[a]-[c]	8	(r)	—	27, 29
Benzo[b]-[d]	5, 8	M.p. 242°–243°	TNF, m.p. 214°–216° (r)	9, 39
Benzo[b]-	8	M.p. 406° (decomp.)	—	31, 39
-5,10-dicarboxylic acid				
5-Bromobenzo[b]-	8	M.p. 112°–113° (y)	—	31
5-Bromo-10-liithiobenzo[b]-[c]	8	—	—	31
5,10-Dibromobenzo[b]-	5	M.p. 222°–223° (y)	—	9, 39
5,10-Dilithiobenzo[b]-[c]	8	—	—	39
5,10-Diphenylbenzo[b]-	5	M.p. 218°–220° (y)	—	6, 23
Dibenzo[a,g]- (?)	2	M.p. 271° (o)	TNF, m.p. 267° (r)	49
Dibenzo[a,l]-	1	M.p. 136.8°–138.9° (r)	TNF, m.p. 205°–206° (r)	28

Dibenzo[b,h]-			
	1, 2, 3	M.p. 376° ± 2° (sealed tube) (subl. 340°–345°) (p-y) fluorescent (bl)	33, 47, 49, 51
5,11(or 6)-Dinitrodibenzo[b,h]-	3	M.p. 384° (subl. 370°) fluorescent (bl)	47
5,6,11,12-Tetraacetoxydibenzo[b,h]-	5, 8	—	14
1,2,3,4-Tetrahydrodibenzo[b,h]-	1	M.p. 300°–302° (y)	49
5,6,11,12-Tetrahydroxydibenzo[b,h]-	5	M.p. 260°–265° (decomp. 185°–245°) (y-g)	See tetraacetyldibenzo[b,h]- 14
5,6,11,12-Tetramethoxydibenzo[b,h]-	5	M.p. 211° (p-y) fluorescent (bl)	14
5,6,11,12-Tetraphenyldibenzo[b,h]-	5	M.p. 350° (o) fluorescent (v)	41

[a] Method 1 is "From 2,2′-Dihalobiaryls," Section B, 1; Method 2 is "Via 1,2- and 2,3-Naphthalynes," Section B, 2; Method 3 is "By the Ullmann Reaction," Section B, 3; Method 4 is "From Organometallic Biaryls," Section B, 4; Method 5 is "Via Reduced Benzobiphenylenes," Section B, 5; Method 6 is "From Non-Aromatic Precursors," Section B, 6; Method 7 is "By Direct Substitution," Section B, 7; and Method 8 is "By Transformation of Functional Groups," Section B, 8.

[b] Key: b (black), b-y (bright yellow), bl (blue), br (brown), d-y (deep yellow), o (orange), p-y (pale yellow), r (red), v (violet), y (yellow), y-g (yellowish green).

[c] Not isolated.

[d] See also tetrahydrodibenzo[b,h]biphenylene, this table.

1140 (w), 1157 (m), 1175 (m), 1205 (w), 1230 (w), 1260 (w), 1306 (m), 1495 (m), 1520 (m), 1595 cm^{-1}.[41]

d. Reactions of the Four-Membered Ring

Dibenzo[b,h]biphenylene is cleaved reductively by Raney nickel to give 2,2'-binaphthyl in 30% yield.[33a] No other reaction of the four-membered ring has been reported. Indeed, because the linearly annellated dibenzo[b,h] isomer has relatively little cyclobutadienoid character, few such reactions can be expected.

5,11(or 6)-Dinitrodibenzo[b,h]biphenylene [(*51a*) (or *b*)] was prepared in low yield (1–3%) by treating 2,3-dibromo-1-nitronaphthalene or 3-bromo-2-iodo-1-nitronaphthalene with copper bronze in refluxing dimethylformamide for 6 hours.[47] The ultraviolet spectrum of the dinitro compound [$\lambda_{max}^{cyclohex}$ 254.5 (log ϵ = 4.58), 262.5 (4.64), 291.5 (4.32), 330 (4.42), 417 (4.06), 446 mμ (4.21)][47] has fewer bands in it than that of the parent hydrocarbon.

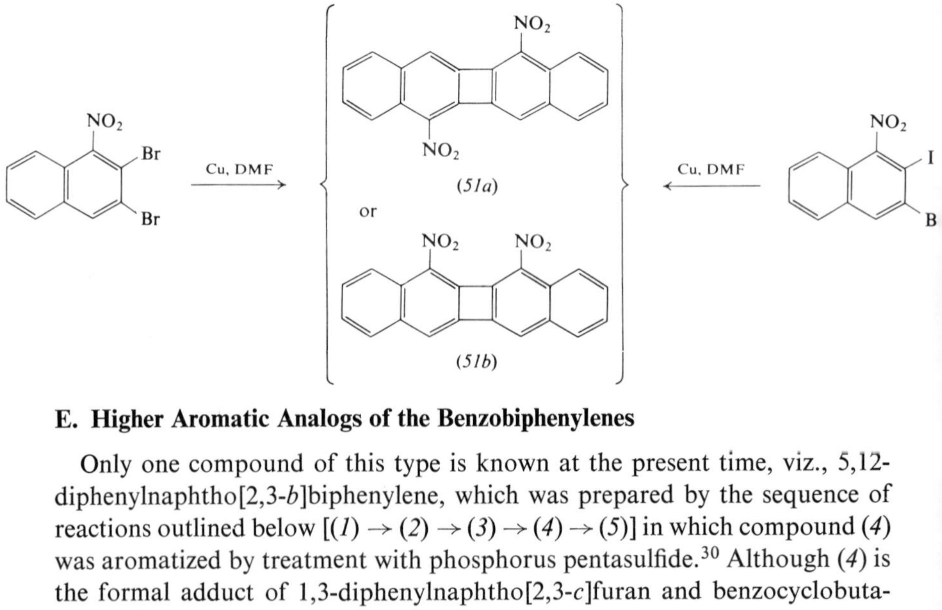

E. Higher Aromatic Analogs of the Benzobiphenylenes

Only one compound of this type is known at the present time, viz., 5,12-diphenylnaphtho[2,3-*b*]biphenylene, which was prepared by the sequence of reactions outlined below [(*1*) → (*2*) → (*3*) → (*4*) → (*5*)] in which compound (*4*) was aromatized by treatment with phosphorus pentasulfide.[30] Although (*4*) is the formal adduct of 1,3-diphenylnaphtho[2,3-*c*]furan and benzocyclobutadiene (*7*), it could not be prepared by generating benzocyclobutadiene in the presence of the naphthofuran. Instead, it was prepared as shown, via 1-bromobenzocyclobutadiene and bromo adduct (*3*).

F. Heterocyclic Benzobiphenylenes

At the present time only two heterocyclic compounds are known among the annellated biphenylenes. 5,10-Diazabenzo[*b*]biphenylene (*1*), long white

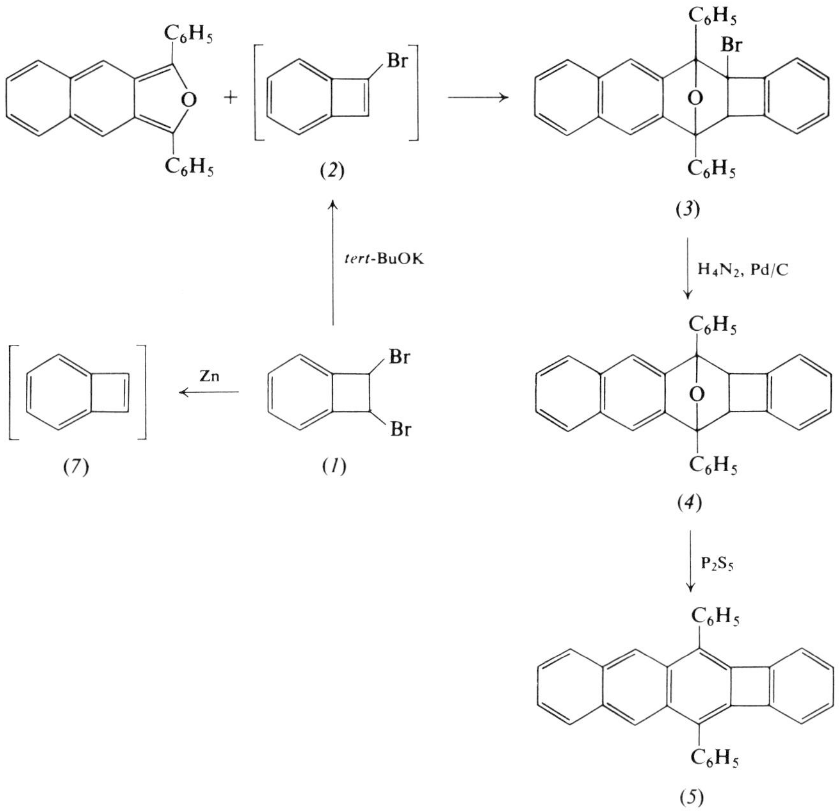

F. HETEROCYCLIC BENZOBIPHENYLENES

needles, m.p. 238°–239°,[22] was obtained in good yield (63%) by treating 1,2-benzocyclobutadienequinone with an equimolar amount o-phenylenediamine.[22] Similarly, benzo[3,4]cyclobuta[1,2-b]phenazine (2), yellow needles, m.p. 324°, was obtained in unspecified yield by treating 2,3-biphenylenequinone with o-phenylenediamine.[12]

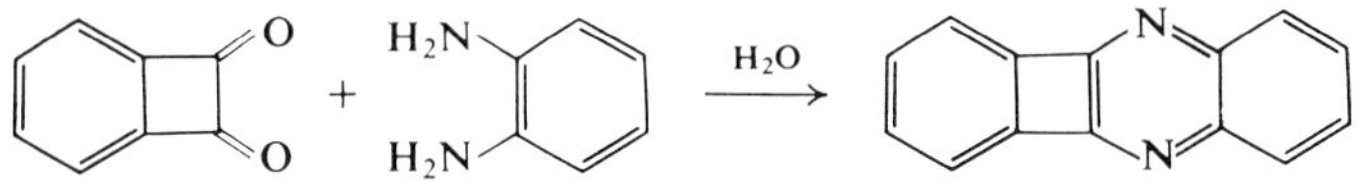

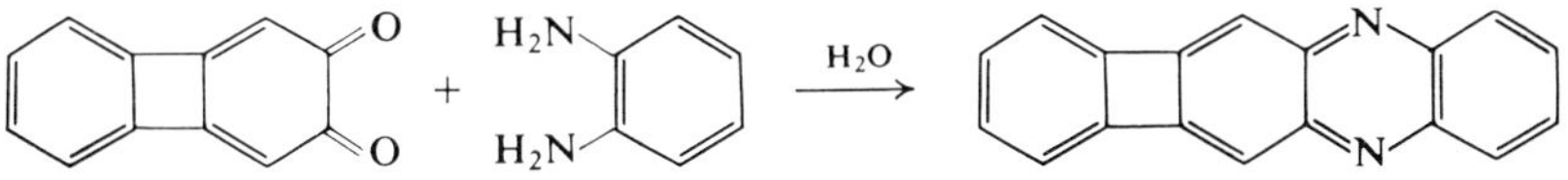

References

1. Ali, M. A., Carey, J. G., Cohen, D., Jones, A. J., Millar, I. T., and Wilson, K. V., *J. Chem. Soc.* 387 (1964).
2. Ali, M. A., and Coulson, C. A., *Tetrahedron* **10**, 41 (1960).
3. Andrade e Silva, M., and Pullman, B., *Compt. Rend.* **242**, 1888 (1956).
4. Avram, M., Dinu, D., Mateescu, G., and Nenitzescu, C. D., *Ber.* **93**, 1789 (1960).
5. Avram, M., Dinulescu, I. G., Dinu, D., Mateescu, G., and Nenitzescu, C. D., *Tetrahedron* 309 (1963).
6. Avram, M., Mateescu, C. D., Dinu, D., Dinulescu, I. G., and Nenitzescu, C. D., *Acad. Rep. Populare Romine, Studii Cercetari Chim.* **9**, 435 (1961).
7. Avram, M., Mateescu, G. D., Dinu, D., Dinulescu, I. G., and Nenitzescu, C. D., *Rev. Chim. Acad. Rep. Populaire Roumaine* **8**, 77 (1963).
8. Baker, W., Barton, J. W., and McOmie, J. F. W., *J. Chem. Soc.* 2666 (1958).
9. Baker, W., Barton, J. W., McOmie, J. F. W., and Searle, R. J. G., *J. Chem. Soc.* 2633 (1962).
10. Baker, W., and McOmie, J. F. W., in Ginsburg, D., "Non-Benzenoid Aromatic Compounds," Interscience, New York, (1959) p. 43.
11. Bell, F., and Hunter, W. H., *J. Chem. Soc.* 2904 (1950).
12. Blatchly, J. M., McOmie, J. F. W., and Thatte, S. D., *J. Chem. Soc.* 5090 (1962).
13. Braun, J. von, and Kirschbaum, G., *Ber.* **54**, 597 (1921).
14. Bruce, J. M., *J. Chem. Soc.* 2782 (1962).
15. Bunnett, J. F., *J. Chem. Ed.* **38**, 278 (1961).
16. Carey, J. G., and Millar, I. T., *J. Chem. Soc.* 3144 (1959).
17. Cava, M. P., and Arora, S. S., unpublished work.
18. Cava, M. P., and Mitchell, M. J., unpublished work. See Mitchell, M. J., Ph.D. diss., The Ohio State Univ. (1960).
19. Cava, M. P., and Muth, K., *J. Org. Chem.* **27**, 757 (1962).
20. (a) Cava, M. P., and Muth, K., *J. Org. Chem.* **27**, 1561 (1962); (b) *ibid.*, *Tetrahedron Letters* No. 4, 140 (1961).
21. (a) Cava, M. P., and Napier, D. R., *J. Am. Chem. Soc.* **79**, 1701 (1957); (b) *ibid.* **78**, 500 (1956).
22. (a) Cava, M. P., Napier, D. R., and Pohl, R. J., *J. Am. Chem. Soc.* **85**, 2076 (1963); (b) Cava, M. P., and Napier, D. R., *ibid.* **79**, 3606 (1957).
23. Cava, M. P., and Pohlke, R., *J. Org. Chem.* **27**, 1564 (1962).
24. Cava, M. P., Pohlke, R., Erickson, B. W., Rose, J. C., and Fraenkel, G., *Tetrahedron* **18**, 1005 (1962).
25. Cava, M. P., Pohlke, R., and Mitchell, M. J., *J. Org. Chem.* **28**, 1861 (1963).
26. Cava, M. P., and Ratts, K. W., *J. Org. Chem.* **27**, 752 (1962).
27. Cava, M. P., Ratts, K. W., and Stucker, J. F., *J. Org. Chem.* **25**, 1101 (1960).
28. (a) Cava, M. P., and Stucker, J. F., *J. Am. Chem. Soc.* **77**, 6022 (1955); (b) *ibid.*, *Chem. Ind. (London)* 446 (1955).
29. Cava, M. P., and Stucker, J. F., *J. Am. Chem. Soc.* **79**, 1706 (1957).
30. Cava, M. P., and Van Meter, J. P., unpublished work. See Van Meter, J. P., Ph.D. diss., The Ohio State Univ. (1963).
31. Coleman, W. E., Ph.D. diss., Univ. of Calif., Berkeley (1960).
32. Crawford, V. A., *Can. J. Chem.* **30**, 47 (1952).
33. (a) Curtis, R. F., and Viswanath, G., *J. Chem. Soc.* 1670 (1959); (b) *ibid.*, *Chem. Ind. (London)* 1174 (1954).
34. Curtis, R. F., *J. Chem. Soc.* 3650 (1959).

35. Dansi, A., and Ferri, C., *Gazz. Chim. Ital.* **71**, 648 (1941).
36. Finkelstein, H., Ph.D. diss., Strasbourg, 1909; *ibid.*, *Ber.* **92**, xxxvii (1959).
37. Gelin, S. *Chim. Mod.* **8**, 241 (1963).
38. Huisgen, R., in Zeiss, H., ed., "Organometallic Chemistry," Reinhold, New York (1960).
39. Jensen, F. R., and Coleman, W. E., *Tetrahedron Letters* No. **20**, 7 (1959).
40. Lee, H. S., *Chemistry (Taipei)* 53 (1963); *ibid* 137 (1962).
41. Nenitzescu, C. D., Avram, M., Dinulescu, I. G., and Mateescu, G., *Ann.* **653**, 79 (1962).
42. Orchin, M., Reggel, L., Friedel, R. A., and Woolfolk, E. O., *U.S. Bureau of Mines Technical Papers* No. **708**, 1 (1948). *Cf.* Orchin, M., and Reggel, L., *J. Am. Chem. Soc.* **69**, 505 (1947); *ibid.* **73**, 436 (1951).
43. Patterson, A. M., Capell, L. T., and Walker, D. F., "The Ring Index," 2nd ed., American Chemical Society, Washington, D.C. (1960), p. 600.
44. Pummerer, R., Lüttringhaus, A., Fick, R., Pfaff, A., Riegelbauer, G., and Rosenhauer, E., *Ber.* **71**, 2569 (1938).
45. Rosenhauer, E., Braun, F., Pummerer, R., and Riegelbauer, G., *Ber.* **70**, 2281 (1937).
46. (a) Schönberg, A., Mustafa, Ahmed, Barakat, M. Z., Latif, N., Moubasher, R., and Mustafa, Akila, *J. Chem. Soc.* 2126 (1948); (b) Schönberg, A., Mustafa, Ahmed, and Barakat, M. Z., *Nature* **160**, 401 (1947).
47. Ward, E. R., and Marriott, J. B., *J. Chem. Soc.* 4999 (1963).
48. Ward, E. R., and Marriott, J. B., private communication.
49. Ward, E. R., and Pearson, B. D., *J. Chem. Soc.* 1676 (1959).
50. Ward, E. R., and Pearson, B. D., *J. Chem. Soc.* 3378 (1959).
51. (a) Ward, E. R., and Pearson, B. D., *J. Chem. Soc.* 515 (1961); (b) Pearson, B. D., *Chem. Ind. (London)* 899 (1960).
52. Wittig, G., Uhlenbrock, W., and Weinhold, P., *Ber.* **95**, 1692 (1962).

CHAPTER 12

Theoretical Aspects of the Cyclobutadiene Problem

H. E. SIMMONS AND A. G. ANASTASSIOU*

Once the nature of the bonds in benzene was recognized, it was natural for chemists to assume that other cyclic hydrocarbons for which equivalent Kekulé structures could be written would exist and possess similar properties associated with the equivocal bonding. Experimentally, it was soon realized that this would not be so for cyclobutadiene. The failure of chemical intuition with respect to such an apparently simple molecule brought the chemist face to face with a new kind of chemical behavior, and subsequently cyclobutadiene has proved as important as a testing ground for theories as it has a subject for experiments.

In this chapter, we review the major papers in the literature dealing with the theoretical aspects of the cyclobutadiene problem. However, because of space limitations, the discussion of many of these papers had to be abridged, and, in addition, only selected theoretical methods have been developed in detail. Section A is in part a chronological survey of studies pertaining to the π-electronic states of cyclobutadiene. In Section B some conclusions and comparisons drawn from these studies are presented, and in Section C framework energies and Jahn–Teller considerations are discussed. Finally, Section D treats substituted cyclobutadienes, cyclobutadiene divalent ions, the metal complexes, the biphenylenes, and the benzobiphenylenes. It is presumed that the reader has some knowledge of simple LCAO–MO or Hückel theory[109, 121] and elementary group theory.[26]

* Central Research Department, Experimental Station, E. I. du Pont de Nemours and Co. Inc., Wilmington, Delaware.

The cyclobutadiene problem has been discussed frequently in the past; in particular, the account of Craig[33] should be referred to.

A. The π-Electronic States of Cyclobutadiene

Chemical experience with planar, cyclic π systems suggests that cyclobutadiene (C_4H_4) is a square molecule, and theoretical studies usually begin with this assumption. It is only practical to study the behavior of the π electrons in detail, relegating all σ electrons to a core which is supposed to provide the field in which the higher energy π electrons move. Since this "π-electron approximation" is generally reliable,[83] there seems to be no reason why its application to C_4H_4 should not be proper.

1. Hückel Molecular Orbital Description

We are interested in the electron configurations that arise when four carbon $2p\pi$ orbitals are arranged with square geometry. By considering a general one-dimensional, one-electron wave, Hückel first gave the π molecular orbitals (MO) for a monocycle of N carbon atoms as

$$\psi_i = \sigma_i \sum_{m=0}^{N-1} \omega^{mi} \chi_m \left[i = 0, \pm 1, \ldots, \begin{array}{l} +N/2 \ (N \text{ even}); \\ \pm(N-1)/2 \ (N \text{ odd}) \end{array} \right] \quad \text{(A-1)}$$

$$\omega = e^{2\pi i/N}$$

where σ_i is a normalizing factor and χ_m are $2p\pi$ atomic orbitals (AO) localized on the carbon atoms m.[59] The atoms are numbered $0, 1, 2, \ldots, N-1$, and i is a "ring quantum number" which labels the MO's and measures a kind of angular momentum about the ring. In the special case of a monocycle, the molecular orbitals ψ_i are determined solely by symmetry considerations.

Although C_4H_4 belongs to the point group D_{4h}, the ψ_i are fully determined under the rotation group C_4.[52] The four $2p\pi$ AO's form a basis for a reducible representation of D_{4h}, $A_{2u} + B_{2u} + E_g$, so that ψ_0 transforms according to a_{2u}, etc.,

$$\psi_0 \subset a_{2u} \quad \psi_{\pm 1} \subset e_g \quad \psi_2 \subset b_{2u} \quad \text{(A-2)}$$

The ground-state configuration must then be $a^2 e^2$, so that the doubly degenerate orbital e is half-filled. Within the scope of its applicability to polyatomic systems, Hund's rule requires that the spins of the two electrons in e be unpaired. Without any calculations we conclude immediately that the lowest state of C_4H_4 is a triplet, provided the energy associated with ψ_0 is less than that with $\psi_{\pm 1}$. In such cases where the highest occupied orbital is not filled, the simple MO's provide a less adequate description of the ground state than they do, e.g., for benzene,[32,132] whose ground-state configuration ($a^2 e^4$) is closed.

We must determine the set of states that arise from the configuration $a^2 e^2$, but first it will be instructive to see what the simple physical model predicts.

Hückel showed that the energies associated with the ψ_i are given by

$$\epsilon_i = \alpha + k_i \beta \tag{A-3}$$

where the k_i are characteristic constants, and α and β are, respectively, the one-electron Coulomb and resonance integrals which occur as elements of the energy matrix $[H_{mn} - \epsilon\, S_{mn}]$.

$$H_{mm} \equiv \alpha = \int \chi_m \mathbf{H} \chi_m \, d\tau \tag{A-4}$$

$$H_{mn} \equiv \beta = \int \chi_m \mathbf{H} \chi_n \, d\tau \tag{A-5}$$

The overlap integrals S_{mn} are neglected ($S_{mn} = \delta_{mn}$) in the usual zero-order calculation where δ_{mn} is the Kronecker symbol. The Hamiltonian operator, $\mathbf{H}$, is unspecified but is supposed to describe the motion of a π electron in the effective field of the core carbon atoms. A calculation performed under these assumptions, which do not explicitly incorporate interelectronic effects, gives the four energy levels shown in Fig. 12.1, along with those of benzene for

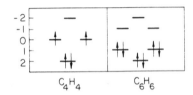

FIG. 12.1. HMO energy levels of C_4H_4 and C_6H_6 (in units of β).

comparison. Not only are $\epsilon_{\pm 1}$ degenerate in C_4H_4, but they have zero relative energy and are nonbonding, thus heightening our suspicion that the ψ_i may be somehow inadequate in this case.

The total π binding energy of C_4H_4 at this level of approximation is $4\alpha + 4\beta$, which equals that of two molecules of ethylene, so that the conventional π delocalization energy is exactly zero. The inclusion of the overlap integral in the calculation cannot remove the degeneracy, nor can any of the other usual refinements of simple LCAO–MO theory. In fact, when overlap is included the delocalization energy becomes negative (-0.53γ where $\gamma = \beta + S\alpha$).[131]

The chemical implications of the Hückel results were early stated.[29, 67, 111] The π-bond order (p_{mn}) measures the relative binding between π electrons on adjacent carbon atoms m and n and is given by

$$p_{mn} = 2 \sum_{\text{occ.} i} \sigma_i^2 \omega^{(m+n)i} \tag{A-6}$$

In C_4H_4, all $p_{mn} = 1/2$, while in C_6H_6, $p_{mn} = 2/3$, clearly suggesting reduced π

binding in cyclobutadiene. Another useful structure index is free valence (f_m) which measures the degree to which a given atom is bonded to its neighbors relative to a theoretical maximum bonding power. Free valence is defined by

$$f_m = \sqrt{3} - \sum_{\text{adj}.n} p_{mn} \qquad \text{(A-7)}$$

and equals 0.73 in C_4H_4, compared to 0.40 in C_6H_6. Thus, the carbon atoms of cyclobutadiene are predicted to have higher chemical reactivity than those of benzene.

2. EMPIRICAL VALENCE-BOND DESCRIPTION

The structure of cyclobutadiene was also studied in early work by the empirical valence-bond method. Calculations with valence-bond wavefunctions, however, gave results in complete disagreement with those obtained from molecular orbital wavefunctions. It was this striking discrepancy in predictions by two theories whose results are normally in agreement that prompted the extensive studies on an unknown molecule that now stretch over three decades.

The valence-bond (VB) method uses a fundamentally different kind of wavefunction to describe the ground states of organic molecules. This method is an extension of one originally developed by Slater[117] for atoms and was first applied to organic molecules by Hückel.[59] In 1934, Wheland[133] studied C_4H_4 by the simple VB and MO theories and compared the results. In essence, VB theory begins by considering the familiar resonance structures of the organic chemist and prescribes a method of constructing wavefunctions, called bond eigenfunctions, to describe each structure. The bond eigenfunctions are then mixed quantum mechanically (resonance) and the energies of the various combinations are determined by the usual application of an energy operator.

In the case of C_4H_4 we can write two Kekulé structures (Chart 12.1, A,B)

CHART 12.1. Kekulé and polar structures for C_4H_4.

and four polar structures (P,Q,R,S). Di-excited structures (e.g., T) are of such high energy that they need not be considered here, but they must not be neglected in accurate calculations.

The bond eigenfunction for a given canonical structure, say A, is given by a linear combination of Slater functions, which are a generalization of the original Heitler–London wavefunction. If we let $a(1)$, $b(2)$, ... represent $2p\pi$ AO's on atoms a, b, ..., where the index in parentheses stands for the coordinates of electrons 1, 2, ..., then $a(1)\alpha(1) \equiv a(\alpha)_1$, is a spin orbital showing occupancy by electron 1 with spin α. The simplest kind of wavefunction we can build from such spin orbitals is a product, which, however, must be antisymmetric to exchange of the electronic coordinates in order to satisfy the Pauli principle. For an n-electron problem we can write 2^n such antisymmetric wavefunctions, each of which is an eigenfunction of the spin operator S_z having eigenvalues $n/2$, $n/2-1$, ..., $-n/2$. The ground state of a π-electron system where n is even can be considered to possess $n/2$ double bonds, so that there will be equal numbers of α and β spins. We will therefore be interested only in those cases where the eigenvalue of S_z is zero.

A Slater function or Slater determinant is the proper antisymmetric combination of products of spin orbitals meeting the above requirements, which in the case of A can be written as

$$\phi_{1A} = (4!)^{-1/2} \begin{vmatrix} (a\alpha)_1 & (b\beta)_1 & (c\alpha)_1 & (d\beta)_1 \\ (a\alpha)_2 & (b\beta)_2 & (c\alpha)_2 & (d\beta)_2 \\ (a\alpha)_3 & (b\beta)_3 & (c\alpha)_3 & (d\beta)_3 \\ (a\alpha)_4 & (b\beta)_4 & (c\alpha)_4 & (d\beta)_4 \end{vmatrix} \quad (A\text{-}8)$$

The important point to note is that in each product of the expanded determinant, atoms a and b are always related by α and β spins, respectively, and so are bonding in the original sense of Heitler and London; atoms c and d are similarly related. Therefore, the simpler and more pertinent notation can be adopted

$$\phi_{1A} \equiv (\alpha\beta\alpha\beta)$$

where we have written only the diagonal term and each position is fixed with respect to the electron coordinates. Inspection shows that three other assignments of spin are compatible with bonds between a and b and between c and d and these exhaust the possibilities.

$$\phi_{2A} = (\alpha\beta\beta\alpha), \quad \phi_{3A} = (\beta\alpha\alpha\beta), \quad \phi_{4A} = (\beta\alpha\beta\alpha)$$

The bond eigenfunction Φ_A is the proper linear combination of ϕ's taken to be antisymmetric to exchange of the spins of two bonded atoms. For structure A, it is easily verified that

$$\Phi_A = (4)^{-1/2}[\phi_{1A} - \phi_{2A} - \phi_{3A} + \phi_{4A}] \quad (A\text{-}9)$$

When an AO is doubly occupied, some other AO must be missing from the

Slater function; the monoexcited structure P, in Chart 12.1, for example, is given by two functions

$$\phi_{1P} = (4!)^{-1/2} \begin{vmatrix} (a\alpha)_1 & (a\beta)_1 & (c\alpha)_1 & (d\beta)_1 \\ & & \text{etc.} & \end{vmatrix} \quad \text{(A-10)}$$

or $\quad \phi_{1P} = (\widehat{\alpha\beta\alpha\beta})_a \qquad \phi_{2P} = (\widehat{\alpha\beta\beta\alpha})_a$

where the linked spins occur on the same AO which is specified in the subscript. In this manner the bond eigenfunctions for all of the structures can be written down. A and B are the only structures which have two π bonds, and any other, such as C,

C

is not independent and can be expressed as a combination of A and B. Furthermore, it can be shown that bond eigenfunctions such as Φ_A, Φ_B are eigenfunctions of the operator S^2 and that structures with different numbers of bonds have different eigenvalues of S^2. Since matrix elements between functions with different eigenvalues of the total spin S^2 vanish and since S^2 commutes with the real Hamiltonian H, we would expect that the ground state of C_4H_4 would be given by solution of a 2×2 secular equation relating A and B only.

In the empirical VB method, the Hamiltonian is unspecified, and resolution of the secular equation (A-11)

$$\begin{vmatrix} H_{AA} - E & H_{AB} - S_{AB}E \\ H_{BA} - S_{BA}E & H_{BB} - E \end{vmatrix} = 0 \quad \text{(A-11)}$$

requires evaluation of the integrals

$$H_{AA} = \int \Phi_A H \Phi_A \, d\tau \qquad H_{BB} = \int \Phi_B H \Phi_B \, d\tau \quad \text{(A-12)}$$

$$H_{AB} = H_{BA} = \int \Phi_A H \Phi_B \, d\tau$$

$$S_{AB} = \int \Phi_A \Phi_B \, d\tau \quad \text{(A-13)}$$

which on expansion give rise to integrals over Slater functions such as,

$$H_{1A,1A} = \int \phi_{1A} H \phi_{1A} \, d\tau \quad \text{(A-14)}$$

These integrals over determinantal wavefunctions are then further expanded[117] as simple product functions of spin orbitals and finally the spin is integrated out. This latter operation causes many elements to vanish, but in general there remain many nonzero integrals. In the case of $H_{1A,1A}$, we find

$$H_{1A,1A} = Q - (ac) - (bd) \quad \text{(A-15)}$$

where

$$Q = \int [a(1)\,b(2)\,c(3)\,d(4)]\,\mathbf{H}[a(1)\,b(2)\,c(3)\,d(4)]\,d\tau \quad \text{(A-16)}$$

$$(ac) = \int [a(1)\,b(2)\,c(3)\,d(4)]\,\mathbf{H}[c(1)\,b(2)\,a(3)\,d(4)]\,d\tau \quad \text{(A-17)}$$

$$(bd) = \int [a(1)\,b(2)\,c(3)\,d(4)]\,\mathbf{H}[a(1)\,d(2)\,c(3)\,b(4)]\,d\tau \quad \text{(A-18)}$$

The notation (ac) and (bd) in Eqs. (A-17) and (A-18) indicates the AO's whose coordinates have been permuted in the signified way. Q is called a Coulomb integral and (ac) and (bd) are single exchange integrals; in this case, symmetry dictates that (ac) and (bd) must be of the same type. Finally, H_{AA} will be a combination of several terms such as in Eq. (A-15).

$$H_{AA} = H_{1A,1A} + H_{2A,2A} + H_{3A,3A} + H_{4A,4A} +$$
$$2[H_{1A,4A} + H_{2A,3A} - H_{1A,2A} - H_{1A,3A} - H_{2A,4A} - H_{3A,4A}]$$

Elaboration of the details of the VB procedure is beyond the scope of this chapter, and the interested reader is referred to the common texts.

When all of the necessary integrals are evaluated symbolically and only single exchange integrals between *adjacent* atoms are retained, the secular determinant for Kekulé structures A and B reduces to

$$\begin{vmatrix} Q + \alpha - E & \tfrac{1}{2}Q + 2\alpha - \tfrac{1}{2}E \\ \tfrac{1}{2}Q + 2\alpha - \tfrac{1}{2}E & Q + \alpha - E \end{vmatrix} = 0 \quad \text{(A-19)}$$

where α has been written for the single exchange integral of the type (ab) or (cd).* The solutions of Eq. (A-19) are $E_G = Q + 2\alpha$ and $E_V = Q - 2\alpha$. Since α is negative the ground state energy of C_4H_4 is given by E_G; E_V then refers to some excited state. The wavefunctions are given by

$$\Phi_G = (2)^{-1/2}[\Phi_A + \Phi_B] \qquad \Phi_V = (2)^{-1/2}[\Phi_A - \Phi_B] \quad \text{(A-20)}$$

and the states G and V are both singlets. Calculation of the triplet-state energies of C_4H_4 by the VB method has been considered by McLachlan.[88]

The energy of a single Kekulé structure (A or B, Chart 12.1) is easily shown to be $Q + \alpha$, so that the resonance energy of C_4H_4 is simply α. The resonance energy of C_6H_6 by a similar calculation is 1.1α. VB theory thus predicts a larger stabilization energy per π electron in cyclobutadiene than in benzene, and this

---

* The definition of α used here should not be confused with the one-electron Coulomb integral which occurs in MO theory.

result is not substantially changed, even when all single exchange integrals are included in the calculation. In this version of the VB method, the basic AO's have been assumed to be orthogonal. This is far from true, and Wheland[130] has shown how nonorthogonality could be taken into account. Even so, with $S_{\pi\pi} = 1/4$, the resonance energy is reduced to 0.78α, a value that is still unexpectedly large.

Actually, the resonance energy is not a fundamental criterion of stability because of its arbitrary definition. Neither Φ_A nor Φ_B alone has physical significance, and only the combinations Φ_G and Φ_V correspond to real states. (See also Section B.) Craig has pointed out that the splitting between Φ_G and Φ_V is the important measure of the resonance interaction, a quantity that retains its significance in any kind of calculation as long as one-electron AO's form the basis. This splitting can be identified with an observable, i.e., a spectroscopic interval, and has fundamental meaning unlike the resonance energy which corresponds to the difference $E_{A(B)} - E_G$.

The simple MO and VB theories are thus in complete disagreement as to the stability of the ground state of C_4H_4, whereas the two theories are generally in close agreement for many known aromatic and olefinic hydrocarbons. The disagreement is even more serious when we inspect the symmetries of the ground-state wavefunctions. The VB wavefunctions $[\Phi_A + \Phi_B]$ and $[\Phi_A - \Phi_B]$ transform like B_{2g} and A_{1g}, respectively, under D_{4h}. Virtually all known hydrocarbons have totally symmetric ground states, i.e., the ground-state wavefunction belongs to an A representation and is unaffected by any of the symmetry operations of the group to which the molecule belongs. Now, it is not physically necessary that a molecule's lowest electronic state be totally symmetric. The wavefunction is not directly related to any physical observable; only its square, which is a measure of the electron density, must remain unchanged under any symmetry operation of the group to which the one-electron basis functions belong. Therefore, the prediction that C_4H_4 has a $^1B_{2g}$ ground state does not violate any physical or theoretical principles; it is a result that must nevertheless be viewed with caution.

Craig[38] has used the symmetry of the ground state as a criterion to classify molecules as aromatic (totally symmetric ground state) and pseudoaromatic (nontotally symmetric ground state). Although "Craig's rule" has proved a useful idea in theoretical discussions, it has no firm physical foundation and is based on an incomplete description of the exact molecular wavefunction—in fact, only the Kekulé forms are considered in its derivation.

3. Comparison of the Simple MO and VB Descriptions of C_4H_4

We can now summarize the viewpoints of the simplest MO and VB pictures. MO theory suggests that C_4H_4 is not stabilized by π-electron delocalization and, further, that it has a triplet ground state. (Triplet states are usually associated with high chemical reactivity akin to that of a diradical.) On the other hand, VB theory predicts exceptionally large stabilization from resonance between the two Kekulé structures, and it too finds a peculiarity in the ground-state wavefunction, in this case a nontotally symmetric singlet.

The dilemma presented by the simple VB and MO descriptions is actually only apparent. If VB wavefunctions are constructed from *all* canonical structures and *all* configurations are included in the MO wavefunctions, the two theories can be shown to be completely equivalent.[77] The reason why so much attention has been paid to the disagreement of the simple theories is that *they normally agree at a level of approximation where their wavefunctions have a close correspondence with the pictorial structures of the chemist* and calculations are trivial to perform. After we have discussed configuration wavefunctions, the two theories will be compared in more detail (Section B).

4. STRAIN ENERGY CONSIDERATIONS

The nonexistence of cyclobutadiene was used in the past to support the MO predictions, but it was early pointed out that the strain energy of four trigonal carbon atoms in a square arrangement might be sufficient to account for its apparent instability.[59, 133]

A calculation by the method of perfect electron pairs was made by Penney[104] to support the view the C_4H_4 would be too highly strained to exist. This method of Heitler–London–Slater–Pauling considers that in some molecules only one "sensible" way of pairing electrons exists, so that one canonical structure is a good approximation to the ground state. When this occurs, the molecule is said to be in a pure valence state. (Examples of pure valence states are found in saturated hydrocarbons and in the σ framework of olefins and aromatics.) The energy of the molecule is then

$$E = \text{constant} - \tfrac{1}{2} \sum_{n>m} J_{mn} + \tfrac{3}{2} \sum_{\text{bonds}} J_{mm'} \qquad \text{(A-21)}$$

where the first sum is over *all* pairs of orbitals and the second is over *bonded* pairs (m bonded to m'), and J is the usual exchange integral. The energy is minimized with respect to the AO's that form the basis for the canonical structure. Unfortunately, Penney considered that perfect overlap would prevail in C_4H_4 and thus found very weak bonds. Coulson and Moffitt[30] later carried out a more detailed calculation which allowed for hybridization and the occurrence of bent bonds. This careful analysis showed that the framework of C_4H_4 was strained by no more than 0.8 eV per CH group, resulting in a total strain of ~ 74 kcal/mole, which is not exceptionally large. A similar study by Weltner[129] with some numerical variations reached the same conclusion. It seems clear that ring strain alone cannot account for the nonexistence of cyclobutadiene, and few authors still consider this a possibility.

There are other effects, however, that may play a role in determining the properties of C_4H_4. Coulson and Moffitt[30] suggest that since MO theory predicted a triplet ground state, Jahn–Teller considerations would predict that a framework with square configuration would be unstable to at least one normal vibration mode. We will consider this in more detail in Section C.

5. ANTISYMMETRIZED MOLECULAR ORBITAL DESCRIPTION

We saw that the ground state of C_4H_4 has an open-shell configuration when the symmetry molecular orbitals are built from linear combinations of $2p\pi$ AO's. When an open shell results from truly degenerate levels, the electronic states cannot be determined by simple inspection of the direct products of the irreducible representations of these levels.[26]

Mulliken[96] has given a method for these cases, which for C_4H_4 starts by considering two π_u electrons in a diatomic molecule. In the imaginary process of splitting the nuclei to form a square molecule, group theory readily shows that π_u^2 of $D_{\infty h}$ goes over to e_u^2 of D_{4h}. If the orbital angular momenta of the electrons of the diatomic are allowed to couple, simple multiplet theory shows that π_u^2 gives rise to the states $^3\Sigma_g^-$, $^1\Delta_g$, $^1\Sigma_g^+$ of $D_{\infty h}$. When the symmetry is reduced to D_{4h}, Mulliken's tables give the correlations shown in Fig. 12.2.

$$
\begin{array}{cc}
D_{\infty h} & D_{4h} \\
\hline
^3\Sigma_g^- & \to \quad ^3A_{2g} \\
^1\Delta_g & \to \quad ^1B_{1g} + {}^1B_{2g} \\
^1\Sigma_g^+ & \to \quad ^1A_{1g}
\end{array}
$$

FIG. 12.2. Correlation of $\pi_u^2(D_{\infty h})$ with $e_u^2(D_{4h})$.

The configuration $a^2 e^2$ gives rise then to four states: $^3A_{2g}$, $^1B_{1g}$, $^1B_{2g}$, $^1A_{1g}$, all of which are degenerate in the semiempirical MO theory. This degeneracy is obviously accidental, and so it is only by the inclusion of electron-repulsion effects that we can hope to learn how the lower states may be split.

In order to consider electron repulsion, a more complete description of the Hamiltonian operator must be made and the wavefunction can be constructed in a manner similar to that used to build antisymmetrized VB wavefunctions from AO's to describe a canonical structure. This is done by taking products of the one-electron wavefunctions, ψ_i, each multiplied by a spin function. The closed-shell ground state of benzene according to Eq. (A-1) is given by the product function

$$\psi_0(1)\,\alpha(1)\,\psi_0(2)\,\beta(2)\,\psi_{+1}(3)\,\alpha(3)\,\psi_{+1}(4)\,\beta(4)\,\psi_{-1}(5)\,\alpha(5)\,\psi_{-1}(6)\,\beta(6)$$

where each spatial factor is now a molecular orbital and contains two electrons with opposed spins to satisfy the Pauli principle. This wavefunction is not sufficiently general due to the indistinguishability of the electrons and must be antisymmetrized by considering all permutations of electrons. The resulting linear combinations of products can then be expressed as the Slater determinantal function [Eq. (A-22)] for the state a^2e^4 of benzene. Depending on the

$$V_0 = (6!)^{-1/2} \begin{vmatrix} \psi_0(1)\alpha(1) & \psi_0(1)\beta(1) & \psi_{+1}(1)\alpha(1) & \cdots & \psi_{-1}(1)\beta(1) \\ \psi_0(2)\alpha(2) & \psi_0(2)\beta(2) & \psi_{+1}(2)\alpha(2) & \cdots & \psi_{-1}(2)\beta(2) \\ \cdot & \cdot & \cdot & \cdots & \cdot \\ \cdot & \cdot & \cdot & \cdots & \cdot \\ \cdot & \cdot & \cdot & \cdots & \cdot \\ \psi_0(6)\alpha(6) & \psi_0(6)\beta(6) & \psi_{+1}(6)\alpha(6) & \cdots & \psi_{-1}(6)\beta(6) \end{vmatrix}$$

(A-22)

nature of the ψ, this single determinant may not be the best wavefunction that can be written for a closed configuration; this will be considered further below. We can use a more compact notation by giving the diagonal element, absorbing the normalization factor, and showing β spin functions by a bar,

$$V_0 = (\,0\bar{0} + 1\overline{+1} - 1\overline{-1}\,)$$

In C_4H_4, the two e electrons can be placed in the degenerate orbitals $\psi_{\pm 1}$ in four distinct ways. The single determinants constructed from imaginary orbitals [Eq. (A-1)] do not form bases for irreducible representations of D_{4h}, but certain linear combinations [Eqs. (A-23)] of those derived from an open-shell configuration are proper. By inspection of the behavior of the four corresponding Slater determinants under the operations of the group D_{4h}, it is easily found that

$$(0\bar{0} + 1\overline{+1}) - (0\bar{0} - 1\overline{-1}) \subset {}^1B_{2g}$$
$$(0\bar{0} + 1\overline{+1}) + (0\bar{0} - 1\overline{-1}) \subset {}^1B_{1g} \quad \text{(A-23)}$$
$$(0\bar{0} + 1\overline{-1}) - (0\bar{0}\overline{+1} - 1) \subset {}^1A_{1g}$$
$$(0\bar{0} + 1\overline{-1}) + (0\bar{0}\overline{+1} - 1) \subset {}^3A_{2g}$$

This limited form of configuration interaction has been called first-order configuration interaction by Moffitt.[93]

The application of a Hamiltonian, which explicitly contains the interelectronic interaction operator, to these wavefunctions can now be considered. The resulting method is the basis for most detailed MO calculations and was first clearly expounded for π-electron systems by Goeppert-Mayer and Sklar.[53] It is usually called the "method of antisymmetrized molecular orbitals"

A. THE π-ELECTRONIC STATES OF CYCLOBUTADIENE

(ASMO), and when the mixing with higher excited configurations is also taken into account, it is called "ASMO including configuration interaction" (ASMO–CI).

Craig[34–38] first studied the cyclobutadiene problem thoroughly by the ASMO–CI method, which will now be developed in some detail. In a closed-shell system containing N π electrons, an excited state occurs by promotion of an electron from a filled orbital i in

$$V_0 = (0\bar{0}1\bar{1}\ldots\tfrac{1}{2}N\tfrac{\overline{1}}{2}N) \tag{A-24}$$

to some unoccupied orbital k'. The resulting singlet (V) and triplet (T) wavefunctions are then given by

$$\begin{matrix}V_{ik'}\\ T_{ik'}\end{matrix} = (2)^{-1/2}[(0\bar{0}\ldots i\bar{k}'\ldots) \mp (0\bar{0}\ldots \bar{i}k'\ldots)] \tag{A-25}$$

where the minus sign goes with the singlet. This form of $V_{ik'}$ and $T_{ik'}$ is required since the wavefunctions must correspond to the same component of the spin quantum number, in this case $S_z = 0$. The ground state of a closed-shell configuration is frequently represented adequately by a single determinantal wavefunction; however, the ground state of an open-shell cannot be decided so simply. We can specify the open-shell configurations by their symmetries, e.g., two of these are

$$V(^1B_{2g}) = (2)^{-1/2}[(0\bar{0} + 1\overline{+1}) - (0\bar{0} - 1\overline{-1})]$$

$$T(^3A_{2g}) = (2)^{-1/2}[(0\bar{0} + 1\overline{-1}) + (0\bar{0} + 1 - 1)]$$

but we cannot determine which is of lower energy without explicit calculation.

A ground state could be approximated in general by Eq. (A-24) and an excited state by Eq. (A-25) and their energies computed. However, a better way to proceed is to write down the wavefunctions of the lower configurations, form linear combinations of the $V_{ik'}$ and of the $T_{ik'}$, and allow these to mix by a variation calculation. This configuration interaction generally causes considerable reduction of the energies of the lower states, and frequently states cross. The problem is simplified since matrix elements between wavefunctions of different multiplicity and of different symmetry vanish.

The π-electron Hamiltonian is given by

$$\mathbf{H} = \mathbf{H}_{\text{core}} + \sum_{\mu > \nu} e^2/r_{\mu\nu} \tag{A-26}$$

$$\mathbf{H}_{\text{core}} = \sum_{\mu} \mathbf{H}_{\text{core}}(\mu) \tag{A-27}$$

where

$$\mathbf{H}_{\text{core}}(\mu) = \mathbf{T}(\mu) + \sum_{m} \mathbf{U}_m(\mu) + \sum_{r} \mathbf{U}_r^*(\mu) \tag{A-28}$$

$\mathbf{T}(\mu)$ is the kinetic energy operator of electron μ, $\mathbf{U}_m(\mu)$ is the potential energy operator of electron μ in the field of the positively charged core atom m, $\mathbf{U}_r{}^*(\mu)$ similarly refers to a neutral atom not explicitly included in the core, and $\mu, \nu, \ldots$ label electrons; $m, n, \ldots$, core atoms; $r, s, \ldots$, atoms uncharged in the core, e.g., hydrogen atoms. For an electronic state of symmetry Λ, configuration wavefunctions are constructed, e.g., for $^1\Lambda$

$$^1\Psi(\Lambda) = \sum_\sigma A_{\Lambda\sigma} V_\sigma \tag{A-29}$$

where the generalized index $\sigma = 0, ik', \ldots$ in Eq. (A-29). A variation calculation using Eq. (A-26) and Eq. (A-29) in Eq. (A-30) is carried out

$$\mathbf{H}\,^1\Psi(\Lambda) = E(^1\Lambda)\,^1\Psi(\Lambda) \tag{A-30}$$

to determine the energies of states with symmetry Λ.

In a straightforward manner integrals over determinantal wavefunctions are expanded first over MO's ψ_i to give a collection of core [Eq. (A-31)] and interelectronic [Eq. (A.32)] integrals. The latter are next expanded in the AO basis,

$$I_{ij} = \int \psi_i^*(\mu)\, \mathbf{H}_{\text{core}}(\mu)\, \psi_j(\mu)\, d\tau(\mu) \tag{A-31}$$

$$(ij|kl) = \int \psi_i^*(\mu)\psi_k^*(\nu)\,(e^2/r_{\mu\nu})\psi_j(\mu)\psi_l(\nu)\, d\tau(\mu)\, d\tau(\nu) \tag{A-32}$$

so that ultimately the energy can be expressed in terms of two one-electron core integrals, α_m and β_{mn},

$$\alpha_m = \int \chi_m^*(\mu)\, \mathbf{H}_{\text{core}}(\mu)\, \chi_m(\mu)\, d\tau(\mu) \tag{A-33}$$

$$\beta_{mn} = \int \chi_m^*(\mu)\, \mathbf{H}_{\text{core}}(\mu)\, \chi_n(\mu)\, d\tau(\mu) \tag{A-34}$$

and the two-electron Coulomb repulsion integrals, $[mn|pq]$.

$$[mn|pq] = \int \chi_m^*(\mu)\chi_p^*(\nu)\,(e^2/r_{\mu\nu})\chi_n(\mu)\chi_q(\nu)\, d\tau(\mu)\, d\tau(\nu) \tag{A-35}$$

The energies and interaction energies of the various configurations can be finally expressed in terms of these basic integrals. Details of the evaluation of the various integrals along with an excellent account of the theory are given by Parr.[101]

It is important to note that although the above description of the ASMO–CI method is easy to set down, the success of its application depends strongly on the values assigned to the various integrals. The number of integrals for a molecule the size of benzene is enormous. Furthermore, *all* excited configurations strictly ought to be taken into account, not simply the singly excited ones. There are reasons to believe, fortunately, that only the latter type need to be considered in many problems. When Craig first carried out the lengthy configuration interaction calculations, there were no high-speed computers available, and, like many works of that period, these calculations required considerable effort.

To perform an actual calculation, the integrals over AO's must eventually be evaluated. The calculation of the electron-repulsion integrals [Eq. (A-32)] is difficult since expansion in terms of Eq. (A-35) generally involves large numbers of integrals over two, three, and four atomic centers. The latter are difficult to evaluate, but it is just their presence that distinguishes the theory and gives hope of properly locating the molecular electronic states. The general problem is so formidable that only a few simple molecules have been treated by considering all repulsion integrals. Cyclobutadiene [37] and benzene [102] were two of the first molecules studied and then many of the integrals had to be approximated. Before giving the results, it is important to pursue this problem a bit further, because the eventual theoretical decision on C_4H_4 will rest on these details.

Whenever the electronic interaction integral $(ij|kl)$ occurs, its expansion over AO's will be given by

$$(ij|kl) = \sum_m \sum_n \sum_p \sum_q c_{mi} c_{nj} c_{pk} c_{ql} [mn|pq] \qquad \text{(A-36)}$$

when the MO's are LCAO wavefunctions

$$\psi_i = \sum_m c_{mi} \chi_m \qquad \text{(A-37)}$$

It is because of Eq. (A-36) that the general calculation founders. In 1953, this impasse was bridged by Pariser and Parr,[98, 99, 100, 103] who introduced a small number of key assumptions that at once gave a tractable semiempirical theory which led to good agreement with experiment for the first time. One approximation was the *neglect of differential overlap*, which says that whenever the charge distribution $\chi_m \chi_n$ $(m \neq n)$ occurs in molecular integrals, these integrals vanish, i.e.,

$$\chi_m \chi_n \equiv 0 \qquad \text{(A-38)}$$

Equation (A-36) now reduces to

$$(ij|kl) = \sum_m \sum_n c_{mi} c_{mj} c_{nk} c_{ql} [mm|nn] \qquad \text{(A-39)}$$

where the two-center Coulomb repulsion integrals $[mm|nn]$ are relatively easy to evaluate. This sweeping simplification is supported by theoretical analysis.

Another important contribution concerned the proper location of the repulsion integrals on a relative energy scale. Evaluation of the one-center integral [11|11], which is the repulsion of two electrons in a single p orbital of a carbon atom, from theoretical formulas using Slater AO's gives values believed to be too large, since changes in the core energy are not allowed for. Pariser[98] showed that the following relation (A-40) involving only experimental quantities could be logically derived,

$$[mm|mm] = I_m - A_m \qquad \text{(A-40)}$$

where I_m is the appropriate valence-state ionization potential and A_m, the corresponding electron affinity.

In 1951, Moffitt[94] also suggested that a semiempirical approach was needed to properly locate excited states and so developed the atoms-in-molecules (AIM) theory. This view sought to relate the states of atoms in a molecule (valence states) with the actual spectroscopic states of the atoms. In this way certain integrals were in effect obtained from experimental data, but the method has not been much used because of certain technical difficulties.

The last three paragraphs sketch the considerations behind the nonempirical and semiempirical calculations that have appeared in the literature. There are those who adhere to one spirit or the other, but it can be said that the semi-empirical approach has been by far the most fruitful. There are, however, many ways of proceeding with a semiempirical calculation. The π-electron literature at this level of complexity is often a nightmare even to the initiated because configuration interaction is invoked to arbitrary extents, different techniques of evaluating integrals are used, and frequently some integrals are neglected entirely. It will be for these reasons that we cannot always compare unequivocally the studies on C_4H_4.

In Craig's studies of C_4H_4, all integrals were evaluated exactly except the three- and four-center repulsions, which had to be approximated, all configurations were taken into account, and no experimental data were employed in the calculations. After first-order configuration interaction the lowest state was found to be the triplet, $^3A_{2g}$, separated by only 0.05 eV from the singlet, $^1B_{2g}$. When full configuration interaction was allowed for, crossing occurred and $^1B_{2g}$ became the ground state with a separation of 0.71 eV. The few lowest states before and after configuration interaction are shown on a relative scale in Fig. 12.3.

Craig considered that the full configuration interaction results were the most significant, but these were obscured for he also found that the triplet was lower than the singlet by 1.00 eV when the estimated three- and four center integrals were omitted from the calculation. Since the more complete calculation agreed with the empirical VB prediction that the ground state was the nontotally symmetric singlet, he favored this picture of C_4H_4 and sought support for this view in his "rule" that was discussed above.

The empirical VB method predicted a much larger separation ($^1B_{2g}$–$^3A_{2g}$) of 3.8 eV compared with 0.7 eV (Fig. 12.3). The first excited singlet separation

A. THE π-ELECTRONIC STATES OF CYCLOBUTADIENE

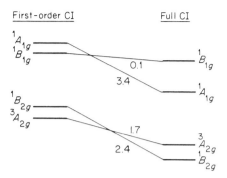

FIG. 12.3. C$_4$H$_4$ state energies after Craig.[37] Lowerings are given in electron volts.

($^1B_{2g}$–$^1A_{1g}$) was 2.5 eV by the nonempirical MO method, compared to the VB value of 7.7 eV. The quantitative agreement of the empirical VB and nonempirical MO calculations was not very good, but Craig considered this reasonable in view of the fact that the empirical theory drew its numerical parts from the experimental data on benzene, a normal aromatic. As we saw above, the ground state–lowest singlet separation is a measure of the "resonance splitting." In these terms, the MO result can be used to estimate a "resonance energy" of $2.5/7.7 \times 32 = 11$ kcal/mole. Craig concluded from his calculations that the results were too sensitive to the various integrals to decide whether $^3A_{2g}$ or $^1B_{2g}$ is the ground state of the square form.

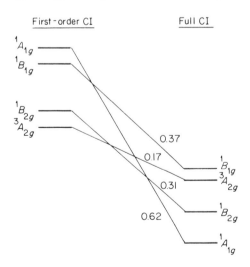

FIG. 12.4. C$_4$H$_4$ state energies after Moffitt and Scanlan.[95] Lowerings are given in electron volts.

Moffitt and Scanlan[95] calculated the energies of the states of C_4H_4 using the atoms-in-molecules approximation (Fig. 12.4). Without configuration interaction the lowest state was $^3A_{2g}$, but configurational mixing gave a $^1A_{1g}$ ground state. It was found that configuration interaction made only slight changes in the energies in contrast with the nonempirical calculations, and this has been found to be characteristic of the semiempirical methods of Moffitt and of Pariser and Parr. Since the total spread in energies of the four lowest states was found to be only 0.2 eV, these results disagree considerably with the ASMO–CI studies of Craig. Wolfsberg[135] earlier used a procedure like that of Moffitt and arrived at similar conclusions.

Shida[113, 114] calculated the ASMO energies of the 36 configurations of C_4H_4, and studied the changes in energy of the π-electron system when the square form is distorted to a rectangle. His calculations employed real molecular orbitals (Table 12.2) and neglected *all* configuration interaction. When the energy was computed using nonempirical integrals, the square form (1.40 Å) was found to be more stable by 0.7 eV than the rectangle (1.33 Å–1.54 Å), whereas the atoms-in-molecules modification suggests that the rectangle is more stable by 0.9 eV. A singlet state of C_4H_4 ($^1A_{1g}$) is so strongly

TABLE 12.1

C_4H_4 CONFIGURATION ENERGIES

State	Wavefunction	Energy (eV) Theoretical[a]	Semiempirical[b]
$^3A_{2g}$	$2^{-1/2}\{|0\bar{0}+1\overline{-1}|+|0\bar{0}+\bar{1}-1|\}$	0	0
$^1B_{1g}$	$2^{-1/2}\{|0\bar{0}+1\overline{+1}|-|0\bar{0}-\bar{1}-1|\}$	0	0
$^1A_{1g}$	$2^{-1/2}\{|0\bar{0}+1\overline{-1}|-|0\bar{0}+\bar{1}-1|\}$	2.74	1.07
$^1B_{1g}$	$2^{-1/2}\{|0\bar{0}+1\overline{+1}|+|0\bar{0}-\bar{1}-1|\}$	2.74	1.07
$^1E_{1u}(1)$	$2^{-1}\{|0\bar{0}-1\bar{2}|+|0+1\overline{+1}\overline{-1}| \\ -|0\bar{0}\overline{-1}2|-|\bar{0}+1\overline{+1}-1|\}$	7.97	5.86
$^3E_{1u}(1)$	$2^{-1}\{|0\bar{0}-1\bar{2}|+|0+1\overline{+1}\overline{-1}| \\ +|0\bar{0}\overline{-1}2|+|\bar{0}+1\overline{+1}-1|\}$	7.97	5.86
$^1E_{1u}(2)$	$2^{-1}\{|0\bar{0}-1\bar{2}|-|0+1\overline{+1}\overline{-1}| \\ -|0\bar{0}\overline{-1}2|+|\bar{0}+1\overline{+1}-1|\}$	13.49	8.03
$^3E_{1u}(2)$	$2^{-1}\{|0\bar{0}-1\bar{2}|-|0+1\overline{+1}\overline{-1}| \\ +|0\bar{0}\overline{-1}2|-|\bar{0}+1\overline{+1}-1|\}$	2.45	3.70

[a] Reference 14.
[b] Reference 73.

depressed that the rectangular form is stabilized relative to the square. The cyclobutadiene wavefunctions are thus unstable to this nuclear displacement, and we will see that this behavior is unlike that found for benzene and normal aromatics.

Longuet-Higgins and McEwen[78] calculated the energies of the lower state of C_4H_4 employing the approximations of Pariser and Parr[98, 99, 100, 103] and including only first-order configuration interaction. Bouman[14] repeated these calculations but evaluated all integrals from theoretical formulas. Their results (Table 12.1) point up the large magnitude of the changes introduced by the semiempirical theory.

A more reasonable test of the semiempirical Pariser–Parr theory would include the singly excited configurations. This has been done[116] employing values of Coulomb and core integrals obtained from the benzene spectrum.[115] The ground state of the square form is $^3A_{2g}$ and the lowest singlet state ($^1B_{2g}$) lies 0.82 eV higher. On deformation to a rectangle whose sides are 1.338 and 1.483 Å (*trans*-butadiene distances), the ground state becomes $^1A_{1g}$, which lies 0.55 eV below the lowest triplet $^3B_{1g}$. Furthermore, inclusion of doubly excited configurations results in degeneracy of the lowest singlet and triplet states of the square form. Full configuration interaction studies were not carried out; however, it seems likely from this trend that the results will be in substantial agreement with the nonempirical calculations of Craig.

6. Nonempirical Valence-Bond Description

In Section A,2, the valence-bond theory, was outlined and the results of empirical calculations for C_4H_4 indicated a singlet ground state of B_{2g} symmetry. McWeeny[90, 91, 92] made a full, nonempirical valence-bond calculation employing all non-, mono-, and diexcited singlet and triplet structures. The theory and calculations are heavy and will not be described here. One essential point of the McWeeny $\overline{VB}$ theory, however, is the consistent use of a set of accurately orthogonal atomic orbitals. *All* integrals were then evaluated theoretically. The results for the state energies agreed exactly with the previously described full ASMO–CI calculations. It can be concluded that the best that π-electron theory can say about C_4H_4 in a purely nonempirical calculation is that the square form has a singlet ground state ($^1B_{2g}$) which has little resonance stabilization in the conventional sense.

Takekiyo[123, 124, 125] repeated the calculation of McWeeny using both orthogonalized and nonorthogonalized atomic orbitals. The former confirmed the McWeeny calculations and the latter showed that neglect of overlap has little effect on the results. Full calculations with orthogonalized orbitals showed the $^1B_{2g}$ ground state to possess 21.5% ionic character in the square form with a carbon–carbon bond order equal to 0.416. (Compare Section A,1.)

B. Some Conclusions and Comparisons

In was pointed out that when all configurations are included in the MO description and all structures in the VB, the two theories must agree exactly. The lack of agreement between the simple theories in the case of C_4H_4 and other pseudoaromatics can only be accounted for by comparing wavefunctions at the different levels of approximation.

As early as 1934, Wheland[133] compared the simple MO and VB results for C_4H_4 and noted that inclusion of electronic interactions would remove the degeneracy predicted by simple MO theory. Wheland[132] considered the proper configuration wavefunctions without making explicit calculations and showed that the primary discrepancy between the two theories arises because the LCAO wavefunctions for C_4H_4 in effect neglect the resonance between the two Kekulé structures while they overemphasize the contribution of excited structures. Resonance among the excited structures is of such a nature that it stabilizes the triplet state more than the singlet. He further showed this is just the reverse of the situation encountered in benzene.

Another comparison, due to Coulson,[27] makes these points particularly clear. If the AO's are numbered as before, then the MO's in real form, their

energies, and their contribution to the bonding and antibonding character of the bonds 0–1, 1–2, 2–3, and 3–0 are given in Table 12.2. The lowest MO

TABLE 12.2

C_4H_4 Wavefunctions

		Bonding per electron	
MO	Energy	0–1	1–2
$\psi_I = \frac{1}{2}(\chi_0 + \chi_1 + \chi_2 + \chi_3)$	$+2\beta$	$+\frac{1}{4}$	$+\frac{1}{4}$
$\psi_{II} = \frac{1}{2}(\chi_0 + \chi_1 - \chi_2 - \chi_3)$	0	$+\frac{1}{4}$	$-\frac{1}{4}$
$\psi_{III} = \frac{1}{2}(\chi_0 - \chi_1 - \chi_2 + \chi_3)$	0	$-\frac{1}{4}$	$+\frac{1}{4}$
$\psi_{IV} = \frac{1}{2}(\chi_0 - \chi_1 + \chi_2 - \chi_3)$	-2β	$-\frac{1}{4}$	$-\frac{1}{4}$

configurations and their equivalent VB structures are given in Fig. 12.5.

B. SOME CONCLUSIONS AND COMPARISONS

Configuration	$\psi_I^2 \psi_{II}^2$	$\psi_I^2 \psi_{III}^2$	$\psi_I^2 \psi_{II} \psi_{III}$	$\psi_I^2 \psi_{II} \psi_{III}$
Multiplicity	Singlet	Singlet	Singlet	Triplet
Bond Structure	□	□	⌐ ¬	⌐ ¬
Energy	4β	4β	4β	4β

FIG. 12.5. Comparison of MO configurations and VB structures in C_4H_4.

We see immediately why the Hückel resonance energy is zero. All the energies are equal to that of two isolated double bonds. Furthermore, if configuration $\psi_I^2 \psi_{II}^2$ or $\psi_I^2 \psi_{III}^2$ is used to calculate the resonance energy, it is evident that only a single Kekulé structure has been taken into account. Since $\psi_I^2 \psi_{II}^2$ and $\psi_I^2 \psi_{III}^2$ do not mix quantum mechanically because of their differing symmetries, we are left in effect with trying to make a single Kekulé structure do the work of two in determining the MO delocalization energy.

When configuration interaction is taken into account, the more complete MO wavefunctions can still be compared with VB structures. Configuration interaction is most important in those states in which nonpolar VB structures are heavily weighted, and this is due to the more complete separation of the electrons in such structures. Triplet states are less influenced by configuration interaction, since the wavefunction already demands that two electrons with parallel spins cannot be found on the same atom, as would occur in an excited VB structure.

Coulson's analysis can be extended by noting the weight of the polar structures in the full MO wavefunctions[36] (Table 12.3). It is seen that any of

TABLE 12.3

COMPARISON OF MO CONFIGURATIONS AND VB STRUCTURES IN C_4H_4

Configuration	Structures		
	Nonpolar	Monoexcited	Diexcited
$\psi_I^2 \psi_{II}^2$	37.5%	50%	12.5%
$\psi_I^2 \psi_{III}^2$	37.5%	50%	12.5%
$\psi_I^2 \psi_{II} \psi_{III}$	37.5%	50%	12.5%
$\psi_I \psi_{II} \psi_{III} \psi_{IV}$	25%	0%	75%
$\psi_{II}^2 \psi_{IV}^2$	37.5%	50%	12.5%
$\psi_{III}^2 \psi_{IV}^2$	37.5%	50%	12.5%
$\psi_{II} \psi_{III} \psi_{IV}^2$	37.5%	50%	12.5%
Minimized	88%	11%	1%

the configurations alone is a very poor approximation to the final mixture for the lowest state. This also says that the empirical VB method which considers only nonpolar structures is already a fairly good approximation to the best that can be done. A similar situation exists with the benzene wavefunctions. Apparently in most cases, whether we are dealing with "normal" or "abnormal" molecules, single configuration MO wavefunctions are rather poor because of the overemphasis on excited structures. The reason why cyclobutadiene must be considered "abnormal" only emerges fully after a complete configuration interaction treatment has been made which reveals to what extent the ground-state degeneracy of a $4n$ π-electron molecule is removed.

C. Jahn–Teller Considerations

Predictions of electronic structures and energies of aromatic molecules are particularly reliable when the σ framework is unstrained. It is not entirely clear, however, that the usual separability of σ and π electrons is valid when the framework is strained as in cyclobutadiene. This was mentioned in previous sections where it was noted that the square-planar form of C_4H_4 may not be an arrangement of minimum potential energy, when only the π electrons are considered explicitly. Intuitively, this is a rather unusual result, for it is difficult at first sight to imagine what forces cause the destruction of symmetry in a molecule possessing equivalent, nonpolar Kekulé structures. Since state functions (wavefunctions) are not directly related to any physical observable, it is surprising to find that it is just the symmetry of state functions that sometimes places stringent conditions on the behavior of a molecule's potential energy surface.

This principle was first clearly stated by Jahn and Teller[62] and can be expressed as follows: A configuration of a polyatomic molecule whose electronic state function is orbitally degenerate cannot be stable with respect to *all* displacements of the nuclei, unless in the original configuration the nuclei all lie on a straight line. This principle was later expanded by Jahn[61] to include spin degeneracy. Since nuclei are very heavy compared to electrons, we might expect a molecule with an orbitally degenerate ground state to behave in three ways: (1) it can dissociate or at least undergo rupture of one bond if it possesses no stable nuclear configuration; (2) it can assume one of several possible shapes having lower symmetry; or (3) if the zero-point energies of the molecular vibrations which destroy the electronic degeneracy are not smaller than the energy the molecule can gain by distorting, then the electronic and nuclear motions can become closely coupled. In the latter case (dynamic Jahn–Teller effect), the molecule remains in a degenerate state and passes rapidly and regularly through a series of distorted structures. If the coupling is weak, it is best to regard the degenerate vibrational levels as being split.

C. JAHN–TELLER CONSIDERATIONS

According to MO theory, cyclobutadiene has an orbitally degenerate ground-state configuration $a^2 e^2$ and thus should be subject to Jahn–Teller effects. Coulson and Moffitt[30] first stressed the possible importance of this principle with respect to C_4H_4 and suggested that its apparent instability was due to an intrinsic decomposition coordinate, such that "any attempts at its synthesis will lead to the formation of its decomposition products, stable unsaturated hydrocarbons."

A simple MO argument given by Coulson[27] (Section B) can be extended by considering the changes in HMO energies that occur when square C_4H_4 is deformed to the rectangle (Fig. 12.6). In general, the energies of MO's $\psi_I \ldots \psi_{IV}$

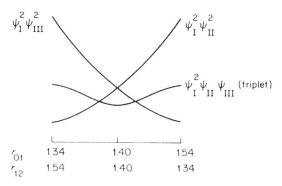

FIG. 12.6. MO configuration energies of C_4H_4 as a function of a continuous rectangular deformation.

are $(\pm \beta_{01} \pm \beta_{12})$ where two resonance integrals for the bonds of unequal length are now needed. It is clear that only the triplet state is stabilized in the square form. If the molecule were in the state $\psi_I^2 \psi_{II}^2$, it would experience forces that tend to shorten bond 0–1 and lengthen bond 1–2. In this one-electron approximation, Coulson has called this behavior a pseudo-Jahn–Teller effect. Configuration interaction generally has the effect of splitting degeneracies, so that Jahn–Teller considerations would no longer apply. In C_4H_4, such splitting would have to be large to make the square form more stable than one of the Kekulé structures, and we saw from Shida's ASMO studies[113, 114] (Section A, 5) that this is not likely. We conclude then that the lowest state will be a rectangular singlet which possesses essentially isolated double bonds and no resonance energy. The lowest excited state will be a square triplet which is stable with respect to nuclear deformation, but this can only be a point of metastable equilibrium at best. Coulson concludes that even if this state were obtained, the molecule "would almost certainly react rapidly to form a dimer or polymer."

In 1937, Lennard-Jones and Turkevich[68] showed how compression of framework σ bonds could be taken into account in LCAO–MO theory.

Recently, Longuet-Higgins and Salem[80] have developed the theory considerably. In essence, the total energy (E) of a conjugated hydrocarbon is the sum of contributions from the π and σ electrons and is given by

$$E = E_\pi + E_\sigma \tag{C-1}$$

where E_π is a function of the resonance integrals of *all* the C—C bonds

$$E_\pi = E_\pi(\beta_1, \beta_2, \ldots, \beta_i, \ldots) \tag{C-2}$$

$$\beta_i = \beta(r_i) \tag{C-3}$$

r is the length of bond i, and E_σ is a sum of independent contributions from the C–C bonds. An expression for the latter contributions $u(r_i)$ was obtained under the assumption of the linear dependence of bond order and bond length.

$$E_\sigma = \sum_i u(r_i) \tag{C-4}$$

Finally, $u(r_i)$ is given[80, 119] in the form

$$u(r_i) = a(r_i - b)\beta(r_i) \tag{C-5}$$

$$\beta(r_i) = \beta_0 \exp[-(r_i - c)/d] \tag{C-6}$$

where the constants a, b, c, d, β_0 can be determined from experiment. In this manner, the total energy can be minimized or can be computed as a function of geometry.

Lennard-Jones applied these ideas to $C_{2n}H_{2n}$ and determined that when n is odd (e.g., benzene) minimum energy occurs with a regular polygonal geometry. When n is even (e.g., cyclobutadiene), however, cyclic molecules will consist of bonds having alternating lengths. He concluded C_4H_4 would contain two isolated single and double bonds of total energy 4β. Hence, even though the minimum energy geometry is no longer square, the prediction is again for zero delocalization energy.

Craig[35] studied another important kind of shape change in C_4H_4. Ignoring changes in the framework, he showed how the energy varies in the rhombus deformation as 1,3-bonding occurs to give bicyclobutadiene.

Empirical VB calculations show that the B_{2g} ground state remains the lowest in energy over the course of this hypothetical process, and the resonance energy slowly decreases. Similar studies using MO wavefunctions show that as soon as 1,3-bonding begins (assignment of an exchange integral), the delocalization energy rises from zero, and the ground state is A_g rather than B_g. Craig

C. JAHN–TELLER CONSIDERATIONS

gives an interesting discussion of the relationship of these changes to wavefunction symmetries. (The original source should be consulted.[33])

Jahn–Teller considerations applied to cyclic molecules having near-degeneracy of several states has also been studied extensively by Liehr.[69–74] Using MO perturbation theory and the parameterization procedure of Lennard-Jones, good agreement was found with Shida's atoms-in-molecules calculations on C_4H_4. The rectangular form was more stable than the square by 0.9 eV (20.9 kcal/mole) and it was shown that the b_{1g} in-plane vibration could contribute to this distortion.

Hobey and McLachlan[55] carried out similar calculations employing the parameterization of Longuet-Higgins and Salem.[80] It was found that the square form with 1.425-Å sides is only 0.5 eV (11.4 kcal/mole) higher in energy than the rectangular form with 1.35- and 1.50-Å sides. They concluded that since the distortion energy is small, the lowest state of C_4H_4 may yet be the square triplet, and so the relative stability of the various planar shapes is not connected with dynamic Jahn–Teller distortion. These same studies suggested that cyclopentadienyl C_5H_5 has a distortion energy of only 0.06 eV (1.38 kcal/mole), and that oscillation between a continuous series of distorted shapes of equal energy will occur.

The results of Liehr[74] and Hobey and McLachlan[55] are supported by the studies of Snyder,[118, 119, 120] who concluded that although the extreme rectangular form is predicted to be more stable than the square modification by 0.9 eV (20.9 kcal/mole), it is itself unstable to a distortion involving concurrent lengthening and compression of the single bonds while the double-bond length remains constant. This distortion does not lead to an energy minimum, but at the bond distances shown in (*1a*), the molecule is predicted to be stable with respect to the rectangular shape (1.35 Å × 1.50 Å) by some 0.4 eV (9.2 kcal/

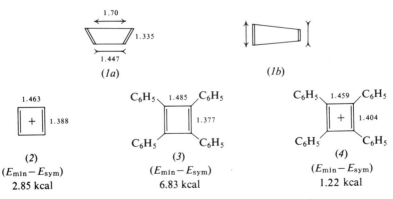

CHART 12.2. Geometry of cyclobutadienes (bond lengths in Ångstroms).

mole). It is of interest that the alternative distortion involving stretching and compressing the double bonds while the single bonds are maintained constant (*1b*) is predicted to be of high energy relative to the uncompressed rectangle. Furthermore, it was shown[118] that introduction of electron repulsion into the Longuet-Higgins and Salem parameterization procedure leads to a greatly reduced instability of the lowest energy singlet of the square with respect to the rectangle. This energy difference is then estimated at only 0.02 eV (0.46 kcal/mole).

From all of these studies we might anticipate that square cyclobutadiene will be a triplet but that the molecule may assume distorted forms. It is also likely that certain distorted forms may represent stable singlet states; however, certain other distortions probably will experience insufficient restoring force so that the molecule will undergo rupture of at least one bond. Since the energies involved in the various distortions are apparently small and the potential wells of any stable singlets are shallow, it can be expected that cyclobutadiene might have transient existence at usual temperatures, but that thermal interconversion of the various forms will be facile. It does not seem unreasonable that as "destructive vibrations" are activated, cyclobutadiene will rapidly drain away as butadienyl diradical and/or acetylene.

$$\text{Square form} \rightleftharpoons \text{Distorted forms}$$

$$HC\overset{HC-CH}{\underset{}{\diagdown}}CH \cdot \quad \text{and/or} \quad \begin{matrix} H & H \\ C & C \\ ||| & + & ||| \\ C & C \\ H & H \end{matrix}$$

Snyder also gave an HMO description of the Jahn–Teller effect on cyclobutadiene monocation $C_4H_4^+$, tetraphenylcyclobutadiene $[(C_6H_5)_4C_4]$, and $[(C_6H_5)_4C_4]^+$. The results are shown in Chart 12.2, where the energy differences are the calculated minimum energies minus those of the square configurations. Introduction of electron repulsion in the $(C_6H_5)_4C_4$ computation resulted in a separation of only 0.02 eV (0.46 kcal/mole).

All of the above treatments rest on the assumption that the Jahn–Teller distortion would operate in the plane of the ring. The possibility of a nonplanar distortion leading to a tetrahedral framework (tetrahedrane), was nevertheless considered by Lipscomb.[75] The configuration of such a species is $a_1^2 t_1^6 e^4$ in which the e level is believed to be slightly bonding. The strain energy of tetrahedrane has been estimated at ~ 90 kcal/mole.[129]

D. Substituted Cyclobutadienes, Condensed Aromatic Cyclobutadienes, and Cyclobutadiene–Metal Complexes

As was previously pointed out, the orbital degeneracy in square-planar cyclobutadiene is a consequence of the number of π electrons and their distribution in degenerate energy levels. The potential instability of such a symmetrical $4n$ π structure is also indicated by ASMO–CI computations, which suggest an uncommonly small energy separation between the lowest singlet and triplet states. The view adopted in the present section is that the instability of square-planar cyclobutadiene can be simply discussed in terms of the orbital degeneracy predicted by HMO theory.

1. SUBSTITUTED CYCLOBUTADIENES

Hückel theory requires that tetra-substituted cyclobutadienes possessing D_{4h} symmetry, such as (5), have an orbitally degenerate ground state irrespective of the nature of the substituent. Tetragonal symmetry on the other hand is not a necessary condition for orbital degeneracy, since theory predicts triplet ground states for substituted cyclobutadienes of any symmetry, provided the substituents are only slightly polar and do not appreciably perturb the uniform electron distribution of the ring. For example, HMO theory requires that systems (6)–(10) have triplet ground states.[111]

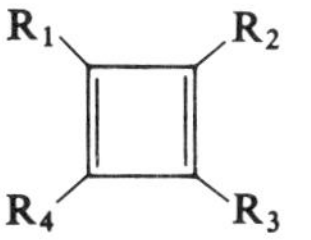

$R_1 = R_2 = R_3 = R_4$
(5)

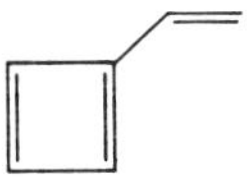

DE = 0.6 β; T
(6)

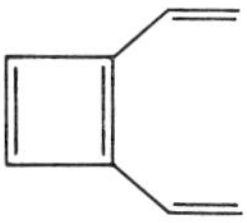

DE = 1.21 β ; T
(7)

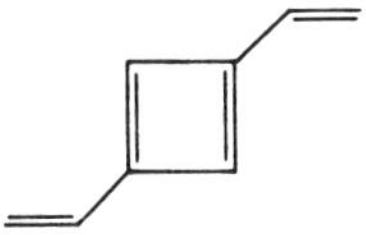

DE = 1.15β; T
(8)

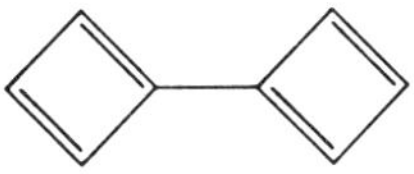

DE = 1.25β; T
(9)

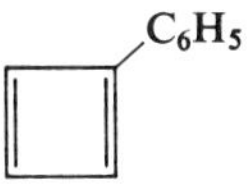

DE = 2.53β; T
(10)

Substituted cyclobutadienes bearing strongly electron-attracting or electron-donating substituents, such as cyano and amino, are predicted to have singlet ground states. Nevertheless, theory suggests high chemical reactivity on account of a vacant bonding molecular orbital (BMO) or nonbonding molecular orbital (NBMO) in cyclobutadienes with electron-attracting groups and to an antibonding molecular orbital (ABMO) or NBMO in cyclobutadienes with electron-donating groups. For 1,2-dicyano- (*11*), 1,3-dicyano- (*12*), 1,2-diamino- (*13*), 1,3-diamino- (*14*), and 1-amino-2-cyanocyclobutadiene (*15*), this is shown in Table 12.4.

TABLE 12.4

HMO Energy Levels of Substituted Cyclobutadienes Bearing Polar Substituents[a]

Cyclobutadiene	Energy levels, $k_i(\beta^{-1})$[b]							
	$i=1$	2	3	4	5	6	7	8
1,2-Dicyano-	2.75	2.62	2.00	0.43	0.33	−0.77	−1.09	−2.29
1,3-Dicyano-	2.74	2.64	2.00	0.61	0	−0.46	−1.25	−2.27
1,2-Diamino-	2.41	1.77	1.29	−0.17	−0.20	−2.09	—	—
1,3-Diamino-	2.39	1.85	1.20	0	−0.35	−2.09	—	—
1-Amino-2-cyano-	2.70	2.24	1.49	0.38	−0.19	−0.92	−2.20	—

[a] The hetero parameters used in the calculations are $\alpha_{NH_2} = \alpha_0 + 1.5\beta_0$; $\alpha_{C \equiv N} = \alpha_0 + 2\beta_0$; $\beta_{C \equiv N} = 1.2\beta_0$.
[b] Equation (A-3).

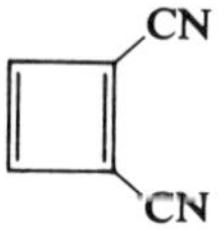

(*11*) 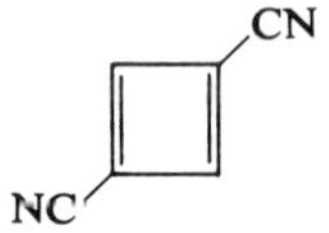 (*12*) 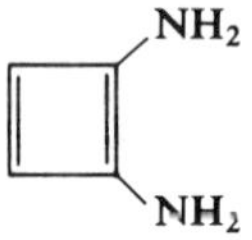 (*13*) 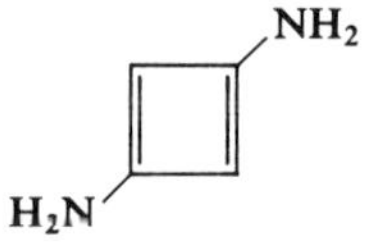 (*14*) 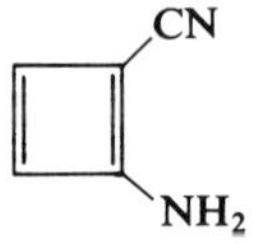 (*15*)

Introduction of both electron-withdrawing and electron-releasing groups in the same molecule gives rise to a stable electronic arrangement such as that shown for 1-amino-2-cyanocyclobutadiene (*15*) in Table 12.4. In terms of resonance theory, the stability of this compound can be rationalized by the important contribution of form (*15a*). The preponderance of this form is also suggested by HMO theory, which predicts the π-bond orders shown in structure (*15a*). The stabilizing effect of the concomitant presence of oppositely polarizable substituents was first suggested by Roberts[110] for 1-(*p*-dimethylaminophenyl)-2-(*p*-nitrophenyl)cyclobutadiene (*16*).

D. SUBSTITUTED CYCLOBUTADIENES   395

(15a)   (16)

2. Cyclobutadiene Divalent Ions

It is generally assumed that cyclobutadiene may be stable as the dication (17) or the dianion (18), since these possess two and six π electrons, respectively, and conform to Hückel's $4n+2$ π-electron requirement for stability.[126]

(17)   (18)

To a first approximation, the ion (17) can be obtained by the removal of two electrons from the doubly degenerate NBMO and the ion (18) by the addition of two electrons to the same NBMO of C_4H_4 without altering the position of the energy levels (Fig. 12.7). This is an oversimplification, however,

FIG. 12.7. HMO and MO–ω energy levels of $C_4H_4^{2+}$ and $C_4H_4^{2-}$.

since it does not provide in any way for the interaction of two like charges, which should be appreciable since they are distributed over only four carbons. If more reasonable Coulomb integrals are used in the computation (for example, those obtained by the ω technique,[121] $\alpha^{2+}=\alpha_0+0.7\beta_0$ and $\alpha^{2-}=\alpha_0-0.7\beta_0$), unstable arrangements are obtained as shown in Fig. 12.7. It is of interest that the instability manifests itself in two different, but related ways, i.e., a doubly degenerate vacant BMO in $C_4H_4^{2+}$ and a doubly degenerate filled ABMO in $C_4H_4^{2-}$. It is not surprising, therefore, that attempts to prepare tetramethylcyclobutadiene dication[63] and the corresponding dianion[1] have failed, whereas $(C_6H_5)_4C_4^{2+}$, in which the charges can be effectively accommodated by the phenyls, has been synthesized successfully.[50]

3. Polycyclic Cyclobutadienes

Polycyclic systems containing fused four-membered rings are usually predicted to possess a singlet ground state. Nevertheless, low stability is predicted for a number of these systems because of (a) excessive skeletal strain, as in tricyclo[3.1.0.0^{2,4}]hexa-1,3,5-triene (*19*) and tricyclo[4.2.0.0^{2,5}]octa-1,3,5,7-tetraene (*20*), (b) bond fixation inhibiting benzenoid resonance, as in benzocyclobutadiene (*21*) and benzotricyclobutadiene (*22*), (c) high free valence, as in dicyclobuta[*a,d*]benzene (*23*) and phenanthro[*l*]cyclobutadiene (*24a*), and (d) a variety of more subtle factors such as the nonbonding nature of the lowest vacant HMO found in tricyclo[5.3.0.0^{2,6}]decapentaene, "bowtiene" (*25*),[5] or the near-degeneracy of the lowest vacant and the highest filled MO's predicted for (*24b*). HMO constants for a large number of polycyclic ions, radicals, and neutral species containing fused four-membered rings have been compiled by Lee.[66]

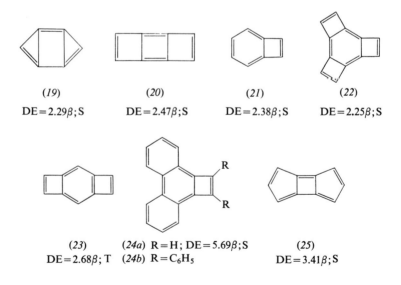

(*19*)   (*20*)   (*21*)   (*22*)
DE=2.29β;S   DE=2.47β;S   DE=2.38β;S   DE=2.25β;S

(*23*)   (*24a*) R=H; DE=5.69β;S   (*25*)
DE=2.68β;T   (*24b*) R=C$_6$H$_5$ DE=3.41β;S

a. Naphtho[b]cyclobutadiene

Recently, Cava[22] reported the synthesis of a substituted naphthocyclobutadiene (26a) which is of considerable theoretical interest, for it represents

TABLE 12.5

HMO CONSTANTS OF 1,2-DIPHENYLNAPHTHO[b]CYCLOBUTADIENE[a]

i	HMO k_i^b	symmetry (C_{2v})	m	f_m^c	π_{mm}^c	mn	p_{mn}^c
1	2.60	—	1	0.309	0.570	2′,3′	0.684
2	2.20	—	1′	0.105	0.331	1′,2′	0.593
3	2.08	—	2′	0.455	0.452	3′,4′	0.654
4	1.87	—	3′	0.394	0.395	1,1′	0.442
5	1.36	—	4′	0.425	0.432	1,8a	0.299
6	1.36	—	6	0.410	0.413	1,2	0.682
7	1.16	—	7	0.453	0.443	2a,8a	0.468
8	1.00	—	7a	0.119	0.345	8,8a	0.750
9	1.00	—	8	0.492	0.489	7a,8	0.491
10	1.00	—	8a	0.215	0.412	7,7a	0.590
11	0.68	a_2	—	—	—	3a,7a	0.532
12	0.25	b_1	—	—	—	6,7	0.690
13	−0.25	a_2	—	—	—	5,6	0.632
14	−0.68	b_1	—	—	—	—	—
15	−1.00	—	—	—	—	—	—
16	−1.00	—	—	—	—	—	—
17	−1.00	—	—	—	—	—	—
18	−1.16	—	—	—	—	—	—
19	−1.36	—	—	—	—	—	—
20	−1.36	—	—	—	—	—	—
21	−1.87	—	—	—	—	—	—
22	−2.08	—	—	—	—	—	—
23	−2.20	—	—	—	—	—	—
24	−2.60	—	—	—	—	—	—

[a] DE = 9.12β.
[b] HMO Energy level, Eq. (A-3), for MO i.
[c] Free valence (f_m), Eq. (A-7), of atom m, atom–atom polarizability (π_{mm}) (ref. 121), and bond order (p_{mn}), Eq. (A-6), of bond mn.

the first example of a stable system possessing a cyclobutadiene ring of which only one pair of neighboring carbons is fused to an aromatic residue.

The observed stability of (26a) is clearly indicated by HMO theory which predicts a closed-shell configuration and a high DE (9.12β).[4] The π-bond orders (Table 12.5) suggest appreciable π-bond fixation constraining the various links as shown in (26). In fact, the pronounced double-bond character of links

(26a) R = C$_6$H$_5$
(26b) R = H

(27)

2a,3 and 8,8a is indicated by the high-field NMR signal due to the protons bound at carbons 3 and 8.[22] Cava's NMR assignment is further supported by the self-atom polarizabilities and free valence indexes which are largest at these positions.[4]

TABLE 12.6

π-Bond Orders of Representative Condensed Aromatic Cyclobutadiene Systems Calculated by the HMO Method and an Approximate SCFMO Method

System	π-Bond orders[a]						Ref.
	a	b	c	d	e	f	
Naphtho[b]cyclobuta-diene (26b)	0.768 (0.806)	0.292 (0.227)	0.842 (0.908)	0.292 (0.227)	0.449 (0.432)	0.768 (0.806)	31, 66
Naphtho[a]cyclo butadiene (27)	0.613 (0.615)	0.194 (0.131)	0.913 (0.957)	0.169 (0.113)	0.615 (0.650)	0.686 (0.664)	31
Benzocyclobutadiene (21)	0.728 (0.747)	0.215 (0.160)	0.897 (0.943)	0.215 (0.160)	0.550 (0.550)	0.728 (0.747)	31, 66, 111
Phenanthro[l]cyclo-butadiene (24a)	0.576	0.144	0.933	0.144	0.682	0.576	66
Tetrabenzobiphenylene (37)	0.544 (0.519)	0.207 (0.149)	0.678 (0.731)	0.207 (0.149)	0.678 (0.731)	0.544 (0.519)	2, 66

[a] Values in parentheses refer to approximate SCFMO bond orders given in the references.

D. SUBSTITUTED CYCLOBUTADIENES

A description of the hypothetical parent molecule (*26b*) has been given by Lee[66] and by Coulson and Poole[31] by the HMO and a β-iterative method, respectively. Comparison of (*26b*) with its position isomer (*27*) and with benzocyclobutadiene (*21*) indicates that (*26b*) is the more stable.[31] The π-bond orders of these systems are shown in Table 12.6.

b. Phenanthro[*l*]cyclobutadiene

Another labile polycyclic system is phenanthro[*l*]cyclobutadiene (*24a*) whose diphenyl derivative (*24b*) was shown recently by Cava and Mangold[23] to have only transient existence and to react readily with oxygen. This led them to suggest that (*24b*) may have a triplet ground state. This conclusion is entirely consistent with the results of HMO calculations (Table 12.7) which

TABLE 12.7

HMO CONSTANTS OF 1,2-DIPHENYLPHENANTHRO[*l*]CYCLOBUTADIENE[a]

i	HMO k_i^b	Symmetry (C_{2v})	m	f_m^c	π_{mm}^c	mn	p_{mn}^c
1	2.67	—	1	0.343	1.107	2′,3′	0.688
2	2.25	—	1′	0.104	0.330	1,2′	0.581
3	2.08	—	2′	0.463	0.494	3′,4′	0.650
4	1.95	—	3′	0.394	0.395	1,1′	0.466
5	1.84	—	4′	0.432	0.472	1,10b	0.191
6	1.32	—	6b	0.158	0.338	1,2	0.732
7	1.26	—	7	0.442	0.443	10a,10b	0.552
8	1.23	—	8	0.420	0.446	2a,10b	0.681
9	1.10	—	9	0.406	0.409	10a,10	0.550
10	1.00	—	10	0.460	0.479	6b,10a	0.504
11	1.00	—	10a	0.126	0.342	9,10	0.722
12	0.79	—	10b	0.308	0.852	8,9	0.603
13	0.78	a_2	—	—	—	7,8	0.708
14	0.064	b_1	—	—	—	6b,7	0.582

TABLE 12.7 (continued)

i	HMO k_i^b	Symmetry (C_{2v})	m	f_m^c	π_{mm}^c	mn	p_{mn}^c
15	−0.064	a_2	—	—	—	6a,6b	0.488
16	−0.78	b_1	—	—	—	—	—
17	−0.79	—	—	—	—	—	—
18	−1.00	—	—	—	—	—	—
19	−1.00	—	—	—	—	—	—
20	−1.10	—	—	—	—	—	—
21	−1.23	—	—	—	—	—	—
22	−1.26	—	—	—	—	—	—
23	−1.32	—	—	—	—	—	—
24	−1.84	—	—	—	—	—	—
25	−1.95	—	—	—	—	—	—
26	−2.08	—	—	—	—	—	—
27	−2.25	—	—	—	—	—	—
28	−2.67	—	—	—	—	—	—

^{a} DE = 10.66β.
^{b} HMO energy level, Eq. (A-3), for MO i.
^{c} Free valence (f_m), Eq. (A-7), of atom m, atom–atom polarizability (π_{mm}) (ref. 121), and bond order (p_{mn}), Eq. (A-6), of bond mn.

indicate that although (24b) has a closed-shell configuration, the highest filled BMO (ψ_{14}) and the lowest unfilled ABMO (ψ_{15}) are separated by only 0.129β, or less than 10 kcal/mole. This suggests the possibility of a thermally populated triplet (also see Appendix). The low stability of (24a) and (24b) is also indicated by the HMO π-bond orders which require the four-membered ring in (24b) to have more cyclobutadiene character than that of (26a).

c. Biphenylene

In 1941, Lothrop[81,82] reported the synthesis of biphenylene, the first compound possessing a totally unsaturated four-membered ring. The proposed structure (28) was immediately questioned by Baker[7] and by Coulson[28] who speculated that the benzopentalene variant (29) was more appropriate on account of its lower strain. Both (28) and (29) are predicted to have comparable HMO delocalization energies.

(28)
DE = 4.506β

(29)
DE = 4.304β

Structure (28) received strong support from electron diffraction and X-ray data published shortly thereafter by Waser et al.[127, 128] The estimated bond distances were 1.41 ± 0.02 Å for all benzenoid bonds and 1.46 ± 0.05 Å for the links connecting the benzene residues (cross-links). The unusual length of the cross-links is also predicted by the HMO π-bond orders (Table 12.10), which predict 1.47 Å[128] and 1.50 Å[18] for the connecting bonds, using Coulson's[29] or Brown's[19] bond order–bond length relation. More recent estimates for the cross-links are 1.49 Å[107] from an SCMO calculation using Coulson's relation, 1.47 Å[2] by an iterative HMO method employing the Longuet-Higgins and Salem parameterization procedure, 1.42 Å[47] from a free-electron MO calculation and 1.45 Å[84] from a VB picture in which all unexcited forms are given equal weight. A recent X-ray crystallographic study by Mak and Trotter[84] gave the values shown in Fig. 12.8. The strain energy

FIG. 12.8. X-ray crystallographic dimensions of biphenylene.[84]

(SE) of biphenylene was originally estimated at ∼100 kcal/mole.[28] This value is now considered to be unrealistically high since it was derived from a Penney straight-bond model.[104] A recent evaluation[10, 21] of SE as the difference between the experimental heat of combustion of the molecule and that calculated by summing bond energy terms,[87] gives 1517.1 − 1441.7 = 75.4 kcal/mole. Comparison of the thermochemical RE of biphenylene (17.1 kcal/mole) to that of biphenyl (81.4 kcal/mole) gives an even lower SE, 64.3 kcal/mole. Furthermore, this latter value should be looked upon as an upper limit since its derivation presupposes that biphenylene and biphenyl possess equal π-electron energies, which may not be the case. In fact, it is conceivable that biphenyl has a lower π energy.[57, 105]

The electronic spectrum of biphenylene has been the subject of considerable speculation. It is easily seen from Table 12.10 that HMO theory predicts the $V_0 \to V_{6,7}$ electronic transition to be of symmetry B_{3g} of D_{2h} and therefore forbidden. In contrast, in other alternant hydrocarbons the $N \to V_1$ excitation which is associated with Clar's p bands is allowed with polarization along the short molecular axis.[121, 106]

The prediction of a forbidden $N \to V_1$ transition in biphenylene was first emphasized by Brown;[17] nevertheless, Clar[25] classified the electronic spectrum in the manner employed for normal polynuclear aromatics. Accordingly, the intense (log ϵ ∼ 4) absorption at 27,910 cm^{-1} was termed the p band. This assignment was employed by Bergman[11] in a correlation between lowest-energy electronic transition and half-wave potential and also more recently by

Layton.[65] The soundness of the HMO prediction was, however, substantiated recently by Hochstrasser,[56] who measured the electronic absorption spectrum of biphenylene in the vapor at a variety of temperatures, in solution at room temperature, and in an EPA rigid glass at $-90°$ K. The results are particularly revealing in that the band at 27,910 cm^{-1} was found to be polarized along the long molecular axis and thus by definition it is not a p band. Furthermore, the weak absorption at $\sim 25,000$ cm^{-1}, considered to be an α band by Clar[25] is probably due to the $N \to V_1$ symmetry-forbidden transition which becomes partly allowed through interaction with vibrations of appropriate symmetry (b_{1u} and b_{2u} for transitions along the short and long molecular axes, respectively).

Recent studies by Hilpern[54b] on the fluorescence and phosphorescence spectrum of biphenylene support the conclusion that the $N \to V_1$ transition is forbidden. Emission to the ground state occurs not from the first excited but the second excited electronic level (B_{1u}).[54b] (However, see the Addendum.)

Quantitative computations of transition energies have been carried out in various degrees of sophistication. The theoretical results together with observed bands of experimentally determined polarization are compiled in Table 12.8.

TABLE 12.8

Electronic Transitions in Biphenylene

Transition	Symmetry	Polarization	Observed[a]	Transition energy (cm^{-1}) Predicted		
				Hush[b]	Brown[c]	Fernandez[d]
6 → 7	B_{3g}	Forbidden	25,480	17,660	26,315	25,000
6 → 8	B_{1u}	Long axis	27,910	25,760	43,500	39,800
5 → 7	B_{1u}	Long axis	—	39,410	35,714	29,940
4 → 7	A_{1g}	Forbidden	—	—	41,666	39,525
6 → 10	B_{2u}	Short axis	—	42,150	—	—
3 → 7	B_{2u}	Short axis	—	43,100	43,500	38,460

[a] Reference 56.
[b] Reference 60.
[c] Reference 17.
[d] Reference 48.

The theoretical treatments which lead to the predictions tabulated above are as follows: Brown employed the LCAO–MO method with incorporation of overlap ($S=0.25$) using $\gamma=3.4$ eV (where γ is the counterpart of β when overlap is explicitly considered). Fernandez performed a near-SCF–MO treatment by successively iterating Coulomb and resonance integrals with incorporation of overlap ($S=0.25$) and using $\gamma=3.4$ eV. Hush and Rowlands used an approximate SCMO method taking into account first-order configuration interaction only. This latter treatment was also employed in computing the electronic spectra of biphenylene monocation radical (b^+), monoanion radical (b^-), and dianion (b^{2-}). The experimental spectra of these ions are also recorded in the literature.[60]

The electron spin resonance (ESR) spectra of b^+ and b^- were determined recently by McDowell and Rowlands.[86] The spectra of both radical ions consisted of five lines with hyperfine splitting constants ($\Delta H = Q\rho_m$) of 4.0 gauss for b^+ and 2.75 gauss for b^-. Theory requires the spectra to consist of five signals, each of which is further split into a quintet, i.e., a total of 25 lines. The observed five-line spectrum is merely a consequence of poor resolution and was ascribed to the set of equivalent protons bound to carbons 2, 3, 6, and 7 (Table 12.10) which are predicted to possess a higher spin density ($\rho_m = c_{im}^2$) than carbons 1, 4, 5, and 8. In HMO theory, the spin density (ρ_m) is synonymous with charge density (q_m). The HMO spin densities are $\rho_1 = 0.0269$ and $\rho_2 = 0.0873$. The different splitting constants observed for b^+ and b^- cannot be rationalized on the basis of simple HMO theory which, due to the alternant nature of biphenylene, predicts that $\rho_m^+ = \rho_m^-$. Qualitative agreement with experiment is obtained, however, when the Coulomb terms of the four-ring are

TABLE 12.9

Experimental and Theoretical Proton Hyperfine Splitting Constants (ΔH) for the Biphenylene Anion and Cation

Method	Total width (gauss)	Hyperfine splitting constants[a,b] (gauss)		
		Position C-1	Position C-2	Position C-4a
Experimental				
Anion	12.28	0.21	2.86	—
Cation	15.60	0.21	3.69	—
HMO ($Q=28$)	12.80	0.76	2.44	—
		(0.027)	(0.087)	(0.136)
HMO ($\beta_{4a,4b} = 2/3\beta_0$; $Q=28$)	11.96	0.36	2.63	—
		(0.013)	(0.094)	(0.143)
SCF ($Q=24.2$)	12.20	0.61	2.44	—
		(−0.025)	(0.101)	(0.174)

[a] Calculated from the relation $\Delta H = Q\rho_m$.
[b] Values in parenthesarees calculated electron spin densities (ρ_m).

assigned a more negative value. Thus for $\alpha = \alpha_0 + 0.2\beta$, ρ_2 is 0.0906 and 0.0853 for b$^+$ and b$^-$, respectively.

More recently Carrington and Santos-Veiga[20] were able to completely resolve the ESR spectra of b$^+$ and b$^-$. As expected five well-separated quintets were observed. The experimental results and the theoretically computed values are reproduced in Table 12.9. The theoretical treatments employed (a) the

TABLE 12.10

HMO CONSTANTS OF BIPHENYLENE

i	HMO k_i^b	Symmetry (D_{2h})	m	f_m^c	π_{mm}^c	mn	p_{mn}^c
1	2.53	b_{3u}	1	0.428	0.419	1, 2	0.621
2	1.80	b_{2g}	2	0.420	0.424	2, 3	0.691
3	1.35	b_{3u}	4a	0.221	0.413	4, 4a	0.683
4	1.25	b_{1g}	—	—	—	4a, 8b	0.565
5	0.88	a_u	—	—	—	4a, 4b	0.263
6	0.44	b_{2g}	—	—	—	—	—
7	−0.44	b_{1g}	—	—	—	—	—
8	−0.88	b_{3u}	—	—	—	—	—
9	−1.25	b_{2g}	—	—	—	—	—
10	−1.35	a_u	—	—	—	—	—
11	−1.80	b_{1g}	—	—	—	—	—
12	−2.53	a_u	—	—	—	—	—

[a] DE = 4.51β.
[b] HMO energy level, Eq. (A-3), for MO i.
[c] Free valence (f_m), Eq. (A-7), of atom m, atom–atom polarizability (π_{mm}) (ref. 121), and bond order (p_{mn}), Eq. (A-6), of bond mn.

simple HMO method, (b) an HMO procedure in which the resonance integrals of the cross-links of b$^+$ and of b$^-$ are assigned the value of $\tfrac{2}{3}\beta_0$, and (c) a semi-empirical SCF method due to McLachlan.[89]

The reactivity of biphenylene has been described on the basis of the various theoretical indexes. In the *localization* approximation, which best describes

D. SUBSTITUTED CYCLOBUTADIENES

transition states occurring late in the reaction coordinate, carbon-2 is predicted to be the most reactive toward electrophiles, nucleophiles, and radicals; for example, Dewar's reactivity numbers [41, 42, 43] are $N_1 = 2.00$ and $N_2 = 1.73$ and the localization energies $L_1 = 2.408$, $L_2 = 2.352$.[17] The situation in the isolated molecule approximation (early transition state) is not as clear cut, since unlike most other alternant hydrocarbons the position of highest free valence ($f_1 = 0.428$, $f_2 = 0.420$)[18, 111] is not that of highest self-atom polarizability ($\pi_{11} = 0.419$, $\pi_{22} = 0.424$).[17] The prediction based on π_{mm} is supported by Brown's reactivity index [16] (Z_m) and by the results of a free-electron treatment,[47] both of which point to carbon-2 as the more reactive toward electrophiles. On the other hand, Fernandez Alonso's iterative treatment [48] leads to a higher electron density at carbon-1 ($q_1 = 1.039$, $q_2 = 1.015$). In fact, all electrophilic substitution reactions such as nitration, halogenation, acylation, etc., occur at carbon-2 exclusively. Streitwieser [122] has recently measured the rate of tritiodeprotonation at carbon-1 and carbon-2 and obtained $k_2/k_1 = 64$. This is greater than the ratio ($k_2/k_1 = 4$) estimated from localization energies.

The orienting effect of an electron-donating substituent (amino, hydroxy, etc.) bound to carbon-2 toward an incoming electrophile is of some interest in view of the disagreement between HMO theory and resonance theory with respect to the most reactive site.[26] If all nonpolar resonance structures of biphenylene are given equal weight, one finds that bonds 1–2 and 2–3 have 3/5 and 2/5 double-bond character, respectively. The molecule would then exist predominantly as in structure (*30*) and accordingly position 1 would be the more reactive. On the other hand, the HMO method requires that the carbanion isoelectronic with (*30*) possesses the NBMO coefficients shown in (*31*), thus making position 3 the more reactive. Recent experiments [8, 9] supported the HMO prediction, since bromination of (*32a*) led to the exclusive formation of (*32b*).

(*30*) R = NH$_2$, OH, ...

(*31*) 9/12, 1/21, CH$_2^\ominus$, 4/21

(*32a*) R$_1$ = R$_2$ = H
(*32b*) R$_1$ = H, R$_2$ = Br

The ionization potential and electron affinity of biphenylene were calculated by Hedges and Matsen [54a] using a semiempirical ASMO method, giving $I = 8.15$ eV and $A = -0.41$ eV.

Calculations [12a, 108] of the magnetic susceptibility (χ_π) of biphenylene afforded negative values in accord with its $4n$ π-electron system. For example, if benzene is assigned a magnetic susceptibility of unity then $4n+2$ hydrocarbons such as naphthalene and azulene have $\chi_\pi > 1$, while pentalene, heptalene, and other $4n$ molecules are predicted to have $\chi_\pi < 0$.

Biphenylene readily forms coordination complexes using one or both of its six-membered rings, but not its four-membered ring.[24] (The metal complexes of cyclobutadiene are discussed in Section D,4.)

d. The Benzobiphenylenes

The two possible monobenzobiphenylenes (*33*) and (*34*) have been known for some time (see Chapter 11, Section D). Theoretical work pertaining to

TABLE 12.11

HMO Constants of Benzo[*b*]biphenylene[a]

i	HMO k_i^b	Symmetry (C_{2v})	m	f_m^c	π_{mm}^c	mn	p_{mn}^c
1	2.57	—	1	0.451	0.441	10,10a	0.517
2	2.15	—	2	0.408	0.409	10,9b	0.736
3	1.62	—	8	0.418	0.421	1,10a	0.575
4	1.37	—	9	0.429	0.420	4a,10a	0.525
5	1.23	—	9a	0.215	0.403	1,2	0.706
6	1.07	—	9b	0.215	0.404	2,3	0.619
7	0.62	a_2	10	0.479	0.463	7,8	0.677
8	0.50	b_1	10a	0.115	0.339	8,9	0.636
9	−0.50	a_2	—	—	—	9,9a	0.667
10	−0.62	b_1	—	—	—	5b,9a	0.556
11	−1.07	—	—	—	—	5a,9b	0.486
12	−1.23	—	—	—	—	9a,9b	0.295
13	−1.37	—	—	—	—	—	—
14	−1.62	—	—	—	—	—	—
15	−2.15	—	—	—	—	—	—
16	−2.57	—	—	—	—	—	—

[a] $DE = 6.25\beta$.
[b] HMO energy level, Eq. (A-3), for MO i.
[c] Free valence (f_m), Eq. (A-7), of atom m, atom–atom polarizability (π_{mm}) (ref. 121), and bond order (p_{mn}), Eq. (A-6), of bond mn.

D. SUBSTITUTED CYCLOBUTADIENES

these systems is limited to a report[66] of the HMO constants of (33) and two papers[11, 65] concerned with the empirical classification of the electronic spectrum of (34). (The two papers contain conflicting assignments.)

Inspection of the HMO constants of (33) and (34) (Tables 12.11 and 12.12) reveals that, like biphenylene, they are predicted to possess appreciable delocalization energies $DE_{(33)} = 6.25\beta$, $DE_{(34)} = 6.14\beta$, closed π-molecular shells and long cross-links. Promotion of an electron from the highest BMO to the lowest ABMO in (33) is symmetry-allowed (B_2) with polarization along the short molecular axis. In (34), all electronic transitions are formally allowed

TABLE 12.12

HMO Constants of Benzo[a]biphenylene[a]

i	HMO k_i^b	m	f_m^c	π_{mm}^c	mn	p_{mn}^c
1	2.59	1	0.456	0.452	5,6	0.679
2	2.10	2	0.409	0.411	6,6a	0.630
3	1.62	3	0.410	0.414	4a,5	0.575
4	1.41	4	0.455	0.448	4,4a	0.548
5	1.20	4a	0.116	0.339	4a,10c	0.494
6	1.00	5	0.478	0.491	3,4	0.729
7	0.83	6	0.423	0.414	2,3	0.592
8	0.32	6a	0.236	0.443	1,2	0.731
9	−0.32	6b	0.228	0.428	1,10c	0.546
10	−0.83	7	0.430	0.421	10b,10c	0.569
11	−1.00	8	0.426	0.434	10a,10b	0.262
12	−1.20	9	0.423	0.428	6a,10b	0.620
13	−1.41	10	0.434	0.427	10,10a	0.688
14	−1.62	10a	0.219	0.416	6b,10a	0.563
15	−2.10	10b	0.280	0.496	9,10	0.610
16	−2.59	10c	0.123	0.339	8,9	0.699
					7,8	0.607
					6b,7	0.695
					6a,6b	0.246

[a] $DE = 6.14\beta$
[b] HMO energy level, Eq. (A-3), for MO i.
[c] Free valence (f_m), Eq. (A-7), of atom m, atom–atom polarizability (π_{mm}) (ref. 121), and bond order (p_{mn}), Eq. (A-6), of bond mn.

408   12. THEROETICAL ASPECTS OF THE CYCLOBUTADIENE PROBLEM

(33)   (34)

TABLE 12.13

HMO Constants of Dibenzo[b,h]biphenylene[a]

i	HMO k_i^b	symmetry (D_{2h})	m	f_m^c	π_{mm}^c	mn	p_{mn}^c
1	2.60	—	2	0.408	0.409	4a,5	0.529
2	2.25	—	4	0.452	0.442	5,5a	0.722
3	2.00	—	4a	0.114	0.338	4,4a	0.569
4	1.41	—	5	0.482	0.467	4a,12a	0.520
5	1.41	—	5a	0.210	0.397	2,4	0.710
6	1.26	—	—	—	—	2,3	0.614
7	1.18	—	—	—	—	5a,5b	0.316
8	0.80	—	—	—	—	5a,11b	0.485
9	0.55	b_{2g}	—	—	—	—	—
10	0.52	a_u	—	—	—	—	—
11	−0.52	b_{3u}	—	—	—	—	—
12	−0.55	b_{1g}	—	—	—	—	—
13	−0.80	—	—	—	—	—	—
14	−1.18	—	—	—	—	—	—
15	−1.26	—	—	—	—	—	—
16	−1.41	—	—	—	—	—	—
17	−1.41	—	—	—	—	—	—
18	−2.00	—	—	—	—	—	—
19	−2.25	—	—	—	—	—	—
20	−2.60	—	—	—	—	—	—

[a] DE = 7.96β.
[b] HMO energy level, Eq. (A-3), for MO i.
[c] Free valence (f_m), Eq. (A-7), of atom m, atom–atom polarizability (π_{mm}) (ref. 121), and bond order (p_{mn}), Eq. (A-6), of bond mn.

D. SUBSTITUTED CYCLOBUTADIENES 409

since the molecule is not subject to any symmetry restrictions. Concerning the reactivity of (33) and (34), it is worth noting that the various indexes (Tables 12.11 and 12.12) are consistent in the choice of the most reactive sites, which are the 5-positions in both (33) and (34).

Of the possible dibenzobiphenylenes (Chapter 11) only dibenzo[b,h]- (35) and dibenzo[a,i]biphenylene (36) are known, the former being the more stable.

TABLE 12.14

HMO CONSTANTS OF DIBENZO[a,i]BIPHENYLENE[a]

i	HMO k_i^b	Symmetry (C_{2v})	m	f_m^c	π_{mm}^c	mn	p_{mn}^c
1	2.63	—	1	0.456	0.454	2,3	0.589
2	2.22	—	2	0.410	0.414	1,2	0.732
3	1.93	—	3	0.411	0.416	3,4	0.732
4	1.49	—	4	0.456	0.451	4,4a	0.544
5	1.45	—	4a	0.117	0.341	4a,5	0.582
6	1.16	—	5	0.486	0.515	4a,12d	0.488
7	1.15	—	6	0.423	0.414	5,6	0.664
8	0.86	—	6a	0.246	0.472	6,6a	0.645
9	0.78	b_1	12c	0.279	0.511	6a,6b	0.222
10	0.23	a_2	12d	0.127	0.342	6a,12c	0.619
11	−0.23	b_1	—	—	—	12c,12d	0.573
12	−0.78	a_2	—	—	—	1,12d	0.544
13	−0.86	—	—	—	—	12b,12c	0.261
14	−1.15	—	—	—	—	—	—
15	−1.16	—	—	—	—	—	—
16	−1.45	—	—	—	—	—	—
17	−1.49	—	—	—	—	—	—
18	−1.93	—	—	—	—	—	—
19	−2.22	—	—	—	—	—	—
20	−2.63	—	—	—	—	—	—

[a] DE = 7.80β.
[b] HMO energy level, Eq. (A-3), for MO i.
[c] Free valence (f_m), Eq. (A-7), of atom m, atom–atom polarizability (π_{mm}) (ref. 121), and bond order (p_{mn}), Eq. (A-6), of bond mn.

This difference in stability was rationalized on the basis of HMO theory by Pullman[6] and by Coulson.[3] Pullman's main arguments are (1) isomer (*35*) possesses a higher DE; $\Delta DE = 0.16\beta$, (2) interaction between the naphthalene moieties should be less pronounced in (*36*) in view of the longer crosslinks, (3) unfavorable steric interactions between protons at C-1 and C-12 in (*36*), and (4) the possibility of a low-lying triplet state in (*36*) due to the relatively low energy of the $N \to V_1$ transition which is 0.46β in (*36*) as compared to 1.04β in (*35*). Similar arguments were employed by Coulson who emphasized the predominance of resonance forms (*35*) and (*36*) suggested by the HMO π-bond orders (Tables 12.13 and 12.14). The importance of these forms is stressed further by the results of an approximate SCMO calculation involving successive iteration of β by the Longuet-Higgins and Salem procedure. The SCMO π-bond orders of (*35*) and (*36*) are shown in (*35a*) and (*36a*), respectively.

(*35*) (*36*)

(*35a*) (*36a*)

The electronic spectrum of dibenzo[*b,h*]biphenylene (*35*) was discussed by Crawford[39] on the basis of HMO theory with inclusion of overlap and employing $\gamma = 23{,}000$ cm^{-1}. He concluded that the $N \to V_1$ transition is allowed with polarization along the long molecular axis. Inspection of the symmetry properties of the MO's in question (Table 12.13) reveals, however, that this conclusion is in error since the $10 \to 11$ promotion is of symmetry B_{3g} and thus forbidden. On the other hand, transitions $9 \to 11$ and $10 \to 12$ are both symmetry-allowed (B_{1u}) with polarization along the long axis.*

The ESR spectrum of the anion radical of (*35*) was determined recently by Carrington and Santos-Veiga.[20] It consists of 91 out of the predicted 125 lines and has a total width of 27.44 gauss. The measured hyperfine splitting constants are 4.31, 1.62, and 0.93 gauss for the protons bound at carbons 1, 2, and 5,

---

* Inclusion of overlap in the calculation does not alter the relative positions of MO's ψ_8 to ψ_{13}.

respectively. Assignments were made on the basis of the HMO spin densities which led to $\Delta H_1 = 3.28$, $\Delta H_3 = 2.21$, and $\Delta H_4 = 0.95$ gauss.

With respect to the reactivity of (35) and (36) it is easily verified that the various diagnostic indexes agree that the most reactive positions should be the 5-positions in both cases.

Recently Ali et al.[2] described an unsuccessful attempt to synthesize the symmetrical tetrabenzobiphenylene (37). The low stability of this system is clearly indicated by its π-bond orders (Table 12.6) which suggest ready dissociation about the cross-links.

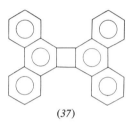

(37)

4. CYCLOBUTADIENE–METAL COMPLEXES

The stabilization of cyclobutadiene derivatives by coordination with transition metals was predicted in 1956 by Longuet-Higgins and Orgel[79] on the basis of HMO theory. Their treatment is summarized in Fig. 12.9 and Table 12.15.

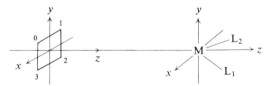

FIG. 12.9. Mononuclear cyclobutadiene–metal complexes ($C_4H_4ML_2$).

TABLE 12.15

MOLECULAR ORBITALS OF MONONUCLEAR CYCLOBUTADIENE–METAL COMPLEXES ($C_4H_4ML_2$)

$\sigma_{xz}{}^a$	$\sigma_{yz}{}^a$	$MO(C_4H_4)^b$	$MO(L_1,L_2)^c$	AO(M)	BMO	NBMO	ABMO
+	+	ψ_I	$1/\sqrt{2}(\chi_1+\chi_2)$	$s, p_z, d_{z^2}, d_{x^2-y^2}$	4	—	2
+	−	ψ_III	$1/\sqrt{2}(\chi_1-\chi_2)$	p_x, d_{xz}	2	—	2
−	+	ψ_II	—	p_y, d_{yz}	1	1	1
−	−	ψ_IV	—	d_{xy}	1	—	1

[a] Reflection operation in indicated planes.
[b] MO's given in Table 12.2.
[c] χ_1 and χ_2 are σ orbitals on ligands L_1 and L_2, respectively, which lie in the xz plane.

12. THEORETICAL ASPECTS OF THE CYCLOBUTADIENE PROBLEM

The $C_4H_4ML_2$ arrangement requires eight BMO's, one NBMO, and six ABMO's which extend over the complex. Longuet-Higgins and Orgel postulated that 16- or 18-electron complexes may be formed depending on the occupancy and relative energy of the critical NBMO, which consists of an additive combination of p_y and d_{yz} only. Furthermore, they recognize that the existence of such complexes would depend on their relative stability compared to that of acetylene complexes to which they might revert. The acetylene complexes, however, are expected to be less stable than the corresponding C_4H_4 complexes due to the high-energy antibonding orbital of acetylene which reduces its acceptor capacity relative to that of cyclobutadiene. This original prediction stimulated a great amount of experimental work that is still continuing.

In 1959, Criegee and Schröder[40] reported the first well-authenticated[45, 51] synthesis of a $C_4H_4ML_2$ compound, tetramethylcyclobutadienenickel chloride (34). Further examples of the same type of complex have been provided since by Malatesta[85] and Freedman[49] with the synthesis of (39) and (40), respectively. In 1959 Hübel[58] reported the synthesis of (41), the first firmly established[44, 51] $C_4H_4ML_3$ complex.

(38) R = CH_3; ML_n = $NiCl_2$
(39) R = C_6H_5; ML_n = $PdCl_2$
(40) R = C_6H_5; ML_n = $NiBr_2$
(41) R = C_6H_5; ML_n = $Fe(CO)_3$

To account for the stability of (41), Brown[15] recently extended the Longuet-Higgins and Orgel treatment to include complexes of type $C_4H_4ML_3$.* In comparing the stability of $C_nH_nML_3$ complexes ranging from C_4 to C_8 π systems, Brown used two criteria. The first involves the relative magnitude of the E_1 group overlap integrals $S(\psi_{\pm 1}, 3d/4p)$ which, by analogy with ferrocene where such interaction accounts for 75% of the binding energy, are predicted to be critical for stability. These integrals are given by Eq. (D-1),

$$S(\psi_{\pm 1}, 3d/4p) = \sqrt{\frac{1}{2}} \left[\frac{\sqrt{3}}{2} \sin 2\omega \cos \omega S_{2p\sigma 3d\sigma} + \cos 2\omega \sin \omega S_{2p\pi 3d\pi} + (S_{2p\sigma 4p\sigma} + S_{2p\pi 4p\pi}) \sin \omega \cos \omega \right] \sum_m c_m \cos \chi_m \quad \text{(D-1)}$$

---

* Brown's classification of the metal AO's and the carbocycle MO's under the D_4 point group contains some errors, e.g., ψ_0 of C_4H_4 is said to belong to the A_1 rather than to the A_2 representation; and the metal AO's p_z and d_{z^2}, are incorrectly assigned to the same representation.

where ω is the angle between the n-fold symmetry axis and the M—C bond, and the summation is taken over the coefficients of ψ_{+1} and ψ_{-1}. Eq. (D-1) gives 0.306, 0.353, 0.361, 0.307, and 0.164 for C_4, C_5, C_6, C_7 and C_8 complexes, respectively.

The second criterion deals with the difference between the Coulomb energies of the doubly-degenerate ring orbital, $\psi_{\pm 1}$, and its symmetry counterparts $(3d/4p)$ on the metal. It is of interest that for the special case where M=Fe, Brown estimates that the smallest energy difference and strongest bonding arise when the π system is cyclobutadiene.

Similar qualitative arguments were applied to the prediction of stable binuclear cyclobutadiene complexes in which each C_4H_4 is flanked symmetrically by two ML_3 groups. Brown postulates, however, that such binuclear complexes should be less stable than their mononuclear analogs since the three-center MO's present in the former should result in stabilization by $\beta/\sqrt{2}$ per metal atom, whereas the two-center MO's involved in the latter result in stabilization by β.

Another, as yet hypothetical, class of C_4H_4-metal complexes is that in which the metal atom is flanked symmetrically by two cyclobutadiene moieties. The symmetry requirements for such an arrangement (Fig. 12.10) are given in Table 12.16,[97] although they will not be discussed further here.

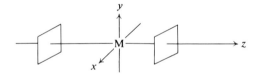

FIG. 12.10. Structure of the hypothetical bis(cyclobutadiene)–metal complexes, $(C_4H_4)_2M$.

TABLE 12.16

MOLECULAR ORBITALS OF THE HYPOTHETICAL BIS(CYCLOBUTADIENE)–METAL COMPLEXES $(C_4H_4)_2M$

Symmetry (C_4)	MO (C_4H_4)[a]	AO (M)	MO (C_4H_4')[a]	Bond symmetry[b]
A	ψ_I	$4s, 3d_{z^2}$	ψ_I	σ_g
		$3p_z$		σ_u
E	ψ_{II}, ψ_{III}	$3d_{xz}, 3d_{yz}$	ψ_{II}, ψ_{III}	π_g
		$3p_x, 3p_y$		π_u
B	ψ_{IV}	$3d_{xy}$	ψ_{IV}	δ_g

[a] MO's are given in Table 12.2.
[b] Subscripts g and u refer, respectively, to symmetry and antisymmetry of the bond with respect to inversion about the center of symmetry of the metal.

References

1. Adam, W., *Tetrahedron Letters* **No. 21**, 1387 (1963).
2. Ali, M. A., Carey, J. G., Cohen, D., Jones, A. J., Millar, I. T., and Wilson, K. V., *J. Chem. Soc.* 387 (1964).
3. Ali, M. A., and Coulson, C. A., *Tetrahedron* **10**, 41 (1960).
4. Anastassiou, A. G., *Chem. and Eng. News* **42**, No. 3, 37 (1964).
5. Anastassiou, A. G., Ph.D. diss., Yale Univ. (1963).
6. Andrade e Silva, M., and Pullman, B., *Compt. Rend.* **242**, 1888 (1956).
7. Baker, W., *Nature* **150**, 210 (1942).
8. Baker, W., and McOmie, J. F. W., *Chem. Soc. Special Publication No. 12*, 49 (1958).
9. Baker, W., McOmie, J. F. W., and Rogers, V., *Chem. Ind.* (*London*) 1236 (1958).
10. Bedford, A. F., Carey, J. G., Millar, I. T., Mortimer, C. T., and Springall, H. D., *J. Chem. Soc.* 3895 (1962).
11. Bergman, I., *Trans. Faraday Soc.* **52**, 690 (1956).
12a. Berthier, G., Mayot, M., and Pullman, B., *J. Phys. Radium* **12**, 717 (1951).
12b. Birks, J. B., Conte, J. M. de C., and Walker, G., *Phys. Letters* **19**, 125 (1965).
13. Blomquist, A. T., and Maitlis, P. M., *J. Am. Chem. Soc.* **84**, 2329 (1962).
14. Bouman, N., *J. Chem. Phys.* **35**, 1661 (1961).
15. Brown, D. A., *J. Inorg. Nucl. Chem.* **10**, 39 (1959).
16. Brown, R. D., *J. Chem. Soc.* 2232 (1959).
17. Brown, R. D., *Trans. Faraday Soc.* **46**, 146 (1950).
18. Brown, R. D., *Trans. Faraday Soc.* **45**, 296 (1949).
19. Brown, R. D., *Trans. Faraday Soc.* **44**, 984 (1948).
20. Carrington, A., and Santos Veiga, J. dos, *Mol. Phys.* **5**, 285 (1962).
21. Cass, R. C., Springall, H. D., and Quincey, P. G., *J. Chem. Soc.* 1188 (1955).
22. Cava, M. P., Hwang, B. Y., and Van Meter, J. P., *J. Am. Chem. Soc.* **85**, 4032 (1963).
23. Cava, M. P., and Mangold, D., *Tetrahedron Letters* **No. 26**, 1751 (1964).
24. Chatt, J., Guy, R. G., and Watson, H. R., *J. Chem. Soc.* 2332 (1961).
25. Clar, E., "Aromatische Kohlenwasserstoffe," Springer, Berlin (1952), p. 279.
26. Cotton, F. A., "Chemical Applications of Group Theory," Interscience, New York (1963).
27. Coulson, C. A., *Chem. Soc. Special Publication No. 12*, 85 (1958).
28. Coulson, C. A., *Nature* **150**, 577 (1942).
29. Coulson, C. A., *Proc. Royal Soc.* (*London*) **A169**, 413 (1939).
30. Coulson, C. A., and Moffitt, W., *Phil. Mag.* **40**, 1 (1949).
31. Coulson, C. A., and Poole, M. D., *Tetrahedron* **20**, 1859 (1964).
32. Coulson, C. A., and Rushbrooke, G. S., *Proc. Cambridge Phil. Soc.* **36**, 193 (1940).
33. Craig, D. P., in Ginsburg, D., ed., "Non-Benzenoid Aromatic Compounds," Interscience, New York (1959), Chap. 1.
34. Craig, D. P., *J. Chem. Phys.* **49**, 143 (1952).
35. Craig, D. P., *J. Chem. Soc.* 3175 (1951).
36. Craig, D. P., *Disc. Faraday Soc.* **9**, 5 (1950).
37. Craig, D. P., *Proc. Royal Soc.* (*London*) **A202**, 498 (1950).
38. Craig, D. P., *Proc. Royal Soc.* (*London*) **A200**, 390 (1950).
39. Crawford, V. A., *Can. J. Chem.* **30**, 47 (1952).
40. (a) Criegee, R., and Schroder, G., *Ann.* **62**, 31 (1959); (b) *ibid.*, *Angew. Chem.* **71**, 70 (1959).
41. Dewar, M. J. S., *Rec. Chem. Prog.* **19**, 1 (1958).
42. Dewar, M. J. S., *J. Am. Chem. Soc.* **74**, 3357 (1952).

43. Dewar, M. J. S., Mole, T., and Warford, E. W. T., *J. Chem. Soc.* 3581 (1956).
44. (a) Dodge, R. P., and Schomaker, V., *Acta Cryst.* **13**-2, 1042 (1960); (b) *ibid.*, *Nature* **186**, 798 (1960).
45. Dunitz, J. D., Mez, H. C., Mills, O. S., and Shearer, H. M. M., *Helv. Chim. Acta* **45**, 647 (1962).
46. Eyring, H., Walter, J., and Kimball, G. E., "Quantum Chemistry," Wiley, New York (1944), Chap. 13.
47. Fernandez Alonso, J. I., and Domingo, R., *Anales Real. Soc. Espan. Fis. y quim.* (*Madrid*) **51B**, 447 (1955).
48. Fernandez Alonso, J. I., and Peradejordi, F., *Anales Real. Soc. Espan. Fis. y Quim.* (*Madrid*) **50B**, 253 (1954).
49. Freedman, H. H., *J. Am. Chem. Soc.* **83**, 2194 (1961).
50. Freedman, H. H., and Young, A. E., *J. Am. Chem. Soc.* **86**, 734 (1964).
51. Fritz, H. P., *Z. Naturforsch.* **16**, 415 (1961).
52. Fumi, F. G., *Nuovo Cimento* **8**, 1 (1950).
53. Goeppert-Mayer, M., and Sklar, A. L., *J. Chem. Phys.* **6**, 645 (1938).
54. (a) Hedges, R. M., and Matsen, F. A., *J. Chem. Phys.* **28**, 950 (1958); (b) Hilpern, J. W., *Trans. Faraday Soc.* **61**, 605 (1965).
55. Hobey, W. D., and McLachlan, A. D., *J. Chem. Phys.* **33**, 1695 (1960).
56. Hochstrasser, R. M., *Can. J. Chem.* **39**, 765 (1961); *ibid.*, *J. Chem. Phys.* **33**, 950 (1960).
57. Hoffmann, R., *J. Chem. Phys.* **39**, 1397 (1963).
58. Hübel, W., Braye, E. H., Clauss, A., Weiss, E., Kruerke, A., Brown, D. A., King, G. S. D., and Hoogzand, C., *Inorg. Nucl. Chem.* **9**, 204 (1959).
59. (a) Hückel, E., *Z. Phys.* **70**, 204 (1931); (b) *ibid.*, **76**, 628 (1932).
60. Hush, N. S., and Rowlands, J. R., *Mol. Phys.* **6**, 317 (1963).
61. Jahn, H. A., *Proc. Royal Soc.* (*London*) **A164**, 117 (1938).
62. Jahn, H. A., and Teller, E., *Proc. Royal Soc.* (*London*) **A161**, 220 (1937).
63. Katz, T. J., Hall, J. R., and Neikam, W. C., *J. Am. Chem. Soc.* **84**, 3199 (1962).
64. Kragten, J., *Chem. Weekblad* **54**, 349 (1958).
65. Layton, E. M., Jr., *J. Mol. Spect.* **5**, 181 (1960).
66. (a) Lee, H. S., *Chemistry* (*Taipei*), 137 (1962); (b) *ibid.*, 22 (1963); (c) *ibid.*, 47 (1963).
67. Lennard-Jones, J. E., *Proc. Royal Soc.* (*London*) **A158**, 280 (1937).
68. Lennard-Jones, J. E., and Turkevich, J., *Proc. Royal Soc.* (*London*) **A158**, 297 (1937).
69. Liehr, A. D., *Ann. Rev. Phys. Chem.* **13**, 41 (1962).
70. Liehr, A. D., *Z. Naturforsch.* **A16**, 641 (1961).
71. Liehr, A. D., *Z. Naturforsch.* **A13**, 429 (1958).
72. Liehr, A. D., *Z. Naturforsch.* **A13**, 311 (1958).
73. Liehr, A. D., *Ann. Phys.* **1**, 221 (1957).
74. Liehr, A. D., *Z. physik. Chem.* (*Frankfurt*) **9**, 338 (1956).
75. Lipscomb, W. N., *Tetrahedron Letters* **18**, 20 (1959).
76. Longuet-Higgins, H. C., *Proc. Chem. Soc.* 157 (1957).
77. Longuet-Higgins, H. C., *Proc. Phys. Soc.* **60**, 270 (1948).
78. Longuet-Higgins, H. C., and McEwen, K. L., *J. Chem. Phys.* **26**, 719 (1957).
79. Longuet-Higgins, H. C., and Orgel, L. E., *J. Chem. Soc.* 1969 (1956). See also ref. 64.
80. Longuet-Higgins, H. C., and Salem, L., *Proc. Royal Soc.* (*London*) **A251**, 172 (1959).
81. Lothrop, W. C., *J. Am. Chem. Soc.* **64**, 1698 (1942).
82. Lothrop, W. C., *J. Am. Chem. Soc.* **63**, 1187 (1941).
83. Lykos, P. G., and Parr, R. G., *J. Chem. Phys.* **24**, 1166 (1956).
84. (a) Mak, T. C. W., and Trotter, J., *J. Chem. Soc.* 1 (1962); (b) *ibid.*, *Proc. Chem. Soc.* 163 (1961).

85. Malatesta, L., Santarella, G., Vallarino, L., and Zingales, F., *Angew. Chem.* **72**, 34 (1960); see also ref. 13.
86. McDowell, C. A., and Rowlands, J. R., *Can. J. Chem.* **38**, 503 (1960).
87. McGinn, C. J., *Tetrahedron* **18**, 311 (1962).
88. McLachlan, A. D., *J. Chem. Phys.* **33**, 663 (1960).
89. McLachlan, A. D., *Mol. Phys.* **3**, 233 (1960).
90. McWeeny, R., *Proc. Royal Soc. (London)* **A223**, 63 (1954).
91. McWeeny, R., *Proc. Royal Soc. (London)* **A223**, 306 (1954).
92. McWeeny, R., *Proc. Royal Soc. (London)* **A227**, 288 (1955).
93. Moffitt, W., *J. Chem. Phys.* **22**, 320 (1954).
94. Moffitt, W., *Proc. Royal Soc. (London)* **A210**, 245 (1951).
95. Moffitt, W., and Scanlan, J., *Proc. Royal Soc. (London)* **A220**, 530 (1953).
96. Mulliken, R. S., *Phys. Rev.* **43**, 279 (1933).
97. Orgel, L. E., "An Introduction to Transition-Metal Chemistry," Wiley, New York (1960), p. 154 *ff*.
98. Pariser, R., *J. Chem. Phys.* **21**, 568 (1953).
99. Pariser, R., and Parr, R. G., *J. Chem. Phys.* **21**, 767 (1953).
100. Pariser, R., and Parr, R. G., *J. Chem. Phys.* **21**, 466 (1953).
101. Parr, R. G., "Quantum Theory of Molecular Electronic Structure," Benjamin, New York (1963).
102. Parr, R. G., Craig, D. P., and Ross, I. G., *J. Chem. Phys.* **18**, 1561 (1950).
103. Parr, R. G., and Pariser, R., *J. Chem. Phys.* **23**, 711 (1955).
104. Penney, W. G., *Proc. Royal Soc. (London)* **A146**, 223 (1934).
105. Peters, D., *J. Chem. Soc.* 1023 (1958).
106. Platt, J. R., "Systematics of the Electronic Spectra of Conjugated Molecules: A Source Book," Wiley, New York (1964).
107. Pritchard, H. O., and Sumner, F. H., *Proc. Royal Soc. (London)* **A126**, 128 (1954).
108. Rebane, T. K., *Soviet Physics, Doklady* **5**, 1246 (1960).
109. Roberts, J. D., "Notes on Molecular Orbital Calculations," Benjamin, New York (1961).
110. Roberts, J. D., *Chem. Soc. Special Publication No. 12*, 111 (1958).
111. Roberts, J. D., Streitwieser, A., Jr., and Regan, C. M., *J. Am. Chem. Soc.* **74**, 4579 (1952).
112. Scherr, C. W., *J. Chem. Phys.* **21**, 1413 (1953).
113. Shida, S., *Bull. Chem. Soc. Japan* **27**, 243 (1954).
114. Shida, S., Kuri, Z., *J. Chem. Soc., Japan, Pure Chem. Section* **76**, 322 (1955).
115. Simmons, H. E., *J. Chem. Phys.* **40**, 3554 (1964).
116. Simmons, H. E., unpublished calculations.
117. Slater, J. C., *Phys. Rev.* **38**, 1109 (1931); see also ref. 46.
118. Snyder, L. C., *J. Phys. Chem.* **66**, 2299 (1962).
119. (a) Snyder, L. C., *Symposium on Molecular Structure and Spectroscopy*, Paper **B-4**, The Ohio State Univ., Columbus, (1960); (b) *ibid.*, Paper **B-7** (1961).
120. Snyder, L. C., *J. Chem. Phys.* **33**, 619 (1960).
121. Streitwieser, A., Jr., "Molecular Orbital Theory for Organic Chemists," Wiley, New York (1961).
122. Streitwieser, A., Jr., and Schwager, I., *J. Am. Chem. Soc.* **85**, 2855 (1963).
123. Takekiyo, S., *Bull. Chem. Soc. Japan* **35**, 463 (1962).
124. Takekiyo, S., *Bull. Chem. Soc. Japan* **35**, 355 (1962).
125. Takekiyo, S., *Bull. Chem. Soc. Japan* **34**, 1686 (1961).
126. Waack, R., *J. Chem. Ed.* **39**, 469 (1962).
127. Waser, J., and Lee, C.-S., *J. Am. Chem. Soc.* **66**, 2035 (1944).

128. Waser, J., and Schomaker, V., *J. Am. Chem. Soc.* **65,** 1451 (1943).
129. Weltner, W., Jr., *J. Am. Chem. Soc.* **75,** 4224 (1953).
130. Wheland, G. W., *J. Chem. Phys.* **23,** 79 (1955).
131. Wheland, G. W., *J. Am. Chem. Soc.* **63,** 2025 (1941).
132. Wheland, G. W., *Proc. Royal Soc.* (*London*) **A164,** 397 (1938).
133. Wheland, G. W., *J. Chem. Phys.* **2,** 474 (1934).
134. Wolfsberg, M., *J. Chem. Phys.* **23,** 793 (1955).
135. Wolfsberg, M., *J. Chem. Phys.* **21,** 943 (1953).

Addendum

Recently Dewar has developed a method for computing ground-state properties of conjugated systems employing empirical bond energies in conjunction with two MO procedures: an SCF variant of the Pariser–Parr method (Pariser–Parr–Pople method) and a similar SCF procedure in which the molecular integrals are modified numerically to account for "vertical" electron correlation (split-p-orbital or S.P.O. method).[5]

By this method, Dewar and co-workers computed that square-planar cyclobutadiene (*42*) has a triplet ground state with a negative resonance energy (-0.53 to -0.82 eV),[6-8] whereas rectangular-planar cyclobutadiene (*43*) possesses a singlet ground state with a slightly positive RE ($+0.034$ eV).[8] However, as has been noted in an earlier section (Section C), singlet (*43*) very probably does not represent an energy minimum with respect to asymmetric in-plane distortions which will ultimately lead to ring opening (see refs. 118–120 in previous list).

Dewar[9] has also employed his method to compute geometries and resonance energies for a number of known and unknown polybenzocyclobutadienes including (*44*)[1] and (*45*)[2] which have been synthesized recently. The results of these calculations indicate that there exists a correlation between the average bond lengths of the four-membered ring and the overall stability of the system, i.e., the greater the cyclobutadiene character of the four-ring the lower the stability of the system.

The reactivity and electron affinity of biphenylene have been the subjects of recent reports. Blatchly and Taylor[4] have measured the relative rates at which positions 1 and 2 undergo tritiodeprotonation in a different medium and temperature from those employed previously (see ref. 122 in previous list), (Section D, 3, c) and obtained $k_2/k_1 = 135$. More recently Dickerman, Milstein, and McOmie[10] studied the homolytic phenylation of biphenylene and found $k_2/k_1 \sim 3$, in good agreement with theory. Concerning the electron affinity of biphenylene Bauld and Banks[3] reported that b⁻ undergoes extensive disproportionation to b⁰ and b²⁻ and attributed the tendency of b²⁻ to form to its $4n+2\pi$ system. However, more recently Waack and co-workers[16] measured the equilibrium for $2b^- \rightleftarrows b + b^{2-}$ and concluded that the tendency of b²⁻ to form is in all probability not greater than that of the dianion of anthracene (a $4n\pi$ system)! This latter result is not unexpected since the orbital arrangement of biphenylene (p. 404) is a stable one and any additional electrons should occupy ABMO's.

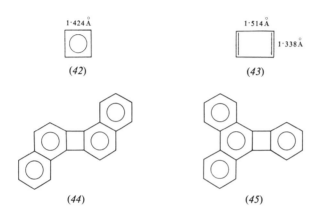

Concerning the emission of biphenylene, recent work by Hochstrasser[13b] indicates that carefully purified material shows neither fluorescence nor phosphorescence in contrast to previous reports (see ref. 54b in previous list).

In view of the observed instability of compound (24b), Hilpern[13a] has recently computed various low-lying states of this molecule by the Pariser–Parr–Pople method and found the lowest triplet configuration ($T_{14,15}$) to be lower in energy by 0.76 eV. than the lowest closed-shell singlet configuration (V_0). Furthermore, $T_{14,15}$ remains the ground state even when the links connecting the phenyls to the four-membered ring are treated as single bonds.

Pettit[11] has recently prepared the iron tricarbonyl complex of cyclobutadiene (46). This interesting compound in addition to being stable, readily undergoes a number of electrophilic substitution reactions.[12]

The ground-state multiplicity of cyclobutadiene has been the subject of two recent chemical investigations. In the first of these, Skell and Petersen[15] have presumably generated free tetramethylcyclobutadiene by the reaction of the cyclobutene (47) with potassium vapor at 245°–255° in a helium atmosphere under conditions which would allow for collisional deactivation of the cyclobutadiene to its ground state. In the absence of a third reactant, the major product was the syn dimer (48), whereas when the reaction was carried out in the presence of triplet methylene (CH_2:), generated *in situ* from CH_2Br_2, the amount of (*48*) decreased markedly and the major product was 1,2-dimethylene-3,4-dimethylcyclobutene (*49*). Skell and Petersen visualize the formation of (*49*) by the simultaneous transfer of two hydrogens from triplet tetramethylcyclobutadiene to triplet CH_2:.

The second investigation was carried out by Pettit and co-workers.[18] Treatment of the complex (*46*) with ceric ion (Ce^{4+}) in the absence of a reactant leads to the syn and anti dimers (*50*) and (*52*) in a ratio of 5 to 1, respectively,[18] whereas "Dewar" benzenes (*51*) are produced when the decomposition of the complex is carried out in the presence of acetylenes.[17] When (*46*) was decomposed in the presence of methyl maleate and methyl fumarate, adducts (*53*) and (*54*) were formed, respectively.[18] Making the unusual assumptions that (*53*) and (*54*) are products of *free* cyclobutadiene in its *ground state* and that

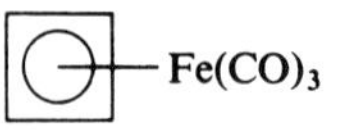

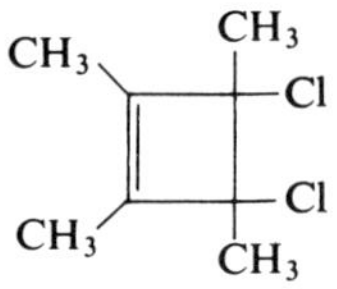

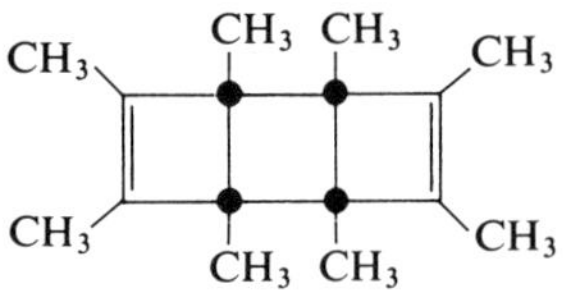

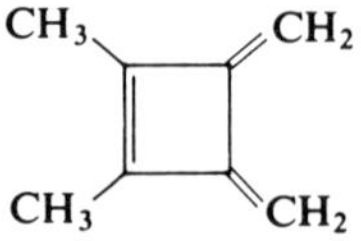

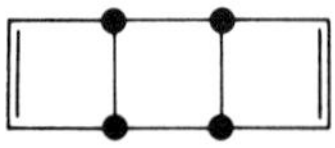

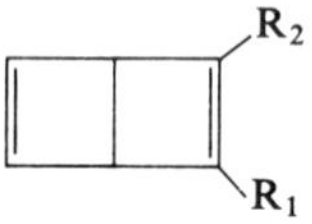

triplet cyclobutadiene is incapable of stereospecific addition,* Pettit and co-workers conclude that the stereospecific formation of (*53*) and (*54*) constitutes strong evidence for a singlet ground state for cyclobutadiene.[18]

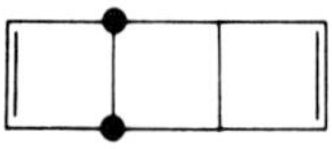

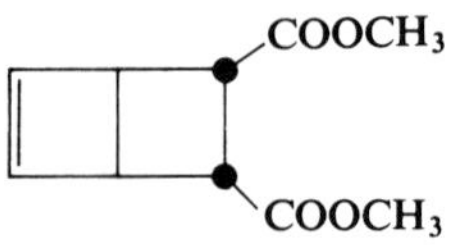

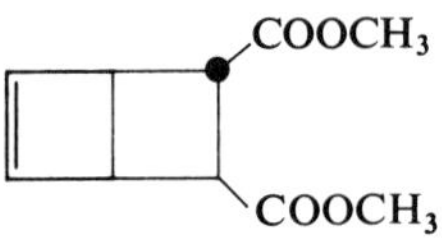

The predominant formation of syn dimers (*48*) and (*50*) over that of the anti counterparts is consistent with the presence of free cyclobutadiene, since

---

* Although this assumption constitutes a rather reliable working hypothesis where additions of methylenes to double bonds are concerned, it does not rest on firm theoretical grounds. See for example, P. P. Gaspar and G. S. Hammond, *in* "Carbene Chemistry," W. Kirmse, ed. Academic Press Inc., New York, New York (1964) pp. 258–261.

Hoffmann and Woodward[14a] have shown recently that for cyclobutadiene syn dimerization ought to be preferred over anti dimerization. On the other hand in view of the very similar energies predicted for the lowest singlet and lowest triplet state of cyclobutadiene, there is no compelling reason why tetramethylcyclobutadiene at 250° and cyclobutadiene in the presence of paramagnetic substances should enter reaction in their electronic ground states.

In view of the inherent deficiencies of most reaction conditions in providing reliable data concerning ground state multiplicity, it appears very likely that a definitive answer could be provided only by a spectroscopic investigation.

Orbital symmetry considerations suggest that cyclobutadiene can act as both a diene toward olefins (4 + 2) and a dienophile toward dienes (2 + 4), but not as an olefin toward olefins (2 + 2) or a diene toward dienes (4 + 4). The transition states for the four possibilities are shown in Fig. 12.11. In each case, the

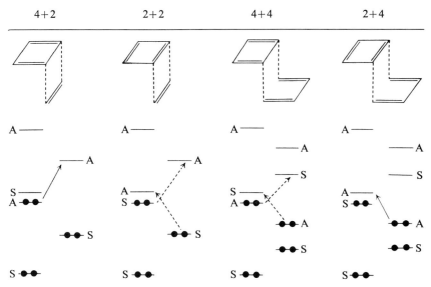

Fig. 12.11. Orbital symmetries for rectangular cyclobutadiene transition states.

MO's of rectangular C_4H_4 are shown on the left, and the symmetric (S) and antisymmetric (A) labels refer to reflection in the unique symmetry plane. The solid arrows correspond to favorable (symmetry allowed) interactions, and the dashed arrows to unfavorable (symmetry forbidden) ones. In accord with a suggestion of Woodward and Hoffmann, the diene component has been depicted as the donor in the two favorable cases.[14b] Similar considerations show that the dimerization of cyclobutadiene must be considered a 2+4 reaction.[14a]

References

1. Barton, J. W., *J. Chem. Soc.* 5161 (1964).
2. Barton, J. W., Rogers, A. M., and Barney, M. E., *J. Chem. Soc.* 5537 (1965).
3. Bauld, N. L., and Banks, D., *J. Am. Chem. Soc.* **87**, 128 (1965).
4. Blatchly, J. M., and Taylor, R., *J. Chem. Soc.* 4641 (1964).
5. For a description of the method see: Chung, A. L. H., Dewar, M. J. S., and Sabelli, N. L., "Molecular Orbitals in Chemistry, Physics and Biology," edited by Per-Olov Löwdin and B. Pullman, Academic Press, New York, New York (1964), pp. 395–404.
6. Chung, A. L. H., and Dewar, M. J. S., *J. Chem. Phys.* **42**, 756 (1965).
7. Dewar, M. J. S., and Gleicher, G. J., *J. Am. Chem. Soc.* **87**, 685 (1965).
8. Dewar, M. J. S., and Gleicher, G. J., *J. Am. Chem. Soc.* **87**, 3255 (1965).
9. Dewar, M. J. S., and Gleicher, G. J., *Tetrahedron* **21**, 1817 (1965).
10. Dickerman, S. C., Milstein, N., and McOmie, J. F. W., *J. Am. Chem. Soc.* **87**, 5521 (1965).
11. Emerson, G. F., Watts, L., and Pettit, R., *J. Am. Chem. Soc.* **87**, 131 (1965).
12. Fitzpatrick, J. D., Watts, L., Emerson, G. F., and Pettit, R., *J. Am. Chem. Soc.* **87**, 3254 (1965).
13a. Hilpern, J. W., *Mol. Phys.* **9**, 295 (1965).
13b. Hochstreasser, R. M., and McAlpine, R. D., *J. Chem. Phys.* **44**, 3325 (1966).
14. (a) Hoffmann, R., and Woodward, R. B., *J. Am. Chem. Soc.* **87**, 4388 (1965); (b) *ibid.*, *J. Am. Chem. Soc.* **87**, 2046 (1965).
15. Skell, P. S., and Petersen, R. J., *J. Am. Chem. Soc.* **86**, 2530 (1964).
16. Waack, R., Doran, M. A., and West, P., *J. Am. Chem. Soc.* **87**, 5508 (1965).
17. Watts, L., Fitzpatrick, J. D., and Pettit, R., *J. Am. Chem. Soc.* **87**, 3253 (1965).
18. Watts, L., Fitzpatrick, J. D., and Pettit, R., *J. Am. Chem. Soc.* **88**, 623 (1966).

APPENDIX

Abstracts of Papers Appearing in the Literature in 1964-1965

Chapter 1

Cyclobutadiene

1.1 Arnett, E. M., and Bollinger, J. M., *J. Am. Chem. Soc.* **86,** 4729 (1964).

The reaction of diisopropylacetylene with dicobalt octacarbonyl affords a mixture of hexaisopropylbenzene (I), tetraisopropylcyclopentadienone (II), 2,7-dimethyl-4,5-diisopropyl-2,4,6-octatriene (III), and 4-isopropylidene-1,2,3-triisopropylcyclobutene (IV). The isolation of compounds III and IV suggests strongly that the reaction products are formed via a cobalt carbonyl complex of tetraisopropylcyclobutadiene (V).

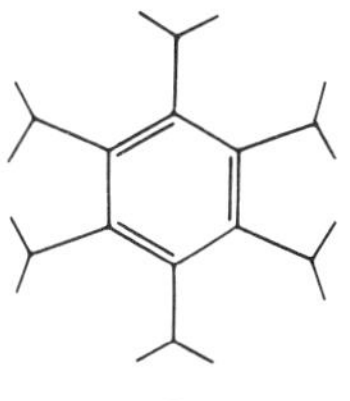

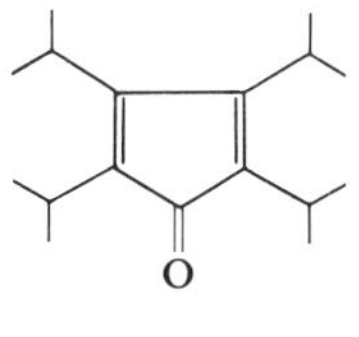

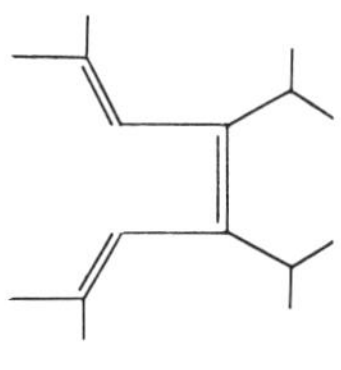

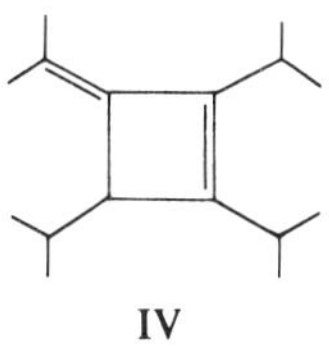

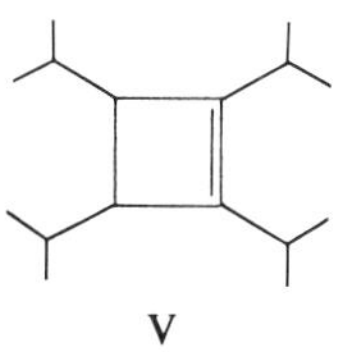

1.2 Avram, M., Dinulescu, I. G., Marica, E., Mateescu, G., Sliam, E., and Nenitzescu, C. D., *Ber.* **97, 382 (1964).**

The generation and dimerization of cyclobutadiene are described. See ref. 5, Chapter 1.

1.3 Berkoff, C. E., Cookson, R. C., Hudec, J., Jones, D. W., and Williams, R. O., *J. Chem. Soc.* **194 (1965).**

Dehalogenation of 3,4-dichloro-1,2,3,4-tetramethylcyclobutene with zinc in the presence of 2-butyne and dimethyl acetylenedicarboxylate gives hexamethylbenzene and dimethyl 3,4,5,6-tetramethylphthalate, respectively, presumably via tetramethylcyclobutadiene. Pyrolysis of tetraphenylthiophene dioxide gives octaphenylcyclooctatetraene, identified by previous workers as a dimer of tetraphenylcyclobutadiene.

1.4 Berry, R. S., Clardy, J., and Schafer, M. E., *Tetrahedron Letters* **1003 (1965).**

Photoinitiated decomposition of benzenediazonium-4-carboxylate yields a surprisingly long-lived species of m/e 76, $[C_6H_4]^{+\cdot}$, for which several structures are proposed, one of them a cyclobutadiene system, bicyclo[2.2.0]hepta-1,3,5-triene (I).

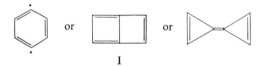

I

1.5 Beynon, J. H., Cookson, R. C., Hill, R. R., Jones, D. W., Saunders, R. A., and Williams, A. E., *J. Chem. Soc.* **7052 (1965).**

Formulation of the high-melting dimer of tetraphenylcyclobutadiene as octaphenylcyclooctatetraene (I) is supported by the formation of diphenylacetylene and hexaphenylbenzene under pyrolytic conditions (670°). The mass spectrum of I is described in detail.

Tetraphenylcyclobutadienepalladium chloride (II) evolves a volatile material of molecular weight 356 when heated to 185° *in vacuo*. Mass spectrometric analysis of the volatile material gives strong indication of the existence of monomeric tetraphenylcyclobutadiene (III) in the gas phase.

1.6 Blomquist, A. T., and LaLancette, E. A., *J. Org. Chem.* **29, 2331 (1964).**
   Abstract 4.1, this appendix.

I: R=C_6H_5

II: R=C_6H_5

III: R=C_6H_5
M.W.=356

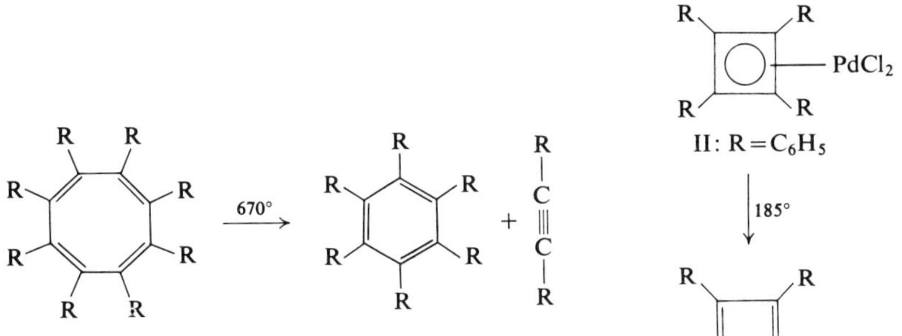

1.7 Breslow, R., Kivelevich, D., Mitchell, M. J., Fabian, W., and Wendel, K., *J. Am. Chem. Soc. 87*, 5132 (1965).

This paper describes a number of unsuccessful approaches to "push–pull" stabilized cyclobutadienes of type I. A full report of the work cited in Chapter 1 (refs. 26, 29, and 30) is given. In addition, the synthesis of 1-dimethylamino-2-nitro-3-phenyl-4-bromo-4-benzoylcyclobutene (III), 1-dimethylamino-2-nitro-3-phenyl-4-iodocyclobutene (II), and 1-dimethylamino-2-nitro-3-phenylcyclobutenone (IV) is described. Ketone IV is resistant to enolization, and attempts to dehydrohalogenate II and III afforded no evidence for cyclobutadienoid intermediates.

I Ia

II: X=I; Y=H
III: X=Br; Y=COC_6H_5

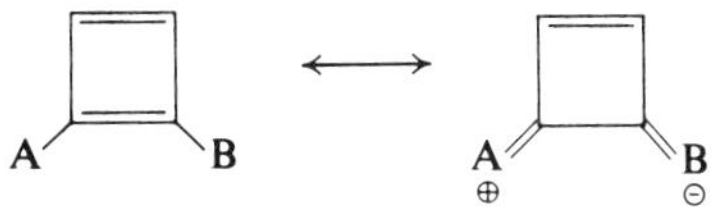

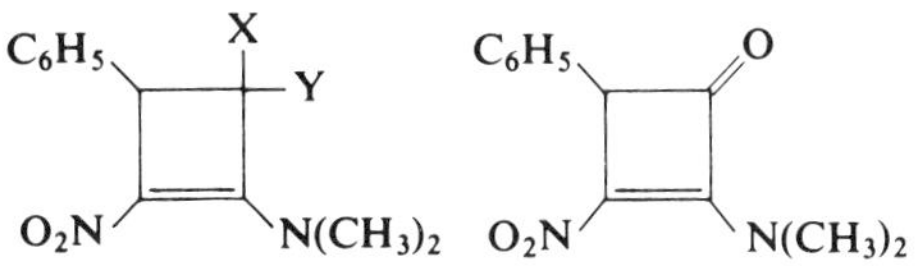

1.8 Burt, G. D., and Pettit, R., *Chem. Commun.* 517 (1965).

Decomposition of cyclobutadieneiron tricarbonyl (I) with ceric ion in the presence of phenylacetylene affords 2-phenylbicyclo[2.2.0]hexa-2,5-diene (III), presumably via cyclobutadiene (II). Hydrocarbon III, referred to as "hemi-Dewar biphenyl," is a colorless oil which isomerizes quantitatively to biphenyl on being heated at 90° for several minutes.

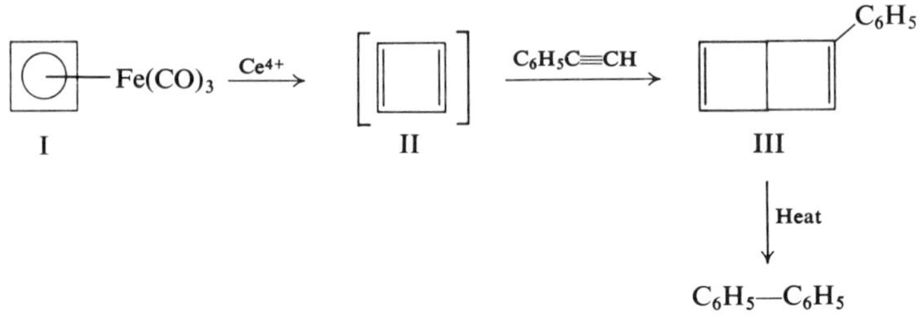

1.9 Childs, R. F., and Johnson, A. W., *Chem. Commun.* 95 (1965).

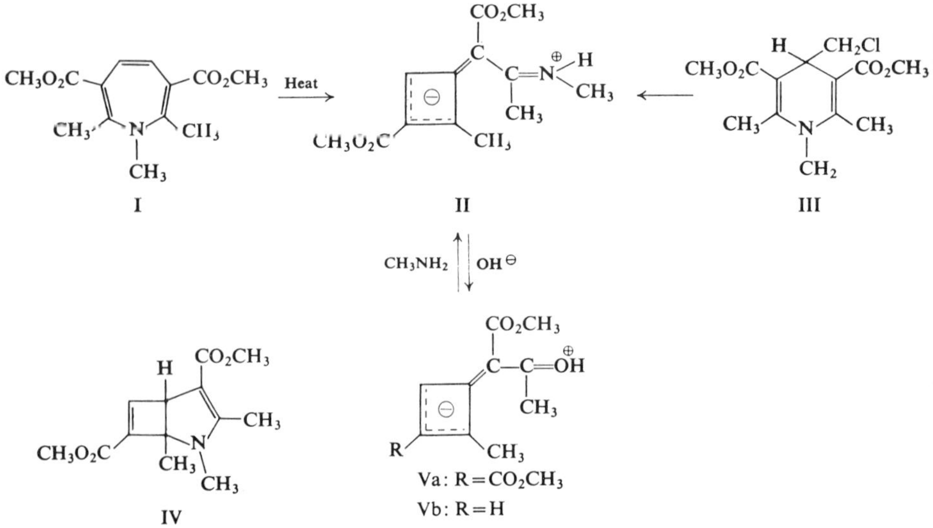

Dimethyl 2,7-dimethylazepine-3,6-dicarboxylate is isomerized quantitatively in boiling benzene to give ylid II which is also obtained by photolysis of dimethyl 4-chloromethyl-2,5-dimethyl-1,4-dihydropyridine-3,5-dicarboxylate (III) and by treating III with sodium methoxide, followed by pyrolysis of the resulting oil. The ylid (II) results from ring-fission and prototropic rearrangement of the valence tautomer (IV) of I. When it was treated with aqueous potassium hydroxide, II gave the oxygen analog, Va, in high yield; Va could be reconverted into II by treatment with hot methanolic methylamine. Three analogs of II were prepared by treating Va with ammonia, *p*-chloroaniline, or dimethylamine. Steam distillation of II from an acid solution gave Vb. The formation of ylids in this series emphasizes the reluctance of cyclobutenyl ions to exist in cyclobutadienoid forms.

1.10 Cookson, R. C., and Jones, D. W., *J. Chem. Soc.* 1881 (1965).

A few of the more salient points in the paper were discussed briefly in Chapter 2 (ref. 12a). In addition: Pyrolysis of tetraphenylcyclobutadienepalladium chloride (I) gives two of the three possible isomers of 1,4-dichloro-1,2,3,4-tetraphenyl-1,3-butadiene (II) and a small amount of 5,10-diphenylindeno[2,1-*a*]indene (III). Extraction of the complex (I) with boiling ethanol gives tetraphenylfuran and a compound formulated as IV. Reduction of the complex (I) with hydrogen in the presence of platinum or with lithium aluminum hydride in ether gives tetraphenylbutadiene (V) and metallic palladium. Reduction with hydrogen and platinum at higher temperatures and pressures gives 1,2,3,4-tetraphenyl-1-butene, probably identical with Freedman's reduction product from tetraphenylcyclobutadienenickel bromide which was tentatively identified as 1,2,3,4-tetraphenylcyclobutene (ref. 25, Chapter 2) but which was subsequently shown to differ from authentic tetraphenylcyclobutene [Freedman and Frantz, *J. Amer. Chem. Soc.* **84**, 4165 (1962)]. Complex I reacts with iron pentacarbonyl to form tetraphenylcyclobutadieneiron tricarbonyl and tetraphenylcyclopentadienone.

Treating I with triphenylphosphine in benzene or chloroform gave green solutions in which the color persisted for several hours in the absence of oxygen and which gave octaphenylcyclooctatetraene, the usual dimer of tetraphenylcyclobutadiene. When air was admitted, the color faded instantly and tetra-

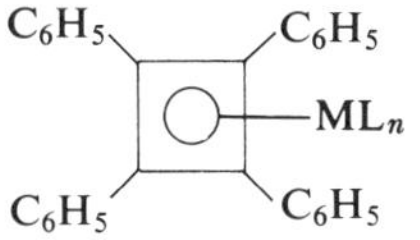

I: $ML_n = PdCl_2$
VI: $ML_n = Fe(CO)_3$

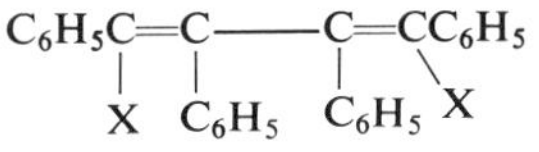

IIa: X = Cl   } two isomers
IIb: X = Cl   }

IV: $X = OC_2H_5$
V: X = H

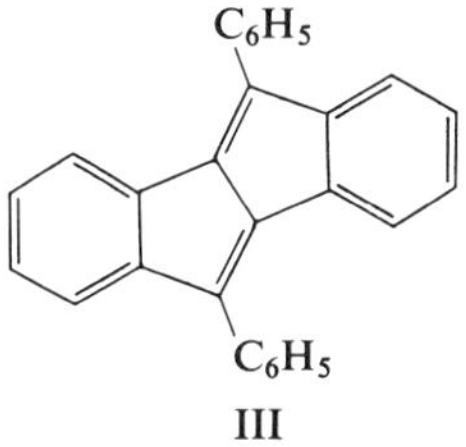

III

phenylfuran was formed in place of the dimer. When tetraphenylcyclobutadiene was generated in the presence of methyl phenylacetylenecarboxylate, the dimer was accompanied by methyl pentaphenylbenzoate. Unlike tetramethylcyclobutadiene, tetraphenylcyclobutadiene does not react with 2-butyne and, hence, is probably less reactive as a dienophile.

Tetrakis(*p*-chlorophenyl)cyclobutadienepalladium chloride was prepared from ethoxytetrakis(*p*-chlorophenyl)cyclobutenylpalladium chloride.

Solutions of tetraphenylcyclobutadiene and tetrakis(*p*-chlorophenyl)cyclobutadiene give ESR signals but the signals are apparently not due entirely to the cyclobutadienes.

1.11 Corey, E. J., and Streith, J., *J. Am. Chem. Soc.* 86, 950 (1964).

The mass spectra of the photoisomers, I and II, of 2-pyrone (III) and *N*-methyl-2-pyridone (IV), respectively, show a peak at m/e 52 corresponding to $C_4H_4^{+\cdot}$. This fragment may well be the cyclobutadiene radical cation (V).

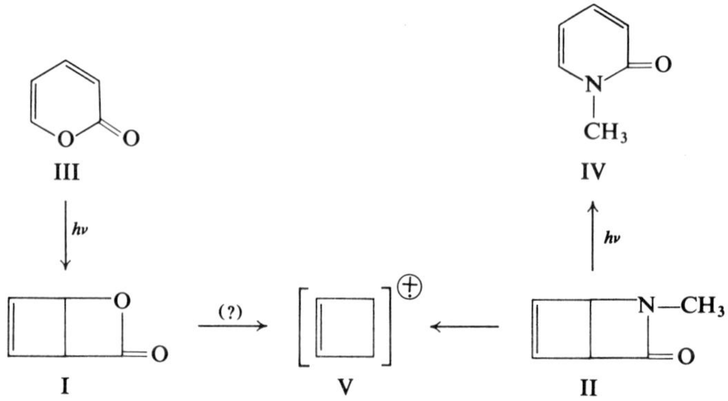

1.12 Criegee, R., and Funke, W., *Ber.* 97, 2934 (1964).

3-Chloro-, 3-hydroxy-, and 3-dimethylamino-1,2,3,4-tetramethylcyclobutene undergo elimination reactions to give products having methylene cyclobutene and butadiene structures and give no tetramethylcyclobutadiene (I).

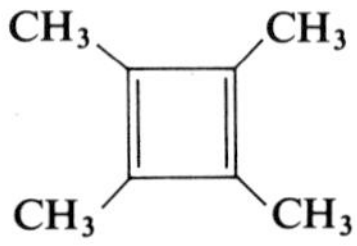

1.13 Criegee, R., Kristinsson, K., Seebach, D., and Zanker, F., *Ber.* 98, 2331 (1965).

Treating *cis*-3,4-diiodo-1,2,3,4-tetramethylcyclobutene with mercury in the presence of maleic anhydride or dimethylmaleic anhydride gave 1,2,3,4-tetramethylbicyclo[2.2.0]hex-2-ene-5,6-dicarboxylic anhydride or the 1,2,3,4,5,6-hexamethyl analog, presumably via tetramethylcyclobutadiene. See also Chapter 1, ref. 45a.

1.14 Criegee, R., and Zanker, F., *Angew. Chem.* 76, 716 (1964). Ref. 56, Chapter 1. A preliminary communication. (The full paper is described in Abstract 1.15.)

1.15 Criegee, R., and Zanker, F., *Ber.* 98, 3838 (1965). (This paper is a full report of the work described in a preliminary communication cited as ref. 56, Chapter 1.)

Dehalogenation of 3,4-diiodo-1,2,3,4-tetramethylcyclobutene (I) with mercury in the presence of 1,2-dihalomaleic anhydrides gives the corresponding *endo*-5,6-dihalo-1,2,3,4-tetramethylbicyclo[2.2.0]hex-2-ene-5,6-dicarboxylic anhydrides (IIa, IIb, IIc), presumably via tetramethylcyclobutadiene.

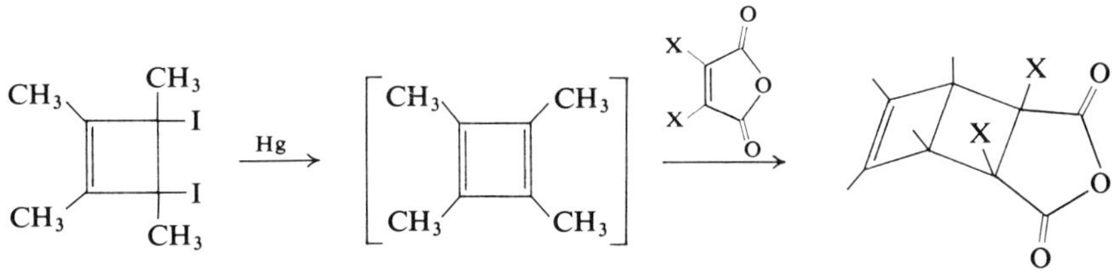

IIa: X = Cl
IIb: X = Br
IIc: X = I

1.16 Drefahl, G., Hoerhold, H. H., and Bretschneider, H., *J. prakt. Chem.* 25, 113 (1964).

Diphenylacetylene undergoes polymerization in the presence of triethylaluminum–titanium tetrachloride to give hexaphenylbenzene and, under the proper reaction conditions, the compound $C_{56}H_{40}$, now known to be octaphenylcyclooctatetraene, the usual dimer of tetraphenylcyclobutadiene (see Chapter 1).

1.17 McOmie, J. F. W., and Bullimore, B. K., *Chem. Commun.* **63 (1965).**

Attempts to prepare tetraphenylcyclobutadiene (I) or a metal complex thereof by thermolysis of 2,3,4,5-tetraphenylthiophene dioxide (II) failed. Thermolysis of the thiophene dioxide in the molten state gave tetraphenylthiophene and tetraphenylfuran; in refluxing di(n-butyl) phthalate, the thiophene dioxide gave 1,2,3-triphenylnaphthalene and 1,2,3-triphenylazulene. Thermolysis of a mixture of the thiophene dioxide and nickel bromide trihydrate at 300°–400° gave triphenylnaphthalene, tetraphenylthiophene, 3,6-diphenyl-1,2,3,4-dibenzopentalene, and a small amount of a white solid, $C_{28}H_{20-22}$; diphenylacetylene is not an intermediate in this transformation. Tetraphenylthiophene dioxide is unaffected by ultraviolet irradiation (Pyrex filter). It is suggested that the intermediate produced in the thermolysis of tetraphenylthiophene dioxide is the same as that produced in the photolysis of diphenylacetylene, possibly triplet tetraphenylcyclobutadiene.

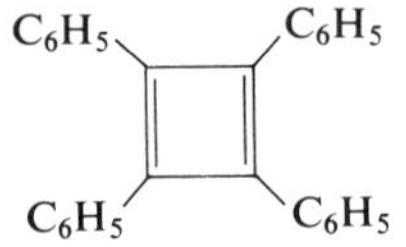

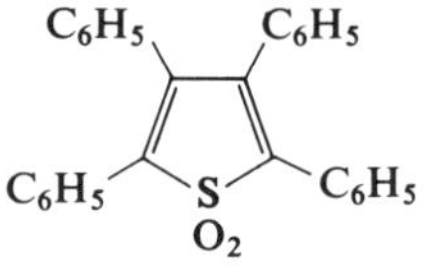

1.18 Nagarajan, K., Caserio, M. C., and Roberts, J. D., *J. Am. Chem. Soc.* **86, 449 (1964).**

The generation and dimerization of 1-fluoro-2,3,4-triphenylcyclobutadiene and 1-fluoro-3-(p-chlorophenyl)-2,4-diphenylcyclobutadiene are described. See ref. 116, Chapter 1.

1.19 Nahon, R., *Dissertation Abstr.* **25, 3846 (1965).** Abstract 5.3, this appendix.

1.20 Pawley, G. S., Lipscomb, W. N., and Freedman, H. H., *J. Am. Chem. Soc.* **86, 4725 (1964).**

An X-ray crystallographic study of the dimer of tetraphenylcyclobutadiene has shown it to be octaphenylcyclooctatetraene. See ref. 119, Chapter 1.

1.21 Skell, P. S., and Petersen, R. J., *J. Am. Chem. Soc.* **86, 2530 (1964).**

The triplet behavior of tetramethylcyclobutadiene was inferred from its dimerization and dismutation reactions. See ref. 35, Chapter 1.

1.22 Throndsen, H. P., and Zeiss, H., *J. Organometal. Chem. 1*, 301 (1964).

Treating phenylacetylene with arylmagnesium compounds gives the usual dimer of tetraphenylcyclobutadiene, i.e., octaphenylcyclooctatetraene (I) (incorrectly identified in this paper as octaphenylcubane). The dimer (I) was not formed when diphenylacetylene was treated with phenyllithium. Analogs of I could not be prepared from 2-butyne or from methylphenylacetylene.

1.23 Watts, L., Fitzpatrick, J. D., and Pettit, R., *J. Am. Chem. Soc. 87*, 3253 (1965). [A few of the more salient points in this paper were cited in Chapter 1 (ref. 48).]

Unambiguous evidence is presented for the first time which indicates that cyclobutadiene (I) has been synthesized. Thus, decomposition of cyclobutadieneiron tricarbonyl (II) with ceric ion at 0° evolves a gas containing I which is a highly reactive substance. Condensation of this gas in liquid nitrogen, followed by reaction of the condensate with methyl propiolate, leads to the formation initially of the "Dewar benzene" derivative, 2-carbomethoxybicyclo[2.2.0]hexa-2,5-diene (III). Compound III isomerizes at 90° to methyl benzoate, which was detected by gas chromatography in low yield.

The "Dewar benzenes," III and IV, are available preparatively by the ceric ion oxidation of complex II in the presence of methyl propiolate and dimethyl acetylenedicarboxylate, respectively. Compounds III and IV, obtained in a state of purity in this way, have been shown to rearrange quantitatively at 90° to methyl benzoate and dimethyl phthalate, respectively.

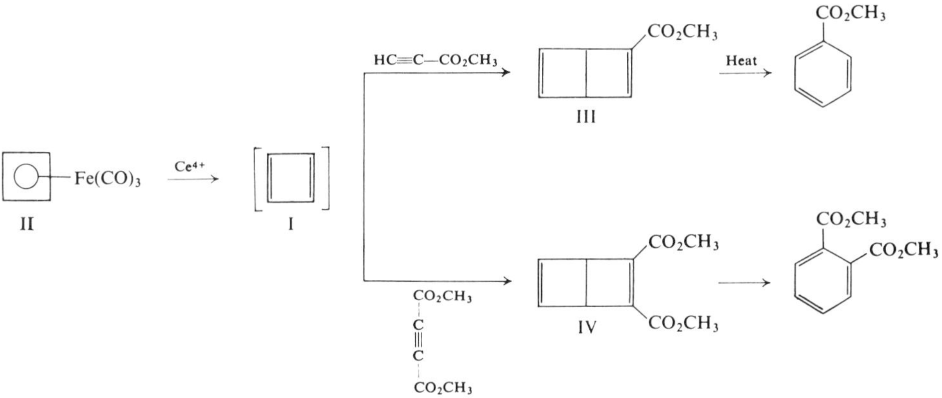

1.24 Watts, L., Fitzpatrick, J. D., and Pettit, R., *J. Am. Chem. Soc. 88*, 623 (1966).

Cyclobutadiene (I), liberated at 0° by the ceric ion oxidation of cyclobutadieneiron tricarbonyl (II), reacts in a *stereospecific* manner both as a diene and a dienophile. This behavior supports the theory that cyclobutadiene has a singlet ground state and, by implication, that it has the diene structure (I) rather than the triplet structure (III) at 0°.

The reaction of I with dimethyl maleate gave only *endo,cis*-5,6-dicarbomethoxybicyclohexene (IV), while the reaction of I with dimethyl fumarate gave only *trans*-5,6-dicarbomethoxybicyclohexene (V). Cyclobutadiene reacts as a dienophile with cyclopentadiene to give compound VI as the sole C_9H_{10} product.

The dimerization of I is not, however, entirely stereospecific. The ratio of the predicted syn dimer to the anti dimer is about 5:1.

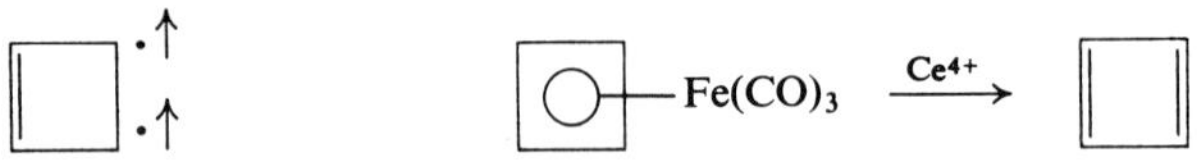

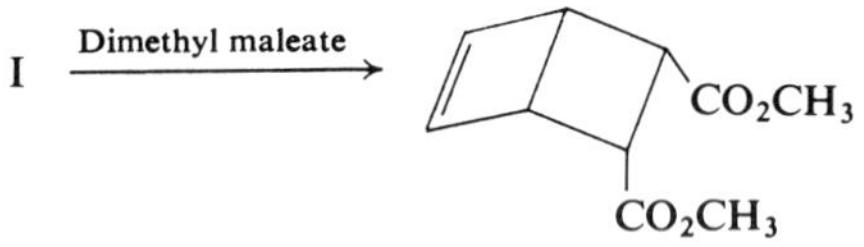

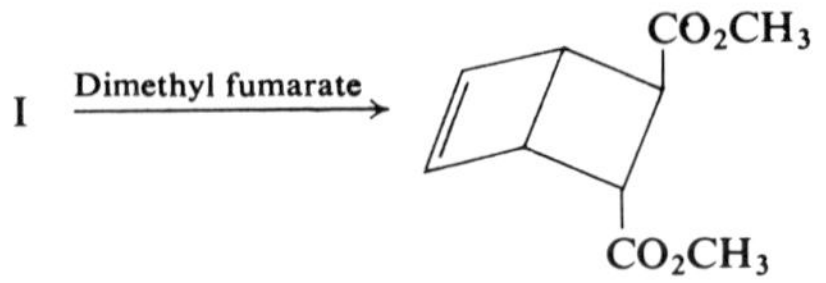

1.25 White, E. H., and Dunathan, H. C., *J. Am. Chem. Soc.* **86**, 453 (1964).

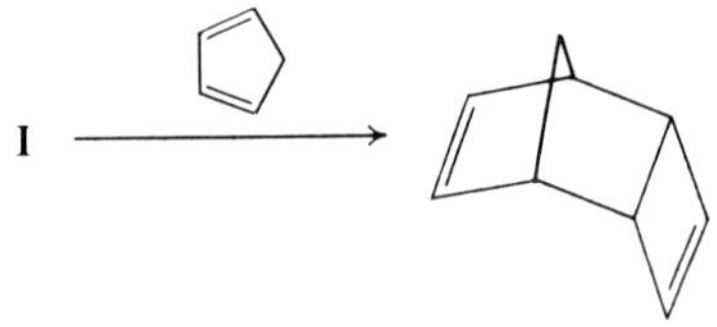

The generation and dimerization of 1,3-diphenylcyclobutadiene are described. See ref. 151, Chapter 1.

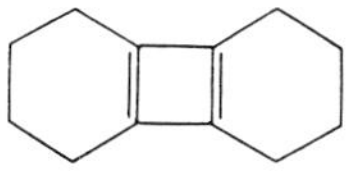

[Structure diagram: octahydrobiphenylene]

I

1.26 Wittig, G., and Weinlich, J., *Ber.* **98,** 471 (1965).

The structure assigned to tetrameric cyclohexyne, formed via octahydrobiphenylene (I) (ref. 158, Chapter 1), is in accord with the Raman spectrum. Transformation reactions of the tetramer are described.

Chapter 2

Cyclobutadiene–Metal Complexes

2.1 Beynon, J. H., Cookson, R. C., Hill, R. R., Jones, D. W., Saunders, R. A., and Williams, A. E., *J. Chem. Soc.* **7052** (1965). Abstract 1.5, this appendix.

2.2 Burt, G. D., and Pettit, R., *Chem. Commun.* **517** (1965). Abstract 1.8, this appendix.

2.3 Cookson, R. C., and Jones, D. W., *J. Chem. Soc.* **1881** (1965). Abstract 1.10, this appendix.

2.4 Criegee, R., Förg, F., Brune, H.-A., and Schönleber, D., *Ber.* **97,** 3461 (1964).

The reaction of tetramethylcyclobutadienenickel chloride with cyclopentadienylsodium is described. See ref. 47, Chapter 1 and ref. 14, Chapter 2.

2.5 Dodge, R. P., and Schomaker, V., *Acta Cryst.* **18,** 614 (1965).

Refinement of the single-crystal X-ray diffraction study reported previously (see Chapter 2) has shown that the carbocycle in tetraphenylcyclobutadieneiron tricarbonyl (I) is square planar with the sides 1.46 Å long.

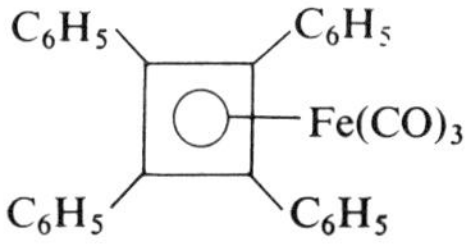

[Structure: tetraphenylcyclobutadiene iron tricarbonyl, C_6H_5 groups at each corner of cyclobutadiene ring with $Fe(CO)_3$]

I

2.6 Emerson, G. F., Watts, L., and Pettit, R., *J. Am. Chem. Soc.* **87,** 131 (1965).

The preparation of the first stable metal complex of cyclobutadiene and of benzocyclobutadiene is reported.

The reaction of *cis*-3,4-dichlorocyclobutene (I) with iron enneacarbonyl at 30° gives cyclobutadieneiron tricarbonyl (II), m.p. 26°, b.p. 68°–70° (3 mm); NMR: $\tau = 6.09$. Oxidation of II with ferric chloride or ceric ion in the presence of excess lithium chloride gives the previously unknown *trans*-3,4-dichlorocyclobutene (III).

The reaction of *trans*-1,2-dibromobenzocyclobutene (IV) (or the corresponding diiodide V) with iron enneacarbonyl at 30° gives benzocyclobutadieneiron tricarbonyl (VI), m.p. 25°, b.p. 73°–78° (0.1 mm); NMR: τ 3.05 (4H), 5.98 (2H). Reaction of VI with triphenylphosphine at 140° affords benzocyclobutadieneiron dicarbonyl triphenylphosphine (VII). Oxidative decomposition of complex VI with silver nitrate gave hydrocarbon IX, m.p. 105°, which is a new dimer of benzocyclobutadiene (X). A suggested mechanism for the formation of IX is shown below.

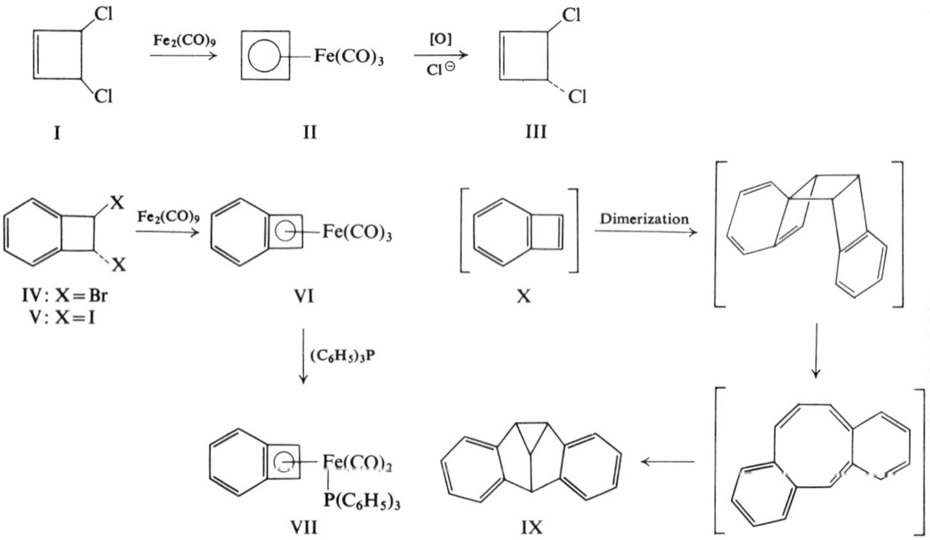

2.7 Fitzpatrick, J. D., Watts, L., Emerson, G. F., and Pettit, R., *J. Am. Chem. Soc.* **87**, 3254 (1965).

Cyclobutadieneiron tricarbonyl (I) can be considered to be a new aromatic system in the sense that it easily undergoes electrophilic substitution reactions. Its reactivity towards electrophiles is comparable to that of ferrocene. The ease of attack of I by an electrophilic species, $G^\oplus$, is rationalized on the basis of the formation of a π-allyliron tricarbonyl cation (II) as a low-energy intermediate in the substitution reaction.

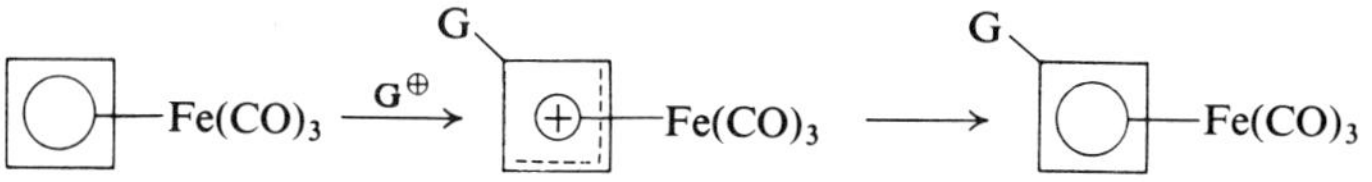

$$\text{I} \xrightarrow{G^{\oplus}} \text{II} \longrightarrow \text{III-IX}$$

I → II →
- III: G = COCH$_3$
- IV: G = COC$_6$H$_5$
- V: G = CHO
- VI: G = CH$_2$Cl
- VII: G = CH$_2$N(CH$_3$)$_2$
- VIII: G = HgCl
- IX: G = D

III or V →
- X: G = CH(OH)CH$_3$ (from III or V)
- XI: G = COOH (from V)
- XII: G = CH$_2$OH (from V)

The above substituted cyclobutadieneiron tricarbonyls were obtained directly from complex I using the reagents indicated in Table 1.

TABLE 1

PREPARATION OF SUBSTITUTED CYCLOBUTADIENEIRON TRICARBONYLS

Substituent	Structure	Reagent used
Acetyl	III	CH$_3$COCl + AlCl$_3$
Benzoyl	IV	C$_6$H$_5$COCl + AlCl$_3$
Formyl	V	C$_6$H$_5$N(CH$_3$)CHO + POCl$_3$
α-Chloromethyl	VI	CH$_2$O + HCl ("chloromethylation")
α-Dimethylaminomethyl	VII	CH$_2$O + HN(CH$_3$)$_2$ + CH$_3$COOH
Chloromercuri	VIII	Hg(OCOCH$_3$)$_2$ + NaCl
Deuterio	IX	CF$_3$COOD

TABLE 2

TRANSFORMATION OF SUBSTITUTED CYCLOBUTADIENEIRON TRICARBONYLS

Starting complex	Reagent	Product
V	CH$_3$MgBr	X (α-Hydroxyethyl-I)
III	NaBH$_4$	X (α-Hydroxyethyl-I)
V	Ag$_2$O	XI (α-Carboxy-I)
V	NaBH$_4$	XII (α-Hydroxymethyl-I)

2.8 Freedman, H. H., and Doorakian, G. A., *Tetrahedron* **20**, 2181 (1964).

Tetraphenylcyclobutadienenickel bromide (I) reacts with pyridine hydrobromide perbromide in methylene chloride to give 3,4-dibromo-1,2,3,4-tetraphenylcyclobutene (II) in 66% yield and a small amount of 2,5-di(p-bromophenyl)-3,4-diphenylfuran (III). Similar results were also obtained by treating I with bromine in the presence of pyridine hydrobromide. Transformation reactions of II and III are reported.

2.9 Fritz, H.-P., in Stone, F. G. A., and West, R. (eds.), *Advan. Organometal. Chem.* **1**, 260 (1964).

[A review of small-ring organometallic complexes (see ref. 27, Chapter 2).]

2.10 Gerloch, M., and Mason, R., *Proc. Royal Soc.* **A279**, 170 (1964).

Correction of the structure of Wilkinson's "tetrakis(trifluoromethyl)cyclobutadiene cyclopentadiene cobalt(III)" (ref. 7, Chapter 2). See ref. 66, Chapter 1.

2.11 Hübel, W., and Merenyi, R., *J. Organometal. Chem.* **2**, 213 (1965).

Diphenylacetylene reacts with molybdenum hexacarbonyl and also with diglyme–molybdenum tricarbonyl complex to give complexes to which tetraphenylcyclobutadiene structures are assigned. Thus, refluxing a solution of diphenylacetylene and molybdenum hexacarbonyl in benzene gives I, II, and III, which are also obtained when diphenylacetylene is heated with diglyme–molybdenum tricarbonyl at 110°–160° (no experimental details) but not at the boiling point of benzene. I, which is formed in 6% yield, is a bright yellow solid, m.p. 255°–262° (decomp.), $\lambda_{max}^{C_2Cl_4}$ 2004, 1961 cm^{-1} only slightly soluble in the usual organic solvents. Complex I is highly stable (the carbonyls are not displaced by triphenylphosphine), but it reacts with lithium aluminum hydride in refluxing tetrahydrofuran to give pentaphenylcyclopentadiene (30%), tetracyclone, and tetraphenylcyclopent-2-en-1-one. II is produced in 45% yield, dark green needles, m.p. 200°–205° (decomp.), $\lambda_{max}^{C_2Cl_4}$ 1988, 1949, 1927 cm^{-1}; it reacts with lithium aluminum hydride at room temperature to give dibenzyl, stilbene, *trans,trans*-tetraphenylbutadiene, tetracyclone, and tetraphenylcyclopent-2-en-1-one. Complex II also reacts with bromine to give *trans*-1,2-dibromostilbene, tetracyclone, and *cis*- and *trans*-dibenzoylstilbene. Thermolysis of II yields hexaphenylbenzene, pentaphenylcyclopentadiene, tetracyclone, and a mixture of diphenylacetylene and stilbene. Similarly, complex III, which is obtained in 6% yield, forms yellow prisms, m.p. 240°–243° (decomp.), $\lambda_{max}^{C_2Cl_4}$ 2012, 1953, and 1618 cm^{-1}. It reacts with lithium aluminum hydride to give *trans,trans*-tetraphenylbutadiene, tetracyclone, and tetraphenylcyclopent-2-en-1-one.

$$C_6H_5C \equiv CC_6H_5 \longrightarrow$$

I

II

III

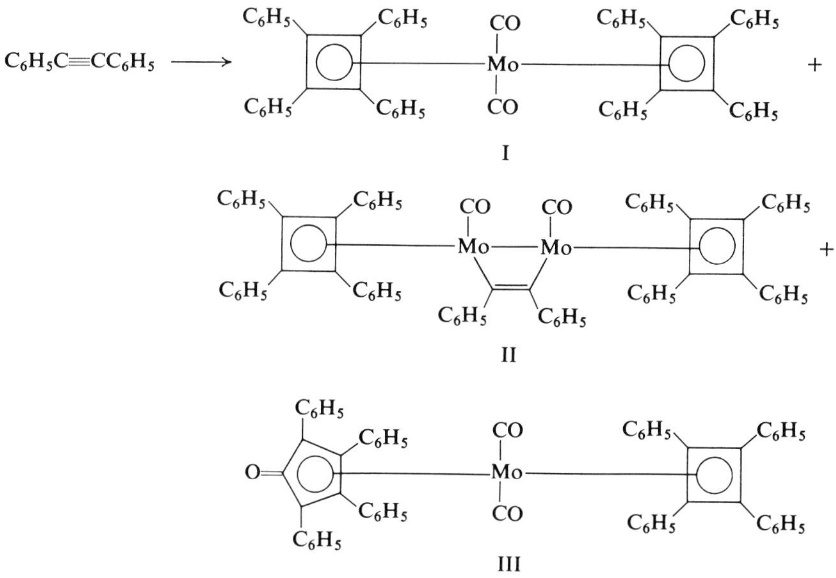

2.12 Hüttel, R., and Neuegebauer, H. J., *Tetrahedron Letters* 3541 (1964).

The most salient points in this paper are reported in Chapter 2 (ref. 33). In addition:

A solution of diphenylacetylene and bisbenzonitrilepalladium chloride in 1:3 ethanol–chloroform deposits a reddish-brown precipitate of tetraphenylcyclobutadienepalladium chloride oligomer $[(C_{28}H_{20})_2(PdCl_2)_3]_m$ (I) and a small amount of hexaphenylbenzene. Complex I absorbs in the ultraviolet at 323 mμ. It gives 1,4-dichloro-1,2,3,4-tetraphenylbutadiene and tetraphenylfuran on heating in a sealed tube at 240°; 1,2,3,4-tetraphenylbutadiene on reduction with sodium borohydride; and *cis*-dibenzoylstilbene on oxidation with nitric acid. Complex I could not be converted into the simpler complex

[(C$_{28}$H$_{20}$)$_2$PdCl$_2$)]$_n$ (II), by treatment with diphenylacetylene in hydrogen chloride–chloroform but when complex I was dissolved in dimethylformamide and treated with concentrated hydrochloric acid, complex II was obtained in 100% yield. Addition of water to dimethylformamide solutions of I or II gives the tetraphenylhydroxycyclobutenyl complex, (C$_{28}$H$_{22}$O)$_2$Pd$_2$Cl$_2$ [L. M. Vallarino and G. Santarella, *Gazz. Chim. Ital.* **94**, 252 (1964)]. When they are dissolved in pyridine, I and II gradually decompose to give after work-up, with ethanol, tetraphenylfuran and 2,5-dihydro-2,5-diethoxy-2,3,4,5-tetraphenylfuran, ascribed to the intermediacy of triplet tetraphenylcyclobutadiene.

2.13 Maitlis, P. M., in *Advan. Organometal. Chem.* 4, 95 (1966).
A review with 95 references.

2.14 Maitlis, P. M., and Efraty, A., *J. Organometal. Chem.* 4, 172 (1965).

Tetraphenylcyclobutadienepalladium halides (I and II) react with cyclopentadienylmolybdenum tricarbonyl dimer to give the corresponding (π-cyclopentadienyl)(π-tetraphenylcyclobutadiene)molybdenum carbonyl halides (III and IV, respectively). The diamagnetic chloride III is stable to hydrogen chloride. Tetraphenylcyclobutadienepalladium chloride (I) reacts also with cyclopentadienyltungsten tricarbonyl dimer, giving (π-cyclopentadienyl)(π-tetraphenylcyclobutadiene)tungsten carbonyl chloride (V), although in very low yield (0.7%).

I: X = Cl
II: X = Br

III: M = Mo; X = Cl
IV: M = Mo; X = Br
V: M = W; X = Cl

2.15 Maitlis, P. M., and Efraty, A., *J. Organometal. Chem.* 4, 175 (1965).

Tetraphenylcyclobutadienepalladium bromide (I) reacts with dicobalt octacarbonyl in homogeneous media to give tetraphenylcyclobutadiene cobalt dicarbonyl bromide (II). Triphenylphosphine reacts with II to give (tetraphenylcyclobutadiene)(triphenylphosphine)cobalt carbonyl bromide (III). The reaction of II with benzene, toluene, or mesitylene in the presence of aluminum chloride gives the corresponding very stable (π-arene)(π-tetraphenylcyclobutadiene)cobalt bromides (IV, V, and VI, respectively). Compounds II and III are paramagnetic whereas the arene complexes IV, V, and VI are diamagnetic.

$$\begin{array}{c}C_6H_5 \underset{C_6H_5}{\overset{C_6H_5}{\square}}-G\\C_6H_5C_6H_5\end{array}$$

I: G = PdBr$_2$
II: G = Co(CO)$_2$Br
III: G = Co(CO)[(C$_6$H$_5$)$_3$P]Br

$$\left[\begin{array}{c}C_6H_5C_6H_5\\\underset{C_6H_5}{\overset{\square}{}}\\C_6H_5C_6H_5\end{array}-Co-\underset{R'}{\overset{R}{\bigcirc}}R'\right]^{\oplus}Br^{\ominus}$$

IV: R = R' = H
V: R = CH$_3$, R' = H
VI: R = R' = CH$_3$

2.16 Maitlis, P. M., Efraty, A., and Games, M. L., *J. Organometal. Chem.* **2**, 284 (1964).

This publication is a preliminary note reporting work described in detail in Abstract 2.17.

2.17 Maitlis, P. M., Efraty, A., and Games, M. L., *J. Am. Chem. Soc.* **87**, 719 (1965).

Tetraphenylcyclobutadienepalladium bromide (I) reacts with cyclopentadienyliron dicarbonyl dimer or cyclopentadienyliron dicarbonyl bromide to give (π-cyclopentadienyl)(π-tetraphenylcyclobutadiene)palladium tetrabromoferrate (II); under the same conditions tetraphenylcyclobutadienenickel bromide (III) is converted into (π-cyclopentadienyl)(π-tetraphenylcyclobutadiene)nickel tetrabromoferrate (IV). Compounds II and IV, which are paramagnetic, are converted by potassium ferrocyanide into the corresponding diamagnetic bromides (V and VI, respectively). Bromide V is reconverted into bromide I by reaction with hydrogen bromide; V reacts with hydroxide, methoxide, and ethoxide ions to give (π-hydroxytetraphenylcyclobutenyl)(π-cyclopentadienyl)palladium (VII) and the analogous methoxy and ethoxy complexes, VIII and IX, respectively. Complexes VII, VIII, and IX all react with hydrogen bromide to give bromide I. The nickel bromide complex, VI, reacts similarly with methoxide ion to give (π-methoxytetraphenylcyclobutenyl)(π-cyclopentadienyl)nickel (X); reaction of X with hydrogen bromide affords VI rather than III.

Tetraphenylcyclobutadienecobalt dicarbonyl bromide (XI) reacts with cyclopentadienyliron dicarbonyl dimer or cyclopentadienyliron dicarbonyl bromide to give (π-cyclopentadienyl)(π-tetraphenylcyclobutadiene)cobalt (XII). Cobalt complex XII is obtained also by the reaction of the palladium complex V with dicobalt octacarbonyl.

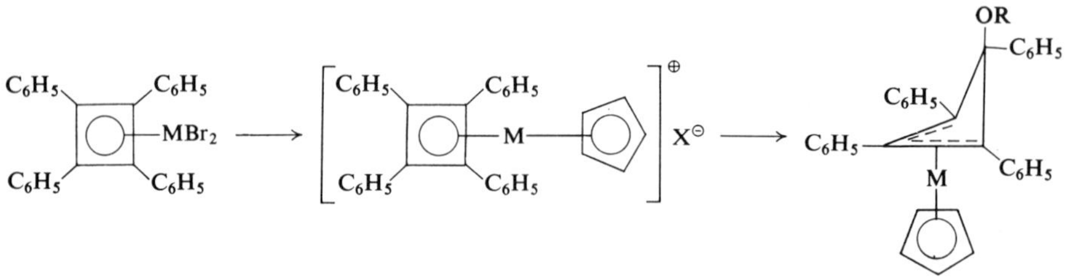

[Structures I–XII depicting tetraphenylcyclobutadiene complexes with Pd, Ni, and Co]

I: M = Pd
III: M = Ni

II: M = Pd; X = FeBr$_4$
IV: M = Ni; X = FeBr$_4$
V: M = Pd; X = Br
VI: M = Ni; X = Br

VII: M = Pd; R = H
VIII: M = Pd; R = CH$_3$
IX: M = Pd; R = C$_2$H$_5$
X: M = Ni; R = CH$_3$

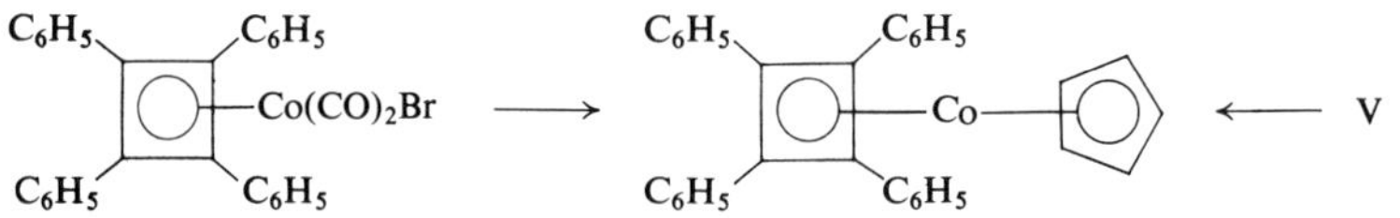

XI

XII

2.18 Maitlis, P. M., and Games, M. L., *Can. J. Chem.* **42,** 182 (1964).

The preparation and chemical behavior of tetraphenylcyclobutadienepalladium bromide and chloride and of tetrakis(*p*-chlorophenyl)cyclobutadienepalladium chloride are reported. See ref. 39, Chapter 2.

2.19 Maitlis, P. M., Pollock, D., Games, M. L., and Pryde, W. J., *Can. J. Chem.* **43,** 470 (1965).

The preparation and properties of the oligomers of tetraphenylcyclobutadienepalladium chloride and of tetrakis(*p*-chlorophenyl)cyclobutadienepalladium chloride are reported.

2.20 McOmie, J. F. W., and Bullimore, B. K., *Chem. Commun.* 63 (1965).
Abstract 1.16, this appendix.

2.21 Temkin, O. N., Brailovskii, S. M., Flid, R. M., Strukova, M. P., Belyanin, V. B., and Zaitseva, M. G., *Kinetika i Kataliz* **5,** 192 (1964).

The correct (noncyclobutadienoid) structure of Erdmann and Koethner's complex is reported. See Chapter 2, Section A, and ref. 61.

2.22 Throndsen, H. P., and Zeiss, H., *J. Organometal. Chem.* **1,** 301 (1964).
Abstract 1.22, this appendix.

2.23 Watts, L., Fitzpatrick, J. D., and Pettit, R., *J. Am. Chem. Soc.* **87**, 3253 (1965). Abstract 1.23, this appendix.

2.24 Watts, L., Fitzpatrick, J. D., and Pettit, R., *J. Am. Chem. Soc.* **88**, 623 (1966). Abstract 1.24, this appendix.

Chapter 3

Cyclobutadiene Divalent Ions

3.1 Bryan, R. F., *J. Am. Chem. Soc.* **86**, 733 (1964).

An X-ray structure analysis demonstrated that Freedman's $C_{28}H_{20}Cl_6Sn$ complex is the pentachlorostannate salt of 4-chlorotetraphenylcyclobutenyl cation. See ref. 2, Chapter 3.

3.2 Farnum, D. G., Heybey, M. A. T., and Webster, B., *J. Am. Chem. Soc.* **86**, 673 (1964).

Attempts to prepare a 3-(cyclobutenonyl) carbonium ion or the cyclobutadienoid dication resonance form thereof, were unsuccessful. See ref. 3, Chapter 3.

3.3 Freedman, H. H., and Young, A. E., *J. Am. Chem. Soc.* **86**, 734 (1964).

The preparation and properties of a methylene chloride solution of a substance believed to be tetraphenylcyclobutadiene dication fluoroborate are described. See ref. 6, Chapter 3.

Chapter 4

Cyclobutadienequinone

4.1 Blomquist, A. T., and LaLancette, E. A., *J. Org. Chem.* **29**, 2331 (1964).

A synthetic study is described in which diphenylcyclobutadienequinone (I) is examined as a possible precursor of 1,2-diphenylcyclobutadiene (II). Thus, lithium aluminum hydride reduction of I affords *cis*-1,2-diphenylcyclobutene-3,4-diol (III) in 23% yield and a small amount (3%) of the corresponding trans diol (IV). Diol III was converted by phosphorus tribromide into *cis*-3,4-dibromo-1,2-diphenylcyclobutene (V). All efforts to obtain evidence for the generation of II by reaction of dibromide V with lithium amalgam, zinc dust, or nickelcarbonyl failed; only intractable polymers were obtained.

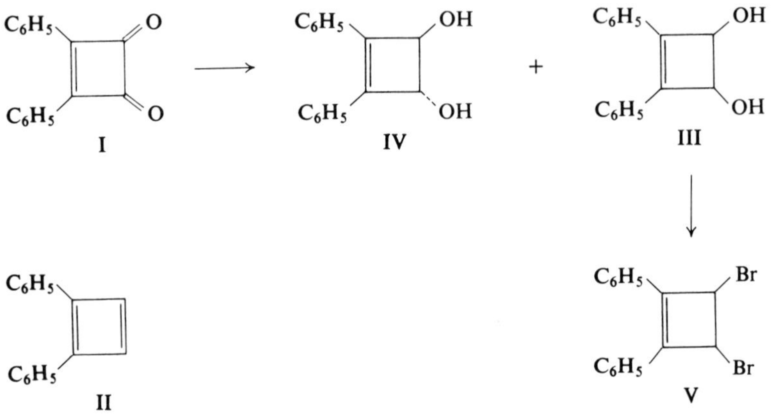

4.2 Blomquist, A. T., and Vierling, R. A., *U.S. 3,169,147*, Feb. 9, 1965.

The preparation of 3,4-dimethylcyclobutadienequinone (I) by hydrolysis of 1,2-dimethyl-3,3,4,4-tetrafluorocyclobutene is described. Physical properties are reported for the dione: b.p. 76°/2 mm, n_D^{24} 1.4893, solidifies at 25°; λ 5.60, 5.67, 5.49, 2.83, and 6.21 μ; λ_{max} 216 (18,800), 340 (26), 355 mμ (23, shoulder); δ 2.40 (one unresolved peak); mono-2,4-dinitrophenylhydrazone, m.p. 205°–208° (ethyl acetate) and ditosylhydrazone, m.p. 190°–191° (decomp.). (See also Chapter 4.)

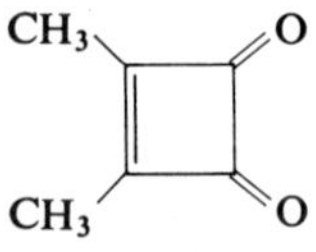

4.3 Breslow, R., Altman, L. J., Krebs, A., Mohacsi, E., Murata, I., Peterson, R. A., and Posner, J., *J. Am. Chem. Soc.* **87**, 1326 (1965).

Hydrolysis of 4,4-dichloro-2,3-dipropylcyclobutenone (I) with 90% sulfuric acid yielded 2,3-di(*n*-propyl)cyclobutadienequinone (II), a yellow liquid, b.p. 70° (0.2 mm); λ_{max}^{EtOH} 218 mμ (ε 18,000), 355sh (28); ν_{max} 1784, 1763, 1589 cm^{-1}; NMR τ 7.25 (triplet), 8.25 (sextuplet), 9.03 (triplet) (all with $J = 7$ cps), obtained in 43% yield. Similarly, hydrolysis of 4,4-dichloro-3-methoxy-2-phenylcyclobutenone gave 2-methoxy-3-phenylcyclobutadienequinone.

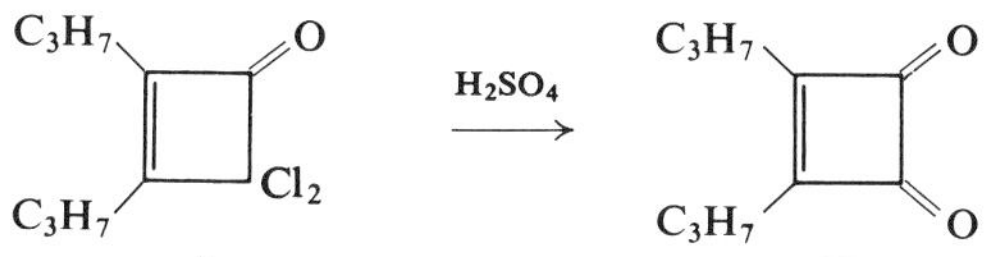

4.4 Chemische Werke Huls A.-G., *Fr. 1,404,928*, July 2, 1965.

Hydrolysis of perchlorocyclobutenone (I) with 20% sulfuric acid for 2 hours gives 3,4-dihydroxycyclobutenedione (II, squaric acid) in good yield (57.1%).

4.5 Farnum, D. G., Heybey, M. A. T., and Webster, B., *J. Am. Chem. Soc. 86*, 673 (1964).

The relationship between the 3-(cyclobutenonyl) carbonium ion, cyclobutadiene dication, and cyclobutadienequinone is discussed. See ref. 5, Chapter 3.

4.6 Karle, I. L., Britts, K., and Brenner, S., *Acta Cryst. 17*, 1506 (1964).

The crystal structure of 3-(1'-cyclohexenyl)cyclobutenedione (I) was determined by the symbolic-addition phase-determination method. One carbon-to-oxygen bond is 0.056 Å longer than the other.

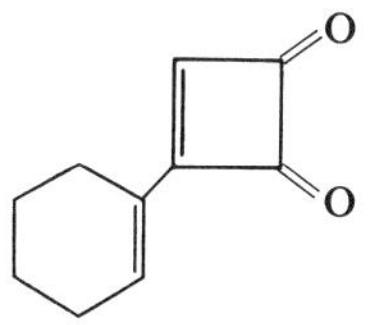

4.7 Maahs, G., *Ann.* **686**, 55 (1965).

Hydrolysis of perchlorocyclobutenone (I) with dilute sulfuric acid gives squaric acid (II), tetrachlorovinylacetic acid, tetrachlorocrotonic acid, and β-formyl-α,β-dichloroacrylic acid. Solvolysis of I with 4 moles of *n*-butanol gives 2,3-dibutoxycyclobutadienequinone (III). When it is treated with 58% oleum at 100°, I gives 2,3-dichlorocyclobutadienequinone (IV) in 43% yield; treating IV with 20% sulfuric acid gives II in 93% yield.

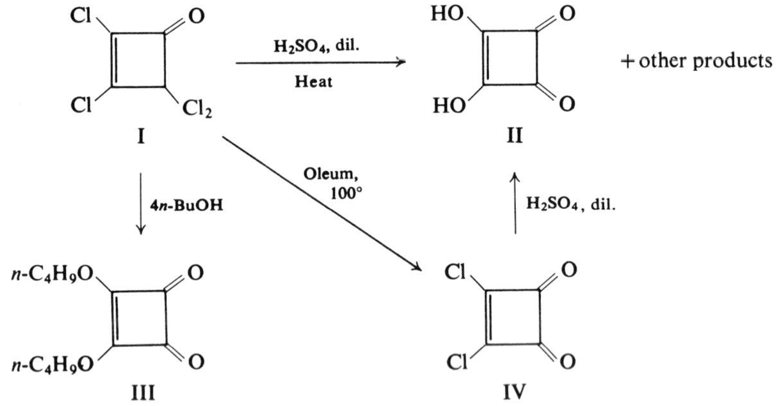

4.8 Müller, E. W., and Korte, F., *Tetrahedron Letters* 3039 (1964).

Hydrolysis of 3-chloro-1-phenyl-3,4,4-trifluorocyclobutene with hot concentrated hydrochloric acid gives phenylcyclobutadienequinone in 71% yield.

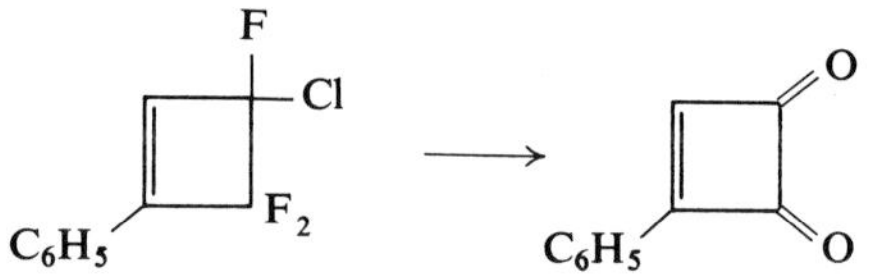

4.9 Treibs, A., and Jacob, K., *Angew. Chem.* **77**, 680 (1965).

The general structure I (or Ia) is proposed for the violet cyclotrimethine dyes formed by condensation of 3,4-dihydroxycyclobutenedione (II) with pyrrole derivatives. II reacted with phloroglucinol to give III.

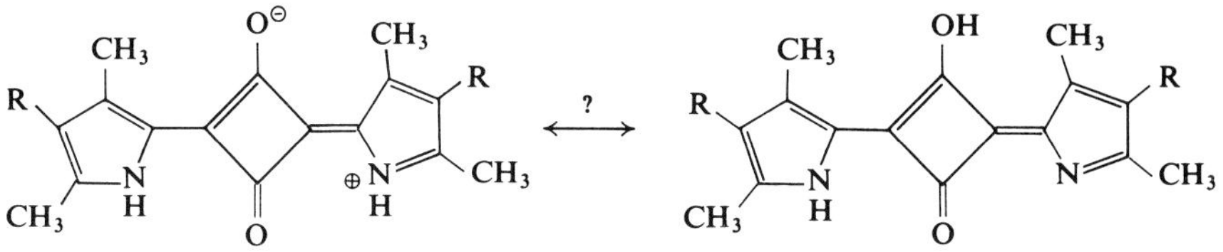

4.10 Wong, C.-H., Marsh, R. E., and Schomaker, V., *Acta Cryst.* 17, 131 (1964).

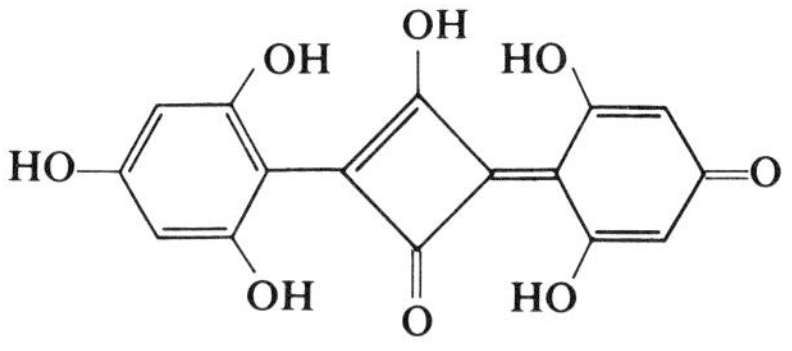

An X-ray diffraction study of the crystal structure of phenylcyclobutadienequinone (I) has shown that the molecule is very nearly planar. The bond distances indicate considerable conjugation throughout the system.

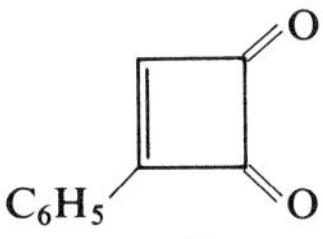

Chapter 5

Methylene Analogs of Cyclobutadienequinone

5.1 Adam, W., *Ber.* 97, 1811 (1964).

In general, thermolysis of 3,4-disubstituted tetramethylcyclobutenes leads to the corresponding butadienes, but in certain cases elimination may compete or predominate. Thus, 3,4-dibromo-1,2,3,4-tetramethylcyclobutene undergoes appreciable elimination while 3,4-diiodo- and 3,4-diacetoxy-1,2,3,4-tetramethylcyclobutene undergo elimination to the complete exclusion of isomerization and afford 1,2-dimethyl-3,4-dimethylenecyclobutene (I).

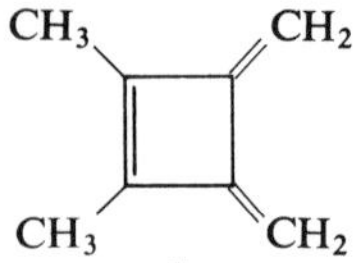

5.2 Fujino, A., Nagata, Y., and Sakan, T., *Bull. Chem. Soc. Japan* **38**, 295 (1965).

Dehalogenation of perchloropropene with aluminum in anhydrous ether gives perchloro(3,4-dimethylenecyclobutene) (I), m.p. 147°–148°, together with perchloro(1,2-dimethylenecyclobutane), the precursor of I. The dimethylenecyclobutene (I) is stable in the solid state but decomposes in ether, benzene, or carbon tetrachloride solution with evolution of *hydrogen chloride*. Prolonged heating with concentrated sulfuric acid converts I into a carboxylic acid of unknown structure; treatment of I with chlorine under various reaction conditions produces one or more of the following: perchloro(1-methylenecyclobutene), perchloro(1,2-dimethylcyclobutene), perchloro(1-methylenecyclobutane), and a dimeric product, $C_{12}Cl_{18}$; oxidation of I with permanganate in acetone, chromic oxide in acetic acid, or fuming nitric acid gave perchlorodimethylene succinic acid. Mild oxidation of I with nitric acid gave a compound, $C_6Cl_6N_2O_4$.

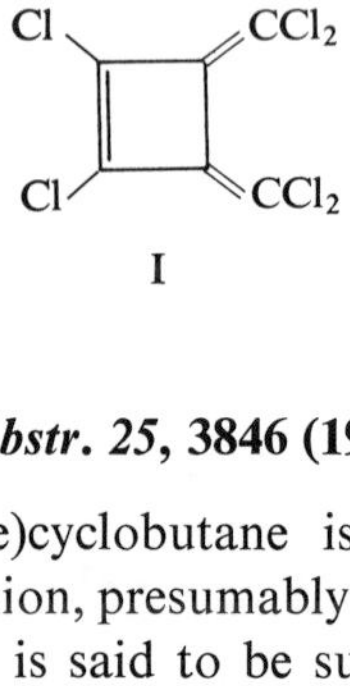

5.3 Nahon, R., *Dissertation Abstr.* **25**, 3846 (1965).

Tetrakis(diphenylmethylene)cyclobutane is cleaved by sodium to give tetraphenyl-2-butyne-1,4-dianion, presumably via a cyclobutadiene tetraanion (I). The cleavage mechanism is said to be supported by the results of MO calculations.

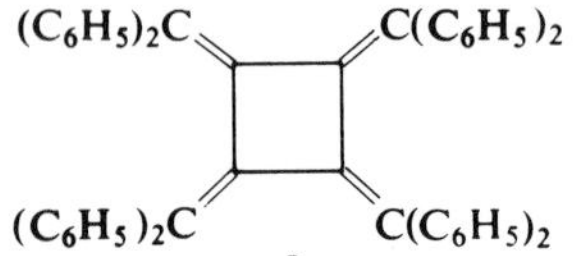

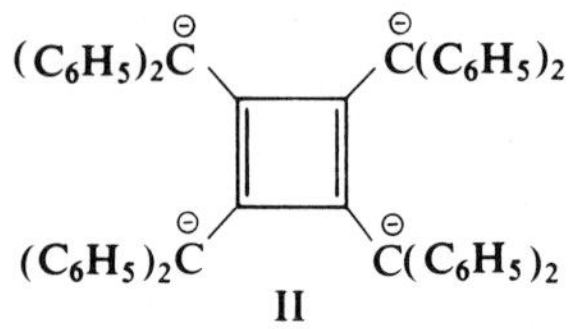

5.4 Roedig, A., Bischoff, F., Heinrich, B., and Märkl, G., *Ann.* 670, 8 (1964).

The preparation of perchloro(3,4-dimethylene)cyclobutene from pentachloropropene is described. See ref. 20a, Chapter 5.

5.5 Roedig, A., Detzer, N., and Friedrich, H. J., *Angew. Chem.* 76, 379 (1964).

Perbromodimethylenecyclobutene (m.p. 165°) was prepared by the dehalogenation of perbromo-1,2-dimethylenecyclobutane with aluminum turnings in ether. The UV spectrum of perbromodimethylenecyclobutene is similar to that of the perchloro compound.

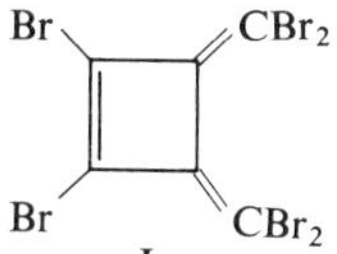

5.6 Roedig, A., and Kohlhaupt, R., *Tetrahedron Letters* 1107 (1964). Abstract 7.4, this appendix.

Chapter 6

Benzocyclobutadiene

6.1 Barton, J. W., Henn, D. E., McLaughlan, K. A., and McOmie, J. F. W., *J. Chem. Soc.* 1622 (1964). Abstract 10.3, this appendix.

6.2 Blomquist, A. T., and Bottomley, C. G., *J. Am. Chem. Soc.* 87, 86 (1965).

This paper, which is concerned with attempts to synthesize 1-methyl-2-phenylbenzocyclobutadiene (I), is a full report of the work cited in Chapter 6, ref. 13.

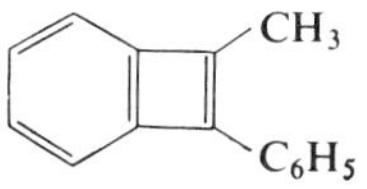

6.3 Boulton, A. J., and McOmie, J. F. W., *J. Chem. Soc.* 2549 (1965). Abstract 10.9, this appendix.

6.4 Emerson, G. F., Watts, L., and Pettit, R., *J. Am. Chem. Soc.* **87**, 131 (1965).
Cited as ref. 27 in Chapter 6. See Abstract 2.6, this appendix.

6.5 Kende, A. S., and MacGregor, P. T., *J. Am. Chem. Soc.* **86**, 2088 (1964).
Abstract 10.19, this appendix.

Chapter 7

Benzocyclobutadienequinone

7.1 Brown, R. F. C., and Solly, R. K., *Chem. and Ind.* (*London*) 1462 (1965).
Abstract 10.10, this appendix.

7.2 Cava, M. P., Mangold, D., and Muth, K., *J. Org. Chem.* **29**, 2947 (1964).

The conversion of 2,2-dibromobenzocyclobutenone (I) into benzocyclobutadienequinone (II) is reported.

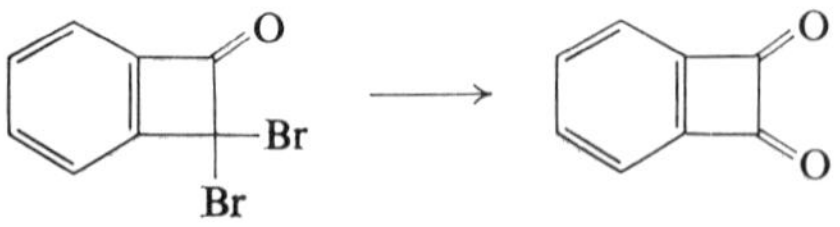

7.3 Geske, D. H., and Balch, A. L., *J. Phys. Chem.* **68**, 3423 (1964).

An interpretation of the ESR spectrum of the anion radical (I) of benzocyclobutadienequinone is presented. The anion radical was obtained by electrolysis of a $1 \times 10^{-3} M$ solution of the dione in acetonitrile.

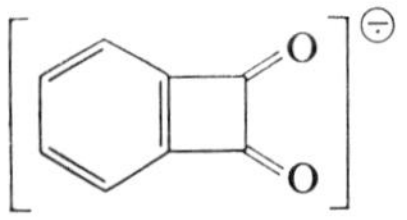

7.4 Roedig, A., and Kohlhaupt, R., *Tetrahedron Letters* 1107 (1964).

Heating the thermolabile dimer of perchlorovinylacetylene, which is believed to have structure I or II, gives perchlorobenzocyclobutene (III).

Hydrolysis of the latter in concentrated sulfuric acid gives perchlorobenzocyclobutadienequinone (IV, m.p. 222°), the structure of which was demonstrated by oxidation to tetrachlorophthalic anhydride (V).

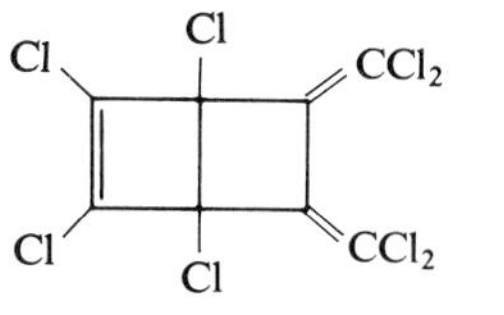

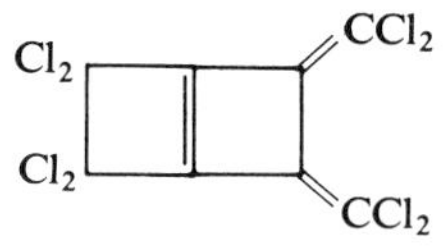

7.5 Stein, R. P., *Dissertation Abstr.* 24, 4411 (1964).

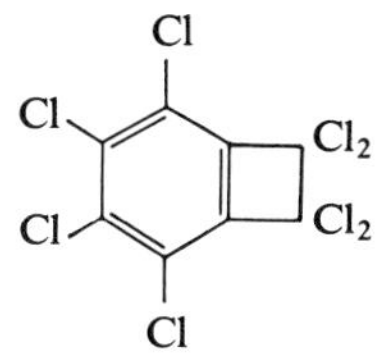

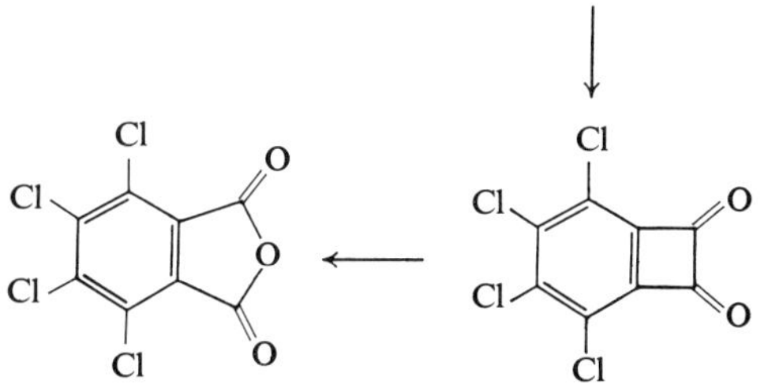

Benzocyclobutadienequinone was subjected to a number of reactions involving addition of reagents to the carbonyl function. (See ref. 6, Chapter 8.)

Chapter 8

Methylene Analogs of 1,2-Benzocyclobutadienequinone

8.1 Blomquist, A. T., and Hruby, V. J., *J. Am. Chem. Soc.* 86, 5041 (1964).

The reaction of *trans*-1,2-dibromobenzocyclobutene with triphenylphosphine, followed by treatment of the resulting bisphosphonium salt with strong base, affords 1,2-bis(triphenylphosphoranyl)benzocyclobutene (I). The deep red bis ylide (I) is fairly stable below $-40°$; it reacts with benzaldehyde to give *trans,trans*-1,2-dibenzylidenebenzocyclobutene (II), m.p. 114°–115°, as well as by-products probably derived from the mono ylide corresponding to I. The 1,2-bis ylide is the first such compound to be prepared. It is probably stabilized through resonance form Ia which is a 10π-electron dianion.

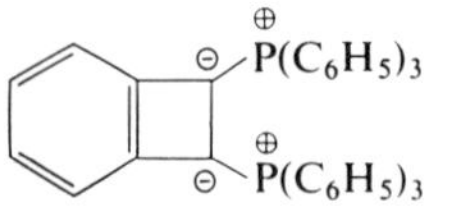

Chapter 9

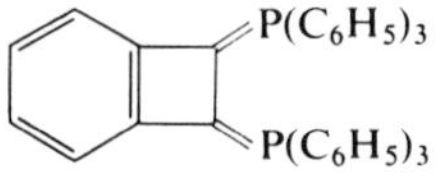

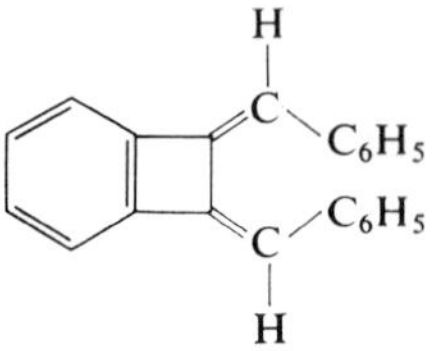

Higher Aromatic Analogs of Benzocyclobutadiene

9.1 Avram, M., Dinulescu, I. G., Elian, M., Farcasiu, M., Marica, E., Mateescu, G., and Nenitzescu, C. D., *Ber.* **97**, 372 (1964).

The generation and dimerization of 3,8-diphenylnaphtho[*b*]cyclobutadiene are described. See ref. 2, Chapter 9.

9.2 Cava, M. P., and Hwang, B. Y., *Tetrahedron Letters* 2297 (1965).

The generation and properties of 3,8-diphenylnaphtho[*b*]cyclobutadiene and 1-bromo-3,8-diphenylnaphtho[*b*]cyclobutadiene are described. 1,2-Dibromo-3,8-diphenylnaphtho[*b*]cyclobutadiene (I) and 3,8-diphenylnaphtho-[*b*]cyclobutadienequinone (II) are both stable compounds. The quinone (II), double m.p. 238° and 248°, was prepared from the corresponding tetrabromide and was reduced with zinc–ammonium chloride to 2-hydroxy-3,8-diphenylnaphtho[*b*]cyclobuten-1-one. See ref. 5, Chapter 9.

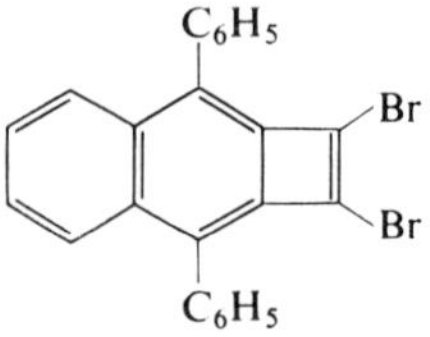

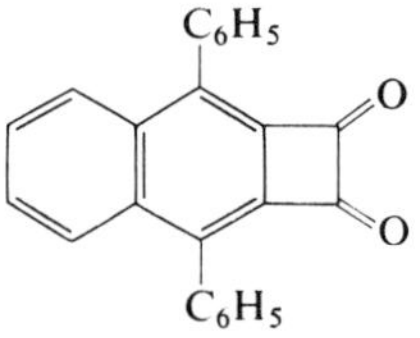

9.3 Cava, M. P., and Mangold, D., *Tetrahedron Letters* 1751 (1964).

The generation and properties of 1,2-diphenylphenanthro[*l*]cyclobutadiene are described. See ref. 8, Chapter 9.

9.4 Kende, A. S., and MacGregor, P. T., *J. Am. Chem. Soc.* **86,** 2088 (1964).
Abstract 10.19, this appendix.

9.5 Wuchter, R. B., *Dissertation Abstr.* **24,** 4413 (1964).

Attempts were made to prepare 1,2-disubstituted naphtho[b]cyclobutenes which are potential precursors of 1,2-disubstituted naphtho[b]cyclobutadienes (I). Dehalogenation of 1,2-dibromo-1,2-diphenylnaphtho[b]cyclobutene gave inconclusive results (but see Chapter 9). An attempted synthesis of 1,2-dimethylnaphtho[b]cyclobutadiene foundered at an early stage as did an attempt to prepare 1,2-dicarbomethoxynaphtho[b]cyclobutadiene.

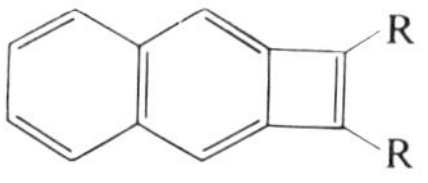

Chapter 10

Biphenylene

10.1 Baker, W., Boulton, A. J., Harrison, C. R., and McOmie, J. F. W., *Proc. Chem. Soc.* **414** (1964).

Treatment of biphenylene with butyllithium at room temperature in ether for 4 days gives 1-lithiobiphenylene (I) and a smaller amount of 2-lithiobiphenylene (II). Carbonation of the mixture of lithio compounds and esterification of the resulting acids with diazomethane gives 1-carbomethoxybiphenylene (III, m.p. 81°, 49%), 2-carbomethoxybiphenylene (IV, m.p. 114°–115°, 2.5%), 1-biphenylenyl-1-pentanone (V, m.p. 63°, 9.5%) and a

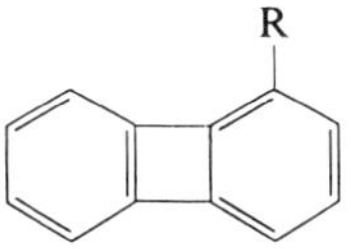

I: R = Li
III: R = CO_2CH_3
V: R = COC_4H_9-n
VII: R = OH
VIII: R = CHO
IX: R = C_6H_5

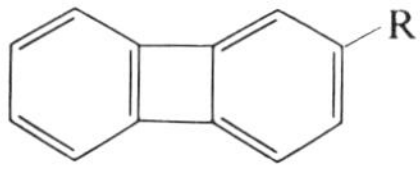

II: R = Li
IV: R = CO_2CH_3
VI: R = COC_4H_9-n
X: R = C_6H_5

trace of 2-biphenylenyl-1-pentanone (VI). 1-Lithiobiphenylene reacts with oxygen to give, after the usual work-up, 1-hydroxybiphenylene (VII, m.p. 132°, 18%) and with dimethylformamide to give biphenylene-1-carboxaldehyde (VIII). Treatment of the mixture of lithium compounds with cyclohexanone, followed by dehydration and dehydrogenation, gave 1-phenylbiphenylene (IX, m.p. 46.5°) and the 2-isomer (X, m.p. 125.5°).

10.2 Baker, W., McLean, N. J., and McOmie, J. F. W., *J. Chem. Soc.* 1067 (1964). (Cited as ref. 17 in Chapter 10.)

1,2,3,6,7,8-Hexamethoxybiphenylene (I) undergoes oxidative demethylation when treated with nitric acid, giving 1,6,7,8-tetramethoxybiphenylene-2,3-quinone (II), m.p. 214°–215°. Quinone II reacts with *o*-phenylenediamine to give a quinoxaline derivative (III), m.p. 186°–187°, and with hydroxylamine to give a monooxime, m.p. 220°–221°, provisionally assigned structure IV. Sodium dithionite reduction of II gives the oily 2,3-dihydroxy-1,6,7,8-tetramethoxybiphenylene (V), which on methylation affords I and which on acetylation affords 2,3-diacetoxy-1,6,7,8-tetramethoxybiphenylene (VI). Bromination of quinone II gives 4-bromo-1,6,7,8-tetramethoxybiphenylene-

I: R = CH_3; X = H
VIII: R = CH_3; X = Br
XII: R = H; X = H
XIII: R = CH_3CO; X = H

II: X = H; Z = O
IV: X = H; Z = NOH
VII: X = Br; Z = O

III

V: R = H
VI: R = CH_3CO

IX: X = Br; R = CH_3; Z = OCH_3
X: X = OCH_3; R = CH_3; Z = H
XI: X = H; R = CH_3; Z = OCH_3
XIV: X = I; R = CH_3CO; Z = OCH_3

XV

XVI

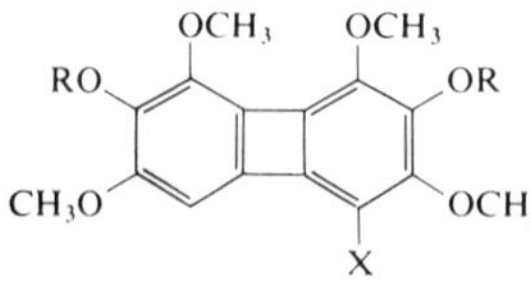

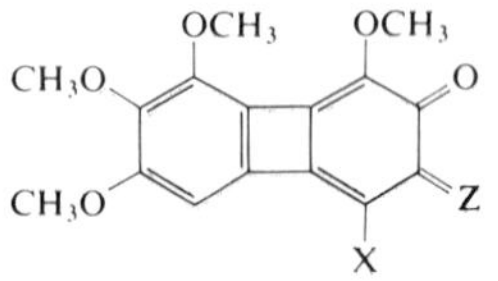

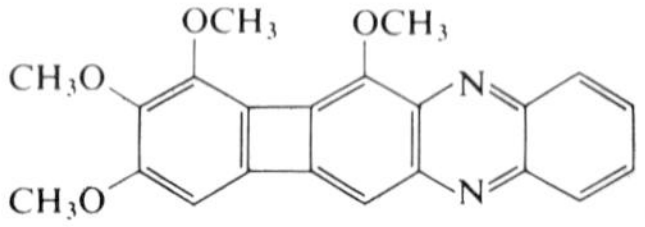

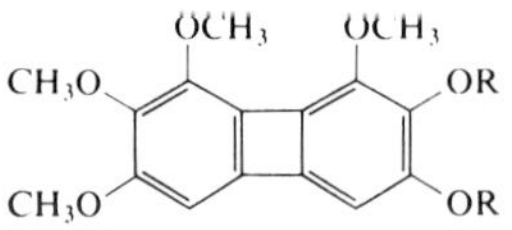

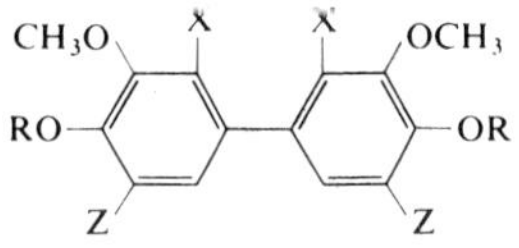

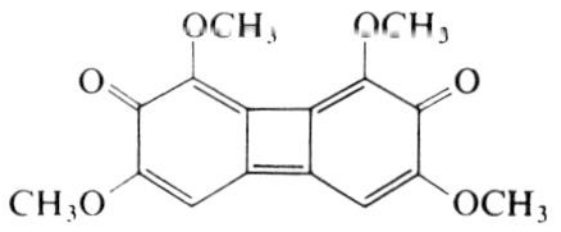

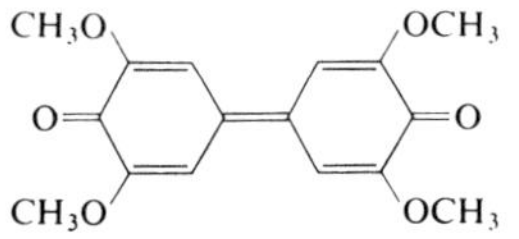

ULTRAVIOLET ABSORPTION DATA FOR NEW BIPHENYLENE DERIVATIVES

Compound	Absorption maxima [mμ (log ϵ)]
II	264, 306, 423 (4.23, 4.36, 4.02)
III	238, 294, 310, 357, 365, 370, 440 (sh), 453 (4.31, 4.76, 4.72, 5.00, 5.01, 5.02, 5.31, 5.34)
VI	269, 346, 365 (4.70, 4.59, 4.57)
VII	289, 318, 436 (4.26, 4.31, 4.04)
VIII	286, 350, 368, 382 (4.64, 4.23, 4.26, 4.31)
IV	235, 301, 419 (4.09, 4.42, 4.07)
XIII	271, 345, 364 (4.59, 4.36, 4.71)
XV	258, 298, 401 (4.17, 4.32, 4.15)

2,3-quinone (VII), m.p. 205°–206°; sodium dithionite reduction of VII followed by methylation *in situ*, gives 4-bromo-1,2,3,6,7,8-hexamethoxybiphenylene (VIII), m.p. 128°–129°. Bromination of 1,2,3,6,7,8-hexamethoxybiphenylene (I) yields a complex mixture containing II, VII, VIII, and 2,2'-dibromo-3,4,5,3',4',5'-hexamethoxybiphenyl (IX). Reductive cleavage of I with Raney nickel in ethanol gives a mixture of almost equal amounts of 2,3,4,2',3',4'-hexamethoxybiphenyl (X) and 3,4,5,3',4',5'-hexamethoxybiphenyl (XI). Demethylation of I by methylmagnesium iodide gives 2,7-dihydroxy-1,3,6,8-tetramethoxybiphenylene (XII), characterized as 2,7-diacetoxy-1,3,6,8-tetramethoxybiphenylene (XIII), m.p. 142°–143°. An attempted Lothrop synthesis of XIII from 4,4'-diacetoxy-2,2'-diiodo-3,5,3',5'-tetramethoxybiphenyl (XIV) was unsuccessful. Air oxidation of diol XII affords the 1,3,6,8-tetramethoxybiphenylene-2,7-quinone (XV), dark mauve crystals, m.p. 265°–266°. Compound XV is the first biphenylene-2,7-quinone to be reported. The structure of XV is confirmed by reductive cleavage with Raney nickel, followed by chromic acid oxidation, to give 3,5,3',5'-tetramethoxybiphenoquinone (coerulignone, XVI).

10.3 Barton, J. W., Henn, D. E., McLaughlan, K. A., and McOmie, J. F. W., *J. Chem. Soc.* 1622 (1964).

The bromination of biphenylene (I) with iodine monobromide gives 2-bromobiphenylene (II) in good yield. The bromination of biphenylene with excess bromine in the absence of a catalyst (preferably in sunlight) gives only a small amount of II, the major products being formed by addition of bromine to I. These addition products are 3,5,6,8-tetrabromo(*trans*-5,6-dihydrobenzocyclooctatetraene) (III), m.p. 139°–140°, and two hexabromides, $C_{12}H_8Br_6$

(IV, m.p. 153°–154.5°, and V, m.p. 183°–185°) of undetermined structure. Bromides III, IV, and V all undergo sodium iodide debromination to give 3,8-dibromobenzocyclooctatetraene (VI), m.p. 93°–94°. Dibromide VI is further debrominated slowly by sodium iodide or zinc to give biphenylene (I). Debromination almost certainly proceeds *via* the valence tautomeric form, VIa. The structures of III and VI are supported by NMR and by X-ray crystallographic studies.

It seems likely that tetrabromide III is identical with the previously reported biphenylene tetrabromide, m.p. 136°–137°, for which structures VII and VIII have been discussed in Chapter 10. (A comparison of the two tetrabromides is not reported in this paper.)

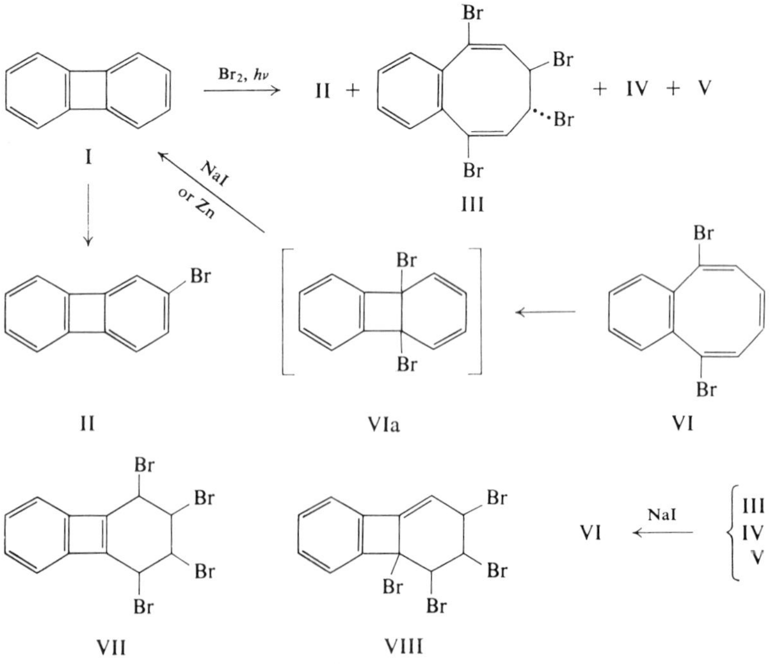

10.4 Barton, J. W., and Whitaker, K. E., *Chem. Commun.* **516 (1965).**

The reaction of biphenylene (I) with either acetyl nitrate or nitrosyl chloride affords 2-nitrobiphenylene (II) as well as products resulting from initial addition of these reagents at the 4a- and 8b-positions of I. The abnormal products formed from acetyl nitrate and nitrosyl chloride are 3-acetoxy-8-nitrobenzocyclooctatetraene (III), m.p. 115°–116°, and 3-chloro-8-nitrobenzocyclooctatetraene (IV), m.p. 111°–113°, respectively. No nitroso compounds were isolated from the reaction of I with nitrosyl chloride.

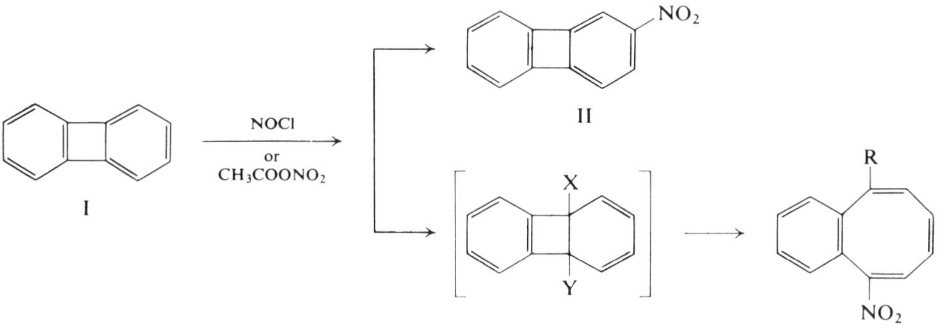

III: R = CH$_3$CO$_2$
IV: R = Cl

10.5 Bauld, N. L., and Banks, D., *J. Am. Chem. Soc.* 87, 128 (1965).

Biphenylene (I) reacts with sodium or potassium in tetrahydrofuran to form two new species, shown to be the biphenylene anion radical (II) and the biphenylene dianion (III). Spectroscopic evidence is reported in support of extensive disproportionation of the anion radical to biphenylene and the dianion (2 II $\rightleftarrows$ I + III). (See, however, Abstract 10.26, this appendix.) The apparent special stability of the dianion III is explained on the basis of the $4n+2$ π-electrons contained in the system.

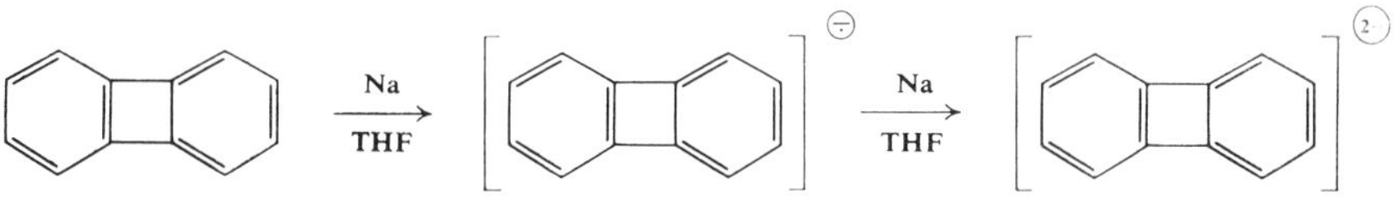

10.6 Birks, J. B., Conte, J. M. de C., and Walker, G., *Phys. Letters* 19, 125 (1965).

Exposure of a 0.1M solution of biphenylene in cyclohexane to selected wavelengths from a high-pressure mercury discharge lamp produces dual fluorescence spectra by exciting both the S_1–S_0 and the S_2–S_0 fluorescences. The quantum yield ratio of the former to the latter is only ~ 0.01 at 293°K, which can be increased by reducing the temperature or by removing dissolved oxygen from the solution.

10.7 Blatchly, J. M., Boulton, A. J., and McOmie, J. F. W., *J. Chem. Soc.* 4930 (1965).

Contrary to previous reports (refs. 11 and 15, Chapter 10), biphenylene undergoes Friedel–Crafts acylation to give mixtures of the 2,6- and 2,7-diacyl derivatives in which the 2,6-diacyl derivatives predominate. The isomer pairs are not interconvertible in the presence of aluminum chloride and hence they are formed directly by competitive reactions. Thus, biphenylene reacts with acetyl chloride and with propionyl chloride in the presence of aluminum chloride to give 2,6- and 2,7-diacetylbiphenylene (I and II, 13% and 4%, respectively) and 2,6- and 2,7-dipropionylbiphenylene (III and IV, 75% and 4%, respectively). Use of a less active sample of aluminum chloride afforded 2-propionylbiphenylene (V, 52%, m.p. 95°), which was converted into the oxime, (VI) m.p. 160°–161°. 2,6-Dipropionylbiphenylene, m.p. 241°–242°

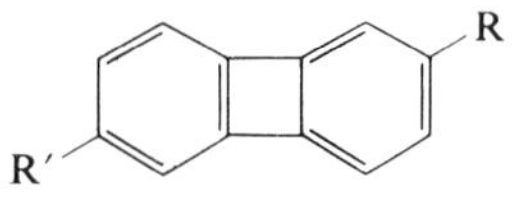

I: R = R′ = COCH$_3$
III: R = R′ = COC$_2$H$_5$
VII: R = R′ = C(NOH)C$_2$H$_5$
VIII: R = R′ = NHCOC$_2$H$_5$
X: R = COCH$_3$, R′ = COC$_2$H$_5$

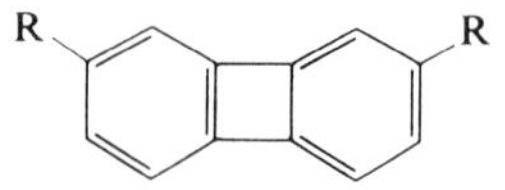

II: R = COCH$_3$
IV: R = COC$_2$H$_5$

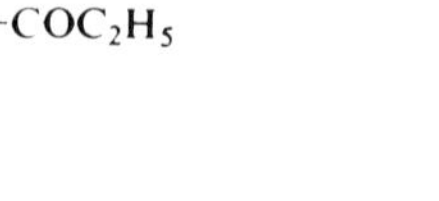

V: R = COC$_2$H$_5$
VI: R = C(NOH)C$_2$H$_5$

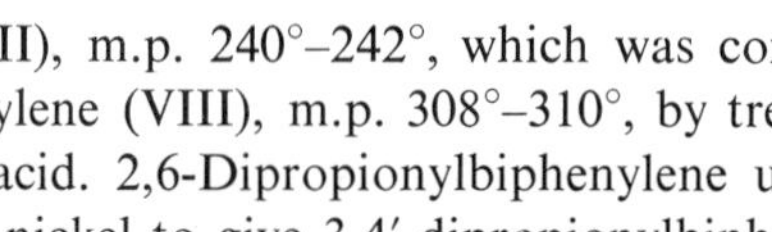

IX

gave the dioxime (VII), m.p. 240°–242°, which was converted into 2,6-dipropionamidobiphenylene (VIII), m.p. 308°–310°, by treatment with boron trifluoride in acetic acid. 2,6-Dipropionylbiphenylene underwent reductive cleavage with Raney nickel to give 3,4′-dipropionylbiphenyl (IX). 2-Acetylbiphenylene reacted with propionyl chloride to give 2-acetyl-6-propionylbiphenylene (X) in 75% yield. The same product was formed in comparable yield (61%) by the reaction of 2-propionylbiphenylene with acetyl chloride. Infrared, ultraviolet, and NMR data are recorded for the biphenylenes.

10.8 Blatchly, J. M., and Taylor, R., *J. Chem. Soc.* 4641, (1964).

Detritiation rate factors, relative to C_6H_6, of 1- and 2-tritiobiphenylene are 14,100 and 104, respectively, in trifluoroacetic acid at 70.1°. The preparation of 2-tritiobiphenylene from 2-lithiobiphenylene (cross-metallation of 2-bromobiphenylene) is described.

10.9 Boulton, A. J., and McOmie, J. F. W., *J. Chem. Soc.* 2549 (1965).

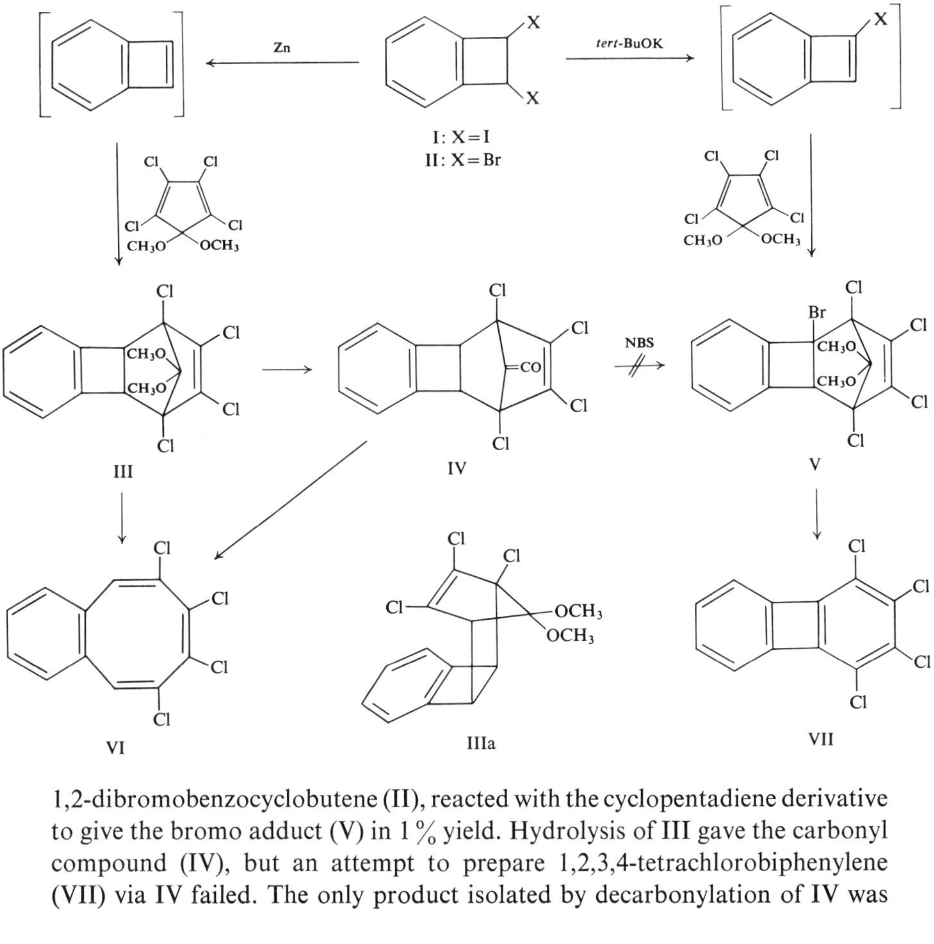

Benzocyclobutadiene, generated by treatment of 1,2-diiodobenzocyclobutene (I) with zinc dust in ethanol, reacted with 1,1-dimethoxy-2,3,4,5-tetrachlorocyclopentadiene to give adduct III in 60% yield. Similarly, 1-bromobenzocyclobutadiene, generated by the action of potassium *tert*-butoxide on 1,2-dibromobenzocyclobutene (II), reacted with the cyclopentadiene derivative to give the bromo adduct (V) in 1% yield. Hydrolysis of III gave the carbonyl compound (IV), but an attempt to prepare 1,2,3,4-tetrachlorobiphenylene (VII) via IV failed. The only product isolated by decarbonylation of IV was tetrachlorobenzocyclooctatetraene (VI), which was also obtained when ketal III was heated with a solution of hydrogen bromide in glacial acetic acid. An attempt to prepare tetrachlorobiphenylene by an indirect route also failed (I → III → IV → V → VII) as did similar attempts to prepare biphenylene and trichlorobiphenylene via the corresponding ketals.

1,2,3,4-Tetrachlorobiphenylene was finally obtained by heating the bromo adduct (V) with the hydrogen bromide in acetic acid (53 % yield). Tetrachlorobiphenylene forms pale yellowish green needles, m.p. 176°: λ_{max} 258sh, 265, 339, 353sh, 357, 373 mμ.

The configuration of the benzocyclobutadiene adducts (III and V) is endo, as shown in IIIa.

10.10 Brown, R. F. C., and Solly, R. K., *Chem. and Ind. (London)* 1462 (1965).

Passage of indanetrione vapor under reduced pressure (0.7 mm) through a silica tube packed with Pyrex helices at 600° gave biphenylene (I) in appreciable yield (23 %), together with smaller amounts of C_{12}–C_{18} aromatic hydrocarbons. At 500°/0.3 mm, 17 % I, 11 % benzocyclobutenedione (II) and 1.2 % fluorenone were formed.

10.11 Campbell, C. D., and Rees, C. W., *Chem. Commun.* 192 (1965).

Oxidation of 1-aminobenzotriazole with lead tetracetate in benzene produces benzyne, which in the absence of a trapping agent affords biphenylene in 83 % yield, together with a trace amount of triphenylene. Similarly, 1-amino-5-methylbenzotriazole yields 67 % 2,6-dimethylbiphenylene. Since the tendency to undergo dimerization is not as marked in benzyne that has been generated by other methods, it is suggested that benzyne generated from 1-aminobenzotriazole exists in the triplet state.

10.12 Dickerman, S. C., Milstein, N., and McOmie, J. F. W., *J. Am. Chem. Soc.* **87**, 5521 (1965).

Partial rate factors have been determined for the homolytic phenylation of biphenylene (I). The formation of 2-phenylbiphenylene (II) takes place about three times faster than the formation of 1-phenylbiphenylene (III); molecular orbital theory predicts a ratio of 4 for k_2/k_1 in the aromatic substitution of biphenylene. Since conflicting values of 64 and 135 were obtained for k_2/k_1 from the tritiodeprotonation reaction, it is suggested that homolytic aromatic substitution (rather than electrophilic aromatic substitution) is the preferred reaction type for evaluating molecular orbital predictions.

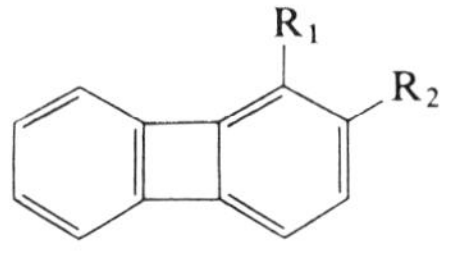

I: $R_1 = R_2 = H$
II: $R_1 = H; R_2 = C_6H_5$
III: $R_1 = C_6H_5; R_2 = H$

10.13 Fawcett, J. K., and Trotter, J., *Acta Cryst.* 20, 87 (1966).

Refinement of the data and calculations obtained in an X-ray crystallographic study of biphenylene (ref. 92 Chapter 10) gave the following values for the bond lengths: 1,2-bond, 1.423 ± 0.003; 2,3-bond, 1.385 ± 0.004; 4,4a-bond, 1.372 ± 0.002; 4a,8b-bond, 1.426 ± 0.003; 4a,4b-bond, 1.514 ± 0.003. The molecule is planar and the angles in the four-membered ring are all 90°, but those in the six-membered rings show significant deviation from 120°.

10.14 Fraenkel, G., Asahi, Y., Mitchell, M. J., and Cava, M. P., *Tetrahedron* 20, 1179 (1964).

NMR spectral data are recorded for biphenylene which indicate that the biphenylene system has two (and not three) ring currents. [See, however, Abstract 10.18, this appendix.]

10.15 Friedman, L., and Seitz, A. H., *Organic Syntheses*, in press.

Biphenylene is obtained in 25–30% yield by rapid decomposition of benzenediazonium-2-carboxylate in ethylene chloride, followed by co-distillation of biphenylene with ethylene glycol.

10.16 Heany, H., and Lees, P., *Tetrahedron Letters* 3049 (1964).

The addition of biphenylene to a reaction mixture which generates benzyne and gives rise to biphenylene and triphenylene does not increase the yield of triphenylene. This result indicates that triphenylene is not formed in such reactions by the addition of benzyne to biphenylene and, by analogy, that the formation of dodecahydrotriphenylene in reactions that generate cyclohexyne does not proceed by addition of cyclohexyne to octahydrobiphenylene.

10.17 Hilpern, J. W., *Trans. Faraday Soc.* **61**, 605 (1965).

The phosphorescence and fluorescence spectra (emission and triplet–triplet absorption) of biphenylene were measured. Phosphorescence and fluorescence stem from the second electronic bands and are probably due to a radiationless selection rule ($u \leftrightarrow u$, $u \not\leftrightarrow g$). Calculation of the electronic bands of biphenylene was carried out, using the approximations of Pariser and Parr and Pople, and the electronic levels of biphenylene were assigned.

10.18 Katritzky, A. R., and Reavill, R. E., *Rec. Trav. Chim.* **38**, 1230 (1964).

Evidence for partial bond fixation in the 1,2- and the 2,3-bonds of biphenylene was deduced from the observed NMR spectrum which has coupling constants $J = 7.1$ and 8.1 cps. The chemical shifts were assigned on the basis of deuteration experiments with d_1-trifluoroacetic acid. The chemical shifts calculated for biphenylene with two ring currents (rather than three) do not agree with the observed values, presumably because of π-electron delocalization in the four-membered ring. [See, however, ref. 64, Chapter 10 (Abstract 10.14, this appendix).]

10.19 Kende, A. S., and MacGregor, P. T., *J. Am. Chem. Soc.* **86**, 2088 (1964).

The reaction of biphenylene (I) with ethyl diazoacetate at 165° affords ethyl fluorene-2-carboxylate (III) in 15% yield. The formation of III is rationalized by assuming an initial attack of carbethoxycarbene at the 2,3-position of I to give II (not isolated), and the subsequent rearrangement of II, as illustrated below:

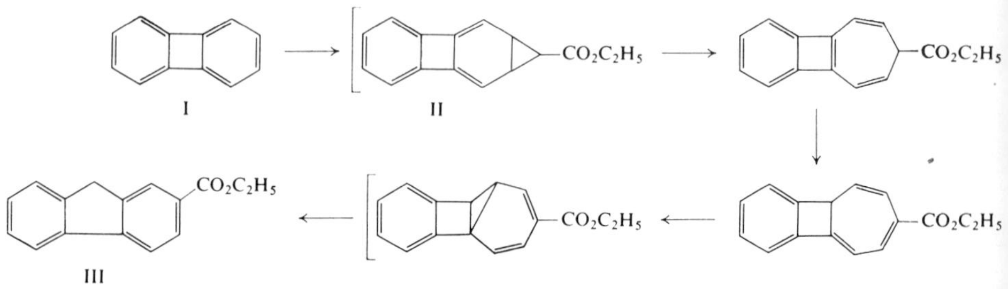

10.20 Kharasch, N., and Alston, T. G., *Chem. and Ind. (London)* **1463 (1965).**

Contrary to a previous report (ref. 84, Chapter 10) biphenylene is not formed by photolysis of solutions of 2-iodobiphenyl or 1,2-diiodobenzene in benzene. This work is cited as ref. 83 in Chapter 10.

10.21 Martin, R. H., Van Trappen, J. P., Defay, N., and McOmie, J. F. W., *Tetrahedron* **20, 2373 (1964).**

The NMR spectra of a number of substituted biphenylenes, benzo[a]- and benzo[b]biphenylenes were recorded at 60 Mc/sec with a view to using NMR evidence in proving the structure of such compounds. NMR spectral data were interpreted for biphenylene, 1-carbomethoxy-, 2-carbomethoxybiphenylene, and 2,3,6-trimethoxybiphenylene; benzo[a]biphenylene, 5-bromo-, 5-iodo-, 5-cyano-, and 5-carbomethoxybenzo[a]biphenylene; 5,10-dibromo-, and 5-bromo-10-carbomethoxybenzo[b]biphenylene; and 1-keto-1,2,3,4-tetrahydrobenzo[b]biphenylene.

10.22 Santos Veiga, J. dos, *Rev. Port. Quim.* **6, 1 (1964).**

The EPR spectra of the blue-violet solution of 2,6-dimethylbiphenylene (I) radical anion (II) and the pink solution of I radical cation (III) in 1,2-dimethoxyethane were recorded. The theoretical values obtained from the hyperconjugative model of Coulson and Crawford were compared with the observed hyperfine splitting constants relative to the aromatic and methyl protons. The anion (II) was obtained by treating I with potassium. The cation (III) was prepared by dissolving I in concentrated sulfuric acid.

$II = I^{\ominus}$
$III = I^{\oplus}$

10.23 Schafer, M. E., and Berry, R. S., *J. Am. Chem. Soc.* **87, 4497 (1965).**

A quantitative study of the photoinitiated decomposition of benzenediazonium-2-carboxylate (I) to benzyne (II), using both ultraviolet spectroscopy and mass spectrometry, affords the necessary data for the calculation of the rate constants of the dimerization of benzyne to biphenylene (III) in the gas phase. The second-order constant has a value greater than or equal to 7×10^8 liter/mole sec and corresponds to a minimum reaction probability per collision of 7×10^{-3}.

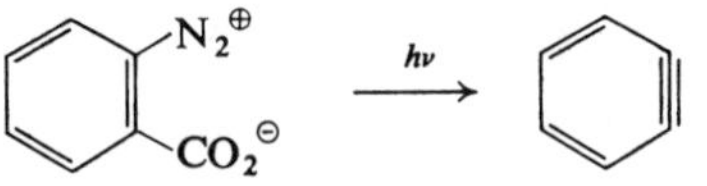

[Scheme: compound I (benzene with N₂⁺ and CO₂⁻) → hv → II (benzene/benzyne); 2 II → III (biphenylene)]

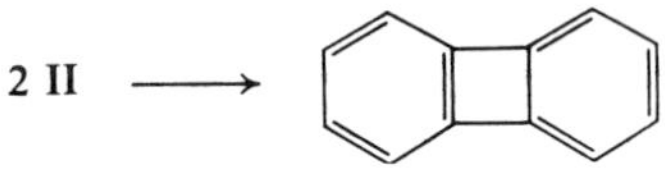

10.24 Vargas Nunez, G., *Bol. Soc. Quim. Peru* 31, 6 (1965).

Thermolysis of 5-methyl-1,2,3-benzothiadiazole 1,1-dioxide (I) gives a product mixture, m.p. 93°–128°, which was separated by zone melting to give 2,6-dimethylbiphenylene (II), m.p. 141°–142°, and a residue believed to contain 2,7-dimethylbiphenylene on the basis of a spectroscopic study of the product mixture. The same product mixture was obtained in 0.33% yield by the dehalogenation of 3-bromo-4-iodotoluene with magnesium in ether.

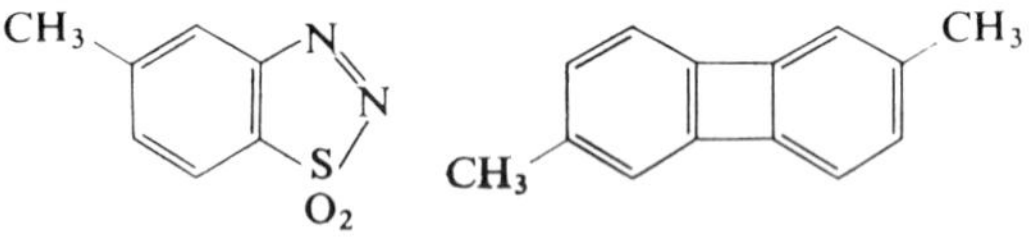

[Structures: I = 5-methyl-1,2,3-benzothiadiazole 1,1-dioxide; II = 2,6-dimethylbiphenylene]

10.25 Waack, R., Doran, M. A., and West, P., *J. Am. Chem. Soc.* 87, 5508 (1965).

A detailed spectroscopic study is reported of the reaction of biphenylene (I) with sodium in tetrahydrofuran to give the corresponding anion radical (II) and dianion (III). The biphenylene anion radical (II) is shown conclusively (by comparison with the anthracene anion radical) to have no special tendency to disproportionate to biphenylene and the biphenylene dianion (III). This work refutes the earlier result of Bauld and Banks (Abstract 10.5, this appendix) and indicates that dianion III does not possess an unusual degree of aromatic stabilization.

The ultraviolet and visible spectra of II and III are reported in detail.

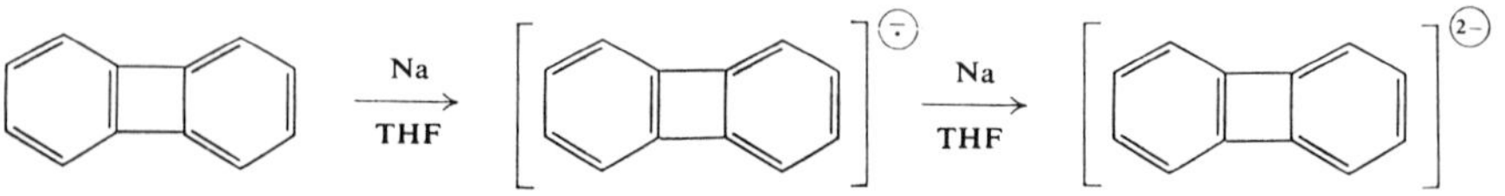

[Scheme: I (biphenylene) → Na/THF → II (anion radical, charge ⊖) → Na/THF → III (dianion, 2⊖)]

10.26 Wolovsky, R., and Sondheimer, F., *J. Am. Chem. Soc.* **87**, 5720 (1965). [Preliminary communication, *ibid.* **84**, 2844 (1962). Cited in Chapter 10 (ref. 137).]

Oxidative coupling of 1,5-hexadiyne (I) under carefully defined conditions afforded the extremely unstable cyclic dimer II in an estimated yield of 9.5%. Rearrangement of II with potassium *tert*-butoxide gave biphenylene (III) as the major reaction product. The overall yield of III from I was about 7.4%.

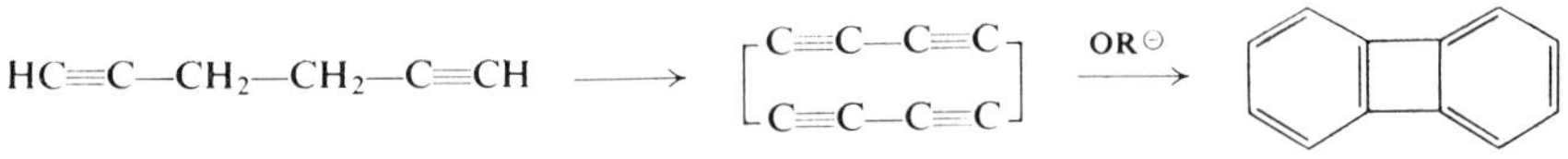

Chapter 11

The Benzobiphenylenes

11.1 Avram, M., Dinulescu, I., Elian, M., Farcasiu, M., Marica, E., Mateescu, G., and Nenitzescu, C. D., *Ber.* **97**, 372 (1964).

5,6,11,12-Tetraphenyldibenzo[*b,h*]biphenylene (I) was obtained by dehydration of 5,6,11,12-tetraphenyl-5,12;6,11-dioxa-5,5a,6,11,11a,11b,12-octahydrodibenzo[*b,h*]biphenylene (II) and by dehydration of 5,6,11,12-tetraphenyl-5,12-oxa-5,5a,11b,12-tetrahydrodibenzo[*b,h*]biphenylene. The dehydration of II gave I, m.p. >350°, in 80% yield. I was identical with a sample prepared previously by another route (see also ref. 41, Chapter 11).

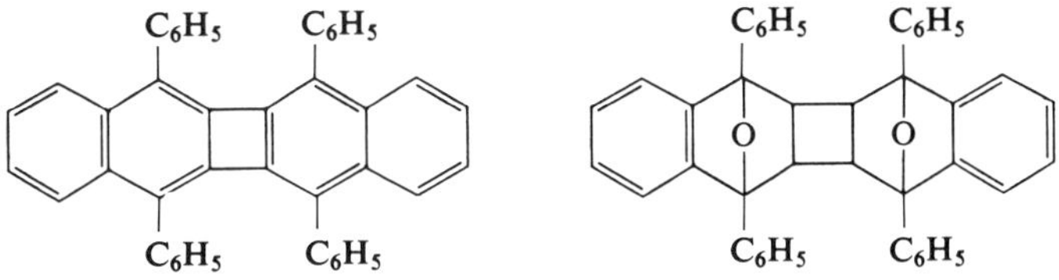

11.2 Barton, J. W., *J. Chem. Soc.* 5161 (1964).

Dibenzo[*a,g*]biphenylene (I) has been synthesized in low yield (9%) by pyrolysis of 1′,2-diiodo-1,2′-binaphthyl (II) with cuprous oxide. Hydrocarbon I forms orange-red plates, m.p. 146°–147.5° (decomp.); λ_{max}^{EtOH} 213, 271.5, 278, 297, 307.5, 324, 381, 394, 420 mμ (log ϵ 4.29, 4.50, 4.65, 4.02, 4.14, 4.12, 3.19, 3.52, 3.59) (corrected data kindly supplied by Dr. J. W. Barton); the 2,4,7-

trinitrofluorenone derivative of I forms dark green needles, m.p. 224°–225°. Raney nickel reductively cleaves I to give 1,2′-binaphthyl (III).

As anticipated on theoretical grounds, I is similar to dibenzo[*a,i*]biphenylene in behaving as a highly unsaturated and unstable molecule. Thus, I decomposes at its melting point and it is destroyed in solution by heating or by exposure to ultraviolet light, giving a yellow amorphous product.

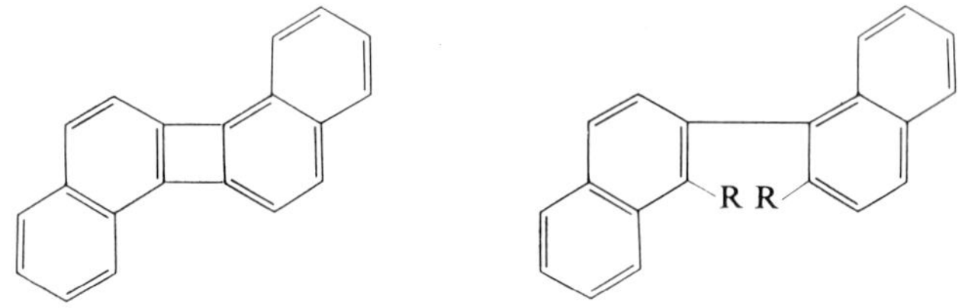

I

II: R = I
III: R = H

11.3 Barton, J. W., Rogers, A. M., and Barney, M. E., *J. Chem. Soc.* **5537 (1965).**

Pyrolysis of 9-iodo-10-(2′-iodophenyl)phenanthrene (I) with cuprous oxide at 240°–340° affords dibenzo[*a,c*]biphenylene (II) in 21% yield. The dibenzobiphenylene (II) was isolated as its purple 2,4,7-trinitrofluorenone complex, m.p. 213°–215°. The free hydrocarbon (II) forms orange-red needles, m.p. 183°–184° (decomp.); λ_{max}^{EtOH} 251.5 (log ε 4.33), 263 (4.25), 267.5 (4.22), 292.5 (3.57), 304 (4.75), 316 (4.73), 337 sh (3.19), 359 (2.97), 400 (3.34), and 420 mμ (3.40). Details of the NMR spectrum of II are also recorded. Hydrocarbon II is stable in the dark, but dilute ethanol solutions are bleached by light. It sublimes unchanged and it appears to be unreactive as a diene, but it adds bromine to give a dibromide (III), which can be debrominated to II with sodium iodide; II is cleaved to 9-phenylphenanthrene (IV) by Raney nickel in ethanol.

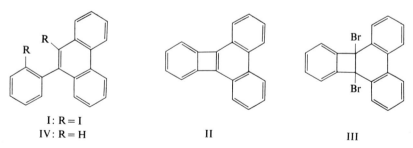

I: R = I
IV: R = H

II

III

11.4 Martin, R. H., Van Trappen, J. P., Defay, N., and McOmie, J. F. W., *Tetrahedron* **20,** 2373 **(1964).** See Abstract 10.21, this appendix.

Author Index

Numbers in parentheses are reference numbers and indicate that an author's work is referred to although his name is not cited in the text. Numbers in italic indicate the page on which the complete reference is cited.

A

Adam, W., 12(1), 52(1), 74(1), 75(1), *84*, 127(1), *127*, 396(1), *414*, *445*
Ali, M. A., 333(1), 334(1), 335(1, 2), *366*, 398(2), 401(2), 410(3), 411, *414*
Allinger, N. L., 36(44), *85*
Alston, T. G., *461*
Altman, L. S., *442*
Anastassiou, A. G., 178(1), *179*, 248(1), *254*, 396(5), 398(4), *414*
Andrade e Silva, M., 335(3), 356(3), 358(3), *366*, 410(6), *414*
Applequist, D. E., 17(2), 48(2), 66(2), 74(2), *84*, *86*
Arnell, E. M., *423*
Arora, S. S., 258(45), *314*, 347(17), *366*
Asahi, Y., 289(64), *314*, *459*, 460(64)
Assony, S. J., 41(4), *84*
Atkinson, E. R., 258(2), 266(1), 267(4), 274(5), 283(3, 4), 284(5), 285(59), 286(59), 290(6), 292(59), 293(59), 303(59), 310(59), 311(59), *313*, *314*
Avram, M., 3(117), 11(5, 9), 14(8, 13), 21(13), 22(13, 118), 34(12), 39(13), 42(7, 9a, 10), 43(6), 47(13), 50(5, 9), 52(5), 61(11, 13, 118), 69(13), 70(8), 71(5, 13), 72(13, 118), 77(5), 79(5), *84*, *87*, 105(1, 2, 3,) *119*, 184(1, 3, 4, 48), 199(1, 2), 202(2, 3, 4, 48), 206(1, 3, 4, 48), 209(48), 210(4), *217*, *218*, 243(2), 245(12), 249(2, 12), 250(2), *254*, 325(6), 327(41), 329(6, 7, 41), 332(4, 5), 337(41), 350(6), 351(6), 356(41), 360(41), 362(6), 363(41), 364(41), *366*, *367*, *424*, *450*, 463(41), *463*

B

Babcock, D., 195(14), *217*
Baker, W., 3, 69(14), *84*, 180, 181(7), 186(5), 196(5), 198(5), 202(8), 208(5), *217*, 255, 256(21), 257(15, 20), 258(10), 259(10, 13, 16), 260(15, 16), 265(13), 267(20), 268(11, 14, 15, 23a), 269(11, 14, 15, 22, 23a), 270(11, 14, 15, 23a), 271(10, 11, 14, 15), 273(15), 274(15), 275(11, 13, 15, 17, 21), 277(7, 15), 278(8, 15, 20), 281(32), 282(11, 15, 21), 283(8, 15, 20, 22), 284(12, 15, 22, 23), 285(15), 286(21), 287, 291(10, 11), 292(11, 14), 293(11, 14, 15), 294(11, 15), 295(11, 14, 15), 296(11, 15), 297(11, 15, 21), 298(11, 15, 23a), 300(11, 15), 301(11, 15), 303(11, 13, 15, 16), 304(10, 13, 15), 305(17), 308(17, 18), 309(10, 11, 12, 14, 15, 22, 23a), 310(10, 11, 13, 14, 15, 23a), 311(10, 11, 14, 15, 16), 312(10, 11, 13, 14, 16, 17), *313*, 318, 325(8, 9), 328(9), 332(9), 334(9), 335(8), 341(9), 345(9), 346(9), 347(9), 348(9), 362(9), *366*, 400, 405(8, 9), *414*, *451*, *452*, 456(11, 15), *456*
Bakunin, M., 180, 181(10), *217*
Balaban, A. T., 41(16), *84*
Balch, A. L., *448*
Ball, W. J., 36(17), *84*
Banks, D., 419, *422*, *455*
Barakat, M. Z., 330(46), 358(46), *367*
Barney, M. E., 418(2), *422*, *464*
Bartlett, P. D., 42(18), 47(18), *84*
Barton, J. W., 186(5), 196(5), 198(5), *217*, 258(10), 259(10, 13), 265(13), 268(11, 14), 269(11, 14), 270(11, 14), 271(10, 11, 14), 273(24), 275(11, 13), 282(11), 284(12), 291(10, 11), 292(11, 14), 293(11, 14), 294(11), 295(11, 14), 296(11), 297(11), 298(11), 300(11), 301(11), 303(11, 13), 304(10, 13), 309(10, 11, 12, 14), 310(10, 11, 13, 14), 311(10, 11, 14), 312(10, 11, 13, 14), *313*, 325(8, 9), 328(9), 332(9), 334(9), 335(8), 341(9), 345(9), 346(9),

347 (9), 348(9), 362(9), *366*, 418(1, 2), *422*, *447*, *453*, *454*, 456(11), *463*, *464*
Battiste, M., 41(27), *84*
Bauld, N. L., 419, *422*, *455*
Baume, E., *257*
Baumgärtel, H., 276(117), *315*
Becker, U., 14(127), *87*
Bedford, A. F., 257(25), 288(25), *313*, 410 (10), *414*
Beesley, R. M., 70(19), *84*
Beineke, T. A., 58(87), *86*
Bell, F., 320(11), 332(11), 333(11), *366*
Belyanin, V. B., 89(61), *121*, *440*
Bergami, O., 6, *86*
Bergman, I., 289(26), *313*, 401, 407(11), *414*
Bergmann, E. D., 89, 106, *119*
Berkoff, C. E., 12(20), 13(20), 39(20), 52 (20), 61(20), 75(20), *84*, *424*
Bernemann, P., 162(19), *179*
Berry, R. S., 261(27), *313*, *424*, *461*
Berthier, G., 290(28), *313*, 405(12a), *414*
Beynon, J. H., 178(2), *179*, 183(11), 217, 272 (29), *313*, *424*, *433*
Bickelhaupt, F., 274(129), *316*
Bieber, T. I., 103, *119*
Biesenbach, T., 10, *87*
Birks, J. B., *414*, *455*
Bischoff, F., 162(20), 164(20a), 165(20a), 166(20a), 176(20a), *179*, *447*
Blatchly, J. M., 187(12), 188(12), *217*, 259 (32), 268(30, 31), 269(31), 270(31), 271 (31), 293(31), 294(31), 295(31), 298(31), 301(31), 303(31), 304(31, 32), 306(31), 307(31), 308(31), 309(30, 31), 310(31, 32), 311(31), 312(31), *313*, 365(12), *366*, 418, *422*, *457*
Blomquist, A. T., 26(21, 23), 77(22), *84*, 95 (6), 98(6), 104(6), 107(6), 109, 111(6), 114(6), 113(6), 118(6), *119*, 129(2), 131 (2, 3), 132(3), 133(2), 137(2), 140(2, 3), 141(2, 3), 143(2, 3), 144(1, 2), 145(1), 146(1), 147(2, 3), 148(1), 149(3), 151 (2, 3), 152(2), 153(3), 155(2, 3), 156(2), *156*, 159(3), 160(4), 162(3), 163(3, 4a, 4c, 5), 164(3, 4a, 4c), 165(4a, 4c), 166(3, 4a, 4c), 167(3, 4), 176(3, 4a, 4c), *179*, 185 (13), 192(13), 198(13), 200(13), 208(13), 209(13), 216(13), *217*, 252(3), *254*, *414*, *424*, *441*, *442*, 447(13), *447*, *449*

Boarland, M. P. V., 257(15), 260(15), 268 (15), 269(15), 270(15), 271(15), 273(15), 274(15), 275(15), 277(15), 278(15), 282 (15), 283(15), 284(15), 285(15), 293(15), 294(15), 295(15), 296(15), 297(15), 298 (15), 300(15), 301(15), 303(15), 304(15), 309(15), 310(15), 311(15), *313*, 456(15)
Bollinger, J. M., *423*
Bordwell, F. G., 195(14), *217*
Boshovsky, W., 32(64), 37(64), *85*
Bosshard, H. H., 268(33), 270(33), 278(33), 281(33), 298(33), 309(33), *313*
Boston, J. L., 92(7), 104(7), 114(7), 118(7), *119*, 436(7)
Bottomley, C. G., 185(13), 192(13), 198(13), 200(13), 208(13), 209(13), 216(13), *217*, 447(13), 449
Boulton, A. J., *447*, *451*, *456*, *457*
Bouman, N., 384(14), 385, *414*
Brailovskii, S. M., 89(61), *121*, *440*
Brand, K., 158, 169(7), *179*
Brass, K., 9(24), *84*
Braun, F., 319(45), 333(45), *367*
Braun, J. von, 318, 332(13), *366*
Braye, E. H., 19(25), 25(25), 59(25), 77(25), *84*, 89(32), 93(31, 32), 100(8, 31, 32), 101(8), 106(31), 110(31), 113(31), 114 (31, 32), *119*, *120*, 412(58), *415*
Brenner, S., *443*
Breslow, R., 27(29), 28(29), 29(29, 30), 41 (26, 27, 30), 42(28), *84*, 129(4), 153(4), *156*, 425(26, 29, 30), *425*, *442*
Bretschneider, H., *429*
Britts, K., *443*
Brown, D. A., 89(32), 93(32), 100(32), 114 (32), *120*, 412(58), *414*, *415*
Brown, Henry C., 27(32), 36(31, 32), *84*
Brown, R. D., 278(35), 282(34), *313*, 401(18), 402(17), 405, *414*
Brown, R. F. C., 225, *448*, *458*
Bruce, J., 47(155), *87*
Bruce, J. M., 267(36), 275(36), *313*, 329(14), 330(14), 331(14), 334(14), 336(14), 358 (14), 363(14), *366*
Brune, H.-A., 23(55), 24(47), 52(55), 75(47), *85*, 103(14), 111(14), *119*, *433*
Bruni, R. J., 258(2), *313*
Bryan, R. F., 125(2), *127*, *441*
Bucher, A., 8(111), 59(111), *86*

AUTHOR INDEX

Buchman, E. R., 4(58), 5, 25(58), 31(58), 34(37), 70(37), *85*, 430(35)
Büchi, G., 18(33), 59(33), 64(33), 66(33), 76(33), *84*
Bullimore, B. K., *430, 440*
Bunnett, J. F., 261(37), *313*, 334(15), *366*
Burckhardt, U., 182(54), 186(54), 190(54), 198(54), 209(54), 214(54), *218*
Burt, G. D., *425, 433*
Busch, M., 273(38), *313*
Buu-Hoï, N. P., 296(39), *314*

C

Cairns, T. L., 103(9, 55), *119, 120*
Campbell, C. D., *458*
Capell, L. T., 322(43), *367*
Caplier, I., 19(25), 25(25), 59(25), 77(25), *84*, 100(8), 101(8), *119*
Carey, J. G., 257(25), 288(25), *313*, 333(1, 16), 334(1), 335(1), *366*, 398(2), 401(2, 10), 411(2), *414*
Carlsohn, H., 7(38), 9(38), *85*
Carr, E. P., 286(40), *314*
Carrington, A., 290, *314*, 404, 410, *414*
Caserio, M. C., 15(91), 16(91, 116), 28(116), 57(91, 116), 58(116), 73(91, 116), *86*, 128(18), 130(18), 131(8, 18a), 133(8, 18), 135(18), 136(18), 137(18), 138(18), 140(18), 141(18), 142(18), 143(18), 144(18), 147(18), 148(18), 151(18a), 152(18a), 153(18), 155(18), *156*, 252(13), *254*, 430(116), *430*
Cass, R. C., 288(42), *314*, 401(21), *414*
Cava, M. P., 180, 181(20), 184(16, 17a, 20, 23), 185(19, 20b), 186(18, 19, 23, 25), 188(21), 189(23), 190(23), 194(18, 19, 25), 195(25), 196(17, 24), 197(17), 198(17, 23), 202(16, 22, 23), 204(23, 26), 205(16), 206(20a), 207(16, 18, 19, 20, 22, 23, 25), 208(17, 18, 23, 24), 209(20), 210(20a), 211(25), 214(20a), *217*, 220, 221(3, 4, 5a), 222(5a), 223(4, 5a), 224(5a), 225(5), 226(5, 7a), 227(7a), 228(2, 6), 229(5), 230(5a), *230, 231*, 233(2a, 2c), 234(1, 2a, 2c, 3), 235(1, 2a, 2b, 2c), 236(2a, 2b), 237(2a, 2b), 238(1, 2a, 2b, 2c), 239(1, 2a, 2b, 2c), 240(2a, 2c), *241*, 244(5, 6, 7), 245(5, 6), 246(4, 6), 249(4), 248(5, 6), 250(5), 251(4, 5, 6), 252(4), 253(8), *254*, 255, 257(46), 258(45), 267(44), 274(44), 277(44), 289(64), *314*, 320(21), 321(28), 322(21, 28, 29), 324(28a), 325(19, 20, 21, 23, 25, 27), 328(19, 20, 21, 25, 27, 29), 329(23, 30), 330(27), 331(27, 29), 332(24, 27), 334(25), 335(20), 337(18, 28), 339(21, 27, 29), 340(20, 27, 29), 341(20, 21), 344(25, 26, 27, 28, 29), 345(27, 29), 346(20, 26, 29), 347(17), 348(25), 350(23, 30), 352(23), 354(28), 355(28), 356(28), 362(19, 20, 21, 23, 26, 27, 28, 29), 363(23), 364(30), 365(22), *366*, 397(22), 398(22), 399, *414, 448, 450, 459*, 460(64)
Chalvet, O., 296(39), *314*
Chatt, J., 272(48), 274(47), 284(48), *314*, 406(24), *414*
Chatterjee, A., 9(39, 40), *85*
Chaudhury, B., 9(39, 40), *85*
Cherchi, F., 9, *85, 86*
Childs, R. F., *426*
Chubb, F., 258(50), *314*
Chung, A. L. H., 418(5, 6), *422*
Cirulis, A., 7(38), 9(38), *85*
Clar, E., 289, *314*, 401, 402, *414*
Clardy, J., *424*
Claus, A., 89(32), 93(32), 100(32), 114(32), *120*, 412(58), *415*
Coates, G. E., 89, *119*
Coffey, C. E., 93(11), 100(11), 114(11), *119*
Cohen, D., 333(1), 334(1), 335(1), *366*, 398(2), 401(2), 411(2), *414*
Cohen, S., 129(13a), 132(13), 134(13), 149(13), 150(13), 151(13), 153(13), 155(13), *156*
Coleman, W. E., 196(39), 198(39), 207(40), 214(40), *218*, 325(39), 328(39), 330(39), 331(31, 39), 332(39), 334(39), 347(31, 39), 348(39), 350(31, 39), 351(39), 352(39), 362(31, 39), *366, 367*
Collette, J., 258(50), *314*
Conte, J. M. de C., *414, 455*
Cookson, R. C., 12(20), 13(20), 22(42), 39(20), 45(41), 52(20), 61(20, 42), 66(41), 75(20), 77(41), *84, 85*, 109(12), 101(12a), *119*, 267(51), *314, 424*, 427(12a), *427, 433*

Cope, A. C., 47(43), *85*
Corbett, J. F., 264, 265(52), 300(52), 311 (52), *314*
Corey, E. J., *428*
Cotton, F. A., 368(26), 377(26), 405(26), *414*
Coulson, C. A., 248(9), *254*, 277(53), 280 (54), *314*, 335(2), *366*, 369(32), 370(29), 376, 377, 386, 389, 398(31), 399, 400, 401(28), 410, *414*
Cox, D. A., 267(51), *314*
Craig, D. P., 369, 375, 379, 381(37, 102), 383, 387(36), 390, 391(33), *414*, *416*
Cram, D. J., 36(44), *85*
Crawford, R., 258(50), *314*
Crawford, V. A., 280(54), *314*, 356(32), *366*, 410, *414*
Criegee, R., 3, 6, 12(49, 52), 13(45, 52, 56), 14(45, 51), 23(45, 53a, 55), 24(45, 46, 47, 50, 54), 25(53), 26(46), 32(48, 49), 52 (45, 49, 52, 54, 55), 53(45), 60(49, 52), 63(45, 56), 74(52), 75(45, 46, 47, 49, 50, 52, 53, 56), *85*, 89, 98(13), 99(13, 17), 103(14, 16), 105(16), 106(13a, 13b, 15, 17), 107(13, 17, 18), 109(17), 110 (17), 111(13, 14, 16), 114(17), 116(17a), 113(17a), 118(17a), *119*, 159(8), 164 (8), 166(8), 167(8), 176(8), *179*, 412, *414*, *428*, 429(45a, 56), *429*, 431(48), *433*
Cullinane, N. M., 274(55), *314*
Curtis, R. F., 260(56), 275(56), 287, 288(56), *314*, 320(33), 321(33), 322(33b), 335 (33, 34), 337(33), 338(33a), 339(33), 354(33a), 356(33a, 34), 357(33), 363 (33), 364(33a), *366*

D

Dahl, L. F., 95(19), 109(19), *119*
Dansi, A., 319(35), 332(35), *367*
Defay, N., *461*, *464*
Dekker, J., 24(46), 26(46), 75(46), *85*, 159 (8), 164(8), 166(8), 167(8), 176(8), *179*
De Puy, C. H., 60(57), *85*
Detzer, N., *447*
Deutsch, D. H., 4(58), 25(58), 31(58), *85*
Dewar, M. J. S., 405, *414*, *415*, 418(5, 6, 7, 8), *422*
Dickelman, T. E., 274(5), 284(5), *313*
Dickerman, S. C., 418, *422*, *458*

Dilthey, W., 242(10), 253(10), *254*
Dinu, D., 3(117), *87*, 184(1, 3, 48), 199(1, 2), 202(2, 3, 48), 206(1, 3, 48), 209(48), *217*, *218*, 325(6), 329(6, 7), 332(4, 5), 350(6), 351(6), 362(6), *366*
Dinulescu, I. G., 11(5), 42(5, 10), 52(5), 71 (5), 77(5), 79(5), *84*, 105(3), *119*, 184(3, 4), 199(2), 202(2, 3, 4), 206(3, 4), 210(4), *217*, 243(2), 245(12), 249(2, 12), 250(2), *254*, 325(6), 327(41), 329(6, 7, 41), 332 (5), 337(41), 350(6), 351(6), 356(41), 360 (41), 362(6), 363(41), 364(41), *366*, *367*, *424*, *450*, 463(41), *463*
Dixit, V. M., 10, *85*
Dobbie, J. J., 256, 273(57, 58), 274(58), 275 (58), 287(58), *314*
Dodge, R. P., 119(20), *119*, 412(44), *415*, *433*
Domingo, R., 272(60), 278(60), *314*, 401 (47), 405(47), *415*
Domnin, N. A., 37(60), *85*
Doorakian, G. A., *436*
Doran, M. A., 419(16), *422*, *462*
Drefahl, G., *429*
Druey, J., 132(7), 140(7), 144(7), *156*
Dunathan, H. C., 22(151), 34(151), 48(151), 55(151), 57(151), 60(151), 72(151), *87*, *432*
Dunicz, B. L., 36(61), *85*
Dunitz, J. D., 118(21), *119*, 412(45), *415*

E

Ebel, H. F., 261(130), 263(130), 266(130), 285(130), *316*
Efraty, A., *438*, *439*
Eicher, T., 42(28), *84*
Eichler, S., 89(60), 92(60), *120*
Ekström, B., 33(63), 36(62, 63), *85*
Elian, M., 243(2), 249(2), 250(2), *254*, *450*, *463*
Emerson, G. F., 88(23), *119*, 180(27), *217*, 419(11, 12), *422*, *434*, *448*,
Engel, W., 23(55), 24(46), 26(46), 52(55), 75 (46), *85*, 159(8), 164(8), 166(8), 167(8), 176(8), *179*
Engelhardt, V. A., 103(9), *119*
Erdmann, H., 89, *119*
Erickson, B. W., 332(24), *366*
Erpelding, T. J., 272(84), *315*, 461(84)

AUTHOR INDEX 469

Evans, M. V., 149(21), 154(21), 155(21), *156*
Eyring, H., *415*

F

Fabian, W., *425*
Farcasiu, M., 22(118), 61(118), 72(118), *87*, 243(2), 249(2), 250(2), *254*, *450*, *463*
Farnum, D. G., 123(4), 124(3), 126(4), *127*, 140(5), *156*, 285(59), 286(59), 292(59), 293(59), 303(59), 310(59), 311(59), *314*, *441*, *443*
Favorskaya, T. A., 4, *85*
Favorskii, A., 4, 32(64), 37(64), *85*
Fawcett, J. K., *459*
Fenton, S. W., 191(28, 36a, 36b), *217*, *218*
Fernández Alonso, J. I., 272(60), 278(60, 61), *314*, 401(47), 402(48), 405(47, 48), *415*
Ferri, C., 319(35), 332(35), *367*
Fick, R., 319(44), 333(44), 350(44), *367*
Finkelstein, H., 181(29, 30), 182(31), 206(29, 30), 207(29, 30), *217*, 320(36), 325(36), 328(36), 340(36), 346(36), *362*(36), *367*
Fischer, H.-G., 24(54), 52(54), *85*, 107(18), *119*
Fischer, W. M., 7(38), 9(38), *85*
Fish, R. W., 123, *127*, 187(38), *218*
Fisher, I. P., 262(62), *314*
Fitzpatrick, J. D., 2(148), *87*, 88(23, 64), *119*, *121*, 419(12), 420(17, 18), *422*, *431*, *434*, *441*
Flid, R. M., 89(41), *121*, *440*
Förg, F., 24(47), 75(47), *85*, 103(14), 111(14), *119*, *433*
Fonken, G. J., 266(63), *314*
Fox, J. J., 256(57, 58), 273(57, 58), 274(58), 275(58), 287(58), *314*
Fraenkel, G., 289(64), *314*, 332(24), *366*, *459*, 460(64)
Frantz, A. M., Jr., 123(6), 124(5), 126(5), *127*, 443(5)
Freedman, H. H., 14(69), 19(67, 69), 20(66, 67), 38(66, 119), 45(67), 46(67), 58(69), 59(68), 64(67), 67(66), 75(69), 76(66, 67), 77(66, 69), *85*, *87*, 98(25), 99(25), 101(24), 104(24), 109(25), 110(25), *120*, 123(6), 124(5), 126(5, 6), *127*, 396(50), 412, *415*, 427(25), *427*, *430*, 436(66), *436*, *441*, 443(5)
Freund, G., 32(70), *85*
Friedel, R. A., 101(65), *121*, 319(42), 332(42), *367*
Friedman, L., 262(65), 263(65), *314*, *459*
Friedrich, H. J., *447*
Fritchie, C., Jr., 57(71), *85*
Fritz, H. P., 42(7, 72), 43(6), *84*, *85*, 89, 105(1, 26, 28), 116(26a), *119*, *120*, 412(51), *415*, *436*
Fujino, A., *446*
Fumi, F. G., 369(52), *415*
Funke, W., *428*
Furukawa, K., 43(80), *86*

G

Gabriel, S., 6, *85*
Games, M. L., 69(104), *86*, 94(42), 95(39), 96(42), 97(40, 41, 42), 98(42), 100(41), 104(41), 105(42), 106(41), 111(42), 112(41), 113(39), 114(39, 40, 42), 117(40), *120*, *439*, *440*
Gaspar, P. P., *420*
Gastaldi, C., 9, *85*, *86*
Gatti, A. R., 283(3), *313*
Gatti, D., 256(93), 273(93, 109), *315*
Gelin, S., 3, *86*, 255, *314*, 318, *367*
Gerloch, M., 104(29), *120*, *436*
Geske, D. H., *448*
Gewanter, H. L., 27(32), 36(32), *84*
Gleicher, G. J., 418(7, 8, 9), *422*
Gochenour, C. I., 37(77), *86*
Goeppert-Mayer, M., 378, *415*
Gold, E. H., 124(8), *127*
Goldish, E., 286(67), *314*
Graham, W. H., 29(78), *86*
Granchelli, F. E., 267(4), 283(4), *313*
Grdenić, D., 266(68), *314*
Griffin, G. W., 26(79), 43(80), *86*, 159(11a), 161(11a), 163(11a), 164(11a), 165(11a), 166(9, 11a), 167(9, 11a), 168(11), 169(11, 12), 170(11), 171(11a), 173(11), 176(11a), 177(11a), 178(9, 10), *179*, 119(32), *217*
Grundmann, C., 9(81), *86*
Guage, A. J. H., 256(57, 58), 273(57, 58), 274(58), 275(58), 287(58), *314*

Günther, H., 261(69), *314*
Guy, R. G., 272(48), 284(48), *314*, 406(24), *414*
Gwynn, D. E., 17(2), 37(82), 48(2), 66(2), 74(2), 79(2), *84*, *86*

H

Haag, A., 182(54), 186(54), 190(54), 198(54), 209(54), 214(54), *218*
Härle, H., 262(131), 264(131), 285(131), 291(131), 292(131), 293, 311(131), 312(131a), *316*
Hager, R. B., 178(10), *179*
Hagihara, N., 92(48), 93(48a), 106(48b), 110(48b), 114(48), 116(48b), *120*
Hahn, E., 261(132), *316*
Hall, J. R., 124(9), *127*, 396(63), *415*
Halvorsen, B. F., 7(96), *86*
Hammond, G. S., *420*
Hancock, J. E. H., 191(33, 34, 35), *217*, *218*
Hanna, M. W., 191(36a, 36b), *218*
Harder, R. J., 36(83), *86*
Hargreaves, A., 286(70), *314*
Harris, J. F., Jr., 36(83), *86*
Harrison, C. R., *451*
Hart, F. A., 262(71), 264(71), 291(71), 292(71), 311(71), *314*
Hart, H., 123, *127*, 187(37, 38), *218*
Hartlage, J., 187(37), *218*
Haszeldine, R. N., 33(84), *86*
Hausmann, J., 6, *86*
Haven, A. C., Jr., 47(43), *85*
Heany, H., 261(72, 73), 267(72), *314*, *459*
Hedges, R. M., 405, *415*
Heinrich, B., 162(20), 164(20a), 165(20a), 166(20a), 176(20a), *179*, *447*
Hellman, H., 262(74), 263(74), 264(74), 291(74), 310(74), *314*
Henn, D. E., *447*, *453*
Henne, A. L., 37(86), *86*
Herwig, W., 266(133), *316*
Heybey, M. A. T., 124(3), *127*, 140(5), *156*, *441*, *443*
Hill, R. R., *424*, *433*
Hilpern, J. W., 69(14), *84*, 402, *415*, 419, *422*, *460*
Hinton, R. C., 264(75), 275(75), 311(75), *314*
Hobey, D. W., 391, *415*

Hochstrasser, R. M., 286(76), 289(76), *314*, 402, *414*, 419, *422*
Hoerhold, H. H., *429*
Höver, H., 32(48), *85*, 431(48)
Hoffmann, R., 401(57), *415*, 421, *422*
Hoffmann, R. W., 261(134), 312(134), *316*
Hoffmeister, E., 256, 272(82), *315*
Hoffmeister, W., *314*
Holmes, E., 42(18), 47(18), *84*
Holt, P. F., 264, 265(52), 300(52), 311(52), *314*
Hoogzand, C., 89(32), 93(32), 100(32), 114(32), *120*, 412(58), *415*
Horst, J. ter, 242(10), 253(10), *254*
Hosaeus, W., 256, *314*
Howton, D. R., 5, *85*, 430(35)
Hruby, V. J., *449*
Hudec, J., 12(20), 13(20), 39(20), 52(20), 61(20), 75(20), *84*, 267(51), *314*, *424*
Hubel, W., 19(25), 25(25), 59(25), 77(25), *84*, 89(32), 93(30, 31, 32), 95(30, 35), 99(30), 100(8, 31, 32, 35), 101(8), 106(31), 109(30), 110(31), 113(31), 114(30, 31, 32), *119*, *120*, 412, *415*, *436*
Hückel, E., 369(59), 371, 376(59), *415*
Hüttel, R., 105(33), 109(33), *120*, *437*
Hughes, E. D., 58(87), *86*
Hughes, E. W., 57(71), *85*
Huisgen, R., 262(79), *315*, 334(38), *367*
Hunter, W. H., 320(11), 332(11), 333(11), *366*
Huntsman, W. D., 162(13, 14), 163(13), 164(14), 166(14), 176(13, 14), *179*
Hush, N. S., 402(60), 403(60), *415*
Hwang, B. Y., 244(5, 6, 7), 245(5, 6), 246(4, 6), 247(4), 248(5, 6), 250(5), 251(4, 5, 6), 252(4), *254*, 397(22), 398(22), *414*, *450*

I

Ilse, F. E., 277(80), *315*
Ipatschi, J., *225*
Ito, M., 132(20), 134(20), 150(20), 154(6), *156*

J

Jackson, H. L., 103(9), *119*
Jacob, K., *444*

Jahn, H. A., 388, *415*
Jenny, E. F., 132(7), 140(7), 144(7), *156*
Jensen, F. R., 196(39), 198(39), 207(40), 214 (40), *218*, 325(39), 328(39), 330(39), 331 (39), 332(39), 334(39), 347(39), 348(39), 350(39), 351(39), 352(39), 362(39), *367*
Johnson, A. W., *426*
Jones, A. J., 333(1), 334(1), 335(1), *366*, 398 (2), 401(2), 411(2), *414*
Jones, D. W., 45(41), 66(41), 77(41), *85*, 109(12), 110(12a), *119*, *424*, 427(12a), *427*, *433*
Jones, E. R. H., 100(34), *120*
Jones, R. C., 5, *86*
Jonescu, M., 7(38), 9(38), *85*

K

Kalb, G. H., 103(9), *119*
Kampmeier, J. A., 272(82), *315*
Karle, I. L., *443*
Katritzky, A. R., *460*
Katz, T. J., 124(8, 9), *127*, 396(63), *415*
Kekulé, A., 3
Keller, H., 42(7), 43(6), *84*, 105(1), *119*
Kende, A. S., *448*, *451*, *460*
Kercher, F., 169(7), *179*
Kerridge, K. A., 258(2), *313*
Kharasch, N., 41(4), *84*, 266(83), 272(83, 84), *315*, 461(84), *461*
Kimball, G. E., *415*
King, G. S. D., 89(32), 93(32), 100(32), 114 (32), *120*, 412(58), *415*
Kipping, F. S., 6, 7(89), 37(90), *86*
Kirschbaum, G., 318, 332(13), *366*
Kitahara, Y., 15(91), 16(91), 57(91), 73 (91), *86*, 131(8), 133(8), *156*
Kivelevich, D., 27(29), 28(29), 29(29), *84*, 425(29), *425*
Klager, K., 39(126), *87*, 89(53), *120*
Kline, G. B., 27(129), *87*
Knoche, R., 5, *86*
Köthner, P., 89, *119*
Kohlhaupt, R., *447*, *448*
Korte, F., 162(17), *179*, *444*
Kozaki, N., 184(53), 202(53), 207(53), 211 (53), *218*
Kragten, J., *415*
Krebs, A., *442*

Kreiter, C. G., 43(6), *84*
Kristinson, H., 13(93), 33(93), 52(93), 75 (93), *86*, *429*
Kruerke, A., 412(58), *415*
Kruerke, U., 89(32), 93(32), 95(35), 100 (32, 35), 114(32), *120*
Kuehne, M. E., 192(41, 42), *218*
Kuprava, Sh. D., 191(43e), *218*
Kuri, Z., 19(94), 71(94), *86*, 384(114), 389 (114), *416*

L

Lacher, J. R., 129(13a), 132(13), 134(13), 149(13), 150(13), 151(13), 153(13), 155 (13), *156*
Lackner, L., 9(132), 89
Lagidze, R. M., 191(43a, 43b, 43c, 43d, 43e), *218*
LaLancette, E. A., 26(21), *84*, 129(2), 131 (2), 133(2), 137(2), 140(1b, 2), 141(2), 143(2), 144(1, 2), 145(1), 146(1), 147 (2), 148(1), 151(2), 152(2), 155(2), 156 (2), *156*, 252(3), *254*, *424*, *441*
Landor, S. R., 36(17), *84*
Lanser, T., 7, *86*
Larson, H. O., 70(97), *86*
Latif, N., 330(46), 358(46), *367*
Lavit, D., 296(39), *314*
Layton, E. M., Jr., 402, 407(65), *415*
Lee, C.-S., 401(127), *416*
Lee, H. S., 242(11), 248(11), 253(11), *254*, 335(40), *367*, 396(66), 398(66), 399, 407(66), *415*
Lees, P., *459*
Le Goff, E., 137(9), *156*, 261(85, 86), *315*
Lehman, G., 273(135), 285(135), 287(135), *316*
Lennard-Jones, J. E., 370(67), 389, *415*
Lester, G. R., 183(11), *217*, 272(29), *313*
Levine, P. L., 274(5), 284(5), *313*
Liebermann, C., 6, *86*
Liehr, A. D., 384(73), 391, *415*
Limpricht, H., 6, *86*
Link, W. J., 15(134), 68(134), *87*
Lipscomb, W. N., 38(119), 58(101), 69(101), *86*, *87*, 191(28), *217*, 392, *415*, *430*
Litten, E., 9(81), *86*
Logullo, F. M., 262(65), 263(65), *314*

Loladze, N. R., 191(43d), *218*
Longuet-Higgins, H. C., 6, *86*, 89, 106, *120*, 281(87), *315*, 376(77), 385, 390, 391, 411, *415*
Lossing, F. P., 262(62), *314*
Lothrop, W. C., 256, 257(88, 89), 259(88, 89), 273(88), 275(89), 276(88), 277(88), 283(88), 284(88), 285(59, 88), 286(59), 291(88, 89), 292(59), 293(59), 303(59), 304(89), 310(59, 89), 311(59, 88, 89), *314*, *315*, 400, *415*
Louis, G., 12(49), 32(49), 52(49), 60(49), 75(49), *85*, 106(15), *119*
Lu, C.-S., 278(123), 285, 292(121), *316*
Ludwig, P., 24(46, 50), 26(46), 75(46, 50), *85*, 103(16), 105(16), 107(37), 111(16), *119*, *120*, 159(8), 164(8), 166(8), 167(8), 176(8), *179*
Lüttringhaus, A., 261(90), *315*, 319(44), 333(44), 350(44), *367*
Lutz, E., 262(106), 271(106), *315*
Lykos, P. G., 369(83), *415*
Lyons, C. E., 60(57), *85*

M

Maahs, G., 129(10b), 132(10a), 135(10), 136(10), 149(10), 155(10), *156*, *444*
McAlpine, R. D., 419(13b), *422*
McDonald, S. G. G., 32(110), *86*, 198(47), *218*
McDowell, C. A., 290(94, 95), *315*, 403, *415*
McEwen, K. L., 385, *415*
McGinn, C. J., 401(87), *416*
McGreer, D., 258(50), *314*
MacGregor, J. R., 272(91), *315*
MacGregor, P. T., *448*, *451*, *460*
McKellin, W. H., 195(14), *217*
McLachlan, A. D., 374, 391, 404, *415*, *416*
McLaughlan, K. A., *447*, *453*
McLean, N. J., 259(16), 260(16), 275(17), 303(16), 305(17), 308(17, 18), 311(16), 312(16, 17), *313*, *452*
McOmie, J. F. W., 3, 42(7, 72), 43(6), 69(14), *84*, *85*, 105(1, 28), *119*, *120*, 178(16), *179*, 180, 181(7), 186(5), 187(12), 188(12), 198(5), 202(8), 208(5), *217*, 236(5), 238(5), *241*, 255, 256(21), 257(15, 20), 258(10), 259(10, 13, 16, 32), 260(15, 16), 261(96), 265(13, 97), 267(20), 268(11, 14, 15, 23a, 30, 31, 99), 269(11, 14, 15, 22, 23a, 31, 96), 270(11, 14, 15, 23a, 31, 98), 271(10, 11, 14, 15, 31, 96, 98, 99), 273(15, 24), 274(15), 275(11, 13, 15, 17, 21, 31, 98), 277(15), 278(15, 20), 281(23, 96), 282(11, 15, 21), 283(15, 20, 22), 284(12, 15, 22, 23), 285(15), 286(21), 287, 291(10, 11), 292(11, 14), 293(11, 14, 15, 31), 294(11, 15, 31), 295(11, 14, 15, 31, 98), 296(11, 15, 22, 23, 98), 297(11, 15, 21), 298(11, 15, 23a, 31), 300(11, 15), 301(11, 15, 31), 303(11, 13, 15, 16, 31), 304(10, 13, 15, 31, 32, 96), 305(17, 96), 306(31, 96), 307(31), 308(17, 18, 31, 96, 99), 309(10, 11, 12, 14, 15, 22, 23a, 30, 31, 98, 99), 310 (10, 11, 13, 14, 15, 23a, 31, 32, 97, 98, 99), 311(10, 11, 14, 15, 16, 31, 97, 98), 312(10, 11, 13, 14, 16, 17, 31, 98), *313*, *315*, 318, 325(8, 9), 328(9), 332(9), 334(9), 335(8), 341(9), 345(9), 346(9), 347(9), 348(9), 362(9), 365(12), *366*, 405(8, 9), *414*, 418, *422*, *430*, *440*, *447*, *451*, *452*, 456(11, 15), *456*, *457*, *458*, *461*, *464*
McWeeny, R., 385, *416*
Märkl, G., 162(20), 164(20a), 165(20a), 166(20a), 176(20a), *179*, *447*
Maier, G., 24(54), 42(103, 152), 52(54), *85*, *86*, *87*, 107(18), *119*, 191(44), *218*
Maitlis, P. M., 25(105), 45(105), 59(105), 69(104), 77(22, 105), *84*, *86*, 94(42), 95(6, 39), 96(42), 97(40, 41, 42), 98(6, 42), 100(41), 104(6, 38, 41), 105(42), 106(41), 107(6), 109(43), 111(6, 42, 43), 112(41), 113(6, 39), 114(6, 39, 40, 42), 117(40), 118(6), *119*, *120*, 159(3), 162(3), 163(3), 164(3), 166(3), 167(3), 176(3), *179*, *414*, *438*, *439*, *440*
Mak, T. C. W., 278(92), 279, 285(92), *315*, 401, *415*, 459(92)
Makowa, O., 89, *120*
Malachowski, R., 4, 31(106), *86*, 181(45), 189(45), *218*
Malatesta, L., 89(45a), 95(45a), 105(45), 109(45), 114(45), *120*, 412, *416*
Mallory, F. B., 147(11), *156*
Manatt, S. L., 138(12), 139(12), 149(12), *156*, 158(15), *179*, 193(46), *218*, 225(8), *231*, 232(4), 235(4), *241*

Mangold, D., 228(2), *230*, 234(1), 235(1), 238(1), 239(1), *241*, 253(8), *254*, 399, *414*, *448*, *450*

Mann, F. G., 261(73), 262(71), 264(71, 75), 275(75), 291(71), 292(71), 311(71, 75), *314*

Manthey, W., 7, *86*

Marica, E., 3(13, 117), 11(5, 9), 14(8, 13), 21(13), 22(13), 39(13), 42(9a, 10), 47(13), 50(5, 9), 52(5), 61(13), 69(13), 70(8), 71 (5, 13), 72 (5, 13), 79(5), *84*, *87*, 105(2, 3), *119*, 243(2), 249(2), 250(2), *254*, *424*, *450*, *463*

Mariella, R. P., 34(108), *86*

Markby, R., 101(65), *121*

Marriott, J. B., 323(47, 48), 324(47), 333(48), 358(47), 363(47), 364(47), *367*

Marsden, J., 22(42), 61(42), *85*

Marsh, R. E., *445*

Martin, R. H., *461*, *464*

Martologue, N., 42(109), *86*, 105(46), *120*

Mascarelli, L., 256(93), 273(93, 109), *315*

Mason, R., 104(29), *120*, *436*

Mateescu, G., 11(5), 22(118), 42(7, 10), 43(6), 50(5), 52(5), 61(11, 118), 71(5), 72(118), 77(5), 79(5), *84*, *87*, 105(1, 3), *119*, 184(1, 4), 199(2), 202(2, 4), 206(1, 4), 210(4), *217*, 243(2), 245(12), 249(2, 12), 250(2), *254*, 325(6), 327(41), 329(6, 7, 41), 332(4, 5), 337(41), 350(6), 351(6), 356(41), 360(41), 362(6), 363(41), 364(41), *366*, *367*, *424*, *450*, 463(41), *463*

Matsen, F. A., 405, *415*

Maxim, M., 3(117), *87*

Mayer, U., 17(158), 37(158), 54(158), 55(158), *64*(158), 69(158), 74(158), *87*, 433(158), *433*

Mayot, M., 290(28), *313*, 405(12a), *414*

Meinwald, Y. C., 26(23), *84*, 160(4), 163(4a, 4c), 164(4a, 4c), 165(4a, 4c), 166(4a, 4c), 167(4), 172(4), 176(4a, 4c), *179*

Merenyi, R., *436*

Merriman, R. W., 8(130), 10(130), *87*

Mez, H. C., 118(21), *119*, 412(45), *415*

Michael, A., 6, 8(111), 57(111), *85*, *86*

Middleton, W. J., 37(112, 113), *86*

Millar, I. T., 257(25), 261(73), 264(75), 275(75), 288(25), 311(75), *313*, *314*, 333(1, 16), 334(1), 335(1), *366*, 398(2), 401(2, 10), 411(2), *414*

Miller, D. B., 31(114), *86*

Mills, O. S., 118(21), *119*, 412(45), *415*

Milstein, N., 418, *422*, *458*

Mitchell, M. J., 29(30), 41(30), *84*, 129(4), 153(4), *156*, 184(16), 188(21), 202(16), 203(16b), 204(16b), 205(16), 207(16), *217*, 221(3), 226(7a), 227(7a), *230*, *231*, 233(2a), 234(2a), 235(2, 2b), 236(2a, 2b), 237(2a, 2b), 238(2a, 2b), 239(2a, 2b), 240(2a), *241*, 267(44), 274(44), 227(44), 289(64), *314*, 325(25), 328(25), 334(25), 337(18), 344(25), 348(25), *366*, 425(30), *425*, *459*, 460(64)

Moffitt, W., 376, 377, 378, 382, 383, 384, 389, *414*, *416*

Mohacsi, E., *442*

Mole, T., 405(43), *415*

Morgan, N. M. E., 274(55), *314*

Mortimer, C. T., 257(25), 288(25), *313*, 401(10), *414*

Moschel, A., 14(51), *85*

Mosl, G., 9(24), *84*

Moubasher, R., 330(46), 358(46), *367*

Müller, E. W., *444*

Mulliken, R. S., 377, *416*

Mumuianu, D., 42(109), *86*, 105(46), *120*

Murata, I., 3, *86*, *442*

Mustafa, Ahmed, 330(46), 358(46), *367*

Mustafa, Akila, 330(46), 358(46), *367*

Muth, K., 184(17a), 186(18), 194(18), 196(17), 197(17), 198(17), 207(18), 208(17, 18), *217*, 221(4), 223(4), *230*, 325(19, 20), 328(19, 20), 335(20), 340(20), 341(20), 346(20), 362(19, 20), *366*, *448*

N

Nagarajan, K., 16(116), 28(116), 57(116), 58(116), 73(116), *86*, 430(116), *430*

Nagata, Y., *446*

Nahon, R., *430*, *446*

Nakamura, A., 91(47), 92(47, 48), 93(47, 48a), 104(47), 106(48b), 110(48b), 114(47, 48), 116(48b), *120*

Napier, D. R., 181(20), 184(20), 185(19, 20b), 186(19), 194(19), 206(20a), 207(19, 20), 209(20), 210(20a), 214(20a),

217, 220, 221(5a), 222(5a), 223(5a), 224(5a), 225(5), 226(5), 229(5), 230(5a), *230*, 320(21), 322(21), 325(21), 328(21), 339(21), 341(21), 362(21), 356(22), *366*

Naylor, P. G., 272(84), *315*, 461(84)

Neikam, W. C., 124(9), *127*, 396(63), *415*

Nenitzescu, C. D., 3(13), 11(5, 9), 14(8, 13), 21(13), 22(13, 118), 34(12), 39(13), 42(7, 9a, 10), 43(6), 47(13), 50(5, 9), 52(5), 61(11, 13, 118), 69(13), 70(8), 71(5, 13), 72(13, 118), 77(5), 79(5), *84, 87*, 105(1, 2, 3), *119*, 184(1, 3, 4, 48), 199(1, 2), 202(2, 3, 4, 48), 206(1, 3, 4, 48), 209(48), 210(4), *217, 218*, 243(2), 245(12), 249(2, 12), 250(2), *254*, 325(6), 327(41), 329(6, 7, 41), 332(4, 5), 337(41), 350(6), 351(6), 356(41), 360(41), 362(6), 363(41), 364(41), *366, 367, 424, 450*, 463(41), *463*

Neuegebauer, H. J., 105(33), 109(33), *120, 437*

Niedenbrück, H., 162(21), *179*

Niekerk, J. N. van, 256(105), 266(105a), 273(105), 285(105a), *315*

Niementowski, S. von, 255, 273(100), *315*

Nierenstein, M., 256(101), 274(101), *315*

Niu, H. Y., 132(20), 134(20), 149(21), 150(19, 20), 153(19), 154(21), 155(21), 156(19), *156*

Noll, K., 12(52), 13(52), 24(46), 26(46), 52(52), 60(52), 74(52), 75(46, 52), *85*, 107(37), *120*, 159(8), 164(8), 166(8), 167(8), 176(8), *179*

Noyori, R., 184(53), 202(53), 207(53), 211(53), *218*

Nozaki, H., 184(53), 202(53), 207(53), 211(53), *218*

O

Oberhansli, W. E., 95(19), 109(19), *119*

Olah, G. A., 267(102), *315*

Orchin, M., 319(42), 332(42), *367*

Orgel, L. E., 6, *86*, 89, 106, *120*, 411, 413(97), *415, 416*

Osborn, J. H., 191(28), *217*

Osborne, J. E., 33(84), *86*

P

Pariser, R., 381, 382, 385, *416*

Park, J. D., 129(13a), 132(13), 134(13), 149(13), 150(13), 151(13), 153(13), 155(13), *156*

Parlati, L., 180, *217*

Parr, R. G., 369(83), 380, 381(102), 385, *415, 416*

Parrack, J. D., 273(103), 274(103), *315*

Patterson, A. M., 322(43), *367*

Pauling, P., 118(21), *119*

Paulus, K. F., 290(94), *315*

Pawley, G. S., 38(119), *87, 430*

Pearson, B. D., 257(121), 259(121), 273(104), 275(121, 122), 312(121), *315, 316*, 321(49), 322(49), 323(49, 51), 332(50, 51), 333(49), 335(49, 51b), 353, 356(49), 357(49), 358(51), 362(49), 363(49, 51), *367*

Penneck, R. J., 259(13), 265(13), 275(13), 303(13), 304(13), 310(13), 312(13), *313*

Penney, W. G., 376, 401(104), *416*

Peradejordi, F., 278(61), *314*, 402(48), 405(48), *415*

Perkin, W. H., Jr., 3(120), *87*

Perry, C. W., 18(33), 59(33), 64(33), 66(33), 76(33), *84*

Peters, D., 401(105), *416*

Petersen, D. R., 14(69), 19(69), 58(69), 75(69), 77(69), *85*

Petersen, L. I., 26(79), *86*, 159(11a), 161(11a), 163(11a), 164(11a), 165(11a), 166(11a), 167(11a), 168(11), 169(11), 170(11), 171(11a), 173(11), 176(11a), 177(11a), *179*

Petersen, R. J., 12(135), 13(135), 23(135), 45(135), 52(135), 75(135), *87*, 419, *422, 430*

Peterson, R. A., *442*

Petrov, A. D., 191(43a), *218*

Pettit, R., 2(148), *87*, 88(23, 64), *119, 121*, 180(27), *217*, 419(12), 420(17, 18), *422, 425, 431, 433, 434, 441, 448*

Pfaff, A., 319(44), 333(44), 350(44), *367*

Pfrommer, J. F., 23(121), *87*, 98(50), 99(50), 107(50), 116(50), *120*

Philipps, F. C., 89, *120*

Pickett, L. W., 286(40), *314*

Pilgram, K., 162(17), *179*

Platt, J. R., 401(106), *416*

Plummer, C. A. J., 274(55), *314*

Pogany, I. I., 11(9), 22(118), 50(9), 61(118), 72(118), *84*, *87*,
Pohl, R. J., 188(21), *217*, 220(5a), 221(5a), 222(5a), 223(5a), 224(5a), 225(5, 9), 226(5, 7a), 227(7a), 228(6), 229(5), 230 (5a), *230*, *231*, 233(2a, 2c), 234(2a, 2c), 235(2a, 2b, 2c), 236(2a, 2b), 237(2a, 2b), 238(2a, 2b, 2c), 239(2a, 2b, 2c), 240(2a, 2c), *241*, 365(22), *366*
Pohlke, R., 184(23), 186(23), 189(23), 190 (23), 198(23), 199(22), 202(22, 23), 204 (23), 207(22, 23), 208(23), *217*, 234(3), *241*, 325(23, 25), 328(25), 329(23), 332 (24), 334(25), 344(25), 348(25), 350(23), 352(23), 362(23), 363(23), *366*
Pohmer, L., 261(136), 263(136), 274(136), *316*
Pollock, D., 94(42), 96(42), 97(42), 98(42), 105(42), 111(42), 114(42), *120*, *440*
Poole, M. D., 248(9), *254*, 398(31), 399, *414*
Posner, J., *442*
Powell, D. L., 149(21), 154(21), 155(21), *156*
Preston, D. R., 268(23a), 269(23a), 270 (23a), 281(23), 284(23), 296(23), 298 (23a), 309(23a), 310(23a), *313*
Pritchard, H. O., 401(107), *416*
Pryde, W. J., 94(42), 96(42), 97(42), 98(42), 105(42), 111(42), 114(42), *120*, *440*
Pullman, B., 290(28), *313*, 335(3), 356(3), 358(3), *366*, 405(12a), 410, *414*
Pummerer, R., 319(44, 45), 333(44, 45), 350 (44), *367*
Purvis, J. E., 8(122), *87*

Q

Quincey, P. G., 288(42), *314*, 401(21), *414*

R

Ramp, E. E., 47(43), *85*
Ranganathan, S., 9(124), 31(123), *87*
Rapson, W. S., 256(105), 266(105a), 273 (105), 285(105a), *315*
Ratts, K. W., 196(24), 208(24), *217*, 325(27), 328(27), 330(27), 331(27), 332(27), 339 (27), 340(27), 344(26, 27), 345(27), 346 (26), 362(26, 27), *366*
Reavill, R. E., *460*
Rebane, T. K., 405(108), *416*

Rees, C. W., *458*
Regan, C. M., 158(18), 168(18), *179*, 192(51), 193(51), *218*, 370(111), 393(111), 398 (111), 405(111), *416*
Reggel, L., 319(42), 332(42), *367*
Reims, A. O., 5, 34(37), 70(37), *85*
Reppe, W., 39(125, 126), *87*, 89, 101, 106 (53b), *120*
Riegelbauer, G., 319(44, 45), 333(44, 45), 350(44), *367*
Riemschneider, R., 14(127), *87*
Robb, E. W., 18(33), 59(33), 64(33), 66(33), 76(33), *84*
Roberts, J. D., 2(128), 15(91), 16(91, 116), 17(133), 26(137), 27(3, 129), 28(116), 29(133), 57(91, 116), 58(116), 66(133), 72(133), 73(91, 116), *84*, *86*, *87*, 128(18), 129(15), 130(18), 131(8, 16, 18a), 132 (16), 133(8, 18), 135(18), 136(18), 137 (16, 18), 138(12, 18), 139(12), 140(16, 18), 141(16, 18), 142(18), 143(16, 18), 144(17, 18), 145(17), 146(17), 147(11, 18), 148(18), 149(12, 18b), 151(16, 18a), 152(18a), 153(16, 18), 155(15, 16, 18), *156*, 158(15, 18), 168(18), *179*, 192(51), 193(46, 49, 51), 198(50), *218*, 223(10), 225(8), *231*, 232(4), 235(4), *241*, 252 (13), *254*, 368(109), 370(111), 393(111), 394, 398(111), 405(111), *416*, 430(116), *430*
Roedig, A., 162(19, 20, 21), 164(20a), 165 (20a), 166(20a), 176(20a), *179*, *447*, *448*
Rogers, A. M., 418(2), *422*, *464*
Rogers, V., 268(23a), 269(23a), 270(23a), 281(23), 284(23), 296(23), 298(23a), 309 (23a), 310(23a), *313*, 405(9), *414*
Rose, J. C., 332(24), *366*
Rosenhauer, E., 319(44, 45), 333(44, 45), 350 (44), *367*
Ross, I. G., 381(102), *416*
Rossman, M. G., 191(28), *217*
Rowlands, J. R., 290(95), *315*, 402(60), 403 (60), *415*, *416*
Ruhemann, S., 8(130), 10(130), *87*
Rushbrooke, G. S., 369(32), *414*

S

Sabelli, N. L., 418(5), *422*
Sakan, T., *446*

Salem, L., 390, 391, *415*
Salfeld, J. C., *257*
Sandin, R. B., 258(50), *314*
Santarella, G., 89(45a), 95(45a), 105(45), 109(45), 114(45), *120*, 412(85), *416*
Santos Veiga, J. dos, 290, *314*, 404, 410, *414*, *461*
Sauer, J. C., 103(9, 55), *119*, *120*
Saunders, R. A., *424*, *433*
Sausen, G. N., 36(83), *86*
Scanlan, J., 383, 384, *416*
Scardiglia, F., 15(91), 16(91), 57(91), 73(91), *86*, 131(8), 133(8), *156*
Schafer, M. E., *424*, *461*
Scherr, C. W., *416*
Scheuchenpflug, D. R., 191(34, 35), *218*
Schiff, H., 6, *87*
Schlatler, M. J., 5, 34(37), 70(37), *85*
Schlichting, O., 39(126), *87*, 89(53), *120*
Schmädel, W. von, 3, 4, 14(156), 37(156), 47(156), 69(156), 71(156), *87*
Schönberg, A., 330(46), 358(46), *367*
Schönleber, D., 24(47), 75(47), *85*, *433*
Schomaker, V., 119(20), *119*, 278(124), *316*, 401(128), 412(44), *415*, *417*, *433*, *445*
Schommer, W., 242(10), 253(10), *254*
Schrauzer, G. N., 89, 92(60), 100(58, 59), 103(56), *120*
Schröder, G., 6, 23(53a), 24(54), 25(53), 52(54), 75(53), *85*, 89(17), 99(17), 103(14), 106(17), 107(17, 18), 109(17), 110(17), 111(14), 113(17a), 114(17), 116(17a), 118(17a), *119*, 412, *414*
Schubert, K., 261(90), *315*
Schüler, H., 262(106, 107), 271(106), 286(106a), *315*
Schumaker, E., 272(108), *315*
Schwager, I., 280(115), 289(114), *315*, 405(122), *416*
Schwechten, H.-W., 256(109, 110), 273(109, 110), *315*
Schweizer, H. R., 305(111), 306(111), *315*
Searle, R. J. G., 186(5), 198(5), 208(5), *217*, 268(14), 269(14), 270(14), 271(14), 292(14), 293(14), 295(14), 309(14), 310(14), 311(14), 312(14), *313*, 325(9), 328(9), 332(9), 334(9), 341(9), 345(9), 346(9), 347(9), 348(9), 362(9), *366*
Seebach, D., *429*
Seitz, A. H., *459*

Seka, R., 9(132), *87*
Selwood, P. W., 290(112), *315*
Sharkey, W. H., 160(26), 164(26), 176(26), *179*
Sharp, D. W. A., 92(7), 104(7), 114(7), 118(7), *119*, 436(7)
Sharts, C. M., 17(133), 29(133), 66(133), 72(133), *87*, 129(15), 131(16), 132(16), 137(16), 140(16), 141(16), 143(16), 151(16), 153(16), 155(15, 16), *156*, 198(50), *218*, 223(10), *231*
Shearer, H. M. M., 118(21), *119*, 412(45), *415*
Shechter, H., 9(124), 15(134), 31(114), 68(134), *86*, *87*, 158, 168(22), 169(24), 170(24), 171(24), 172(24), 173(24), 174(24), 175(22, 24), 177(24), 178(24), *179*
Sheppard, N., 42(7, 72), 43(6), *84*, *85*, 105(1, 28), *119*, *120*
Shida, S., 19(94), 71(94), *86*, 384, 389, *416*
Shuttleworth, R. G., 256(105), 266(105a), 273(105), 285(105a), *315*
Simmons, H. E., Jr., 27(129), *87*, 202(52), *218*, 385(115, 116), *416*
Simon, I., 290(6), *313*
Sisido, K., 184(53), 202(53), 207(53), 211(53), *218*
Skattebøl, L., 144(17), 145(17), 146(17), *156*
Skell, P. S., 12(135), 13(135), 23(135), 45(135), 52(135), 75(135), *87*, 419, *422*, *430*
Sklar, A. L., 378, *415*
Slater, J. C., 371, 373(117), *416*
Sliam, E., 11(5), 50(5), 52(5), 71(5), 77(5), 79(5), *84*, *424*
Small, G., Jr., 27(159), *87*
Smirnov-Zamkov, I. W., 14(136), *87*
Smith, J. P., 265(97), 310(97), 311(97), *315*
Smutny, E. J., 26(137), *87*, 128(18), 130(18), 131(18a), 133(18), 135(18), 136(18), 137(18), 138(18), 140(18), 141(18), 142(18), 143(18), 144(18), 147(18), 148(18), 149(18b), 151(18a), 152(18a), 153(18), 155(18), *156*, 252(13), *254*
Snyder, L. C., 390(119), 391, 392(118), *416*
Solly, R. K., *225*, *448*, *458*
Sondheimer, F., 268(137), *316*, *463*
Spokes, G. N., 261(27), *313*
Springall, H. D., 257(25), 288(25, 42), *313*, *314*, 401(10, 21), *414*

Srimany, S. K., 9(40), *85*
Staab, H. A., *225*
Stein, G., 258(45), *314*
Stein, R. P., 226(11), 228(11), 229(11), 230 (11), *231*, 233(6), 234(6), 237(6), *241*, *449*
Stern, R. L., 22(153), *87*
Sternberg, H. W., 101(65), *121*
Stiebler, P., 5, *87*
Stiles, R. M., 182(54), 186(54), 190(54), 198 (54), 209(54), 214(54), *218*, 261(27), *313*
Stobbe, H., 7, 8(139), 9(140), *87*
Stone, F. G. A., 25(105), 45(105), 59(105), 77(105), *86*, 109(43), 111(43), *120*
Strömer, R., 10, *87*
Streith, J., *428*
Streitwieser, A., Jr., 158(18), 168(18), *179*, 192(51), 193(51), *218*, 280(113, 115), 283(116), 289(114), *315*, 368(121), 370 (111), 393(111), 396(121), 397(121), 398 (111), 400(121), 401(121), 404(121), 405 (111), 406(121), 407(121), 408(121), 409 (121), *416*
Strübell, W., 276(117), *315*
Stucker, J. F., 186(25), 194(25), 195(25), 196(24), 205(25), 208(24), 211(25), *217*, 257(46), *314*, 321(28), 322(28, 29), 324 (28a), 325(27), 328(27, 29), 330(27), 331 (27, 29), 332(27), 335, 337(28), 339(27, 29), 340(27, 29), 344(27, 28, 29), 345 (27, 29), 346(29), 354(28), 355(28), 356 (28), 362(27, 28, 29), *366*
Sturkova, M. P., 89(61), *121, 440*
Sumner, F. H., 401(107), *416*
Suzuki, S., 283(116), *315*
Swan, G. A., 273(103), 274(103), *315*

T

Taber, H. W., 191(34), *218*
Takekiyo, S., 385, *416*
Taylor, R., 418, *422, 457*
Teller, E., 388, *415*
Temkin, O. N., 89(61), *121, 440*
Thatte, S. D., 187(12), 188(12), *217*, 268(31), 269(31), 270(31, 98), 271(31, 98), 275 (31, 98), 293(31), 294(31), 295(31, 98), 296(98), 298(31), 301(31), 303(31), 304 (31), 306(31), 307(31), 308(31), 309(31, 98), 310(31, 98), 311(31, 98), 312(31, 98), *313, 315*, 365(12), *366*
Thiele, J., 4, *87*
Thorpe, J. F., 70(19), *84*
Throndsen, H. P., *431, 440*
Tiers, G. V. D., 15(134), 68(134), *87*, 158, 169(24), 170(24), 171(24), 172(24), 173 (24), 174(24), 175(24), 177(24), 178(24), *179*
Tochtermann, W., 261(132), *316*
Toepel, T., 39(126), *87*, 89(53), *120*
Tokes, L., 272(84), *315*, 461(84)
Tolgyesi, W. S., 267(102), *315*
Triebs, A., *444*
Trotter, J., 278(92), 279, 285(92), *315*, 401, 415, 459(92), *459*
Trumbull, E. R., 47(43), *85*
Tsutsui, M., 19(143), 38(143), 58(143), 59 (143), 76(143), *87*, 92(63), 101(62), 103 (63), 104(63), *121*
Turkevich, J., 389, *415*

U

Uhlenbrock, W., 333(52), 334(32), *367*
Uhler, R. O., 158, 169(24), 170(24), 171(23, 24), 172(24), 173(23, 24), 174(23, 24), 175(24), 177(23, 24), 178(24), *179*
Unseld, W., 262(74), 263(74), 264(74), 291 (74), 310(74), *314*

V

Vallarino, L., 89(45a), 95(45a), 105(45), 109 (45), 114(45), *120*, 412(85), *416*
Van Meter, J. P., 202(26), *217*, 244(6, 7), 245 (6), 246(12), 248(6), 251(6), *254*, 329 (30), 350(30), 364(30), *366*, 397(22), 398 (22), *414*
Van Trappen, J. P., *461, 464*
Vargas Nunez, G., *462*
Veber, D. F., 178(10), *179*
Vellturo, A. F., 43(80), *86*, 169(12, 25), *179*
Veraguth, H., 181(55), 188(55), 189(55), *218*
Verdol, J. A., 163(5), *179*
Vetter, H., 101, *120*
Vierling, R. A., 131(3), 132(3), 140(3), 141 (3), 143(3), 144(3b), 147(3), 149(3), 151

(3), 153(3), 155(3), *156*, 252(14), *254*, *442*
Viswanath, G., 260(56), 275(56), 287(56), 288(56), *314*, 320(33), 321(33b), 322(33), 335(33), 337(33), 338(33a), 339(33), 354(33a), 356(33a), 357(33), 363(33), 364(33a), *366*
Vitale, T., 181(10), *217*
Vogel, E., 3, 69(145), *87*
Vol'pin, M. E., 3, *87*, 255, 307(119), *316*
Voris, D., 286(40), *314*

W

Waack, R., 122(10), *127*, 395(126), *416*, 419, *422*, *462*
Wailes, P. C., 100(34), *120*
Walker, D. F., 322(43), *367*
Walker, G., *414*, *455*
Wallenberger, T. F., 289, *316*
Walter, J., *415*
Walton, T. R., 37(86), *86*
Ward, E. R., 257(121), 259(121), 275(121, 122), 312(121), *316*, 321(49), 322(49), 323(47, 48, 49, 51), 324(47), 332(50, 51), 333(48, 49), 335(49), 353, 356(49), 357(49), 358(47, 51), 362(49), 363(47, 49, 51), 364(47), *367*
Warford, E. W. T., 405(43), *415*
Waser, E., 4(157), *87*
Waser, J., 278(123, 124), 285, 292(121), *316*, 401(127, 128), *416*, *417*
Wasserman, E. R., 9(147), *87*
Watson, H. R., 272(48), 284(48), *314*, 406(24), *414*
Watts, L., 2(148), *87*, 88(23, 64), *119*, *121*, 180(27), *217*, 419(11, 12), 420(17, 18), *422*, *431*, *434*, *441*, *448*
Watts, M. L., 259(13, 32), 265(13), 268(99), 271(99), 275(13), 303(13), 304(13, 32), 308(99), 309(99), 310(13, 32, 99), 312(13), *313*, *315*
Weber, D. F., 199(32), *217*
Weber, W., 273(38), *313*
Webster, B., 123(4), 124(3), 126(4), *127*, 140(5), *156*, *441*, *443*
Weinhold, P., 333(52), 334(52), *367*
Weinlich, J., *433*
Weiss, E., 89(32), 93(32), 100(32), 114(32), *120*, 412(58), *415*

Weltner, W., Jr., 69(149), *87*, 376, 392(129), *417*
Wendel, K., *425*
Wender, I., 101, *121*
West, P., 419(16), *422*, *462*
West, R., 132(20), 134(20), 149(21), 150(19, 20), 153(19), 154(6, 21), 155(21), 156(19), *156*
West, W. W., 285(125), *316*
Whaley, W. M., 258(2), *313*
Wheland, G. W., 369(132), 370(131), 371, 375, 376(133), 386, *417*
Whitaker, K. E., *454*
White, D. M., 27(32), 36(32), *84*
White, E. H., 22(151, 153), 34(151), 42(152), 48(151), 55(150, 151), 57(151), 60(151), 70(150), 72(151), *87*, *432*
Whiting, M. C., 100(34), *120*
Wilkinson, G., 92(7), 104(7, 66), 114(7), 118(7), *119*, *121*, 436(7)
Williams, A. E., 178(2), *179*, 183(11), *217*, 272(29), *313*, *424*, *433*
Williams, J. K., 160(26), 164(26), 176(26), *179*
Williams, R. O., 12(20), 13(20), 39(20), 52(20), 61(20), 75(20), *84*, *424*
Willstätter, R., 3, 4, 5, 14(156), 37(156), 47(155, 156), 69(156), 71(156), *87*, 181(55), 188(55), 189(55), *218*
Wilson, K. V., 333(1), 334(1), 335(1), *366*, 398(2), 401(2), 411(2), *414*
Winkler, H. J. S., 262(126), *316*
Wirth, W.-D., 23(55), 52(55), *85*
Wittig, G., 17(158), 37(158), 54(158), 55(158), 64(158), 69(158), 74(158), *87*, 261 (127, 128, 130, 132, 134, 136), 262(126, 131), 263(130, 136), 264(131), 266(130, 133), 273(135), 274(129, 136), 285(130, 131, 135), 287(291(131), 292(131), 293, 304(127), 311(131), 312(131a, 134), *316*, 333(52), 334(52), *367*, 433(158), *433*
Wolf, W., 272(84), *315*, 461(84)
Wolfsberg, M., 384, *417*
Wolovsky, R., 268(137), *316*, *463*
Wong, C.-H., *445*
Woods, W. G., 27(32), 36(32), *84*
Woodward, R. B., 27(159), 70(97), *86*, *87*, 421, *422*
Woolfolk, E. O., 319(42), 332(42), *367*

Wreden, F., 180, *218*
Wright, D., 273(103), 274(103), *315*
Wristers, H. J., 162(14), 164(14), 166(14), 176(14), *179*
Wuchter, R. B., *451*

Y

Young, A. E., 126(6), *127*, 396(50), *415*, *441*

Z

Zaitseva, M. G., 89(61), *121*, *440*
Zanker, F., 13(56), 63(56), 75(56), *85*, 429(56), *429*
Zeiss, H., 89, 92(63), 103(63), 104(63), *121*, *431*, *440*
Zingales, F., 89(45a), 95(45a), 105(45, 68), 109(45), 114(45), *120*, *121*, 412(85), *416*
Zollinger, H., 268(33), 270(33), 278(33), 281(33), 298(33), 309(33), *313*
Zschoch, F., 7(140), 9(140), 87

Subject Index

Compounds are indexed on the basis of parent compound first and substituent groups second. The latter are written in alphabetical order, e.g., "4,5-dinitro-1-fluoro-" and "6-bromo-2-chloro-" but not "1-fluoro-4,5-dinitro-" or "2-chloro-6-bromo-." Methylene-substituted four-membered carbocyclic compounds having only trigonally hybridized carbon atoms in the ring are treated as an exception to this rule and are entered in the index under "dimethylenecyclobutene," "methylenecyclobutanetrione," etc. (not under "cyclobutene," "cyclobutanetrione," etc.).

A question mark (?) within or following a compound name means that the assigned structure is in doubt; an asterisk (*) signifies an unsuccessful synthesis, an unsuccessful transformation, or an incorrect structure assignment in the original literature; and a dagger (†) indicates that an alternate naming or numbering system unsuitable for indexing purposes has been used in the body of the monograph.

Page numbers in boldface type refer to important or definitive passages, while those in italic refer to tabular matter, e.g., product yields and physical properties.

A

Acebenzylen, see Benzocyclobutadiene
Acetylene, 69
  from cyclobutadiene, 14
  metal complexes, 91–95
  dimerization of, 19
—, diphenyl-, 18
—, vinyl-, 14
Acrylic acid, β-(2-biphenylenyl)-, 271, 295, 296, 312
Acrylophenone, β-(2-biphenylenyl)-,271,312
AIM method, 382, 384, 391
  ASMO–CI method and, 384
Anthra[b]cyclobutadiene, 243
Anthraquinone, 23
  ASMO–CI method, 380, 385
  AIM method and, 384
Azulene, 1,2,3-triphenyl-, 18–20, 38, 39, **41**, 67, *76*, *82*

B

Benz[a]anthra-7,12-quinone, 23
Benzene, 23
—, hexaisopropyl-, 423
—, hexamethyl-, 13, 63, *75*, *80*, 424
—, hexaphenyl-, 18, 64, 91, 101, 104
5,10-(o-Benzeno)benzo[b]biphenylene, 5,5a,9b,10-tetrahydro-, 202, *211*

Benzobiphenylene, 317–367, see also Tetrahydrodibenzobiphenylene
  preparation of, 321–334
  stability, 335
Benzo[a]biphenylene, 317
  chemistry, 339–340
  addition reactions, 339, **344**, 346
  Friedel–Crafts reaction,* 332
  oxidation, 344
  photolysis, 337, 354
  reductive cleavage, 344
  delocalization energy, 407
  HMO constants, *407*
  preparation of, 332,* **339**, *362*
    from bromobenzo[a]biphenylene, *331*
    via benzo[a]biphenylenemagnesium iodide, 339
    from benzo[b]naphth[1,2-d]iodolium iodide, *322*, 339
    from benzo[b]naphtho[1,2-d]mercurole, *324*, 339
    from dihydrobenzo[a]biphenylene, 325, *328*, 339
    from iodophenyliodonaphthalene, *322*, 339
    via lithiobenzo[a]biphenylene, 339
  properties, 337
  half-wave potential, 289
  IR data, 338

Benzo[a]biphenylene—cont.
properties—cont.
NMR data, 461
UV data, 289, **337**, **347**, 354–356
trinitrofluorenone complex, 337, *362*
—, 5-acetyl-, 330
preparation of, *331*, 332,* 343–344
properties, *362*
UV data, *345*
—, 5-bromo-, 256, 320
chemistry, 325, *331*, 335, 343–344, **346**
preparation of, 194, *322*, **325**, *328*, 340
proof of structure, 341
properties, *211*, *362*
IR data, 344
NMR data, 461
UV data, *345*
—, 5-cyano-
chemistry, *331*, 343–344
preparation of, *331*, 343–344
properties, *362*
NMR data, 461
UV data, *345*
—, 5,6-dibromo-
bromination, 335, **346**
preparation of, 196, 197, 325, 330, *328*, 334,* **340–343**, *362*
properties, *362*
UV data, *345*
—, 6a,10b-dihydro-, 194, 202, *209*
—, 5-iodo-
chemistry, *331*, 343–344
preparation of, **195**, 325, *328*, *331*, 340
properties, *212*, *362*
NMR data, 461
UV data, *345*
—, 5-lithio-
chemistry, *331*, 343–344
preparation of, *331*, 343–344, *362*
properties, *362*
Benzo[a]biphenylene-5-carbonyl chloride
chemistry, *331*, 343–344
preparation of, *331*, 343–344
properties, *362*
Benzo[a]biphenylene-5-carboxamide, *331*, 343–344, *362*
Benzo[a]biphenylene-5-carboxylic acid
chemistry, *331*, 343–344
methyl ester, *331*, 343–344
NMR data, 461

Benzo[a]biphenylene-5-carboxylic acid—cont.
preparation of, *331*, 343–344
properties, 337, *362*
UV data, *345*
Benzo[a]biphenylene-5-magnesium bromide, *331*, 343–344
Benzo[a]biphenylene-5-magnesium iodide,† *362*
Benzo[b]biphenylene, 317, *see also* Diazabenzo[b]biphenylene *and* Tetrahydrodibenzo[b,h]biphenylene
chemistry, 348–353
addition reactions, 351
oxidation, 352
reductive cleavage, 351
delocalization energy, 406–407
HMO constants, *406*
preparation of, *332*,* *362*
from dibromobenzo[b]biphenylene, 330, *331*, 348
from dihydro- and tetrahydrobenzo[b]biphenylene, *328*, 348
properties, 347, *362*
half-wave potential, 289
IR data, 348
UV data, 347
trinitrofluorenone complex, 347
—, 5-bromo-
preparation of, *331*, 350
properties, *362*
UV data, 350
—, 5-bromo-10-lithio-, 36
chemistry, *331*
preparation of, *331*, 350
properties, *362*
—, 5a-bromo-5,10-diphenyl-5,10-oxido-5,5a,9b,10-tetrahydro-, 203–205, *211*
—, 5,10-dibromo-
chemistry, 330, *331*, 348, 351–352
preparation of, **196–197**, *212*, 325,* *328*, 330, *332*,* 348
properties, *362*
NMR data, 461
UV data, 350
—, 5a,9b-dibromo-5,10-diphenyl-5,10-oxido-5,5a,9b,10-tetrahydro-, *212*
—, 1,2-dihydro-, *270*, *292*, *310*, *328*
—, 5,10-dihydro-5,10-diphenyl-5,10-oxido-,* 189, 190

Benzo[b]biphenylene, 1-iodo-,—cont.
—, 5,10-dilithio-
 chemistry, 348, 350
 preparation of, 330, *331*, 350
 properties, *362*
—, 5,10-diphenyl-
 chemistry, 234, 350,* 352
 preparation of, 190, **325**, **328**, 350
 properties, *362*
 UV data, 351
—, 5,10-diphenyl-5a-iodo-5,10-oxido-5,5a,
  9b,10-tetrahydro-, 203–205, 212
—, 5,10-diphenyl-5,10-oxido-5,5a,9b,10-
  tetrahydro-, 201, *210*
—, 1-hydroxy-1,2,3,4-tetrahydro-, *270, 271*,
  311
—, 1-keto-1,2,3,4-tetrahydro-
 preparation of, 269, 293, *311*, 348
 properties
  NMR data, 461
  UV data, *295*
 reduction, 271
—, 1,4-oxido-1,4,4a,8b-tetrahydro-, 202, *211*
—, 1,2,3,4-tetrahydro-, *271, 312, 328*
 UV data, *292*
—, 5,5a,9b10,-tetrahydro-, 202, *210*
Benzo[b]biphenylene-10-carboxylic acid,
  5-bromo-, methyl ester, 461
 NMR data, 461
Benzo[b]biphenylene-5,10-dicarboxylic acid,
  330, *331*, 350, *362*
Benzocyclobutadiene, 180–218, see also
  Dicyclobuta[a,d]benzene, 1,2-Di-
  hydrobiphenylene, 5,10-Dihydro-
  5,10-oxidobenzo[b]biphenylene, 3,6-
  Dioxabenzocyclobutadiene, 1,4-
  Ethanobiphenylene-2,3-dicar-
  boxylic anhydride, Pentacyclo-
  [10.2.0.0$^{2, 11}$.0$^{4, 9}$.0$^{5, 8}$]tetradecapen-
  taene, Spiro{tricyclo[6.2.0.0$^{2, 7}$]deca-
  1(8), 2,4,6-tetraene-9,1'-cyclo-
  hexane}, 1,2,3,4-Tetrahydrobipheny-
  lene, Tetrahydro-5,10-oxidobenzo-
  [b]biphenylene, Tricyclobuta[a,c,e]-
  benzene, Tricyclo[6.2.0.0$^{2, 7}$]decatet-
  raene, Tricyclo[7.4.0.0$^{2, 8}$]trideca-
  2(8),3,6,9,11,13-hexaene, Tricyclo-
  [6.3.0.0$^{2, 7}$]undeca-1(8),2,4,6,9-pen-
  taene
 adducts, configuration of, 458

Benzocyclobutadiene—cont.
 bond fixation in, 396
 bond orders, *398*
 chemistry
  addition of halogen, 214
  adduct formation, **201–205**, 350, 364,
   457
  dimerization, **194–195**, 198–200, 434
  photolysis, 225
  polymerization, 183, 186, **193**, 194
 delocalization energy, 192, **396**
 free-valence index, 193
 generation of, 182–187, *206–207*
  from benzocyclobutene, 186, 194
  from bis(dimethylamino)-o-xylene
   dimethohydroxide,* 181
  from bromobenzocyclobutene, **185**
  from dibromobenzocyclobutene, 181–
   182, **183–184**
  from dibromo-o-xylene,* 181, 188
  from diiodobenzocyclobutene, **184**, 457
 as resonance contributor, 305
 stability, 219, 242, 248, 399
—, 1-bromo-
 chemistry, 192–216, *328*
 adduct formation, **203–204**, 457
 dimerization, 194, **195**, 328
 polymerization, 194
 generation of, *207*
  from bromobenzocyclobutene (?), 186,
   194
  from dibromobenzocyclobutene, 181–
   182, **185**, 457
—, 1,2-dibromo-
 adduct formation, 203–205
 dimerization, **196**, *328*, 330
 generation of, *208*
  from tetrabromo-o-xylene, **196–197**
  from tetrabromobenzocyclobutene, 184
—, 1-dimethylamino-,* 191
—, 1,2-dimethyl-4,4,5,5-tetracyano-3,4,5,6-
  tetrahydro-,* 26, 167
—, 1,2-diphenyl-,* 190
 delocalization energy, 193
—, 1,2-diphenyl-4,4,5,5-tetracyano-3,4,5,6-
  tetrahydro-,* 26, 167
—, 1-ethoxy-,* 190
—, 1-hydroxy-3,4,5,6-tetramethyl-,* 187
—, 1-iodo-
 adduct formation, 203,* 203–204

Benzocylobutadiene, 1-iodo-,—*cont.*
  dimerization, **195–196**, *328*
  generation of, from diiodobenzocyclobutene, **185**, 191,* *208*
—, 1-methyl-2-phenyl-
  dimerization, 198
  generation of, *208*
  via benzyne, 186
  from bromomethylphenylbenzocyclobutene, **185**, 185,* 192,* 447
  rearrangement, 214
—, 3-nitro-1-phenyl-,* 181
—, 1-phenyl-
  delocalization energy, 193
  dimerization, 198
  generation of, 186, *209*
  rearrangement, 214
—, 4,4,5,5-tetracyano-3,4,5,6-tetrahydro-,* 167
Benzocyclobutadiene-4,5-dicarboximide, 3,4,5,6-tetrahydro-*N*,1,2-triphenyl-,* 167
Benzocyclobutadiene-4,5-dicarboxylic acid
—, 3,6-dihydro-1,2-diphenyl-, dimethyl ester,* 167
—, 1,2-diphenyl-3,4,5,6-tetrahydro-,* anhydride, 167
Benzocyclobutadieneiron dicarbonyl triphenylphosphine, 434
Benzocyclobutadieneiron tricarbonyl, 434
1,2-Benzocyclobutadienequinone, 219–231, *see also* Methylenebenzocyclobutenone *and* Tetrahydronaphthalenequinone
  anion radical, ESR spectrum, 448
  chemistry, 225, 449
    addition reactions, 226–228
    condensation reactions, 229, 365
    dimerization,* 221
    ketal formation, 226
    oxidation, 225
    reaction with base, 226
      with phosphoranes, 235, 237*
    reduction, 226
    ring cleavage, 143
  delocalization energy, 225
  derivatives, 228,* 229, 229,* *230*
  history, 220
  oxime, in Beckmann rearrangement, 228–229

1,2-Benzocyclobutadienequinone—*cont.*
  preparation of, 220–222
    from benzocyclobutenediol dinitrate, 220
    from bis(trifluoroacetoxy)benzocyclobutenone, 222
    from dibromobenzocyclobutenone, 448
    from dihydrophthaloyl chloride,* 223
    from indanetrione, 458
    from tetrabromobenzocyclobutene, **221**
    from tetrakis(trifluoroacetoxy)benzocyclobutene, 221
  properties, 223–224
  stability, 225
—, 3,4,5,6-tetrachloro-, 448–449
—, 3,4,5,6-tetrahydro-,* 137
1,3-Benzocyclobutadienequinone, 219
1,5-Benzocyclobutadienequinone, 219
3,6-Benzocyclobutadienequinone, 219
4,5-Benzocyclobutadienequinone, 219
Benzo[3,4]cyclobuta[1,2-*b*]phenazine, 307, 365
—, 7,8-dimethoxy-, 308
—, 6,7,8,9-tetramethoxy-, 308, 452
  UV data, 453
Benzocyclobutene
—, 1,2-bis(triphenylphosphoranyl)-, 449
—, 1-[1′(1)-cyclohexenyl]-2-isopropylidenyl-,* 192
—, 1,2-diiodo-, *210*
Benzocyclobutene-1,2-dione, *see* 1,2-Benzocyclobutadienequinone
Benzo[*a,d*]dicyclobutene, *see* Dicyclobuta[*a,d*]benzene
Benzoic acid, 2,3,4,5-tetraphenyl-, methyl ester, 77, 83
Benzotricyclobutadiene, *see* Tricyclobuta[*a,c,e*]benzene
Benzynes
  polarization of, 262
  preparation of biphenylenes via, **260–264**, 458, 462
Bicyclobutadiene, 390
Bicyclo[6.4.0]dodeca-1,7-diene
—, 3-methylene-2,6,6,9,9,12,12-heptamethyl-, 18, 66, *74*, *79*
—, 3,3,6,6,9,9,12,12-octamethyl-, 18, 48, *74*, *79*

Bicyclo[3.2.0]hepta-5,7-diene-2,4-dione-3,3-dicarboxylic acid, 6,7-diphenyl-, diethyl ester,* 10
Bicyclo[2.2.0]hexa-2,5-diene, 2-phenyl-, 426
Bicyclo[2.2.0]hexa-2,5-diene-2-carboxylic acid, methyl ester,† 431
Bicyclo[2.2.0]hexa-2,5-diene-2,3-dicarboxylic acid, dimethyl ester, 431
Bicyclo[2.2.0]hexa-1,3,5-triene, 424
Bicyclo[2.2.0]hex-2-ene-5,6-dicarboxylic acid, dimethyl ester,† 432
—, 5,6-dibromo-1,2,3,4-tetramethyl-, anhydride, 13, 64, 75, 80, **429**
—, 5,6-dichloro-1,2,3,4-tetramethyl-, anhydride, 429
—, 5,6-diiodo-1,2,3,4-tetramethyl-, anhydride, 429
—, 1,2,3,4,5,6-hexamethyl-, anhydride, 13, 63, 75, 80, 429
—, 1,2,3,4-tetramethyl-, anhydride, 13, 63, 75, 80, 429
1,1'-Bicyclohexenyl
—, 2,6,6,3',3',6',6'-heptamethyl-3-methylene-, 18, 66
—, 3,3,6,6,3',3',6',6'-octamethyl-, 18, 74, 79
Bicyclo[4.3.0]nona-2,4,7-triene, 2,3,4,5-tetraphenyl-, 77, 83
Bicyclo[4.2.0]octa-2,7-diene, 1,2,3,4,5,6,8-heptamethyl-7-methylene-, 23, 75, 80, 106–107
Bicyclo[4.2.0]octa-3,7-diene, 1,3,4,5,6,7,8-heptamethyl-2-methylene-, 23, 75, 80, 106–107
Bicyclo[4.2.0]octa-1,3,5,7-tetraene, see Benzocyclobutadiene
1:2-Binaphthylene, see Dibenzo[a,i]biphenylene
2:3-Binaphthylene, see Dibenzo[b,h]biphenylene
Biphenyl, 2,2'-diiodo, 257–258
Biphenylene, 2, 180, 255, see also 1-Benzoyl-2-(2'-biphenylenyl)ethylene, Acrylic acid, Acrylophenone, Biphenylenequinone, Bisbiphenylenyl, Dihydrobenzobiphenylene, Glyoxylic acid, Octahydrodibenzobiphenylene, Tetrahydrobenzobiphenylene
anion radical, 419, 455
chemistry, 455, 462

Biphenylene—cont.
anion radical—cont.
ESR spectrum, 290, 403
UV data, 462
anisotropy, 290
bond fixation in, 460
bond lengths, 279, 286, 401, 459
cation radical, 272
ESR spectrum, 290, 403
chemistry, 278–285
acetoxylation, 268, 282
acylation, 280, **293**, **456**
addition reactions, 282–285, 454–455, 460
adduct formation, 284,* 459*
bromination, 268, 282,* 283, 296, **453–454**
carbene addition, 284,* 460
chlorination, 269, 296
cleavage, by bromine, 453–454
iodination, 269
mercuration, 268, 301
metallation, 283, 451, 455, 462
nitration, 268, 269, 282, 298, 454–455
nucleophilic substitution, 282
oxidation, 276, 284
ozonolysis, 284
phenylation, 418, 458–459
reaction with benzyne,* 459
with complexing agents, 284
reduction, 276, 282, 283*
reductive cleavage, 276, 278, 283
substitution reactions (rate factors), 458–459
sulfonation, 269
tritiodeprotonation, 418, 458–459
chromium tricarbonyl complex,* 274
co-distillation, 459
complexes, 284, **285**, 293, 406
π-cyclopentadienenickel complex,* 274
dianion, 419, 455
UV spectrum, 462
delocalization energy, 288, **401**
electron affinity, 405, 419
electron diffraction study, 278
electronic transitions, 402
electrophilic substitution in, 280–282
formation, rate of, from benzyne, 462
free-radical substitution, 282, 297*
free-valence indices, 279, 296, **405**

Biphenylene—cont.
in Friedel–Crafts reaction, 268, 269, **456**
history, 255–256
HMO constants, *404*
Hückel calculations, 290
iron tricarbonyl complex,* *274*
nickel carbonyl complex,* *274*
orienting effect, 280–282, 405, 456
platinum bis(triphenylphosphine) complex,* *274*
preparation of, 256–272, *273–275**
from aminobenzotriazole, 458
from benzene, 261–262
from benzenediazonium carboxylate, 261, *263*,* 459
from biphenyl, 272
from biphenylenemercury, 265–266
from biphenylene reduction products, 267
from biphenylbis(diazonium chloride),* 255
from biphenyliodoniumcarboxylate,† 261
from biphenylylbis(magnesium bromide),† 256, 266
from bis(iodophenyl)mercury, 261
from bromofluorobenzene, 261
from bromoiodobenzene, 261
from bromodibenzoiodolium iodide,† 258
from cyclododecatetraenediyne, 267–268, *268**
from cyclododecatetrayne, 463
from dibenzofuran,* 256
from dibenzoiodolium iodide,† 257
from dibenzothiadiazacycloheptatrienedioxide, 261
from dibromobenzene, *256*,* 261
from dibromobenzocyclooctatetraene, 453–454
from dibromobiphenyl, 257
from diiodobenzene, 260,* 261, 262, *461**
from diiodobiphenyl, 257
from *o*-dilithiobenzene, 262
from fluorobiphenyl, 271
from indanetrione, 458
from iodobiphenyl, 272,* *461**
from iodobiphenylene, 269

Biphenylene—cont.
preparation of—cont.
from molybdenum tricarbonyl complexes, 272
from octahydrobiphenylene, 267
from oxidotetrahydrobiphenylene,* 202, 267, 337
from purpurotanin,* 256, *274*
from phenylacetylene, 262
from *o*-phenylenemercury, 266
from phthaloyl peroxide, 261
from *m*-terphenyl, 272
from tetrabromotetrahydrobiphenylene, 267, 454
proof of structure, 276
properties, 285–290
electrochemical and electromagnetic data, 288–290
fluorescence, 455, **419**, 460
half-wave potential, 289
infrared data, 288
ionization potential, 405
magnetic susceptibility, 289, 405
NMR data, 289, **459**, **460**, 461
phosphorescence, **419**, 460
thermochemical data, 288
UV data, 238, 287, *292*, **293**, 301, 338, 401–402
reactivities, 335, **405**, 457
stability, 285, 335
strain energy, 276–277, 401
tetramethylenecyclobutane as resonance contributor to, 286–287
trinitrobenzene complex, 263
X-ray crystallographic analysis, 278, 285
—, 2-acetamido-
chemistry, *268*, *270*, 278,* 281, 294, **298**, 405
preparation of, *270*, 294, 298
properties, *309*
UV data, *300*
—, 2-acetamido-3-bromo-
chemistry, *270*, 284, **298**, 405
preparation of, *268*, 281, 298
properties, *309*
—, 2-acetamido-3-phenylazo-, *270*, 278, 309
—, 2-acetoxy-
chemistry, *268*, *271*, 301
preparation of, *268*, *275*,* 282, 294, *304**
properties, *309*

Biphenylene—*cont.*
—, 2-acetoxymercuri-
  chemistry, *270*, *271*, 301
  preparation of, *268*
  properties, *309*
—, 2-acetyl-
  acylation, *282*, 293, **456**
  in Bayer–Villiger reaction,* 294, 304
  condensation reactions, *270*, 296
  dinitrophenylhydrazone, *309*
    UV data, *295*
  oxidation, *270*, 294
  oxime, *270*, 294, 298, *309*
    UV data, *295*
  preparation of, *268*, 275,* 280, **293**, 456
  properties, *309*
    UV data, *295*, **297**, 301
  reductive cleavage, 278
  in Schmidt reaction, 298
—, 2-acetyl-3-hydroxy-, *268*, *309*
—, 2-acetyl-2-hydroxy-7-methoxy-, *268*, *309*
—, 2-acetyl-6-propionyl-, 456
—, 2-amino-
  coupling reaction, with diazonium salts, *268*, 298
  diazotization of, 298
  hydrochloride, *300*
    UV data, 301
  orienting effect, 405
  preparation of, *270*, 275,* 298
  properties, *309*
    UV data, **297**, *300*, 301
  in Sandemeyer reaction,* 294, 301
—, 2-amino-3-bromo-
  chemistry, *270*, 298
  preparation of, *270*, 298
  properties, *309*
—, 2-amino-3-hydroxy-, *271*, *309*
  sulfate dihydrate, *270*
—, 2-amino-3-phenylazo-, *268*, *270*, 298, *309*
—, 2-benzamido-, *270*, 298, *309*
—, 2-benzoyl-
  chemistry, *270*, 294, 304
  oxime, *270*, 298
  preparation of, 268, *270*, 293
  properties, *309*
    UV data, *295*
—, 2-benzoyloxy-
  chemistry, *271*, 301
  preparation of, *270*, 294

Biphenylene, 2-benzoyloxy-,—*cont.*
  properties, *309*
    UV data, *303*
—, 2-(2′-biphenylenyl)-, *268*, *270*, 282, *309*
    UV data, *292*
—, 2,6(or 7)-bis(iodomethyl)-,† *269*, 271, 291, *310*
—, bromo-, preparation of,* 275
—, 2-bromo-
  preparation of, 282*
    from acetoxymercuribiphenylene, *270*, 301
    from aminobromobiphenylene, 298
    from biphenylene, *268*, 296, 297,* 453–454
    from bromodibenzoiodolium iodide, † 258, 259
  properties, *309*
    UV data, *298*
—, 2-bromo-3-diazonium chloride, 298
—, 1-bromo-2,3,4,5,6,7-hexamethoxy-, † 452, 453
    UV data, 453
—, 2-(β-carboxypropionyl)-
  chemistry, *270*, *271*
  methyl ester, 268, *270*, 293, 348
  preparation of, *268*, *270*, 293
  properties, *309*
    UV data, *295*
—, 2-(γ-carboxypropyl)-, see Butyric acid
—, 2-chloro-
  preparation of, *275**
    from 2-acetoxymercuribiphenylene, *270*, 296, **301**
    from biphenylene, *269*, 296, 296*
  properties, *309*
    UV data, *298*
—, 2-(p-chlorophenylazo)-3-hydroxy-, *268*, 304, *309*
—, 2-cinnamoyl-, *270*, 296, *310*
    UV data, *295*
—, 2-cyano-, *270*, *310*
    UV data, *300*
—, 2,6-diacetamido-
  chemistry, *269*, *270*, 298
  preparation of, *270*, 298
  properties, *310*
—, 2,6-diacetamido-3-bromo-, *269*, 298, *310*
—, 2,6-diacetamido-3,7-dibromo-, *269*, 298, *310*

SUBJECT INDEX 487

Biphenylene—*cont*.
—, 1,3-diacetoxy-1,2-dihydro-2-keto-,* 188, 307
—, 2,3-diacetoxy-1,6,7,8-tetramethoxy-, 452–453
UV data, 453
—, 2,7-diacetoxy-1,3,6,8-tetramethoxy-, 452–453
UV data, 453
—, 2,6-diacetyl-
chemistry, *270*, 278, **298**, 456*
preparation of, *269*, 293, 456
properties, *310*
UV data, *295*
—, 2,7-diacetyl-, 456
—, 2,6-diamino-
chemistry, 298
preparation of, *270*, 298
properties, *310*
—, dibromo-,* 275
—, dichloro-, *269*, 296
—, 2,3-dihydroxy-, *270*, *271*, 304, *310*
—, 2,7-dihydroxy-, *275*,* 304*
—, 2,3-dihydroxy-4,5,6,7-tetramethoxy-, † 452
UV data, *453*
—, 2,7-dihydroxy-1,3,6,8-tetramethoxy-, 453
—, 1,2-di(iodomethyl)-, *see* 1,2-Bis-(iodomethyl)biphenylene
—, 1,5(or 8)-diisopropyl-4,8(or 5)-dimethyl-,* 272
—, 1,8-dimethoxy-, *259*, 304, *310*
UV data, *303*
—, 2,3-dimethoxy-, *259*, *271*, 304, *310*
UV data, *303*
—, 2,7-dimethoxy-
chemistry, *268*, 304
preparation of, *259*, *275*,* 304
properties, *310*
UV data, *303*
—, 3,7(or 6)-dimethoxy-1,5(or 8)-dinitro-, *265*, *310*
—, 2,7-dimethoxy-1,3,6,8-tetramethyl-, *259*, 304, *310*
UV data, *303*
—, 1,5-dimethyl-, 291
—, 1,5(or 8)-dimethyl-, 262, 263, ***264***, *310*
—, 1,8-dimethyl-, *259*, 291, 293, *311*
UV data, *292*

Biphenylene—*cont*.
—, 2,6-dimethyl-, 291
ion radicals, EPR spectra, 461
preparation of, *264*, *275*,* 458, 462
properties, 291, *311*
UV data, *292*
—, 2,7-dimethyl-, 262
preparation of, *259*, 262, 277, 462
properties, 291, 293, *311*
UV data, *292*
—, 3,7(or 6)-dimethyl-1,5(or 8)-dinitro-, 264–265, 265,* *311*
UV data, *300*
—, 1,5(or 8)-dimethyl-1,2,3,4,5,6,7,8-octahydro-,* 37
—, 1,8-dimethyl-3,4,5,6-tetramethoxy-, *311*
UV data, *303*
—, dinitro-,* 275
—, 2,6-dinitro-, *259*, *269*, 298, *311*
UV data, *300*
—, 2,6-dipropionamido-, 456
—, 2,6-dipropionyl-, 456
—, 2,7-dipropionyl-, 456
—, 1-formyl-,† 452
—, 2-formyl- †
condensation reactions, *271*
preparation of, *271*, *275*,* 294, 296
properties, *311*
UV data, *295*
—, 1,2,3,6,7,8-hexamethoxy-
chemistry, *271*, 308, 453
preparation of, *259*, 260, **452**
properties, *311*
UV data, *303*
—, 1,2,4,5,7,8-hexamethoxy-, *259*, *311*
UV data, *303*
—, 1,2(or 3),4,5,6,8-hexamethyl-, 262, *264*
properties, 291, *311*
UV data, *292*, 293
—, 1-hydroxy-, 452
—, 2-hydroxy-
chemistry, *269*, *271*, 281, 304, 405
preparation of, *271*, *275*,* 294, 301, 304
properties, *311*
UV data, 297, *303*
—, 2-hydroxy-3-phenylazo-, *269*, 304, *311*
—, 2-hydroxy-3-(*p*-sulfophenyl)azo-, *269*, *270*, 304, *311*
—, 2-iodo-
chemistry, *269*, *270*

Biphenylene, 2-iodo-,—*cont.*
 preparation of, *269, 271,* 297,* 301
 properties, *311*
  UV data, *298*
—, 1-lithio-, 451–452
—, 2-lithio-, 451–452
—, 1-methoxy-, *259, 311*
  UV data, *303*
—, 2-methoxy-
  chemistry, *271,* 301
  preparation, *259,* 275*
  properties, *311*
  UV data, *303*
—, 1-methyl-, *259,* 291, *311*
  UV data, *292*
—, 2-methyl-, *259,* 291, *311*
  UV data, *292*
—, 2-nitro-,
  chemistry, *270,* 282
  preparation of, *259,* 261, *269, 275,** 298, 454–455
  properties, *312*
  UV data, *300*
—, 1,2,3,4,5,6,7,8-octahydro-
  dimer, 18, **54,** *74, 78,* 432–433
  generation of, **17,** 36,* 37,* 54, 64, *74*
  reaction with cyclohexyne, 17, 64, **459–460**
—, 1,2,3,4,5,6,7,8-octahydro-1,1,4,4,5,5,8,8-octamethyl-,
  generation of, **17,** 48, *74*
  rearrangement of, 66
—, 1,2,3,4,5,6,7,8-octamethoxy-, *259, 312*
  UV data, *303*
—, 1-(*n*-pentanoyl)-,† 451
—, 2-(*n*-pentanoyl)-,† 451–452
—, 1-phenyl-, 452, 458–459
—, 2-phenyl-, 452, 458–459
—, 2-propionyl-, 456
—, 1,4,5,8-tetraacetoxy-2,3,6,7-tetra-methyl-,* 275
—, 1,2,3,4-tetrabromo-1,2,3,4-tetrahydro-,* 283, 453–454
—, 1,2,3,4-tetrachloro-, 457, 457*
  UV data, 458
—, 2,2,3,3-tetracyano-1,2,3,4-tetrahydro-,* 188, 239–240
—, 1,2,3,4-tetrahydro-,* 192
—, 1,4,4a,8b-tetrahydro-,* 202
—, 1,4,5,8-tetrahydroxy-2,3,6,7-tetra-methyl-,* 267, 336

Biphenylene—*cont.*
—, 2,3,6,7-tetramethoxy-
  chemistry, *271,* 308
  preparation of, *259, 265, 275,** 304, 304*
  properties, *312*
—, 2,3,6,7-tetramethyl-
  preparation of, *269, 271*
  properties, 291, *312*
  UV data, *292*
—, tetranitro-,* 275
—, 1,2,3-triacetoxy-,* 275
—, trichloro-,* 457–458
—, 2,3,6-trimethoxy-, 461
Biphenylene-d_1, 457, 460
Biphenylene-t_1, 457
2,3,6,7-Biphenylenebis(quinone),* 178, 305
Biphenylenebis(molybdenum tricarbonyl),† 284
Biphenylene-1-carboxylic acid, methyl ester,† 451, 461
  NMR data, 461
Biphenylene-2-carboxylic acid methyl ester,† 451
  NMR data, 461
  preparation of, *270,* 294
  properties, *309*
  UV data, *295*
  reductive cleavage, 278
Biphenylene-2-diazonium chloride,* 275
Biphenylene-1,4-dicarboxylic acid, 2,2,3,3-tetracyano-1,2,3,4-tetrahydro-, dimethyl ester,* 240
2,6(?)-Benzenedisulfonic acid, *269,* 301, *311*
Biphenyleneiodonium iodide, *see* Dibenzo-iodolium iodide
Biphenylenemolybdenum tricarbonyl,† 284
1,2-, 1,4-, 1,6-, and 1,8-Biphenylenequinone, 305
  delocalization energy, 306
  stability, 304–306
2,3-Biphenylenequinone
  chemistry
    acetylation, 187, **307**
    bromination, *269,* 308
    condensation with *o*-phenylenediamine, 307
    reaction with maleic anhydride,* 188
    reduction, *270,* 304
    stability, 304–306
    substitution reactions, 306

2,3-Biphenylenequinone—*cont.*
  preparation of, *271, 275,** 304, 306
  properties, *312*
  UV data, 308
—, 1-bromo-2,3,4,5,6,7-hexamethoxy-, 453
—, 1-bromo-4,5,6,7-tetramethoxy,† 452–453
—, 1,4-dibromo-
  acetylation,* 308
  preparation of, *269*, 308
  properties, 310
  UV data, 308
—, 6,7-dimethoxy-, *271*, 308, *310*
—, 1,6,7,8-hexamethoxy-, *259*, 260
—, 1,6,7,8-tetramethoxy-†
  chemistry, 308, 452–453
  oxime, 236, 308, 453
  preparation of, *271*, 308, 452–453
  properties, *312*
  UV data, 453
2,7-Biphenylenequinone, 305
  delocalization energy, 306
—, 1,3,6,8-tetramethoxy-, 305, 452–453
  UV data, *453*
β-(2-Biphenylenyl)acrylic acid, *see* Acrylic acid
2,2′-Bis(biphenylenyl), 298
Bis(cyclobutadiene)-metal complexes, MO's, 413
Bis(cyclobutadiene)molybdenum dicarbonyl, octaphenyl-, 436–437
Bis(cyclobutadienemolybdenum carbonyl)-(diphenylacetylene), 437
Bis(cyclobutadiene)nickel, 1,2,3,4,1′,2′,3′,4′-octaphenyl-, 91, 103
Bond orders
  bond lengths and, 390
  in condensed cyclobutadiene aromatic systems, *398*
Bonds, bent, 376
Butadiene, 14, 21, 22, 23, 47
—, 1,4-dichloro-1,2,3,4-tetraphenyl-, 25, *76, 83*, 107, 427
—, 1,4-diethoxy-1,2,3,4-tetraphenyl-, 427
—, 1,3-diphenyl-, 21, 48
—, 1,2,3,4-tetramethyl-, 23, 106
—, 1,2,3,4-tetraphenyl-, 15, 19, 68, *76, 82*
—, 1,2,3,4-tetraphenyl-1-thiophenoxy-, *76, 83*

8b,12b-Butanobenzo[3′,4′]cyclobuta-[1′,2′:3,4]cyclobuta[1,2-*e*]biphenylene, 1,2,3,4,5,6,7,8,9,10,11,12-dodecahydro-, *see* Octahydrobiphenylene dimer
Butyric acid, γ-(2-biphenylenyl)-,† *271, 292, 293, 312*, 348

C

Charge distribution in cyclobutadiene, 381
Compound of unknown structure
  $C_{28}H_{18}N_4$, 144
  $C_{28}H_{20}$, 41
  $C_{28}H_{20-22}$, 430
  $C_{28}H_{20}N_4$, 144
  $(C_{28}H_{20}Ni)_x$, 104
  $C_{28}H_{22}$, 19, 38
Configuration (electronic) of cyclobutadiene, 382
  closed shell, 377, 379
  effect of bond lengths on, 384
  excited state, 380, 385
  ground state, 369, 377
  open shell, 377, 378, 379
  wavefunctions, 380
Configuration energies of cyclobutadiene, 384
Configuration interaction in cyclobutadiene, 378–380, 382, 384, 388
  first order type, 382, 385
  Jahn–Teller effect and, 389
  nonpolar structures and, 387
Craig's rule, 375, 382
Cross-ring resonance interaction
  in cyclobutadienequinone, 140
  in dimethylenecyclobutenes, 163
  in 3-hydroxy-4-phenylcyclobutadiene, 148
Cubane, octaphenyl-, *see* Octaphenylcyclooctatetraene
Cycloaddition
  of benzocyclobutadiene, 198
  of cyclobutadiene, 60
Cyclobutadiene, **1–87**, 180, 219, 255, 423–433, *see also* Bicycloheptadienedione, Bicyclo[2.2.0]hexatriene, Cyclobutadiene–metal complexes, Cyclobutadiindene, Cyclobutadiindenequinone, Cyclobuta[*a*]indenone, Cyclobutadienyl, Dihydro-

Cyclobutadiene—cont.
  benzocyclobutadiene, Dioxabenzobenzocyclobutadiene, Octahydrobiphenylene, Tetrahydrobenzocyclobutadiene, Tricyclo[5.3.0.0.^{2,6}]-decadiene, Tricyclo[8.6.0.0^{2,9}]-hexadecadiene, Tricyclo[7.5.0.0.^{2,8}]-tetradecadiene
  adducts, 71, 426, 431
  cation radical, 428
  chemistry, 49–70
    adduct formation, stereospecificity of, 431–432
    addition to aromatic compounds, 61
      to dienes, 60, 420
      to dienophiles, 59–61, 63, 420
    cycloaddition, 50, 60
    dimerization, 11, 12, 14, **50**, 71, 420
      stereospecificity of, 432
    fissioning, 69
    hydrogen abstraction, 14, 21, **47**, 68
    polymerization, 70
    reaction with transition metal compounds, 100
  rearrangement, 66
    ring cleavage, 4,* 14, **392**
  dianion,* 44, **122**, 123, 395, 396, 441
  dication, 44, **122**, 123, 139, 443
  dimer, 50, 71, 77
  distortion, 391–392
  erroneous structure assignments, 6–10
  electron density in, 375
  π-electronic states, 369
  energy levels, 370
  generation of, 2, **11–25**, 26–43,* 71
    from acetoxycyclobutene,* 34
    from acetylene, 19, 36
    from 1,3-bis(dimethylamino)cyclobutane dimethiodide,* 21
    from 1,2-bis(dimethylamino)cyclobutane dimethobromide,* 34
    from 1,2-bis(dimethylamino)cyclobutane dimethohydroxide,* 5, 34
    from 1,3-bis(dimethylamino)cyclobutane dimethohydroxide,* 5, 21, 47
    from cyclobutadieneiron tricarbonyl, 420, 425–426, **431–432**
    from cyclobutanetetracarboxylic anhydride,* 43
    from cyclobutene,* 70

Cyclobutadiene—cont.
  generation of—cont.
    from cyclobutenyltrimethylammonium hydroxide,* 5
    from diacetoxytetrahydroethenocyclobuta[b]naphthalene, 23
    from dibromocyclobutane, 3, 14,* 32,* 47
    from dicarbomethoxytricyclo[4.2-.2.0^{2,5}]deca-3,7,9-triene,* 39
    from dichlorocyclobutene, 11, 12
    from diethenooctahydrodicyclobuteno-[b,i]anthraquinone,* 39
    from dimethoxytetrahydroethenocyclobuta[b]naphthalene, 23, 47
    from dimethyl tricyclo[4.2.2.0^{2,5}]-decatrienedicarboxylate, 22
    from ethenotetrahydrocyclobuteno[b]-anthraquinone,* 39
    from methylenecyclobutyltrimethylammonium hydroxide,* 5
    from silver nitrate complex,* 43
    from tetrabromocyclobutane, 11, 12
    from tetrahydroethenocyclobuta[b]anthraquinone, 47
    from tricyclo[4.2.2.0^{2,5}]decatriene, 23
  ground state, 375, 419
  history, 3–10
  Hückel description, **369**, 371
    delocalization energy and, 386
  mercury complex,* 104
  nickel carbonyl complex,* 103
  nickel complex,* 105
  rectangular form, 391
    ground state, 385
  resonance contributors, 371
  ring strain, 391
  silver complexes,* 42, 105
  singlet ground state, 420, 431–432
  square form, 391
    ground state, 385
    triplet stabilization and, 389
  stability, 2, 248, 425
  theory, 368–416
  triplet ground state, **43–49**, 50, 60, **369**, 374, 376, 418, 430, 438
  valence tautomerization,* 46, 69
  VB description, 371
  ylids, 426–427
—, 1-acetoxy-2,4-dichloro-3-phenyl-,* 27

Cyclobutadiene—cont.
—, 1-amino-2-cyano-, energy levels, 394
—, 1-benzoyl-2-dimethylamino-3-nitro-4-phenyl-,* 425
—, 1,2-bis(trimethylsilyl)-3,4-diphenyl-, cobalt carbonyl complex, 100
—, 1-bromo-2-dimethylamino-3-nitro-4-phenyl,* 30
—, 1-(1'-cyclohexenyl)-2,4-dichloro-3-fluoro-, 17, 66, 72
—, 3-(1'-cyclohexenyl)-1,2-dihydroxy-,* 141
—, 1-chloro-2-dimethylamino-3-nitro-4-phenyl-,* 29
—, 1-chloro-2-fluoro-3-nitro-4-phenyl-,* 29
—, 1-chloro-2-fluoro-4-phenyl-,* 29
—, 1-chloro-2-fluoro-4-vinyl-,* 29
—, 1-chloro-2-hydroxy-3-nitro-4-phenyl-,* 28
—, 1-(p-chlorophenyl)-2,4-diphenyl-3-fluoro-, 430
  dimer, 16, 58, 73, 78
  generation of, 16, 58, 73
—, 1-cyano-2-dimethylamino-3,4-diphenyl-,* 42
—, 1,2-diamino-, energy levels, 394
—, 1,3-diamino-, energy levels, 394
—, 1,3-dichloro-2-hydroxy-4-phenyl-,* 27
—, 1,2-dicyano-, energy levels, 394
—, 1,3-dicyano-, energy levels, 394
—, 1,2-difluoro-3,4-diphenyl-,* 33
—, 1,3-dihydroxy-2,4-dimethyl-,* 27
  dication,* 124
—, 1,2-dihydroxy-3,4-diphenyl-,* 141
—, 1,3-dihydroxy-2,4-diphenyl-, 123, 126
—, 1,2-dihydroxy-3-phenyl-,* 26
—, 1,2-dimethyl-, nickel carbonyl complex,* 102–103
—, 1-dimethylamino-2-hydroxy-4-nitro-3-phenyl-,* 425
—, 1-dimethylamino-2-nitro-3-phenyl-,* 425
—, 1-(p-dimethylaminophenyl)-2-(p-nitrophenyl)-,* 394
—, dimethyldiphenyl-,* 431
—, 1,3-dinitro-2,4-diphenyl-,* 31
—, 1,2-diphenyl-,* 42, 70, 100, 441–442
—, 1,3-diphenyl, 9,* 10,* 432
  dimerization, 21, 55, 72, 78
  generation of, 21, 34,* 42,* 48, 55, 70,* 72
  hydrogen abstraction, 48, 68

Cyclobutadiene—cont.
—, 1,2-divinyl-, ground state, 393
—, 1,3-divinyl-, ground state, 393
—, 1-fluoro-2,3,4-triphenyl-, 430
  dimerization, 15, **57**
  generation of, **15**, 28,* **57**, *73*
—, 1-formyl-2-hydroxy-,* 6
—, 1-hydroxy-,* 6
—, 1-phenyl-, ground state, 393
—, 1,2,3,4-tetrafluoro-,* 33, 37
—, 1,2,3,4-tetraisopropyl-, 423
—, 1,2,3,4-tetraisopropyl-, 423
—, 1,2,3,4-tetrakis(benzhydryl)-, tetranion, 446
—, 1,2,3,4-tetrakis(trifluoromethyl)-,* 36
—, 1,2,3,4-tetramethyl-, 107, 286, 419, 424, 428
  adduct formation, 13, **63**, 64, 424, 429
  (π-cyclopentadiene)nickel complex, 105
  dianion,* 127
  dication,* 124, 396
  dimerization, 12, 13, 23, **52**, *74*, *79*, 419, 430
  dismutation, 430
  generation of, *74*, 428,* 431*
    from dibromotetramethylcyclobutene, 13
    from dichlorotetramethylcyclobutene, 12, 32, 44, 61, 99
    from diiodotetramethylcyclobutene, 13, 32,* **63**, 429
    from dimethyl octamethyltricyclo-[4.2.2.0^{2,5}]decatrienedicarboxylate,* 39
    from tetramethylcyclobutadienenickel chloride, 23, 25
    from tetramethylcyclobutanediol,* 42
    from tetramethylcyclobutanedione ditosylhydrazone,* 42
  hydroxylation of, 25
  reaction with nickel tetracarbonyl, 99
    with triplet methylene, **44**, 419
  triplet character, 44, 430
—, 1,2,3,4-tetraphenyl-, 100, 107, 429, 437–438
  adduct formation, 20, **64**, 65, 66, 428
  butadienyl valence tautomer, 46
  and compound $C_{28}H_{20}$, m.p. 154°–155°, 41
  copper complex,* 104

Cyclobutadiene, tetraphenyl-,—*cont.*
dication, 123, 125, 396, 441
dimer, 14, 18–20, 25, 42, *58*, *75*, *76*, *81*, 97,* 424, 430, 431
ESR data,* 428
generation of, 75
from bromoiodotetraphenylbutadiene, 19, 38
from bromotetraphenylbutadienyldimethyltin bromide, 19, 45, 64, 67
from dibromotetraphenylbutadiene,* 38
from dibromotetraphenylcyclobutene, 14, 58
from dichlorotetraphenylcyclobutane,* 15, 68, 69
from diiodotetraphenylbutadiene, 19
from dilithiotetraphenylbutadiene, 19
from diphenylacetylene, 18, 66
from iodotetraphenylbutadienyldimethyltin iodide, 20, 67
from octaphenylcyclooctatetraene,* 38
from tetraphenylcyclobutadienepalladium bromide, 25, 109
chloride, 45, 66, 427, 438
from tetraphenylcyclobutane,* 32
from tetraphenylmercurole, 25
from tetraphenylthiophene dioxide,* 430
from triphenylcyclopropenyl bromide,* 39
magnesium complex,* 101
mercury complex,* 101
monomer, 424
reaction with air, 46, 109, 427
with thiophenol, 46
rearrangement, 19, 67
triplet ground state, 430
—, 1-vinyl-, ground state, 393
Cyclobutadiene(π-benzene)cobalt bromide, 1,2,3,4-tetraphenyl-, 438
Cyclobutadienebis(π-cyclopentadiene)nickel, 1,2,3,4-tetramethyl-,† 101
Cyclobutadienebis(quinone),* 150, 157, 173, *see also* Dimethylenecyclobutanedione, Methylenecyclobutanetrione, Trimethylenecyclobutanone
Cyclobutadienecarboxylic acid, 3
—, 2-acetoxy-4-(*p*-acetoxyphenyl)-,* 10
—, 2-acetoxy-4-(*p*-anisyl)-,* 10

Cyclobutadienecarboxylic acid—*cont.*
—, 2-acetoxy-4-(3'-methyl-4'-methoxyphenyl)-,* 10
—, 2-acetyl-3-methyl-4-phenyl-,* 8
—, 2-chloro-4-(*p*-anisyl)-,* 10
—, 2-chloro-4-(*p*-hydroxyphenyl)-,* 10
—, 2-chloro-4-(3'-methyl-4'-methoxyphenyl)-,* 10
Cyclobutadienecobalt carbonyl triphenylphosphine bromide, 1,2,3,4-tetraphenyl-, 438
Cyclobutadienecobalt dicarbonyl bromide, 1,2,3,4-tetraphenyl-, 438, 439
Cyclobutadienecobalt mercury carbonyl
—, 1,2(or 3)-di(*tert*-butyl)- (?), 95
—, 1,2(or 3)-diphenyl- (?), 95
Cyclobutadiene(π-cyclopentadiene)cobalt
—, 1,2,3,4-tetrakis(trifluoromethyl)-,* 104, 436
—, 1,2,3,4-tetraphenyl-, *114*
chemistry, 106, 110
preparation of, 92, 97, 439
UV data, *116*
Cyclobutadiene(π-cyclopentadiene)molybdenum carbonyl bromide, 1,2,3,4-tetraphenyl-,† 438
Cyclobutadiene(π-cyclopentadiene)molybdenum carbonyl chloride, 1,2,3,4-tetraphenyl-,† 438
Cyclobutadiene(π-cyclopentadiene)nickel bromide, 1,2,3,4-tetraphenyl-, 439
Cyclobutadiene(π-cyclopentadiene)nickel tetrabromoferrate, 1,2,3,4-tetraphenyl-, 439
Cyclobutadiene(π-cyclopentadiene)palladium bromide, 1,2,3,4-tetraphenyl-, 439
Cyclobutadiene(π-cyclopentadiene)palladium tetrabromoferrate, 1,2,3,4-tetraphenyl-, 439
Cyclobutadiene(π-cyclopentadiene)tungsten carbonyl chloride, 1,2,3,4-tetraphenyl-, 438
Cyclobutadiene(π-cyclopentadienone)molybdenum carbonyl, 1,2,3,4,2',3',4',5' octaphenyl-, 438
Cyclobutadiene-1,2-dicarboxylic acid,* 3
—, 3,4-diphenyl-, 7,8
Cyclobutadiene-1,3-dicarboxylic acid,* 4, 31
—, 2,4-diphenyl-,* 7, 8

Cyclobutadiene(π-diphenylacetylene)molybdenum carbonyl, 1,2,3,4-tetraphenyl-, 436, 437
Cyclobutadieneiron tricarbonyl
  electrophilic substitution reactions, 419, 434–435
  NMR data, 434
  oxidation, 425, 434
  preparation of, 433–435
—, 1-acetyl-, 435
—, 1-benzoyl-, 435
—, 1-chloromercuri-, 435
—, 1-chloromethyl-,† 435
—, 1-dimethylaminomethyl-, 435
—, 1-formyl-, 435
  reaction with methylmagnesium bromide, 435
  reduction, 435
—, 1-hydroxymethyl-, 435
—, 1,2,3,4-tetrakis(p-chlorophenyl)-,† 93, 97, *114*
  UV data, *116*
—, 1,2,3,4-tetraphenyl-, 89, 106
  preparation of, 104*
    from dilithiotetraphenylbutadiene, 100
    from diphenylacetylene, 92, 93, 104, 104*
    from tetraphenylcyclobutadienepalladium bromide and corresponding chloride, 97
  properties, *114*
    IR data, 113
    UV data, *116*
  reduction, 110
  stability, 412
  X-ray crystallographic analysis, 118, 433
Cyclobutadiene-d_1-iron tricarbonyl, 435
Cyclobutadiene(iron tricarbonyl)-1,2(or 3)-dicarboxylic acid, 3(or 2),4-diphenyl, dimethyl ester, 94
Cyclobutadiene(π-mesitylene)cobalt bromide, 1,2,3,4-tetraphenyl-, 438–439
Cyclobutadiene–metal complexes, 6, **88–121**, 433–441, *see also* Bis(cyclobutadiene)–metal complexes *and specific metal complexes of the substituted cyclobutadienes*
  anion exchange, 110
  decomposition of, 23

Cyclobutadiene–metal complexes—*cont.*
  diamagnetic and paramagnetic behavior, 438
  electronic configuration (metal), 110–111
  history, 89
  ligand transfer, 97
  MO description, 411
  monomeric, dimeric, and oligomeric species, 98, 111
  in polymerization of acetylenes, 89, 92, 423
  preparation of, 90–100, 104–105,* 430,* 438
  properties, 112–119
    IR and UV data, 113
  reaction with complexing agents, 110
  reduction, 436–437
  stability, 438
Cyclobutadienemolybdenum tricarbonyl bromide, 1,2,3,4-tetraphenyl-, 97, *114*, 117
Cyclobutadienemolybdenum tricarbonyl iodide, 1,2,3,4-tetraphenyl-, 97, *114*, 117
Cyclobutadienenickel, *see* Bis(cyclobutadiene)nickel
Cyclobutadienenickel acetate, 1,2,3,4-tetraphenyl-, 98, 109
Cyclobutadienenickel azide, 1,2,3,4-tetramethyl-, 98, 107
Cyclobutadienenickel bromide
—, 1,2,3,4-tetramethyl-, 98, 99, *116*
—, 1,2,3,4-tetraphenyl-, 104, 412, 427
  bromination, 436
  carbonylation, 112
  oxidation, 109
  preparation of, 97, 99
  properties, 109, *116*, 117
  reaction with base, 109
    with cyclopentadienyliron dicarbonyl, 439
  reduction, 110
  solvolysis, 109
Cyclobutadienenickel chloride
—, 1,2,3,4-tetramethyl-, 2, 110, 412
  anion exchange, 98
  hydrate, 107
  oxidation, 109
  preparation of, 99
  properties, *114*
    IR data, 117

Cyclobutadienenickel chloride, 1,2,3,4-tetramethyl-,—*cont.*
  properties—*cont.*
    spectral data, 112, *116*, 118
    reaction with cyclopentadienyl sodium, 24, 101, 105, 111, 433
      with dimethylglyoxime,* 106
      with phenanthroline, 98, 111
      with triphenylphosphine, 98, 111
    reduction, 110
    solvolysis, 107, 109
    thermolysis, 23, 52, **106**
    X-ray crystallographic analysis, 118
—, 1,2,3,4-tetraphenyl-, 99
  UV data, 113, *116*
Cyclobutadienenickel chloride phenanthroline, 1,2,3,4-tetramethyl-, 23, 52, 98, 106
Cyclobutadienenickel chloride triphenylphosphine, 1,2,3,4-tetramethyl-, 23, 98, 106
Cyclobutadienenickel fluoride, 1,2,3,4-tetramethyl-, 98
Cyclobutadienenickel hydroxide, 1,2,3,4-tetraphenyl-,* 98, 109
Cyclobutadienenickel iodide, 1,2,3,4-tetramethyl-, 99, *116*
Cyclobutadienenickel nitrate, 1,2,3,4-tetramethyl-, 98
Cyclobutadienenickel sulfate, 1,2,3,4-tetramethyl-, 98
Cyclobutadienepalladium bromide
—, 1,2,3,4-tetrakis(*p*-chlorophenyl)-
  carbonylation, 112
  preparation of, 96, 98, 111
  properties, *114*
  reaction with iron pentacarbonyl, 97
—, 1,2,3,4-tetraphenyl-
  carbonylation, 100, 112
  dimer,* 98
  preparation of, **95**, 98,* 439, 440
  properties, 113, *114*
  reaction with cyclopentadienyliron complexes, 439
    with cyclopentadienylmolybdenum tricarbonyl dimer, 438
    with dicobalt octacarbonyl, 438
    with iron pentacarbonyl, 97, 104, 104*
    with nickel tetracarbonyl, 97

Cyclobutadienepalladium chloride
—, 1,2(or 3)-diethyl-3(or 2),4-diphenyl-,* 105
—, 1,2(or 3)-dimethyl-3(or 2),4-diphenyl-,* 105
—, 1,2,3,4-tetrakis(*p*-chlorophenyl)-
  oligomers, 440
  preparation of, 428, 440
  properties, *114*
  reaction with hydrogen bromide, 98, 111
    with iron pentacarbonyl, 97
—, 1,2,3,4-tetraphenyl-, 88, 110, 412, 440
  demetallation, 25
  dimer, 105,* 109, 437,* 438
  oligomers, 105,* 109, 437, 440
  oxidation, 437
  preparation of, **94**, 95, 104,* 109
  properties, *114*
    IR data, 117
    NMR data, 18
    UV data, *116*, 437
  reaction with cobaltocene, 97
    with cyclopentadienemolybdenum tricarbonyl dimer, 438
    with cyclopentadienetungsten tricarbonyl dimer, 438
    with dimethylformamide, 109
    with iron pentacarbonyl, 97
    with molybdenum hexacarbonyl, 97
    with trimethyl phosphite, 45, 109, 111
    with triphenylphosphine, 45, 109, 111, 427
    with tungsten hexacarbonyl, 97
  reduction, 427, 437
  solubility, 98, 109, **111**
  solvolysis, 109, 438
  thermolysis, 25, 107, 424, 427, 437
Cyclobutadienepalladium iodide, 1,2,3,4-tetraphenyl-, 95, 97
Cyclobutadienequinone, 128–156, 219, 441–445, *see also* Tetrahydrobenzocyclobutadienequinone
  delocalization energy, 138
  history, 128
  resonance contributors, 139
—, 3-amino-4-phenyl-
  hydrolysis, 137, 147
  preparation of, *136*, **137**, 147
  properties, 147, 150, *155*
  UV data, *152*

Cyclobutadienequinone—*cont.*
—, 3-bromo-4-(1′-cyclohexenyl)-,* 138, 142
—, 3-bromo-4-phenyl-
  displacement reactions, 137, 147
  preparation of, **135**, *136*, 137,* 138,* 141
  properties, *155*
    UV data, 152
  reduction, 147
—, 3-chloro-4-phenyl-
  displacement reactions, 137, 147
  preparation of, 135, *136*, 141
  properties, *155*
    UV data, 151
  reduction, 147
—, 3-(1′-cyclohexenyl)-
  bond lengths, 443
  chemistry, 137–138, 141
  preparation of, 131, *132*
  properties, 143, *155*, 156
    IR data, 151
    UV data, *152*
  stability, 140
—, 3-cyclohexyl-,* 137–138, 141
—, 3-(1′,2′-dibromocyclohexyl)-,* 142
—, 3,4-di(*n*-butoxy)-, *132*, 135, *136*, 149, 444
—, 3,4-dichloro-, 444
—, 3,4-dihydroxy-†
  chemistry, 149, 444–445
  dianion, 149
  phenylhydrazone, 149
  preparation of, *132*, **443**
    from chlorodifluorotetramethoxycyclobutane, 134
    from chlorodifluorotriethoxyclobutene, 134
    from chlorodifluorotrimethoxycyclobutene, 134
    from dichlorocyclobutadienequinone, 444
    from diethoxytetrafluorocyclobutene, 134
    from perchlorocyclobutenone, 135, 444
  properties, 150, *155*
    acid strength, 149
    IR data, 153, 154
    UV data, *152*, 154
  salts, properties, **150**, 154, 156
—, 3,4-dimethyl-
  chemistry, **140**, 144, 147
  derivatives, **143**, *155*, 442

Cyclobutadienequinone, dimethyl-,—*cont.*
  preparation of, **131**, *132*, 442
  properties, *155*
    IR data, 153, 442
    NMR data, 156, 442
    UV data, *152*, 442
—, 3,4-diphenyl-
  cleavage, 144, 146
  condensation with *o*-phenylenediamine, 144
  delocalization energy, 139
  derivatives, 143, *155*
  oxidation, 141
  photolysis, 26, 140
  preparation of, 131, *133*, 137*
  properties, 155
    IR data, 153
    NMR data, 156
    UV data, *152*
  reduction, 441
  stability, 140
—, 3,4-di(*n*-propyl)-, 442–443
—, 3-hydroxy-4-phenyl-
  chemistry, 137, 148
  preparation of, *136*, 137, 147
  properties, 150, *155*
    UV data, *152*, 153
—, 3-iodo-4-phenyl-, *136*, 137, *147*, 155
    UV data, *152*
—, 3-methoxy-4-phenyl-, 443
  hydrolysis, 137, 147
  preparation of, *136*, 137, 147, 148
  properties, 150, *155*
    UV data, *152*
—, 3-phenyl-, 128
  cleavage, **143–144**, 147
  condensation with *o*-phenylenediamine, 143
  delocalization energy, 139
  derivatives, 143, *155*
  halogenation, 135, 138,* 141
  oxidation, 141
  polarographic behavior, 141
  preparation of, *133*
    from chlorofluorophenylcyclobutene, 130
    from chlorodifluoroethoxyphenylcyclobutene, 131
    from chlorophenyltrifluorocyclobutene, 129, 444

Cyclobutadienequinone, 3-phenyl-,—cont.
  preparation of—cont.
    from cyclohexenylcyclobutadienequinone, 137, 138, 143
    from diethoxydifluorophenylcyclobutene, 131
    from difluorophenylcyclobutenone, 130, 131
    from phenyltetrafluorocyclobutene, 130
  properties, 140, 155
    IR data, 153
    UV data, 151
  reduction,* 26, 141
  X-ray crystallographic analysis, 445
—, 3-vinyl-, delocalization energy, 139
Cyclobutadiene-1,2,3,4-tetracarboxylic acid,* 32
Cyclobutadiene(π-toluene)cobalt bromide, 1,2,3,4-tetraphenyl-, 438
Cyclobutadienetungsten carbonyl bromide, 1,2,3,4-tetraphenyl-, 97
Cyclobutadienetungsten carbonyl iodide, 1,2,3,4-tetraphenyl-, 97
Cyclobutadienyl carbinyl radical (and ions),* 27
Cyclobuta[1,2-c:3,4-c']difuran, 1,3,4,6-tetraphenyl-,* 178
Cyclobuta[1,2-a:3,4-a']diindene, 5,10-dihydro-,* 6
Cyclobuta[1,2-a:4,3-a']diindene, 9,10-dihydro-,* 9
5,10-Cyclobuta[1,2-a:3,4-a']diindenequinone,* 7
9,10-Cyclobuta[1,2-a:4,3-a']diindenequinone,* 9
7H-Cyclobuta[a]inden-7-one, 1-acetyl-2-methyl-,* 8
Cyclobutadienoid character, 2
  of cyclobutenyl ions, 426
  stability and, 418
Cyclobutane, tetrakis(benzhydrylene)-, see Tetramethylenecyclobutane
Cyclobutene
—, 3,4-dichloro-1,2,3,4-tetramethyl-, 14
—, 3,4-dimethylene-, see dimethylenecyclobutene
—, 3-isopropylidenyl-1,2,4-triisopropyl-, 423
—, 3-methylene-1,2,4-trimethyl-, 24, 44,† 75, 79,† 107

Cyclobutene—cont.
—, 1,2,3,4-tetramethyl-, 12, 13, 44, 75, 79, 81
Cyclobutene(π-cyclopentadiene)nickel, 3-1'(or 3),2-cyclopentadienyl-1,2,3,4-tetramethyl-,† 75, 81, 101
Cyclobutenediol, 1,2,3,4-tetramethyl-, 25, 109
Cyclobutenedione, see Cyclobutadienequinone
Cyclooctatetraene, 44, 89, 92, 106
—, 1,2,3,4,5,6,7,8-octamethyl-, 12, 13, 75, 80, 106
—, 1,2,3,4,5,6,7,8-octaphenyl-, 19, 20, 25, 38, 44, 52, 58, 59, 66, 76, 82, 101, 111, 424, 427, 429, 431
  X-ray crystallographic analysis, 430
—, 1,2,4,7-tetraphenyl-,† 22, 56
—, 1,3,5,7-tetraphenyl-, 22, 56, 72, 78
Cyclopentadiene, pentaphenyl-, 41
Cyclopentadienone, 2,3,4,5-tetraisopropyl-, 423

D

Degeneracy
  in cyclobutadiene, 370
  ground state, 388
  spin, 388
Delocalization energy of cyclobutadienoid systems
  benzobiphenylenes, 406, 407
  benzocyclobutadiene, 192, 396
  benzocyclobutadienequinone, 225
  biphenylene, 288, 401
  biphenylenequinones, 305
  cyclobutadiene, 370, 387, **418**
    divalent ions, 122
  cyclobutadienequinone, 138, 168
  dibenzobiphenylenes, 408, 409
  dicyclobutabenzene, 396
  dicyclobutadienyl, 393
  dihydroxycyclobutadienequinone dianion 149
  dimethylenebenzocyclobutene, 235
  dimethylenecyclobutene, 158, 168
  diphenylbenzocyclobutadiene, 193
  diphenylcyclobutadienequinone, 139
  diphenylnaphthocyclobutadiene, 398
  diphenylphenanthrocyclobutadiene, 400
  divinylcyclobutadiene, 393

Delocalization energy of cyclobutadienoid systems—*cont*.
  methylenebenzocyclobutenone, 232
  naphthocyclobutadiene, 248
  phenanthrocyclobutadiene, 396
  phenylbenzocyclobutadiene, 193
  phenylcyclobutadiene, 393
  phenylcyclobutadienequinone, 139
  tetramethylenecyclobutane, 168
  tricyclobuta[*a,c,e*]benzene, 396
  tricyclo[5.3.0.0^{2,6}]decapentaene, 396
  tricyclo[3.1.0.0^{2,4}]hexatriene, 396
  tricyclo[4.2.0.0^{2,5}]octatetraene, 396
  vinylcyclobutadiene, 393
  vinylcyclobutadienequinone, 139
5,10-Diazabenzo[*b*]biphenylene, 229, *230*,† 364–365
3,8-Diazanaphtho[*b*]cyclobutadiene
  1,2-dimethyl-,* 144, 252
  1,2-diphenyl-,* 144, 252
  1-phenyl-,* 143, 252
2,3:6,7-Dibenzobiphenylene, *see* Dibenzo-[*b,h*]biphenylene
Dibenzo[*a,c*]biphenylene, 317, 464
Dibenzo[*a,g*]biphenylene, 317, 323,* 362*
  preparation of, *332*,* *353*,* **463**
  reductive cleavage, 464
  stability, 464
  UV data, 353, 463–464
Dibenzo[*a,g*(or *i*)]biphenylene,* 318, *332*, 335
Dibenzo[*a,h*]biphenylene, 317
Dibenzo[*a*(or *b*),*h*]biphenylene,* *332*
Dibenzo[*a,h*(or *i*)]biphenylene,* *332*
Dibenzo[*a,i*]biphenylene, 317, 321, *322*
  delocalization energy, 409
  HMO constants, *409*
  photolysis, 337, 354, 356
  preparation of, 333,* 355, *362*
  properties, 354
    half-wave potential, 289
    IR data, 354
    UV data, 289, 354, **355**, 356, **357**
  reactivity, 411
  stability, 335, 354, 464
  trinitrofluorenone complex, 354
Dibenzo[*b,h*]biphenylene, 317, 319*
  addition reactions, 364
  anion radical, ESR data, 410
  delocalization energy, 408

Dibenzo[*b,h*]biphenylene—*cont*.
  HMO constants, *408*
  nitration,* *333*
  preparation of,* 333
    from bromoiodonaphthalene, 323
    from dibromonaphthalene, 322–323, 357–358
    from diiodonaphthalene, 323
    dinaphtho[2,3-*b*:2′,3′-*d*]iodolium iodide, 357
    octahydrodinaphtho[2,3-*b*:2′,3′-*d*]-iodolium iodide, 258
    from octahydrobinaphthylbis(diazonium chloride),† 321
    sulfate, 321*
  properties, 356, *363*
    IR data, 357
    UV data, 356, **357**, 359, 364, 410
  reactivity, 411
  stability, 335, 356
—, dinitro-,* 330
—, 1,7(or 10)-dinitro-,* 323, *333*
—, 5,6(or 11)-dinitro-,† 323, *363*, 364
—, 1,2,3,4,7,8,9,10-octahydro-, *312*, 321*
  preparation of, 258,* 259,* 275,* 357*
  UV data, *292*
—, 5,6,11,12-tetraacetoxy-, *328*, *331*, 334, 358, *363*
  spectral data, *359*
—, 1,2,3,4-tetrahydro-, *322*
—, 5,6,11,12-tetrahydroxy-, 358, *362*, *363*
  preparation of, 327, *328*, 330, 334, 336, 358
  spectral data, *359*
—, 5,6,11,12-tetramethoxy-, *363*
  preparation of, *328*, 334, 358
  spectral data, *359*
—, 5,6,11,12-tetraphenyl-, *363*
  preparation of, 325–327, *328*, 337, 358, 360, 463
  spectral data, 360
—, 5,6,11,12-tetraphenyl-, 360
Dibenzo[*b,h*]biphenylene-5,6,11,12-bis-(quinone),* 319, *333*
Dibenzo[*a,e*]cyclooctatetraene
—, 5,6-dimethyl-11,12-diphenyl-, 186, 198, *212*
—, 5,6-diphenyl-, 186, 198, *213*
—, 5,6,11,12-tetrabromo-, 197
Dibenzoiodolium iodide,† 257–258

Dibenzo[c,g]tricyclo[4.2.0.0^{2,5}]octadiene,† 199, *209*
—, 5,6(or 11)-dimethyl-11(or 6),12-diphenyl-, 199–200, *213*
—, 5,6,11,12-tetrabromo-,* 32, 197
Dibenzo[c,h]tricyclo[5.3.0.0^{2,6}]decapentaene,* 178
Dibenzo[c,i]tricyclo[5.3.0.0^{2,6}]decapentaene,* 178
Dicyclobuta[a,d]benzene
 delocalization energy, 396
—, 1,2,4,5-tetramethylene-, *see* Tetramethylenedicyclobuta[a,d]benzene
Dicyclobutadienyl, ground state, 393
1,1'-Dicyclohexenyl, *see* Bicyclohexenyl
1,2-Dimethylenebenzocyclobutene, 232
 adduct formation,* 188
 delocalization energy, 235
 preparation of, from benzocyclobutadienequinone,* 227, 237
  from dimethylbenzocyclobutenediol diacetate, 236
  from 1,2-dimethylenebenzocyclobutene-α,α'-dicarboxylic acid,* 236
 properties, 237, *238*
  IR data, 238
  UV data, 239
 reduction, 240
 as resonance contributor, 281, 305
—, α,α'-diphenyl-,† 228, 235, *238*, 239, 449
—, α,α,α',α'-tetrachloro-,* 227, 237
—, 4,4,5,5-tetracyano-3,4,5,6-tetrahydro-, 172–173
1,2-Dimethylenebenzocyclobutene-α'-carboxylic acid, α-cyano-α',3,4,5-tetramethoxy-,† 236, *238*, 308
1,2-Dimethylenebenzocyclobutene-4,5-dicarboximide, *N*-phenyl-3,4,5,6-tetrahydro-, 172–173
1,2-Dimethylenebenzocyclobutene-α,α'-dicarboxylic acid, *238*, 240
 decarboxylation,* 237
 dimethyl ester, 227, 305
  preparation of, from acid, 240
   from benzocyclobutadienequinone, 235
   from methyl 2-methylenebenzocyclobutenone-α-carboxylate, 234, 235
  UV data, 238–239

1,2-Dimethylenebenzocyclobutene-α,α'-dicarboxylic acid—*cont.*
 reduction, 240
 stability, 239
5,6-Dimethylenebicyclo[2.2.0]hex-1(4)-ene, α,α',2,2,3,3-hexachloro-, 449
7,8-Dimethylenebicyclo[4.2.0]oct-1(6)-ene, 3,3,4,4-tetracyano-
 preparation of, 161
 properties, *176*
  spectral characteristics, *164*
 reaction with air, 166
7,8-Dimethylenebicyclo[4.2.0]oct-1(6)-ene-3,4-dicarboximide, *N*-phenyl-
 preparation of, 161
 properties, *176*
  spectral characteristics, *164*
 reaction with air, 166
1,2-Dimethylene-2,4-cyclobutanedione,* 157, 173
—, α,α,α',α'-tetraphenyl-
 preparation of, 172–173
 properties, 174, *175*, *177*, 178
1,3-Dimethylene-2,4-cyclobutanedione,* 157, 173
—, α,α,α',α'-tetraphenyl-derivatives,* 174, 178
 preparation of, 171, 174
 properties, *175*, *177*, 178
 reduction, 174, 174*
3,4-Dimethylenecyclobutene, 157, 445–447, *see also* Dimethylenebicyclo[4.2.0]octene, Tetramethylenetricyclodecadiene
 adduct formation,* 167
 polymerization, 165–166
 preparation of, 159, **161–162**
 properties, 162–163, *176*
  spectral characteristics, 163, *164*
 reduction, 167
 resonance contributor, 248
—, 1,2-dichloro-α,α,α',α'-tetraphenyl-,* 162
—, 1,2-dimethyl-
 adduct formation,* 26, 167
 preparation of, 45, **159**, 162, 445
 properties, *176*
  NMR data, 163, 165
  UV data, *164*
 reaction with hydrogen chloride, 166
 reduction, 166–167

SUBJECT INDEX 499

3,4-Dimethylenecyclobutene—*cont.*
—, 1,2-diphenyl-
  adduct formation,* 26, 167
  bromination, 166
  delocalization energy, 158
  ozonolysis, 165
  preparation of, 159–160
  properties, *176*
    spectral characteristics, 163, *164*
  reduction, 167
—, $\alpha,\alpha,\alpha',\alpha'$,1,2-hexabromo-, 447
—, $\alpha,\alpha,\alpha',\alpha'$,1,2-hexachloro-†
  chlorination, 166
  oxidation, 165
  preparation of, 162, **446**, 447
  properties, *176*
    spectral characteristics, 164
—, 1-methyl-
  preparation of, 160–162
  properties, *176*
    spectral characteristics, *164*
—, 1-phenyl-, delocalization energy, 158
—, 1-vinyl-, delocalization energy, 158
3,6-Dioxabenzocyclobutadiene, 1,2,4,5-tetraphenyl-3,4,5,6-tetrahydro-,* 26

E

Eigenfunctions for cyclobutadiene, 371
Electron
  affinity, 382
  angular momentum, 377
  density, 375
  interaction, 381
  repulsion, 377, 381
    in three and four centers, 382
  states, 381
Energy (ground-state and triplet) in cyclobutadiene, 374
Energy levels, *see also* Configuration energies
  of cyclobutadiene, 370, **382–383**
    ground-state and excited, 379
  of cyclobutadiene divalent ions, 395
  of substituted cyclobutadienes, 394
5,10-Ethanobenzo[*b*]biphenylene, 5,5a,9b,10-tetrahydro-, 202, *210*
1,4-Ethanobiphenylene-2,3-dicarboxylic anhydride, 9,10-diketo-1,2,3,4-tetrahydro-,* 188

Ethylene, 1-benzoyl-2-(2'-biphenylenyl)-,† 296

F

Free valence, 371
Free-valence index
  of benzocyclobutadiene, 193
  of cyclobutadiene, 371
  of dimethylenecyclobutene, 158
  of tetramethylenecyclobutane, 168
Furan, 2,3,4,5-tetraphenyl-, 427

G

Glyoxylic acid, 2-biphenylenyl-, *271*, 296, 311
Ground state, *see also* Configuration (electronic), Pure valence state
  of cyclobutadiene, 382
  of square cyclobutadiene, 383
  of substituted cyclobutadienes, 393

H

Hamiltonian
  core, 379
  operator, 370, 378
Heitler–London–Slater–Pauling method, 376
Heitler–London wavefunction, 372
2,4,6-Heptatrienonitrile, 103
2,4-Hexadiene, 3,4-dimethyl-, 75, *81*
Hückel rule, 1, 44, **306–307**
Hund's rule, 369
Hybridization (in cyclobutadiene), 376

I

Indeno[2,1-*a*]indene, 5,10-diphenyl-, 427
Integral
  core, 380
  Coulomb, 370, 374, 381
  exchange, 374, 376
  interaction (electronic), 380–381
  one-center, 381
  overlap, 370
  repulsion (electronic), 380–381
  resonance, 370, 390

Integral—*cont.*
  three- and four-center, 382
  vanishing, 381
Interelectronic effects, 370
Ionization potential, valence state, 382

J

Jahn–Teller effect, 388–392
  in cyclobutadiene dication, 123
  in cyclobutadiene monocation, 392
  dynamic, 388, 391
  pseudo, 389
  in tetrahedrane, 392

K

Kronecker symbol, 370

L

LCAO–MO theory, 370
  σ bonds and, 390
Ligand transfer reactions, 97, 104,* 110

M

Metal complexes, *see* Benzocyclobutadiene, Biphenylene, Cyclobutadiene, *etc.*
1,4-Methanobiphenylene
—, 9-isopropylidenyl-1,4,4a,8b-tetrahydro-, 201, *210*
—, 1,4,4a,8b-tetrahydro-, 201, *209*
2-Methylenebenzocyclobutenone, delocalization energy, 232
—, α,α-dichloro-,† 227, 233, *234,* 237*
—, α-phenyl-α-(o-benzoylphenyl)-, 234
2-Methylenebenzocyclobutenone-α-carboxylic acid,† 233, *234*
  methyl ester, 227, 233, 234, *234*
4-Methylene-1,2,3-cyclobutanetrione, 158, 173*
4-Methylenecyclobutenone, 158
Molecular orbital (MO) description, *see also* Penney electron pairs method
  of cyclobutadiene
    antisymmetrized, 377
    nonempirical, 382
    perturbation theory and, 391
    semiempirical, 377

Molecular orbital (MO) description—*cont.*
  of cyclobutadiene—*cont.*
    valence-bond description and, **375**, 386
  of cyclobutadienyl carbinyl ions and radical, 27
  of octaphenyltetramethylenecyclobutene, 446
  of 1,2,3,4-tetrakis(benzhydryl)cyclobutadiene tetraanion, 446
Mulliken's correlation tables, 377

N

Naphthalene, 23
—, 1,4-diacetoxy-, 23
—, 1,3-dichloro-2-fluoro-5,6,7,8-tetrahydro-, 16, 66, *72, 77*
—, 1,2-dihydro-, 22
—, 1,4-dimethoxy-, 23
—, 1,2,3-triphenyl-, 18, 67, *76, 82*
—, 1,2,4-triphenyl-, 41
Naphthalene-1,2-dicarboximide
—, 1,2-dihydro-N-phenyl-, 204–205, *211,* 214–215
—, N-phenyl-, 204–205, 214–215
Naphthalene-2,6-dicarboxylic acid
  dimethyl ester, 22, 61
—, 1,2-dihydro-, dimethyl ester, 22, 61
—, 3,4-dihydro-, dimethyl ester, 22, 61
—, 4a,8a-dihydro-, dimethyl ester, 61
Naphtho[2,3-b]biphenylene
—, 4b-bromo-5,12-diphenyl-5,12-oxido-4b,5,12,12a-tetrahydro-, 204–205
—, 5,12-diphenyl-, 364
Naphthocyclobutadiene, *see also* Diazanaphthocyclobutadiene
Naphtho[a]cyclobutadiene, 243
  π-bond orders, *398*
  stability, 399
Naphtho[b]cyclobutadiene,
  π-bond orders, 248, *398*
  delocalization energy, 248
  generation of, 243–245
  stability, 242, 399
—, 1-bromo-3,8-diphenyl-, 245
  generation of, **245**, 250, 450
  reaction with diphenylisobenzofuran, 250
—, 1,2-dibromo-3,8-diphenyl-, 244, 246, 248, 251, 450
—, 1,2-dimethyl-,* 451

Naphtho[*b*]cyclobutadiene—*cont.*
—, 1,2-diphenyl-, 2
  adduct formation, 250–251
  bromination, 251
  delocalization energy, 248, 397
  HMO constants, *397*
  oxidation, 251
  photodimerization, 252
  preparation of, 244, 451*
  properties, 245
    IR data, 247
    NMR data, 246
    UV data, 247
  reduction, 251
  stability, 248
  trinitrofluorenone complex, 245
—, 3,8-diphenyl-, 245, 450
  adduct formation, 243, 245–246, **249**
  dimerization,* 249, 450
  generation of, **243**, 245, 249–250, 450
  hydrogen abstraction,* 249
  polymerization, 249
  stability, 243
Naphtho[*b*]cyclobutadiene-1,2-dicarboxylic acid, dimethyl ester,* 451
Naphtho[*a*]cyclobutadienequinone, 5,6,7,8-tetrahydro-,* 223
Normalization factor, 378

O

2,4,6-Octatriene, 4,5-diisopropyl-2,7-dimethyl-, 423
Omega technique, 396
Operator, *see also* Hamiltonian
  kinetic energy, 380
Orbitals, in cyclobutadiene
  degenerate, 378
  nonorthogonal and orthogonal, 385

P

Parameterization procedure, 391
Pariser–Parr approximation, 381, 385
Pariser–Parr–Pople method, 418
Pauli principle, 372, 378
Penney method (electron pairs), 376
Pentacyclo[10.4.0.0^{2,11}.0^{3,10}.0^{4,9}]hexadeca-4,6,8,12,14,16-hexaene, *see* Dibenzotricyclo[4.2.0.0^{2,5}]octadiene

Pentacyclo[10.2.0.0^{2,11}.0^{4,9}.0^{5,8}]tetradeca-1(12),2,4(9),5(8),10-pentaene,* 191–192
Phenanthrene
—, 1,4-diacetoxy-, 23
—, 1,4-dimethoxy-, 23
Phenanthro[*l*]cyclobutadiene
  π-bond orders, *398*
  delocalization energy, 253, 396
  free valence, 396
  stability, 242
—, 1,2-diphenyl-, 450
  delocalization energy, 400
  electronic configuration, 419
  generation of, 252
  HMO constants, *398–399*
  stability, 396
Phthalic acid
  dimethyl ester, 22
—, 1,2-dihydro-3,4,5,6-tetraphenyl-
  diethyl ester, 76, *82*
  dimethyl ester, 65
—, 3,4,5,6-tetramethyl-, dimethyl ester, 13, 39, 63, *75, 80*, 424
—, 3,4,5,6-tetraphenyl-, dimethyl ester, 65, *76, 82*
Pi-binding energy in cyclobutadiene, 370
Pure valence state, 376

Q

Quantum number (spin), 379
Quinoxaline, *see* Benzocyclobutaphenazine, 3,8-Diazanaphtho[*b*]cyclobutadiene

R

Reactivity of cyclobutadiene, 371
Representation
  irreducible, 377, 378
  reducible, 369
Resonance energy (in cyclobutadiene), 374, 383, 386, 390, *see also* Delocalization energy
Resonance splitting, 383
Ring strain (in cyclobutadiene), 376, 377

S

Secular determinant, 374
Secular equation, 373

Slater function, 372, 378
Spin operator, 372
Spiro[cyclopropane-1,9'-(1,4-methanobiphenylene)], 201, *210*
Spiro{tricyclo[6.2.0.0^{2,7}]deca-1(8),2,4,6-tetraene-9,1'-cyclohexane}, 10,10-dimethyl-,* 192
Squaric acid, *see* Dihydroxycyclobutadienequinone
Stabilization
  push-pull, 2, 29, 394, 425
  symmetry and, 393
Strain energy, *see* Ring strain (in cyclobutadiene)

T

Tetrabenzo[*a,c,g,i*]biphenylene,* 318, *333*, 334, *398*, 411
Tetracyclo[10.1.0.0^{3,10}.0^{4,9}]trideca-2,4,6,8,10-pentaene-13-carboxylic acid, ethyl ester, 460
Tetrahedrane, *see* Tricyclo[1.1.0.0^{2,4}]butane
Tetralin, 23
1,2,3,4-Tetramethylenecyclobutane, 157
  adduct formation, 161, **172–173**
  biphenylene and, 286
  delocalization energy, 168
  dimerization, 161, **172–173**
  free-valence index, 168
  preparation of, 168, 169*
  properties, *177*
    spectral data, 170–171
  reduction, 172–173
  stability, 168, 172–173
—, octa(*p*-anisyl)- (?), 169
—, octaphenyl-
  cleavage, 446
  ozonolysis, 171, 173–174
  preparation of, 158, 169, 169*
  properties, *177*
    spectral data, 170, *175*
  stability, 172–173
  thermolysis, 171
—, α,α',α'',α'''-tetraphenyl-,* 169
1,2,3,4-Tetramethylenecyclobutane-α,α',α'',α'''-tetracarboxylic acid,* 178
5,6,11,12-Tetramethylenetricyclo[8.2.0.0^{4,7}]dodeca-1(10),4(7)-diene, 161, 164, 166, 172–173, *176*

Tribenzo[*a,c,h*]biphenylene, 318
Tricyclo[1.1.0.0^{2,4}]butane, 69
  Jahn–Teller effect in, 392
  as valence tautomer of cyclobutadiene, 46, 49
—, 1,3-diphenyl-,* 42
Tricyclo[1.1.0.0^{2,4}]butane-2,3,4-tricarboxylic acid, 1-methyl-,* 69
Tricyclobuta[*a,c,e*]benzene,† delocalization energy, 396
Tricyclo[5.3.0.0^{2,6}]deca-1,6-diene,* 36, 37
Tricyclo[5.3.0.0^{2,6}]deca-1,3,5,7,9-pentaene,* 178
  delocalization energy, 396
Tricyclo[6.2.0.0^{2,7}]deca-1(8),2,4,6-tetraene, 9,9,10,10-tetramethyl-,* 192
Tricyclo[6.4.0.0^{2,7}]dodeca-1(8),2(7)-diene, *see* Octahydrobiphenylene
Tricyclo[8.6.0.0^{2,9}]hexadeca-1,9-diene, 36, 37
Tricyclo[8.6.0.0^{2,9}]hexadeca-1(10),2(9)-diene,* 36, 37
Tricyclo[3.1.0.0^{2,4}]hexa-1,3,5-triene, delocalization energy, 396
Tricyclo[4.2.1.0^{2,5}]nona-3,7-diene, 432
Tricyclo[4.2.0.0^{2,5}]octa-3,7-diene, *anti*-, 11, 52, 420, *see* Cyclobutadiene dimer (*for syn isomer*)
—, 1,2,3,4,5,6,7,8-octamethyl-, *anti*-, 23, 107
Tricyclo[4.2.0.0^{2,5}]octa-1,3,5,7-tetraene, delocalization energy, 396
Tricyclo[7.5.0.0^{2,8}]tetradeca-1,8-diene,* 36, 37
Tricyclo[7.5.0.0^{2,8}]tetradeca-1(9),2(8)-diene,* 36, 37
Tricyclo[7.4.0.0^{2,8}]trideca-2(8),3,6,9,11,13-hexaene-5-carboxylic acid, ethyl ester, 460
Tricyclo[6.3.0.0^{2,7}]undeca-1(8),2,4,6,9-pentaenyl ion, 183
Trimesic acid, triphenyl-, *see* Diphenylcyclobutadienedicarboxylic acid
2,3,4-Trimethylenecyclobutanone,* 157, 173
—, hexaphenyl-,† 171, 173–174, *175*, *177*, 178
Triphenylene, 1,2,3,4,5,6,7,8,9,10,11,12-dodecahydro-, 17, 64
Truxene, *see* Dihydrocyclobutadiindene

V

Valence-bond (VB) description of cyclobutadiene, 371
  empirical, 382, 388, 390
  MO description and, **375**, 386
  nonempirical, 385
Variation calculation, 379

W

Wavefunctions (for cyclobutadiene), *386*

Wavefunctions (for cyclobutadiene)—*cont.*
  antisymmetrized, 372, 378
  determinantal, 373
  LCAO, 381
  singlet and triplet, 379
Woodward–Hoffmann rules, 61, **420**

Z

Zero-order calculation, 370